Advances in Experimental Medicine and Biology

Series Editors

Wim E. Crusio, Institut de Neurosciences Cognitives et Intégratives d'Aquitaine,
CNRS and University of Bordeaux, Pessac Cedex, France

Haidong Dong, Departments of Urology and Immunology
Mayo Clinic, Rochester, MN, USA

Heinfried H. Radeke, Institute of Pharmacology & Toxicology
Clinic of the Goethe University Frankfurt Main, Frankfurt am Main,
Hessen, Germany

Nima Rezaei, Research Center for Immunodeficiencies, Children's Medical Center
Tehran University of Medical Sciences, Tehran, Iran

Ortrud Steinlein, Institute of Human Genetics
LMU University Hospital, Munich, Germany

Junjie Xiao, Cardiac Regeneration and Ageing Lab, Institute of Cardiovascular
Science, School of Life Science, Shanghai University, Shanghai, China

Advances in Experimental Medicine and Biology provides a platform for scientific contributions in the main disciplines of the biomedicine and the life sciences. This series publishes thematic volumes on contemporary research in the areas of microbiology, immunology, neurosciences, biochemistry, biomedical engineering, genetics, physiology, and cancer research. Covering emerging topics and techniques in basic and clinical science, it brings together clinicians and researchers from various fields.

Advances in Experimental Medicine and Biology has been publishing exceptional works in the field for over 40 years, and is indexed in SCOPUS, Medline (PubMed), Journal Citation Reports/Science Edition, Science Citation Index Expanded (SciSearch, Web of Science), EMBASE, BIOSIS, Reaxys, EMBiology, the Chemical Abstracts Service (CAS), and Pathway Studio.
2020 Impact Factor: 2.622

Kavindra Kumar Kesari
Shubhadeep Roychoudhury
Editors

Oxidative Stress and Toxicity in Reproductive Biology and Medicine

A Comprehensive Update on Male Infertility- Volume One

Editors
Kavindra Kumar Kesari [iD]
Department of Bioproducts and Biosystems,
School of Chemical Engineering
Aalto University
Espoo, Finland

Shubhadeep Roychoudhury
Department of Life Science and
Bioinformatics
Assam University
Silchar, India

Responsible Editor
Carolyn Spence

ISSN 0065-2598   ISSN 2214-8019 (electronic)
Advances in Experimental Medicine and Biology
ISBN 978-3-030-89339-2   ISBN 978-3-030-89340-8 (eBook)
https://doi.org/10.1007/978-3-030-89340-8

© The Editor(s) (if applicable) and The Author(s), under exclusive license to Springer Nature Switzerland AG 2022
This work is subject to copyright. All rights are solely and exclusively licensed by the Publisher, whether the whole or part of the material is concerned, specifically the rights of translation, reprinting, reuse of illustrations, recitation, broadcasting, reproduction on microfilms or in any other physical way, and transmission or information storage and retrieval, electronic adaptation, computer software, or by similar or dissimilar methodology now known or hereafter developed.
The use of general descriptive names, registered names, trademarks, service marks, etc. in this publication does not imply, even in the absence of a specific statement, that such names are exempt from the relevant protective laws and regulations and therefore free for general use.
The publisher, the authors and the editors are safe to assume that the advice and information in this book are believed to be true and accurate at the date of publication. Neither the publisher nor the authors or the editors give a warranty, expressed or implied, with respect to the material contained herein or for any errors or omissions that may have been made. The publisher remains neutral with regard to jurisdictional claims in published maps and institutional affiliations.

This Springer imprint is published by the registered company Springer Nature Switzerland AG
The registered company address is: Gewerbestrasse 11, 6330 Cham, Switzerland

Foreword

The book by Dr. Kavindra Kumar Kesari and Dr. Shubhadeep Roychoudhury, ***Oxidative Stress and Toxicity in Reproductive Biology and Medicine: A Comprehensive Update on Male Infertility***, highlights the strong connection between oxidative stress and male infertility. This topic has gained prominence with new knowledge about the impact of lifestyle and mutagenic factors on male infertility. Exposure to several environmental toxicants, such as chemicals, radiations, and viral and microbial infections, represents risk factors for male infertility. The book's title itself defines the theme of the book, wherein the authors have considered relevant oxidative stress–induced toxicity in male fertility. This book contains 15 chapters which cover most of the recent progress in the field of free radical biology in cellular toxicology and clinical manifestations of various issues related to men's health and infertility along with the therapeutic use of herbal and natural medicines to control the oxidative stress. The introductory chapter by Prof. Ralf Henkel highlights the relationship between oxidative stress and male infertility with historical perspectives. Several chapters in this book focus on the physiological and pathological role of oxidative stress in male reproduction by providing appropriate pathways. The follow-up chapters mostly explore the role of oxidative stress at the molecular level, for example, Chapter 5 unravels the molecular impact of sperm DNA damage and repair mechanism along with certain other types of damage relevant to male reproduction.

The scientific value of this book is that most of the exiting factors which are associated with our daily life exposures are well covered. Several important chapters on uropathogenic factors (bacterial and viral infection) have been discussed to explore the role in male infertility. Especially Chapter 7 discusses the role of oxidative stress in bacteriospermia vis-a-vis male infertility. Chapters 9 and 10 on inflammation and varicocele describe how different types of lifestyle and environmental factors induce oxidative stress and in turn cause infertility. Last but not least, Chapter 15 presents an interesting study on sperm redox system equilibrium and its implication in fertilization and male fertility. Understanding the sperm redox system will allow us to explore the mechanism of free radical biology in reproductive toxicology. The book presents an update on the etiology of infertility, where a number of chapters defining the known and probable causes of male infertility and therapeutic measures against oxidative stress are delineated.

The editors should be congratulated for providing chapters with concepts and mechanisms using illustrations, flow charts, and tables. The language is kept simple in this book so that an inquisitive reader with a scientific or nonscientific background may easily understand the subject matter. This book will be useful for general readers as it covers many important public issues. The editors of this book deserve special thanks for assembling a great team of contributors and for editing such an important book on a clinical topical issue.

Director, American Center for Reproductive Medicine   Ashok Agarwal
Cleveland Clinic Foundation
Cleveland, OH, USA
Director, Andrology Center
Cleveland Clinic Foundation
Cleveland, OH, USA
Professor of Surgery (Urology), Lerner College of Medicine
Cleveland Clinic Foundation
Cleveland, OH, USA

Preface and Acknowledgments

There are several factors in the indoor and outdoor environment which may affect reproductive health. Potential threats to reproduction and fertility that are encountered by men in everyday life emanate from chemical, physical, and biological sources. Most common etiology is oxidative stress which may induce cellular toxicity and enhance free radical production within the cells which often overwhelm the scavenging capacity by antioxidants due to excessive production of reactive oxygen species (ROS). Oxidative stress is a result of the imbalance between ROS and antioxidants in the body which may lead to sperm damage (DNA or count), deformity, and, eventually, male infertility. Environmental toxicants, such as radiations (ionizing and non-ionizing), radiations from radiotherapy, chemicals, toxic metals, fine particles, lifestyle factors, medications or drugs consumption, and microbial and viral contaminations, and other factors including climate change and pollution affect male fertility. This book presents various state-of-the-art chapters on the recent progress in the field of cellular toxicology and clinical manifestations of various issues related to men's health and fertility. This book also provides a way towards therapeutic use of natural antioxidants to control or manage oxidative stress for the prevention of male infertility outcome. This book has value-added collection of 15 chapters as Volume I, which discuss **an up-to-date review on the impact of oxidative stress factors in male reproduction with** multidisciplinary approaches with a focus on environmental toxicity. Chapter 1 by Prof. Ralf Henkel mainly introduces the role of oxidative stress and toxicity in reproductive biology and medicine. This chapter primarily points out the factors that are responsible for inducing oxidative stress, which may lead to male infertility, and these are presented with historical perspectives. Chapter 2 discusses the role of ROS and antioxidants in both male and female reproduction. This chapter largely highlights the origin of ROS in the female and male reproductive tracts with possible mechanisms of their generation. It also describes the specific physiological roles of the reactive molecules in upholding the structural integrity and functionality of both the reproductive systems. Chapter 3 discusses the pathological roles of ROS in male reproductive problems which are linked to infertility, erectile dysfunction, and prostate cancer. The chapter mainly emphasizes the role of lipid peroxidation, DNA damage, and apoptosis induced by oxidative stress, with an insight into the probable mechanism through which ROS exerts its pathological impact. The effect of varicocele-induced ROS on infertility is value added discussion towards male in fertility in this chapter. In this continuation, Chapter 4 represents the molecular interactions which are directly or indirectly associated with oxidative stress–mediated male infertility. The chapter mainly highlights the association of proteins with dysregulated molecular pathways in sperm and seminal plasma due to oxidative stress. Chapter 4 also discusses the molecular interactions between proteins associated with oxidative stress and their potential role leading to male infertility with the help of advanced proteomic techniques and bioinformatic tools. Chapter 5 by Prof. Ashok Agarwal and his group, one of the leading scientists in the field of oxidative stress and male infertility research, focuses on the role of ROS in causing sperm DNA damage, along with certain other types of damages. This chapter elaborates the molecular mechanisms involved in DNA repair during spermatogenesis or after fertilization of the oocyte. The impact of sperm DNA damage on reproductive outcomes, such as fertilization and pregnancy rates, has been discussed with a focus on animal and human studies. Chapter 6 discusses the role of leukocytes and infection in male infertility. There are several types of infections that may affect human health, especially male urogenital infection, involving bacterial, viral, protozoal, and fungal infections. These infections are more or less asymptomatic in nature and pass on to sexual partner through intercourse which may lead to fertilization failure, pregnancy loss, and even development of illness in the offspring. Among these, leukocytospermia emerges as one of the most likely causes of ROS and oxidative stress-induced sperm dysfunction, poor fertilization, and male infertility. In this connection, Chapter 7 elaborates the role of oxidative stress towards bacteriospermia and male infertility. In most of the infection cases, bacteria attack the urological system, and these uropathogens are involved in infection and inflammation of the male urogenital tract. Bacteriospermia may induce infertility by interacting directly with sperm cells and start producing ROS and lead to impaired sperm parameters. Chapter 7 also discusses the protective measures by supplementing herbal medicines, antioxidants, and plant extracts. The role of herbal medicines against viral or microbial infections have been emphasized in recent years. Therefore, this book provides a collection of oxidative stress–induced infertility prevention and fertility protection measures. Chapter 8 discusses the oxidant-sensitive inflammatory pathways in male reproduction, where inflammation has been considered as a major signal of oxidative stress. Oxidative stress–induced male infertility and its mechanisms are extensively documented in this chapter along with several important fertility parameters. In this series, Chapter 9 discusses more about oxidative stress with emphasis on various factors of idiopathic male infertility and the mechanism by which several endogenous as well as exogenous factors may induce idiopathic male infertility. This chapter covers several important aspects such as inflammation, varicocele, obesity, and lifestyle factors which may induce oxidative stress and lead to infertility. It also elaborates the pathophysiology of male oxidative stress infertility with a focus on its diagnostic and therapeutic aspects. In this connection, Chapter 10 discusses an important clinical male infertility aspect – varicocele. This has been reported as one of the most common causes of male subfertility, although the pathophysiology of varicocele remains largely unknown. This chapter mainly explores the correlation among oxidative stress, varicocele, and male infertility with focus on various fertility parameters. It also discusses several therapeutic aspects to explore the role of interventions and to reduce oxidative stress in men with varicocele-associated infertility. In a way, Chapters 1, 2, 3, 4, 5, 6, 7, 8, 9, and 10 pose oxidative stress to be the responsible factor behind elevated levels of ROS, and as a result may also be considered as a hallmark for diabetes-induced male infertility, as has been reported in Chapter 11. For better understanding of the underlying mechanisms, Chapter 11 discusses the central role of oxidative stress, which may play a major role as mediator of obesity, metabolic syndrome, and type 2 diabetes mellitus in men. This chapter mainly covers the pathophysiological role of oxidative stress in the management and mechanism of reproductive dysfunctions. Although, management of oxidative stress is not well defined, however, improved nutrition and exercise has been recommended for the improvement in male fertility parameters under the influence of oxidative stress. Chapter 12 discusses the key causes of decreased sperm motility and some of the muted genes and metabolic causes, wherein athenozoospermia or poor sperm motility has been reported as causes of infertility in the male. This chapter covers several important aspects related to the metabolic dysregulation with focus on signaling pathways and genes involved in sperm motility. This chapter explores the central role of oxidative stress and ROS production in association with sperm motility, molecular pathways, and lifestyle factors. Interestingly, a therapeutic approach for athenozoospermia has also been proposed. Chapter 13 discusses as to how viruses invade the male genital system thus in turn leading to detrimental consequences on male fertility. This chapter uncovers the facts associated with the tropism of various viruses in the male genital organs and explores their sexual transmissibility. It summarizes the functional and mechanistic approaches employed by the viruses in inducing oxidative stress inside spermatozoon thus leading to male infertility. Additionally, this chapter also highlights the various antiviral therapies that have been studied so far in order to ameliorate viral infections and to combat the harmful consequences leading to male infertility. This chapter would be highly interesting considering the current scenario of COVID-19 viral infections affecting human health. The male reproductive system is one of the attainable targets of many viruses including immunodeficiency virus, Zika virus, adenovirus, cytomegalovirus, and potentially severe acute respiratory syndrome coronavirus 2 (SARS-CoV-2), and infection with such viruses may cause serious health issues which has been reported in Chapter 14. This chapter discusses the pathogenesis of viral infections and possible roles in male reproductive health. This is an evidence-based study where authors explore how the viruses affect the male reproductive system, including their distribution in tissues and body fluids, possible targets as well as the effects on the endocrine system. This chapter also covers several important therapeutic options to protect cells from the viruses. A last Chapter 15, the final chapter, mainly discusses the sperm redox system equilibrium and its implications for fertilization and male fertility. The role of ROS which has been responsible for cellular oxidative damage is discussed in this chapter. The fundamental questions like how glutathione transferases, glutathione peroxidases, the mammalian testis thioredoxin system and their respective substrates and cofactors, specifically glutathione and selenium, facilitate the protection, successful development, and maturation of mammalian spermatozoa have been outlined in this chapter. This chapter shows high impact towards redox system and free radical research.

All the chapters presented in this book are interlinked and gradually lead to the conclusion that oxidative stress–induced ROS formation is a responsible factor behind male infertility. This book also focuses on the signaling pathways and molecular factors associated with environmental toxicity in understanding the mechanisms of toxicity associated with free radical generation and male infertility. One who could read all the chapters from 1 to 15 will inculcate a better understanding of environmental challenges for male reproduction and possible protective measures in the form of herbal medicine, bioactive compounds, therapeutic approaches, and antioxidant intakes. We hope this book will serve as both an excellent review and a valuable reference for formulating suitable measures against oxidative stress and for promoting the science involved in the male reproductive system.

Finally, we would like to dedicate this book to all the frontline workers globally, who are fighting 24/7 against COVID-19 to save our lives, and also to COVID-19 warriors who lost their lives saving humanity. We would like to thank all the authors who contributed to this book. Last but not least, our special thanks go to the series editors, Dr. W. E. Crusio, Dr. H. Dong, Dr. H. H. Radeke, Dr. N. Rezaei, Dr. O. Steinlein, and Dr. J. Xiao, and the entire Springer editorial team for their sincere assistance and support. Our special thanks go to Dr. Carolyn Spence for her continuous support and suggestions throughout the book editing. The editors are also thankful to the entire production team including Mr. Vishnu Prakash and Ms. A. Meenahkumary for their support.

Aalto UniversityKavindra Kumar Kesari
Espoo, Finland
Assam UniversityShubhadeep Roychoudhury
Silchar, India

Contents

1 **Oxidative Stress and Toxicity in Reproductive Biology and Medicine: Historical Perspectives and Future Horizons in Male Fertility**.. 1
  Ralf Henkel

2 **Reactive Oxygen Species in the Reproductive System: Sources and Physiological Roles** 9
  Anandan Das and Shubhadeep Roychoudhury

3 **Pathological Roles of Reactive Oxygen Species in Male Reproduction**....................................... 41
  Saptaparna Chakraborty and Shubhadeep Roychoudhury

4 **Molecular Interactions Associated with Oxidative Stress-Mediated Male Infertility: Sperm and Seminal Plasma Proteomics**.......... 63
  Manesh Kumar Panner Selvam, Damayanthi Durairajanayagam, and Suresh C. Sikka

5 **Unraveling the Molecular Impact of Sperm DNA Damage on Human Reproduction** 77
  Renata Finelli, Bruno P. Moreira, Marco G. Alves, and Ashok Agarwal

6 **Role of Infection and Leukocytes in Male Infertility**.............. 115
  Sandipan Das, Shubhadeep Roychoudhury, Shatabhisha Roychoudhury, Ashok Agarwal, and Ralf Henkel

7 **Bacteriospermia and Male Infertility: Role of Oxidative Stress** 141
  Sandipan Das, Shubhadeep Roychoudhury, Anwesha Dey, Niraj Kumar Jha, Dhruv Kumar, Shatabhisha Roychoudhury, Petr Slama, and Kavindra Kumar Kesari

8	**Oxidant-Sensitive Inflammatory Pathways and Male Reproductive Functions**	165
	Sulagna Dutta, Pallav Sengupta, and Srikumar Chakravarthi	
9	**Oxidative Stress and Idiopathic Male Infertility**	181
	Pallav Sengupta, Shubhadeep Roychoudhury, Monika Nath, and Sulagna Dutta	
10	**Oxidative Stress and Varicocele-Associated Male Infertility**	205
	Terence Chun-Ting Lai, Shubhadeep Roychoudhury, and Chak-Lam Cho	
11	**Oxidative Stress in Men with Obesity, Metabolic Syndrome and Type 2 Diabetes Mellitus: Mechanisms and Management of Reproductive Dysfunction**	237
	Kristian Leisegang	
12	**Metabolic Dysregulation and Sperm Motility in Male Infertility** ...	257
	Sujata Maurya, Kavindra Kumar Kesari, Shubhadeep Roychoudhury, Jayaramulu Kolleboyina, Niraj Kumar Jha, Saurabh Kumar Jha, Ankur Sharma, Arun Kumar, Brijesh Rathi, and Dhruv Kumar	
13	**Tale of Viruses in Male Infertility**	275
	Shreya Das, Arunima Mondal, Jayeeta Samanta, Santanu Chakraborty, and Arunima Sengupta	
14	**Pathogenesis of Viral Infections and Male Reproductive Health: An Evidence-Based Study**	325
	Diptendu Sarkar, Shubham Dutta, Shubhadeep Roychoudhury, Preethi Poduval, Niraj Kumar Jha, Paltu Kumar Dhal, Shatabhisha Roychoudhury, and Kavindra Kumar Kesari	
15	**Sperm Redox System Equilibrium: Implications for Fertilization and Male Fertility**	345
	Lauren E. Hamilton, Richard Oko, Antonio Miranda-Vizuete, and Peter Sutovsky	

Index 369

Chapter 1
Oxidative Stress and Toxicity in Reproductive Biology and Medicine: Historical Perspectives and Future Horizons in Male Fertility

Ralf Henkel

Abstract Since the discovery by John MacLeod in 1943 that spermatozoa produce small amounts of hydrogen peroxide, a member of the so-called reactive oxygen species (ROS), the importance and functions of these highly reactive oxygen derivatives in physiology and pathology are a subject of numerous studies. It has been shown that they play essential roles, not only in causing oxidative stress if their concentration is excessively high, but also in triggering crucial cellular functions if their concentration is low. On the other hand, antioxidants counterbalance the action of ROS to maintain a fine balance between oxidation and reduction as an excessive amount of antioxidants leads to a condition called reductive stress and is as harmful as oxidative stress. This book *"Oxidative Stress and Toxicity in Reproductive Biology and Medicine – A Comprehensive Update on Male Infertility"* authoritatively summarizes the current knowledge of various causes of oxidative stress including various andrological conditions and environmental pollution as well as the physiological effects of ROS. Moreover, this book expands into the treatment of oxidative stress with antioxidants and phytomedicine, a rapidly developing area. As a first of its kind, this book also sheds light on the effects of the redox potential during the fertilization process and thus highlights the importance of the correct balance of oxidants and antioxidants, even in the culture medium in assisted reproduction. The editors have brought together an impressive group of renowned

R. Henkel (✉)
Department of Metabolism, Digestion and Reproduction, Imperial College London, London, UK

Department of Medical Bioscience, University of the Western Cape, Bellville, South Africa

American Center for Reproductive Medicine, Cleveland Clinic, Cleveland, OH, USA

LogixX Pharma, Theale, Reading, UK
e-mail: ralf.henkel@logixxpharma.com

experts to share their knowledge on the topic of oxidative stress and its clinical management in andrology and assisted reproduction.

Keywords Oxidative stress · Reductive stress · Antioxidants · Sperm functions · Fertilization

1.1 Introduction

It is almost 80 years ago that John MacLeod [45] made the pioneering discovery that the primary role of oxygen in spermatozoa is not for metabolism and motility as they cannot oxidize glucose, lactate, and pyruvate. MacLeod's observation went even further by indicating that high partial pressure of oxygen is not only inhibiting motility but also producing small amounts of hydrogen peroxide (H_2O_2), a highly oxidizing chemical compound that is now counted among the so-called reactive oxygen species (ROS). Since that time, scientists conducted a lot of research to understand not only the impact of extrinsic oxygen and highly reactive oxygen derivatives such as ROS but also cells' own intrinsic ROS production on cellular and organismal physiology and development and also their pathological impact on disease. By now, we have learnt much about the chemistry and biochemistry of oxygen [12] and its highly reactive derivatives with reaction times in the nano- to millisecond range [29], as well as the metabolic role of H_2O_2 in cellular signaling cascades and oxidative stress [53]. It is clear now that for normal physiologic function of cells, including reproduction in general and spermatozoa in specific, ROS at very low concentrations are essential because these molecules do not only trigger essential sperm functions such as capacitation, acrosome reaction, and spermatozoa-oocyte binding [10, 17, 49, 50] but also cause oxidative stress at higher concentrations [11].

Oxidative stress is a concept that was formulated by Helmut Sies in 1985 [52]. Groundbreaking work by John Aitken [9] and Ashok Agarwal [3] lead to the recognition of oxidative stress as major cause of male [2, 51] and female infertility [4, 5, 44] leading to miscarriage or birth defects [56]. Later it became clear that due to the dual effect of ROS on sperm function [18, 19] there must be a very fine balance between ROS essential for proper physiological activity and antioxidant protection from cellular oxidative damage. Hence, not only the ROS but also the antioxidants have to be regulated to counterbalance oxidative stress. Too high concentrations of antioxidants have also been found to be detrimental, as this condition can cause reductive stress [28, 34, 43] and is as harmful as oxidative stress [13]. In light of the recent, rapid developments in both the fields, redox-biology and human reproduction, it is time for a comprehensive update.

Oxidative Stress and Toxicity in Reproductive Biology and Medicine – A Comprehensive Update on Male Infertility provides such an update at the right time highlighting the important topics. This book covers human fertility, including assisted reproduction, in general, the special emphasis is on male reproduction and provides a unique collection of chapters spanning from the description of various sources of ROS in the reproductive system and their physiological roles in the generation of ROS to their pathological influence on fertility in a variety of diseases. Since not only diseases or aging but also environmental pollution can cause increased oxidative stress, the book also describes a number of environmental toxicants as well as radiation and their effects on fertility. This is followed by a section dedicated to the treatment of oxidative stress with herbal medicine and antioxidants. The topics covered are then rounded off with chapters on the impact of male oxidative stress on recurrent pregnancy loss, the treatment of oxidative stress in embryo culture as well as the impact of reductive stress on male fertility.

The sources of ROS causing oxidative stress in the reproductive system are manifold and include physiological production of ROS in all aerobic living cells. These, small amounts of ROS have been shown to trigger essential physiological functions such as capacitation [17]. On the other hand, excessive production or exposure of cells to excessive amounts of ROS caused by adverse lifestyle [22, 26, 47] and related conditions such as obesity [42], diabetes mellitus [1, 24], or poor nutrition negatively [54, 62] affect fertility in men and women and modification of the lifestyle conditions can have positive effects on fertility [62]. In addition, genital tract infections/inflammations [33, 35], scrotal hyperthermia due to tight underwear [38, 48], sedentary position [25, 55], varicocele [37] or cancer [57] and its treatment with radiation [8, 15] and chemotherapy [20, 36] all have significant negative effects on fertility. On the other hand, infertility is also regarded as a surrogate marker for cancer development [40].

In addition to the aforementioned production of high ROS levels due to medical reasons, in the past century, environmental pollution has become a significant cause of infertility, of which parts of the causes are oxidative stress-driven where the pollution is directly or indirectly triggering high ROS production. The resultant oxidative stress does not only have a direct negative effect on the gametes [21, 60] but also detrimental effects on the testes [32, 46]. Furthermore, the detrimental effects are not only direct but also indirect if pregnant women are exposed to environmental toxicants. Among other problems, this can lead to disturbances in the sex determination.

About 80% of the global population is dependent on traditional and herbal medicine for their primary health care [27] including reproductive issues. This book addresses this globally important issue by including four chapters on the role of herbs and phytomedicine. Many plants are used not only due to their contents of pharmacologically active compounds such as atropine, triterpene saponins, and lactones, or flavonoids and tannins, but also due to the high concentrations of antioxidants such as lycopene and various vitamins. Interestingly, about 25% of modern prescription medicines contain bioactive plant compounds. In Western medicine, still little is known about herbal remedies to treat diseases [14]. Therefore, it is important to bring these aspects of the usage of herbal medicine as primary or supplementary therapy option to the attention of the readers [16].

Since ROS play such important physiological roles in all bodily functions, shedding light on redox imbalances in reproductive events as a whole is important, including embryo development [31, 59]. The oxygen concentration in the embryo surrounding environment is only about 8% in the fallopian tube, about 5% or less in the uterus [58] and is becoming hypoxic or anoxic during early implantation [41]. This situation makes a metabolic shift from oxidative phosphorylation to glycolysis necessary [31, 59]. Hence, keeping the redox balance between oxidants and antioxidants is of utmost importance for normal embryonic development as too high concentrations of either ROS or antioxidants detrimental because essential transcription factors such as Nrf2, NFκB, or AP1 are redox-sensitive and if not properly triggered can interfere in signaling pathways [23, 59] and lead to embryonic death and teratogenesis [30, 61]. Therefore, this book, as the first of its kind in the field of human reproduction, throws more light not only on the pathology of male factors leading to recurrent pregnancy loss but also on the essential functions of the finely balanced redox systems needed for successful fertilization with its implications even in embryo culture systems.

Last, but not least, this book on oxidative stress and its toxicity in reproductive biology also addresses the use of antioxidants, which are not only freely available over the counter but also often prescribed by clinicians to treat oxidative stress-related conditions of infertility. In most cases, this is done without having properly diagnosed the patient for oxidative stress. Oral antioxidants provide an excellent safety and are cost-effective [6, 7, 39]. On the other hand, in respect to their effects, the bivalent actions of both ROS and antioxidants to trigger essential physiological reactions and to cause harm are widely unknown. Whereas for ROS the detrimental effects are generally known, the detrimental effects of antioxidants are either not known or are ignored. However, as ROS have beneficial effects in triggering e.g. sperm capacitation, antioxidants have detrimental effects, particularly if they are available at too high concentrations when they are causing reductive stress [34], a condition which is as harmful as oxidative stress [13].

Since the discovery of ROS and the recognition of oxidative stress as major contributing factor to numerous diseases including infertility, we have learnt a lot about oxidative stress and generally associate negative events with it. Opposed to this, antioxidants are thought to be 'good' and 'healthy' and neutralize the "bad" effects of oxidative stress. With the increase of knowledge, however, we began to realize that such "black and white painting" is incorrect. It is now clear that ROS are also crucial triggers for essential physiological functions. On the other hand, for antioxidants, we are just at the beginning of understanding that they can also have serious detrimental effects. This includes not only their functions but also the realization of reductive stress as cause of disease and infertility. The lesson that we have to learn is that only a finely regulated balance between oxidation and reduction provides a healthy environment for cellular function. Therefore, it is essential to properly diagnose the redox status and the causes of a patient's infertility before starting with a treatment which could worsen the fertility status. Hence, a personalized diagnosis with subsequent adjusted treatment is of utmost importance. The problem that we are facing at this stage is that we do not know what is "normal," not only for male infertility but also for female infertility. For embryo development, there are even no data available yet.

To come to this point has been a long journey, and with this limited information available for human infertility treatment, this book provides the reader with an authoritative overview of redox stress in human reproduction. The comprehensive knowledge provided in this book will assist not only andrologists and urologists in better comprehension of male infertility but is also a source of information for gynecologists, embryologists, and basic scientists to further their understanding of fertilization as a highly complex process. With all the up-to-date evidence provided, this book will be an invaluable source of information to improve patient management and laboratory procedures.

References

1. Abbasihormozi S, Babapour V, Kouhkan A, Naslji AN, Afraz K, Zolfaghary Z, Shahverdi A. Stress hormone and oxidative stress biomarkers link obesity and diabetes with reduced fertility potential. Cell J. 2019;21:307–13.
2. Agarwal A, Said TM. Oxidative stress, DNA damage and apoptosis in male infertility: a clinical approach. BJU Int. 2005;95:503–7.
3. Agarwal A, Ikemoto I, Loughlin KR. Relationship of sperm parameters with levels of reactive oxygen species in semen specimens. J Urol. 1994;152:107–10.
4. Agarwal A, Gupta S, Sharma RK. Role of oxidative stress in female reproduction. Reprod Biol Endocrinol. 2005;14:28.
5. Agarwal A, Aponte-Mellado A, Premkuma BJ, Shaman A, Gupta S. The effects of oxidative stress on female reproduction: a review. Reprod Biol Endocrinol. 2012;10:49.
6. Agarwal A, Leisegang K, Majzoub A, Henkel R, Finelli R, Panner Selvam MK, Tadros N, Parekh N, Ko EY, Cho CL, Arafa M, Alves MG, Oliveira PF, Alvarez JG, Shah R. Utility of antioxidants in the treatment of male infertility: clinical guidelines based on a systematic review and analysis of evidence. World J Mens Health. 2021a;39:233–90.
7. Agarwal A, Finelli R, Panner Selvam MK, et al. A global survey of reproductive specialists to determine the clinical utility of oxidative stress testing and antioxidant use in male infertility. World J Mens Health. 2021b;39:e17.
8. Ahmad G, Agarwal A. Ionizing radiation and male fertility. In: Gunasekaran K, Pandiyan N, editors. Male Infertility. Springer; 2017. p. 185–96.
9. Aitken RJ, Clarkson JS. Cellular basis of defective sperm function and its association with the genesis of reactive oxygen species by human spermatozoa. J Reprod Fert. 1987;81:459–69.
10. Aitken RJ, Irvine DS, Wu FC. Prospective analysis of sperm-oocyte fusion and reactive oxygen species generation as criteria for the diagnosis of infertility. Am J Obstet Gynecol. 1991;164:542–51.
11. Aitken RJ, Baker MA, Nixon B. Are sperm capacitation and apoptosis the opposite ends of a continuum driven by oxidative stress? Asian J Androl. 2015;17:633–9.
12. Cadenas E. Biochemistry of oxygen toxicity. Annu Rev Biochem. 1989;58:79–110.
13. Castagne V, Lefevre K, Natero R, Clarke PG, Bedker DA. An optimal redox status for the survival of axotomized ganglion cells in the developing retina. Neuroscience. 1999;93:313–20.
14. Castleman M. Healing Herbs. New York: Bantam Books; 1995. p. 1–5.
15. Centola GM, Keller JW, Henzler M, Rubin P. Effect of low-dose testicular irradiation on sperm count and fertility in patients with testicular seminoma. J Androl. 1994;15:608–13.

16. Dada R, Sabharwal P, Sharma A, Henkel R. Use of herbal medicine as primary or supplementary treatments. In: Henkel R, Agarwal A, editors. Herbal Medicine in Andrology – An Evidence-Based Update. London: Elsevier Academic Press; 2020. p. 9–15.
17. de Lamirande E, Gagnon C. Human sperm hyperactivation and capacitation as parts of an oxidative process. Free Radical Biol Med. 1993;14:157–66.
18. de Lamirande E, Gagnon C. Impact of reactive oxygen species on spermatozoa: a balancing act between beneficial and detrimental effects. Hum Reprod. 1995;10(Suppl 1):15–21.
19. de Lamirande E, Jiang H, Zini A, Kodama H, Gagnon C. Reactive oxygen species and sperm physiology. Rev Reprod. 1997;2:48–54.
20. Delbes G, Hales BF, Robaire B. Effects of the chemotherapy cocktail used to treat testicular cancer on sperm chromatin integrity. J Androl. 2007;28:241–9.
21. Di Nisio A, Foresta C. Water and soil pollution as determinant of water and food quality/contamination and its impact on male fertility. Reprod Biol Endocrinol. 2019;17:4.
22. Durairajanayagam D. Lifestyle causes of male infertility. Arab J Urol. 2018;16:10–20.
23. Filomeni G, Rotilio G, Ciriolo MR. Disulfide relays and phosphorylative cascades: partners in redox-mediated signaling pathways. Cell Death Differ. 2005;12:1555–63.
24. Gandhi J, Dagur G, Warren K, Smith NL, Sheynkin YR, Zumbo A, Khan SA. The role of diabetes mellitus in sexual and reproductive health: An overview of pathogenesis, evaluation, and management. Curr Diabetes Rev. 2017;13:573–81.
25. Gill K, Jakubik J, Kups M, Rosiak-Gill A, Kurzawa R, Kurpisz M, Fraczek M, Piasecka M. The impact of sedentary work on sperm nuclear DNA integrity. Folia Histochem Cytobiol. 2019;57:15–22.
26. Gupta S, Fedor J, Biedenharn K, Agarwal A. Lifestyle factors and oxidative stress in female infertility: is there an evidence base to support the linkage? Expert Rev Obstet Gynecol. 2013;8:607–24.
27. Gurib-Fakim A. Medicinal plants: traditions of yesterday and drugs of tomorrow. Mol Aspects Med. 2006;27:1–93.
28. Gutteridge JMC, Halliwell B. Mini-review: oxidative stress, redox stress or redox success? Biochem Biophys Res Commun. 2018;502:183–6.
29. Halliwell B, Gutteridge JMC. Free Radicals in Biology and Medicine. 2nd ed. Oxford: Clarendon Press; 1989.
30. Hansen JM. Oxidative stress as a mechanism of teratogenesis. Birth Defects Res C Embryo Today. 2006;78:293–307.
31. Harvey AJ, Kind KL, Thompson JG. REDOX regulation of early embryo development. Reproduction. 2002;123:479–86.
32. Henkel R. Environmental Contamination and Testicular Function. In: Sikka SC, Hellstrom WJG, editors. Handbook of Bioenvironmental Toxicology: Men's Reproductive & Sexual Health. Elsevier; 2018. p. 191–208.
33. Henkel R. Infection in Infertility. In: Parekattil SJ, Esteves SC, Agarwal A, editors. Male Infertility: Contemporary Clinical Approaches, Andrology, ART and Antioxidants. 2nd ed. Cham: Springer; 2020. p. 409–24.
34. Henkel R, Sandhu IS, Agarwal A. The excessive use of antioxidant therapy: a possible cause of male infertility? Andrologia. 2019;51:e13162.
35. Henkel R, Offor U, Fisher D. The role of infections and leukocytes in male infertility. Andrologia. 2020 Jul 21;e13743. doi: https://doi.org/10.1111/and.13743. Online ahead of print.
36. Howell SJ, Shalet SM. Testicular function following chemotherapy. Hum Reprod Update. 2001;7:363–9.
37. Jensen CFS, Østergren P, Dupree JM, Ohl DA, Sønksen J, Fode M. Varicocele and male infertility. Nat Rev Urol. 2017;14:523–33.
38. Jung A, Schuppe HC. Influence of genital heat stress on semen quality in humans. Andrologia. 2007;39:203–15.

39. Kuchakulla M, Soni Y, Patel P, Parekh N, Ramasamy R. A systematic review and evidence-based analysis of ingredients in popular male fertility supplements. Urology. 2020;136:133–41.
40. Lao M, Honig SC. Male infertility and subsequent risk of cancer development. J Men's Health. 2015;11:19–28.
41. Leese HJ. Metabolic control during preimplantation mammalian development. Hum Reprod Update. 1995;1:63–72.
42. Leisegang K, Henkel R, Agarwal A. Obesity and metabolic syndrome associated systemic inflammation and the impact on the male reproductive system. Am J Reprod Immunol. 2019;82:e13178.
43. Lipinski B. Evidence in support of a concept of reductive stress. Br J Nutr. 2002;87:93–4.
44. Luti S, Fiaschi T, Magherini F, Modesti PA, Piomboni P, Semplici B, Morgante G, Amoresano A, Illiano A, Pinto G, Modesti A, Gamberi T. Follicular microenvironment: Oxidative stress and adiponectin correlated with steroids hormones in women undergoing in vitro fertilization. Mol Reprod Dev. 2021;88:175–84.
45. MacLeod J. The role of oxygen in the metabolism and motility of human spermatozoa. Am J Physiol. 1943;138:512–8.
46. Mathur PP, Huang L, Kashou A, Vaithinathan S, Agarwal A. Environmental toxicants and testicular apoptosis. Open Reprod Sci J. 2011;3:114–24.
47. Mendiola J, Torres-Cantero AM, Agarwal A. Lifestyle factors and male infertility: an evidence-based review. Arch Med Sci. 2009;5(1A):S3–S12.
48. Minguez-Alarcon L, Gaskins AJ, Chiu YH, Messerlian C, Williams PL, Ford JB, Souter I, Hauser R, Chavarro JE. Type of underwear worn and markers of testicular function among men attending a fertility center. Hum Reprod. 2018;33:1749–56.
49. O'Flaherty C. Peroxiredoxins: hidden players in the antioxidant defence of human spermatozoa. Basic Clin Androl. 2014;24:4.
50. O'Flaherty CM, Beorlegui NB, Beconi MT. Reactive oxygen species requirements for bovine sperm capacitation and acrosome reaction. Theriogenology. 1999;52:289–301.
51. Sharma RK, Agarwal A. Role of reactive oxygen species in male infertility. Urology. 1996;48:835–50.
52. Sies H. Oxidative Stress. London: Academic Press; 1985. p. 1–507.
53. Sies H. Role of metabolic H_2O_2 generation: redox signaling and oxidative stress. J Biol Chem. 2014;289:8735–41.
54. Skoracka K, Eder P, Lykowska-Szuber L, Dobrowolska A, Krela-Kazmierczak I. Diet and nutritional factors in male (in)fertility-underestimated factors. J Clin Med. 2020;9:1400.
55. Song GS, Seo JT. Changes in the scrotal temperature of subjects in a sedentary posture over a heated floor. Int J Androl. 2006;29:446–57.
56. Toboła-Wróbel K, Pietryga M, Dydowicz P, Napierała M, Brązert J, Florek E. Association of oxidative stress on pregnancy. Oxid Med Cell Longev. 2020;2020:6398520.
57. Tvrda E, Agarwal A, Alkuhaimi N. Male reproductive cancers and infertility: a mutual relationship. Int J Mol Sci. 2015;16:7230–60.
58. Ufer C, Wang CC. The roles of glutathione peroxidases during embryo development. Front Mol Neurosci. 2011;4:12.
59. Ufer C, Wang CC, Borchert A, Heydeck D, Kuhn H. Redox control in mammalian embryo development. Antioxid Redox Signal. 2010;13:833–75.
60. Varghese AC, Ly KD, Corbin C, Mendiola J, Agarwal A. Oocyte developmental competence and embryo development: impact of lifestyle and environmental risk factors. Reprod Biomed Online. 2011;22:410–20.
61. Wang CC, Rogers MS. Oxidative stress and fetal hypoxia. In: Laszlo G, editor. Reactive Oxygen Species and Disease. 282: Research Signpost, 257; 2007.
62. Wright C, Milne S, Leeson H. Sperm DNA damage caused by oxidative stress: modifiable clinical, lifestyle and nutritional factors in male infertility. Reprod Biomed Online. 2014;28:684–703.

Chapter 2
Reactive Oxygen Species in the Reproductive System: Sources and Physiological Roles

Anandan Das and Shubhadeep Roychoudhury

Abstract Reactive oxygen species (ROS) are oxygen-containing molecules which are reactive in nature and are capable of independent existence in the body. ROS comprise mostly of free radicals that contain at least one unpaired electron. Endogenous sources are the foremost birthplaces of ROS, which include mitochondrial electron transport chain, endoplasmic reticulum and peroxisome. Conversely, numerous enzymatic pathways such as xanthine oxidase and cyclooxygenase systems are among the prominent generators of cellular ROS. Major sources of ROS in the female reproductive tract include Graafian follicles, follicular fluid, fallopian tube, peritoneal cavity and endometrium. On the contrary, leukocytes, immature spermatozoa and varicocele are the key originators of ROS in the male reproductive system. For the sake of maintaining a proper oxidative balance, cells have evolved a variety of antioxidative molecules. From the physiological perspective, ROS and antioxidants are actively involved in the regulation of myriad female reproductive processes, such as cyclic luteal and endometrial changes, follicular development, ovulation, fertilization, embryonic implantation, maintenance of pregnancy and parturition. Similarly, physiological amounts of ROS are crucial in the accomplishment of various male reproductive functions as well, which include spermatozoa maturation, capacitation, hyperactivation and acrosome reaction. This chapter highlights the birthplaces of ROS in the female and male reproductive tract along with mechanisms of their production. This chapter will also put forward specific physiological roles of these reactive molecules in upholding the structural integrity and functionality of both the reproductive systems.

Keywords Oxidative stress · Free radicals · Female · Endometrium · Ovulation · Implantation · Male · Spermatozoa · Hyperactivation · Acrosome reaction

---

A. Das · S. Roychoudhury (✉)
Department of Life Science and Bioinformatics, Assam University, Silchar, India

2.1 Introduction

Reactive oxygen species (ROS) are extremely reactive oxidizing molecules which are endogenously produced during mitochondrial oxidative phosphorylation of the cellular aerobic metabolism [1]. Among the endogenous sources, mitochondria are the most prominent sources of ROS [2]. Apart from the mitochondrial electron transport chain (ETC), the major intracellular sources of ROS production include endoplasmic reticulum and peroxisomes [3]. Moreover, enzymatic pathways such as reduced nicotinamide adenine dinucleotide phosphate (NADPH) oxidase, xanthine oxidase, lipoxygenase, cyclooxygenase and the cytochrome P450 systems serve as prominent sources of ROS generators in the cell [4]. ROS may also arise exogenously from interactions with external sources like environmental pollutants, ionizing radiations, ultraviolet light, chemotherapeutic agents, inflammatory cytokines, xenobiotic compounds and bacterial invasion [1, 5]. ROS are mostly free radicals in nature, most significant of which are superoxide ($O_2^{\bullet-}$) and hydroxyl anions ($OH^{\bullet}$) [6]. Apart from free radicals, non-radicals such as hydrogen peroxide (H_2O_2) and singlet oxygen ($^1O_2$) are also considered as important forms of ROS [7].

ROS, irrespective of its class, possess the fundamental chemical properties necessary for conferring reactivity to different biological targets. ROS also play the role of signalling molecules which help in regulating biological and physiological processes [8, 9]. From evolutionary biology studies, it is quite evident that nature had selected ROS as signal transducers for the allowance of adaptation to changes in nutrient availability and the oxidative environment [10]. When present in reasonable strength, ROS play vital roles in maintaining a large number of physiological activities in both female and male reproductive systems [11]. In the female, ROS are known to modulate several physiological processes ranging from ovulation to fertilization. They also play a critical role during gestation and normal parturition as well as initiation of pre-term labour [3]. In men, ROS are actively involved in the normal physiology of the spermatozoa, including facilitation of the maturational stages, thereby, maximizing their fertilizing ability [12].

As already mentioned, ROS serve as key signalling molecules in physiological processes when present in optimum concentration, but when the balance is interrupted towards a surplus of ROS, it gives rise to various pathological conditions. Due to their ability to play both physiological and pathological roles, ROS are often regarded as a double-edged sword [3]. In battling the excess ROS there exists an antioxidant defence mechanism in the body, comprising enzymes or non-enzymes which are capable of neutralizing the reactive molecules. Antioxidant defence is surpassed when ROS concentration rise beyond its normal capacity, thereby disrupting the balance, which leads to a condition called oxidative stress. Oxidative stress can result in disruption of cellular functioning or even cell death through several mechanisms including lipid peroxidation, nucleic acid, and protein damages [13].

This chapter highlights some of the important free radicals and non-radicals that exert biological effects on the reproductive system. Along with this, some of the major endogenous sources responsible for production of ROS in the female and the male reproductive tracts have been delineated. Finally, it sheds light on various ROS-mediated reproductive physiologies and their complicated mechanisms of action.

2.2 Reactive Oxygen Species (ROS)

ROS is a recurrently utilized term in biological sciences and medicine. Aerobic organisms utilize molecular oxygen (O_2) which results in the generation of a large number of oxygen-containing reactive molecules that are collectively known as ROS [14]. ROS is a cumulative term to include $O_2^{\bullet-}$, H_2O_2, $OH^\bullet$, $^1O_2$, peroxyl radical ($LOO^\bullet$), alkoxyl radical ($LO^\bullet$), lipid hydroperoxide (LOOH), peroxynitrite ($ONOO^-$), hypochlorous acid (HOCl) and ozone (O_3), among others [15]. The oxides of nitrogen such as nitric oxide ($NO^\bullet$), $ONOO^-$ and nitrogen dioxide ($NO_2^\bullet$) are generally included under the category of reactive nitrogen species (RNS). RNS are a family of nitrogen moieties associated with O_2, which are produced during the reaction of NO with $O_2^{\bullet-}$ and H_2O_2. Similarly, chlorine-containing reactive species, such as HOCl, that are capable of chlorinating and oxidizing other molecules are called reactive chlorine species (RCS). However, these are not a frequent matter of concern in biology and medicine, as compared to ROS and RNS [14].

Some of the O_2-containing reactive molecules have unpaired electrons and are thus called O_2-free radicals. On the other hand, those ROS which are devoid of unpaired electrons are called non-radical ROS [14]. Some of the major free radicals and non-radicals are discussed below.

2.2.1 Free Radicals

Free radicals are defined as atoms or molecules containing one or more unpaired electrons in the outer electronic orbital that are capable of existing independently in the body. Their high reactive nature renders them the ability to abstract electrons from other compounds to attain stability. The molecule which is attacked eventually loses its electrons and becomes a free radical itself, which then initiates a chain of reaction cascade [16].

2.2.2 Superoxide Anion ($O_2^{\bullet-}$)

Superoxide ($O_2^{\bullet-}$) is an anionic radical which is generated during the one-electron reduction of O_2 [17]. This reaction can occur non-enzymatically in the mitochondrial ETC or by the activity of NADPH oxidase and xanthine oxidase enzymes [18].

The concentration of $O_2^{\bullet-}$ are kept in check by effective compartmentalization of the sequential decrease of O_2 and by widespread expressions of an enzyme called superoxide dismutase (SOD), which converts these anions into non-radical species such as H_2O_2 and $^1O_2$ [19, 20].

Superoxide ($O_2^{\bullet-}$) is fairly non-reactive to most of the biological molecules, even though the low level of $O_2^{\bullet-}$ tolerable in cells and tissues indicate that restraining cell exposures to $O_2^{\bullet-}$ is a significant selector for survival [21].

2.2.3 Hydroxyl Radical (OH•)

Hydroxyl radicals (OH•) are generated in situ by either oxidation of water or hydroxide ions. The interaction between H_2O_2 and reduced transition metals such as copper and iron may also lead to the production of OH•. Moreover, water molecules when exposed to ionizing radiation can give rise to OH• [17, 18, 22].

Hydroxyl radicals (OH•) are highly reactive species and most of the organic molecules are vulnerable to its attack. The oxidation potential of these molecules renders them highly oxidizing in nature [23]. Furthermore, due to the non-selective nature of OH•, many susceptible organic molecules in the body such as acids, aromatics and ketones get degraded by its action [24].

2.2.4 Nitric Oxide (NO•)

Nitric oxide (NO•) is a gas and a diatomic free radical which is synthesized in vivo from L-arginine and O_2 by nitric oxide synthase (NOS). Heme-containing proteins are accountable for biological synthesis of NO•, which is generated by two successive monooxygenase reactions at the NOS heme active site [25].

Nitric oxide (NO•) mediates myriad biological functions through stimulation of its primary cellular receptor haemoprotein-soluble guanylate cyclase, which in turn, produces the second messenger molecule cyclic-guanosine monophosphate (cGMP) [25]. The bioavailability and actions of NO• are mediated by its reaction with $O_2^{\bullet-}$, which yields reactive peroxides and $ONOO^-$, thereby merging O_2 radicals and NO• pathways [26]. NO• regulates endothelium-dependent regulation of blood flow and pressure as well as inhibits the activation of blood platelets, which holds particular importance in achieving a penile erection. Moreover, NO• is recognized as a neurotransmitter in certain types of nerves [27].

2.3 Non-radicals

Non-radicals are the atoms or molecules which, unlike the free radicals, are devoid of unpaired electrons [14]. The non-radical species mediates their biological effects by initiating free radical reactions in the living organisms [28].

2.3.1 Hydrogen Peroxide (H_2O_2)

It is primarily produced as a metabolic by-product of aerobic respiration taking place in the mitochondria, where partial reduction of O_2 results in the generation of $O_2^{\bullet-}$. In addition, enzymes such as SOD helps in the production of H_2O_2 from $O_2^{\bullet-}$ [29]. Moreover, H_2O_2 may also be generated endogenously by p66[Shc] enzyme activity, a member of the Src homology/collagen protein family [30]. H_2O_2 may get decomposed to water in cytosol and mitochondria by the activity of scavenger enzyme glutathione peroxidase (GPx) or in the peroxisome by the action of catalase [31].

It has been considered a toxic molecule for living systems for a long time [32]. However, recent findings have elucidated the physiological roles of H_2O_2 and has recognized it as a ubiquitous endogenous molecule. H_2O_2 has been suspected to play the role of second messenger system with a pro-survival activity in several physiological processes [33].

2.3.2 Singlet Oxygen ($^1O_2$)

Singlet oxygen ($^1O_2$) is a non-radical that represents an excited state of O_2. In humans, $^1O_2$ is produced by enzymatic activation of O_2 and during phagocytosis by macrophages [34, 35]. $^1O_2$ is not as stable as the ground state O_2 and tends to change back fairly quickly at the cost of surrounding molecule, thereby inducing a chemical effect [34].

Being a strong oxidant, $^1O_2$ possesses the capability of readily oxidizing lipids, proteins and nucleic acids present in the cell [36]. Though, unlike radical O_2 species, $^1O_2$ is rather mild and non-toxic to mammalian tissues [37]. $^1O_2$ exerts its biological effects by acting as a cell signal messenger through the activation of Ca^{2+}-regulated K^+ channels [38].

Table 2.1 Major ROS in human body along with their biosynthetic pathways

	ROS	Mechanism of biosynthesis	References
Free radicals	Superoxide ($O_2^{\bullet-}$)	Enzymatic or non-enzymatic reduction of molecular oxygen (O_2)	[17,18]
	Hydroxyl ($OH^{\bullet}$)	Oxidation or ionization of water molecules	[17,18]
	Nitric oxide ($NO^{\bullet}$)	Reaction of L-arginine and molecular oxygen (O_2) in presence of NOS	[25]
Non-radicals	Hydrogen peroxide (H_2O_2)	Mostly by the partial reduction of molecular oxygen (O_2) and subsequent action of SOD during aerobic metabolism	[29]
	Singlet oxygen ($^1O_2$)	Enzymatic activation of oxygen (O_2) and during phagocytosis	[35]
	Ozone (O_3)	Decomposition of dihydrogen trioxide (H_2O_3) during antibody-catalyzed water oxidation pathway	[39]

2.3.3 Ozone (O_3)

Ozone (O_3) is best known as a gaseous compound occurring in the stratosphere where it shields the entry of harmful solar radiations. However, it has been suspected that O_3 is also produced endogenously in biological systems through the process of antibody-catalyzed water oxidation pathway [39]. This reaction occurs in a hydrophobic site in the antibody molecule, where dihydrogen trioxide (H_2O_3) is formed during the initial reaction of water with $^1O_2$, which subsequently decomposes to H_2O_2 and O_3 [40].

Ozone (O_3) is a highly reactive gaseous allotrope of O_2 and it can have pathogenetic consequences by cleaving the unsaturated lipids and oxidizing the sulphur and nitrogen atoms of proteins [41]. However, judicious concentration of O_3 can reduce oxidative stress by upregulating intracellular antioxidant enzymes and it can also provoke the generation of ROS acting as natural physiological activators of various biological functions [42]. The major ROS produced in the human body along with their respective biosynthetic pathways are summarized in Table 2.1.

2.4 Sources of ROS in the Female Reproductive System

ROS have been linked with various physiological and pathological functions of the female reproductive system [3]. The female reproductive tract is thought to be a potent site of ROS production. ROS are mostly generated in the ovarian cells such as the Graafian follicles and in other non-ovarian tissues like the fallopian tube, the endometrium and the peritoneum [7].

2.4.1 Graafian Follicle

Graafian follicles have been claimed to be a possible source of ROS because of the oocyte and various immune cells such as macrophages and neutrophils. The oocyte produces ROS by various mechanisms – the most prominent being oxidative phosphorylation that takes place in the oocyte for production of ATP by utilizing O_2. Another pathway is the xanthine oxidase system of the follicle, which generates ROS during the elimination of nitrogenous wastes mainly in the form of uric acid. Furthermore, immune cells produce ROS in the form of HOCl by the NADPH oxidase system to protect Graafian follicles against microbial attack [43].

The role of NADPH oxidase along with SOD type 3 (SOD3) has been studied extensively in the follicular cells of *Drosophila*. An NADPH oxidase and an extracellular SOD3 functions in follicular cells of *Drosophila* egg chamber to produce H_2O_2, which is critical for follicular rupture and ovulation. Human follicles also express these two enzymes and thus could perform similar roles in women [44]. The mammalian NADPH oxidase family comprises of seven members, of which, the expression of NADPH oxidase 4 and 5 (NOX4 and NOX5) are found in human granulosa cells [45]. NOX4 and NOX5 allocate an electron across the granulosa cell membrane from NADPH in the cytosol to O_2 in the luminal or extracellular space. This intercellular transportation of electrons eventually generates $O_2^{\bullet-}$, which in turn, is rapidly converted to H_2O_2 by SODs [44]. ROS can also generate from the cumulous oophorus, which originates from undifferentiated granulosa cells [46].

Generation of native ROS is essential for cellular physiology while widespread ROS may be harmful [44]. Oxidative stress initiating from excessive production of ROS in the granulosa cells and cumulous oophorus is detrimental to oocyte functioning and viability and can even lead to DNA damage [47].

2.4.2 Follicular Fluid

The production of follicular fluid is undertaken by granulosa cells and theca interna cells, which are the likely producers of ROS [48]. Follicular fluid contains various steroid hormones, growth factors, leukocytes and cytokines, all of which are the potent generators of ROS [49]. During the ovulatory process, ROS are produced in the follicular fluid physiologically and serve as key signalling molecules in regulating various reproductive functions such as maturation of oocyte, steroid production by ovary, and functions of the corpus luteum, as well as the development of fertilization and embryo [50].

Follicular fluid creates a unique microenvironment necessary for maintaining the quality of oocyte, implantation, and the development of early embryo. Overproduction of ROS in follicular fluid may alter the fine balance of distinctive composition which are critical for female fertility [51]. For this reason, follicular fluid also houses various antioxidant enzymes, which under physiological conditions, avoid

ROS production and forage prevailing free radicals, thereby protecting the ovarian somatic cells, and maintaining a healthy oocyte physiology [52].

2.4.3 Fallopian Tube

Some animal and human studies have confirmed the presence of ROS in the fallopian tubes – the most noteworthy is NO• [53–55]. An endogenous ROS generation system exists in the fallopian tube which has been demonstrated by positive NADPH diaphorase staining [53]. The formation of NO• is catalyzed by NOS which exists in the human tubal cells. It has already been mentioned that NO• arises from O_2 and L-arginine by the action of NOS. NO• further reacts with $O_2^{•-}$ to give rise to ONOO⁻, which is necessary for the process of signal transduction [25].

There are three isoforms of NOS: endothelial (eNOS), neuronal (nNOS) and inducible (iNOS) [3]. The three isoforms of NOS are produced differently in various tissues of the human fallopian tube [56]. However, only tubal epithelial cells produce all the isoforms of NOS, signifying that the tubal epithelial cells are potentially significant determinant of the effect of NO• in the fallopian tube [57]. eNOS and nNOS are accountable for incessant basal release of NO• and both require calcium/calmodulin system for activation [58]. The third isoform iNOS is an inducible form independent of calcium and is released in reply to inflammatory cytokines and lipopolysaccharides [59]. A study conducted on knockout mouse model has indicated that the activity of NO• is sustained at appropriate levels by the action of NOS isoforms [60].

Studies have shown that NO• could affect the contraction of fallopian tube smooth muscle and ciliary beat frequency present in epithelial cells of airway via cGMP pathway, which is necessary for smooth passage of ovum from the ovary [55, 61]. However, at high concentrations, it can lead to immune reactions and tissue damage [57]. Thus, to limit ROS toxicity, antioxidant enzymes are supplied in abundance to the fallopian tubes [3]. It is noteworthy that decreased level of NO• can have negative effects on the ovum [47].

2.4.4 Peritoneal Cavity

The space between the parietal peritoneum and visceral peritoneum is called peritoneal cavity. It largely houses the visceral organs such as liver, stomach, small and large intestines, and associated smaller organs. The female reproductive organs such as uterus, fallopian tubes and ovaries also project into the peritoneal cavity [62, 63].

The main origin points of ROS in the peritoneal cavity include macrophages, apoptotic endometrial cells, red blood cells (RBC) and menstrual-reflux debris [64, 65]. Various pro-oxidant and pro-inflammatory factors are released by RBCs, such as haemoglobin and its by-products heme and Fe^{2+}, into the peritoneal conditions.

Heme and Fe^{2+} are indispensable for human cells but play a crucial role in the generation of lethal ROS, unless appropriately chelated [66].

Disruption in the balance of ROS and antioxidant enzymes in the peritoneal cavity is suspected to cause endometriosis, a major causal factor of female infertility [67]. Increased level of ROS may also regulate the expression of genes encoding immunoregulators, cytokines and cell-adhesion molecules implicated in the induction of endometriosis [64].

2.4.5 Endometrium

The endometrium is the innermost epithelial lining of the uterus, which is highly organized in nature and is under cyclical control of ovarian steroids and pituitary hormones [68]. Various physiological processes related to the endometrium are controlled by ROS present in physiological concentrations. Endometrial stromal cells are speculated to be the potential source of ROS. The endoplasmic reticulum and the electron transport system both in mitochondria and in the nucleus of these stromal cells are the intracellular sources of ROS [69]. The production of ROS increases during the late secretory phase of the endometrium, which is believed to carry out a vital role in endometrial shedding, prior to menstruation [70].

A fine balance exists between ROS and the various defence systems against oxidative stress in the endometrium. SODs present in cytosol and mitochondria metabolize $O_2^{\bullet-}$, whereas GPx metabolize H_2O_2. Disruption of oxidative balance in the endometrium results in abnormalities of normal reproductive physiological functions such as menstruation and decidualization [69].

Various endogenous sources of ROS in the female reproductive tract have been elucidated in Fig. 2.1.

2.5 Sources of ROS in the Male Reproductive System

Human spermatozoa are the active production sites of ROS and $O_2^{\bullet-}$ is considered the most abundant ROS [71, 72]. In fact, $O_2^{\bullet-}$ is the precursor of other ROS types in spermatozoa. The reduced product of $O_2^{\bullet-}$ reacts with itself via dismutation to generate H_2O_2. In presence of Fe^{2+} and Cu^{2+}, H_2O_2 further reacts with $O_2^{\bullet-}$ to undergo the Haber-Weiss reaction which generates the highly reactive $OH^{\bullet}$ [73, 74]. The generation of ROS in spermatozoa occur via two methods: (a) the NADPH oxidase system at the level of spermatozoa plasma membrane, and (b) the reduced nicotinamide adenine dinucleotide (NADH)-dependent oxidoreductase system at the level of mitochondria [75].

NADPH oxidases are a set of enzymes located in the plasma membrane of spermatozoa, which catalyze the conversion of bivalent O_2 to $O_2^{\bullet-}$. NADPH molecules produced by the hexose monophosphate shunt act as the substrate for this reaction

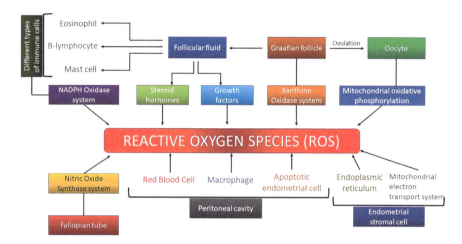

Fig. 2.1 Endogenous sources of ROS in the female reproductive system. ROS may be produced in the Graafian follicle via the xanthine oxidase system or by the oxidative phosphorylation that takes place in the mitochondria of the oocyte. The follicular fluid present inside the Graafian follicle harbours various steroid hormones and growth factors which may also give rise to ROS. Immune cells such as eosinophils, mast cells and B-cells present in the follicular fluid also produce ROS via the reduced nicotinamide adenine dinucleotide phosphate (NADPH) oxidase system. The nitric oxide synthase system operating in the tubal cells of fallopian tube contributes to the generation of ROS. Red blood cells, macrophages and apoptotic endometrial cells present in the peritoneal cavity also release ROS. Endoplasmic reticulum and the mitochondrial electron transport chain present in the endometrial stromal cells are also a potent generation site of ROS.

[76]. NOX5 is a member of the NADPH oxidase family, the presence of which has been confirmed in the acrosome and the midpiece region of spermatozoa [77]. NOX5 is stimulated when Ca^{2+} attaches to its cytosolic N-terminal domain. Eventually, several conformational changes in the cell initiate the production of $O_2^{\bullet-}$ [72].

The mitochondrial oxidoreductase system contributes to the bulk of ROS in the human spermatozoa. This is due to the abundance of mitochondria, which is required to provide relentless supply of energy for the sake of spermatozoa motility [75]. Respiring spermatozoa are an active foundation of ROS, in which the NADH oxidoreductase catalyzes the transfer of electrons from NADH to coenzyme Q10 in the mitochondrial ETC. During this process of inverse transfer of electron, electron leakage occurs, which reduces O_2 to $O_2^{\bullet-}$ [78]. The upsurge of the respiratory rate of spermatozoa increases the concentration of O_2 in the mitochondria, which in turn results in escalated release of $O_2^{\bullet-}$ [79].

ROS are mainly detected in the seminal plasma, which are endogenously originated from different sources. The human ejaculate contains diverse types of round cells from different developmental stages of spermatozoa, leukocytes and epithelial cells. Leukocytes mainly neutrophils and macrophages, and immature spermatozoa are considered the chief endogenous generators of ROS [80, 81].

2.5.1 Leukocytes

Leukocytes have been shown to be a prime generator of ROS in the human semen. Activated leukocytes, which are the peroxidase-positive variants, yield huge amounts of ROS [7]. Peroxidase-positive leukocytes include polymorphonuclear leukocytes (PMNs) and macrophages [82]. These leukocytes are generated in a large proportion by the prostate and seminal vesicles, which later make way to the seminal fluid [83]. Several intracellular and extracellular inducers such as infection and inflammation activate these leukocytes, thereby exciting them to release up to 100 times more ROS than normal [84, 85].

Neutrophils containing myeloperoxidase use $O_2^{\bullet-}$ to oxidize Cl^-, which in turn, results in the generation of HOCl. Additionally, leukocytes intensify the synthesis of NADPH by the hexose monophosphate shunt, which increases the action of NADPH oxidases, thereby upregulating the synthesis of $O_2^{\bullet-}$ [7].

There is an association between male accessory gland infection (MAGI) and seminal ROS levels. Prostatitis is the most common form of MAGI, with prostato-vesiculitis and prostato-vesiculo-epididymitis being the other relevant forms. MAGI generate immune reactions which results in an increased infiltration of leukocytes in the male genital tract, thereby leading to ROS production [76]. Bacterial prostatitis has also been linked with autoimmune response to prostate antigens with excessive ROS production. People diagnosed with chronic prostatitis who do not contain leukocytes may also be found to have raised levels of ROS, indicating some other sources of ROS in the prostate gland [86].

Optimum concentrations of ROS and antioxidants are to be maintained for the sustenance of a healthy state. An increase in the pro-inflammatory cytokines such as interleukin (IL)-8 and a reduction in the SOD concentrations can initiate respiratory burst, resulting in increased ROS productivity and ultimately, oxidative stress. If the seminal leukocyte levels are unusually increased, then oxidative stress tends to cause damage to spermatozoa [87].

2.5.2 Immature Spermatozoa

The quantity of ROS production varies among spermatozoa of different stages of development, which indicates that the level of maturation influences the ROS synthesis [83]. For the sake of fertilization, developing spermatozoa extrude their cytoplasm during spermatogenesis. However, damaged or immature spermatozoa may experience a condition in which they maintain excess residual cytoplasm (ERC) around the midpiece due to a seizure in spermiogenesis. ERC encompasses large quantity of glucose-6-phosphate dehydrogenase, a cytosolic enzyme that exploits the hexose monophosphate shunt to generate excessive NADPH. NADPH being the substrate, results in the production of $O_2^{\bullet-}$ by the activity of NADPH oxidase [88].

Spermatozoa morphology may also be considered as a factor for determining the level of ROS production in the semen. Morphologically abnormal spermatozoa are found to produce more ROS than their normal counterparts. According to a study, spermatozoa having normal or borderline morphology are found to have lower levels of ROS in them. However, spermatozoa having amorphous heads, defective tail, excessive cytoplasm and abnormal midpiece are characterized by increased production of ROS [89].

2.5.3 Varicocele

Varicocele is an abnormality in the vascular drainage system. It is manifested by the enlargement of veins in pampiniform or cremasteric venous plexus around the spermatic cord [90, 91]. The grade of varicocele determines the level of ROS production. The higher the varicocele grade, the greater is the ROS generation [92].

Scrotal hyperthermia has been hypothesized to be one of the leading causes of varicocele-induced oxidative damage. Varicocele is thought to elevate temperature in the scrotum by reflux of warm abdominal blood through incompetent valves of the internal spermatic veins and cremastic veins into the pampiniform plexus [91]. The increase in temperature has been positively correlated with upregulated production of mitochondrial ROS, mediated through the transfer of electrons to O_2. Moreover, ROS synthesis from cell membrane, cytoplasm and peroxisome also increases during heat-stress [93, 94].

Varicocele is often associated with oxidative stress-induced male infertility. A close relationship persists between varicocele and oxidative stress, which is confirmed by the higher levels of ROS including NO• and the products of lipid peroxidation in infertile men with varicocele in comparison to infertile men not having varicocele [95]. It has been established that fertile men with varicocele are more likely to experience oxidative stress-induced reproductive complications in comparison to their varicocele-free counterparts [96].

The various endogenous production sites of ROS in the male reproductive tract have been elucidated in Fig. 2.2.

2.6 Physiological Role of ROS in Female Reproductive System

Modest amounts of ROS are vital for sustaining various physiological processes of the female reproductive system. ROS are positively associated with physiological processes such as follicular growth, ovarian steroidogenesis, ovulation, corpus luteum formation, luteolysis, germ cell functioning, proper continuation of pregnancy and parturition [11].

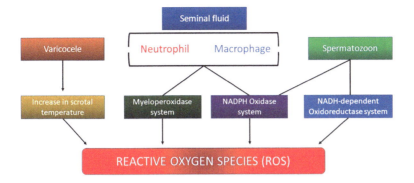

Fig. 2.2 Endogenous sources of ROS in the male reproductive system. Leukocytes such as neutrophils and macrophages present in the seminal fluid produce ROS via the myeloperoxidase system and the reduced nicotinamide adenine dinucleotide phosphate (NADPH) oxidase system. The developing spermatozoa also contributes to the generation of ROS via the NADPH oxidase system and the reduced nicotinamide adenine dinucleotide (NADH)-dependent oxidoreductase system. Increase in scrotal temperature due to varicocele also produces ROS in the testicular tissues.

2.6.1 Folliculogenesis

Folliculogenesis is defined as the process of progressive growth of ovarian follicles starting from a reserve state of dormant follicles set up in early life, whose fate is either ovulation or atresia. During the follicular stage of menstrual cycle, primordial follicles present in the ovary undergo the developmental and maturational processes, often resulting in numerous dominant mature follicles. These mature follicles are intended for ovulation whereas immature follicles end up being dead following follicular atresia [97].

Follicular angiogenesis is considered vital for follicular growth and development, which is due to the stimulation of hypoxia in granulosa cells [98]. ROS function as intracellular messengers of this angiogenic process and also serve the process of signal transduction [99]. Additionally, ovary possesses a defensive mechanism for cleansing and defence against excessive ROS. These comprise of enzymes such as catalase and SOD as well as antioxidants such as vitamins C and E [100]. A study conducted on genetically modified mouse suggests that these enzymes and antioxidants may perform vital functions in the development and survival of ovarian follicles [101].

About 99% of the ovarian follicles undergo a deteriorating procedure called follicular atresia, which is initiated by apoptosis of granulosa cells [102]. ROS may assist as an initiator of atresia by causing granulosa cell death in the antral follicles [101, 103]. Metabolic processes such as ovarian steroidogenesis generate $O_2^{\bullet-}$ as by-products. In case the concentration of SOD is inadequate to forage $O_2^{\bullet-}$, the latter may cause atresia or death of follicles [104].

2.6.2 Ovarian Steroidogenesis

Steroidogenic cells significantly uplift the generation of ROS in the ovary [104]. Controlled and adequate ROS concentration along with the transcripts of different antioxidant enzymes in the female reproductive tract may involve in ovarian steroidogenesis through feedback regulation [105, 106]. Peroxidase activity is found to increase 200-fold in the presence of estradiol, which is basically due to the existence of eosinophils in the leukocyte infiltrate. Peroxidase plays an important role by maintaining the antibacterial environment in the uterus. Moreover, peroxidase controls the level of oestrogen by a negative feedback mechanism [11].

NADPH oxidase present in the apex region of the epithelial membrane of endometrium causes dismutation of $O_2^{\bullet-}$, which in turn, results in the production of H_2O_2 [107]. H_2O_2 helps in the generation of prostaglandins through enzymatic activity of cyclooxygenase, as well as other non-enzymatic pathways [11, 108, 109].

Eicosanoids are signalling molecules which act as hormone-like substances and give rise to other compounds such as prostaglandins. Arachidonic acid is an essential unsaturated fatty acid which functions as a precursor of eicosanoid group of compounds. ROS trigger the reaction of O_2 and arachidonic acid, which is followed by the production of different prostaglandins by cyclooxygenase activity. This plays an important role in chemotaxis, implantation and degradation of corpus luteum [110].

2.6.3 Ovulation

Ovulation is an essential pre-requisite to successful reproduction, which is triggered by the pre-ovulatory upwelling of the pituitary luteinizing hormone (LH) [111]. The flow of LH induces the expression of a number of genes associated with inflammatory response in the pre-ovulatory follicles [112]. A positive correlation of ovulation with a heightened inflammatory response speaks for a role of ROS in this process [113]. Massive recruitment of inflammatory cells such as macrophages and neutrophils after the LH surge to the ovary are mainly responsible for production of ROS and their depletion greatly hampers ovulation [114, 115]. More specifically, the level of $O_2^{\bullet-}$ is found to increase when ovulation takes place [100]. It has been assumed that the $O_2^{\bullet-}$ is a causal factor of membrane damage which then commence the synthesis of prostaglandins mediated by the cyclooxygenase system via the activation of phospholipase. These prostaglandins are considered indispensable for ovulation [11]. H_2O_2 can mimic the effect of LH and helps to retain, at least partially, the ovulatory events in an LH-independent situation [113]. Earlier studies have demonstrated the ability of H_2O_2 to trigger p42/44 MAPK and EGFR, both of which are important factors of the LH-induced signalling pathway in the ovary [116, 117].

An in vitro study conducted on perfused rabbit ovary preparation has shown that SOD supresses the process of ovulation [118]. Another study has shown that broad-spectrum foragers of oxidative species such as butylated hydroxyanisole (BHA) and N-acetyl-L-cysteine (NAC) when administered in vivo in the ovarian bursa of mice treated with equine choriomic gonadotropin/human chorionic gonadotropin (eCG/hCG) considerably downregulate the rate of ovulation. Cumulous mucification/expansion, which is a prelude to ovulation, was also unnoticeable in these mice, [119]. Progesterone is associated with ovulation, and it has been revealed that antioxidants substantially reduce progesterone production in isolated follicles exposed to LH [113]. Moreover, oxidative stress induced by free radicals in the ovarian follicles provoke apoptosis of granulosa cells and instigate ovulation, while ROS foragers have a suppressive effect on the same [113, 120].

2.6.4 Formation and Luteolysis of Corpus Luteum

The corpus luteum is an impermanent endocrine gland which originates from the differentiated granulosa and thecal cells. It is known to produce oestrogen and progesterone following the completion of ovulation [121]. It has been shown that LH displays luteotropic activity in the conservation of high level of vitamins in the corpus luteum. Due to the abundance antioxidant vitamins, it has been hypothesized that ROS may have purposeful implication in the corpus luteum [110, 120]. The corpus luteum is considered to be significantly exposed to locally produced ROS due to its high blood vasculature and elevated metabolic activity. There is also a relation between ROS and progesterone. Reduced concentration of ROS employ luteotropic effects and help in the maintenance of corpus luteum. Additionally, ROS activity decreases and antioxidant level increases at the time of development of corpus luteum [106]. Moreover, cellular antioxidants play important roles in the rescue of corpus luteum from regression when pregnancy ensues [122].

In case of pregnancy failure, luteolysis occurs, which is a two-step regression process comprising of functional luteolysis and structural luteolysis [121]. Functional luteolysis is the depletion of progesterone production, whereas structural luteolysis is characterized by physical involution. Functional luteolysis holds particular importance because fast deterioration in the synthesis of progesterone is vital for follicular development in the subsequent cycle of reproduction [104]. During this process, an increase in ROS generation takes place coupled with a decrease in SOD activity – this serves to exert luteotropic effect on the corpus luteum [106]. The increase in ROS is followed by the production of cyclooxygenase-2, which triggers the expression of nuclear factor-kappa B (NF-κB) and a subsequent surge in the production of prostaglandin F-2-alpha ($PGF_{2\alpha}$) synthesis, which provokes luteal cells and phagocytic leukocytes to upregulate $O_2^{\bullet-}$ [123, 124].

Studies conducted on pregnant and pseudo-pregnant rat models have shown that functional luteolysis is the main driver of the regression phase of the corpus luteum. Both the models yielded similar results of having increased activity of copper/

zinc-SOD (Cu/Zn-SOD) from early to mid-luteal phase, which then gradually tends to decrease during the regression phase. This change in Cu/Zn-SOD activity is in parallel with the serum progesterone concentrations which points towards significant correlation between them [125, 126]. Furthermore, studies conducted on $PGF_{2\alpha}$-induced functional luteolytic pseudo-pregnant rat models showed that ROS (including $O_2^{\bullet-}$ and H_2O_2) rises in the corpus luteum at the time of regression phase and is directly involved in luteolysis [110, 127].

Hydrogen peroxide carries out a vital function in ROS-induced relapse of corpus luteum. It restrains LH-dependent cyclic adenosine monophosphate (cAMP) and progesterone synthesis, following which it depletes the ATP within the cells. This is an implication of peroxide-induced DNA damage [128]. H_2O_2 also opposes the transport of cholesterol across the outer mitochondrial membrane, thereby inhibiting progesterone metabolism [129]. H_2O_2 is also found to increase the production of prostaglandins by triggering lipid peroxidation [109]. This suggests that H_2O_2 synthesizes $PGF_{2\alpha}$ by positive feedback mechanism of ROS production [121].

2.6.5 Implantation

A study conducted on rat model by cytochemical staining procedure has confirmed the presence of NADPH oxidase in the membrane of trophoblast cell surface. This implies that trophoblasts release ROS into the extracellular medium and by virtue of this ability, they can phagocytose maternal cells during the embryo implantation [130]. Xanthine oxidase functioning has also been detected in the pre-implantation embryo of mouse. Xanthine is the major end-product of purine metabolism during the pre-implantation development. Downregulation of xanthine oxidase reduces the ROS activity in the embryo, while hypoxanthine causes a 2-cell block in mouse embryo [131]. Furthermore, ROS has probable implications in regulating the pace of pre-implantation embryo development [132]. Also, $NO^{\bullet}$, which is synthesized by oocytes and embryos, is thought to regulate embryonic growth and implantation [133].

2.6.6 Maintenance of Pregnancy

In case of regular pregnancy, oxidative stress in the placenta is existent during all three trimesters and is essential to maintain routine cell functions. $O_2^{\bullet-}$ concentrations remain stable throughout the pregnancy and are found to be present in much higher levels than in a non-pregnant woman [134]. ROS plays a vital function in the signal transduction process and has the capability to employ vasoactive consequences on the placenta [135]. Recently, it has been elucidated that oxidative stress and hypoxia induce placental apoptosis of the trophoblast, which is necessary for the developmental processes of embryo and placenta [4]. During implantation, decidual macrophages are concerned with the apoptotic activity, which is a necessary prerequisite for the development of embryo during pregnancy [136]. In addition to the regulation of normal placental apoptosis, ROS also bring about changes in the maternal vasculature which is necessary for foeto-placental interactions [137].

Healthy pregnancy is characterized by profound alterations in the maternal cardiovascular system. Vascular resistance decreases to meet the requirements of the growing foeto-placental unit, due to which vascular firmness is lowered during healthy pregnancy [137]. In the third trimester of pregnancy, a significant relationship persists between oxidative stress and vascular stiffness, which may imply that when oxidative stress increases above a certain threshold, the vasculature may get damaged, which may lead to vascular stiffness [138]. $NO^{\bullet}$ has been reported to regulate gene transcription and downstream activities such as vasodilation, trophoblast proliferation, invasion, anti-inflammation, antithrombosis and pro-angiogenesis in the placenta [139, 140]. Moreover, oxidative stress takes part in regulating autophagy and apoptosis, which are interconnected processes necessary in maintaining placental homeostasis [140].

Despite all the beneficial effects, an increase in the level of oxidative stress has a devastating effect on pregnancy. Increased oxidative stress during early stages of placental development may give rise to an imbalance between protective and destructive mechanisms of autophagy and apoptosis, which in turn, may lead to miscarriage, intrauterine growth restriction, pre-eclampsia and embryo resorption [135, 140]. Premature pregnancy loss has been linked with flawed antioxidant system in the placenta and is suggestive of the fact that proper balance of ROS and antioxidant enzymes is a critical factor for maintaining a healthy pregnancy [141].

2.6.7 Parturition

Parturition is an event which is characterized by active muscle contractions and is under the positive feedback control of ROS and inflammatory agents such as prostaglandins and cytokines [142, 143]. During parturition, inflammatory events take place, which are mediated by the invasion of neutrophils and monocytes [144]. Increase in the migration of leukocytes to the cervix due to the upregulation of cytokine expression is held responsible for production of $O_2^{\bullet-}$, by the reaction of hypoxanthine and xanthine oxidase. Generation of $O_2^{\bullet-}$ increases the concentration of intracellular Ca^{2+} thus leading to myometrial contraction [145]. Furthermore, H_2O_2 participates in a number of regulatory mechanisms involving calcium release and uptake by sarcoplasmic reticulum along with the modulation of myofibrillar calcium sensitivity, thereby acting as a modulator of smooth muscle contraction [146].

Reduction in the level of progesterone and elevation of oestrogen level is necessary for parturition to occur. However, oestrogen-mediated enhancement of ROS production brings about metabolic and enzymatic changes, which work in conjunction to bring about myometrial contractions and uterine involutions, which plays an essential function in the contraction of uterus after parturition [143]. Also, the last phase of gestation is depicted by an enormous upsurge in antioxidant enzyme reserve for the sake of preparation of life in an O_2-abundant condition [147].

2.7 Physiological Roles of ROS in the Male Reproductive System

ROS has been elucidated to have positive implications on the physiological processes of spermatozoa. Physiological concentrations of ROS are critical for the accomplishment of most of the developmental stages of spermatozoa and is particularly important for attaining their fertilizing potential [12]. However, spermatozoa are vulnerable to increased oxidative stress as they lack adequate cellular repair mechanism and antioxidant defence system. Therefore, maintaining a healthy equilibrium of ROS and antioxidants is critical for appropriate reproductive functionality of a man [148].

2.7.1 Spermatogenesis and Sperm Maturation

Spermatogenesis is a multifaceted procedure of male germ cell propagation and development from diploid spermatogonia to haploid spermatozoa, by the process of meiosis. These replicative processes guide to an intensification in the amount of mitochondrial O_2 consumption and ROS generation [149]. In the testis, ROS may prove to be indispensable in the complex process of male germ cell proliferation and maturation [150]. Spermatogenetic processes such as proliferation of spermatogonia and the differentiation of spermatocytes to spermatozoa needs to be supported by many humoral factors. Some receptors for these humoral factors are the members of the receptor-tyrosine kinase (RTK) family, and it has been seen in other somatic cells that ROS modulate these RTK-mediated signalling pathways. Thus, ROS play a seemingly beneficial role by regulating phosphorylation signals via controlling phosphatases [151].

ROS is also hypothesized to participate in transcriptional control of spermatogenesis. There are various regulatory transcription factors, which control stage-specific gene expression at the cellular level [152]. Along with other transcription factors, pituitary gland secretes LH and follicle stimulating hormone (FSH), which are involved in the regulation of spermatogenesis [153]. FSH acts on the Sertoli cells by activating signalling process, encoding various genes, aiding in the secretion of peptides and other signalling molecules [154]. The cAMP response element binding protein (CREB) present in the Sertoli cells is an important transcription factor for FSH signal transduction. Overproduction of ROS is thought to be a vital inducer of gene expression involving CREB family of receptors [155]. However, the functional relationship between different transcription factors and overproduction of ROS is under the regulatory control of various oestrogen receptors [156].

ROS are also engaged in maturational stages of spermatozoa during the epididymal transit. This stage is important for hyperactivation of spermatozoa, as they undergo remodelling of the cellular membrane and nuclear material, instalment of cell signalling pathways and enzymatic machinery [148]. Unlike somatic cells, spermatozoa are devoid of the defensive DNA repair systems owing to their insufficiency of histone packaging to uphold the biochemical integrity stored in their genes [157]. Therefore, additional stability must be ensured to the spermatozoon's nuclear genome to avoid any damage. Genetic durability is conferred by the oxidation of small nuclear thiol group proteins present in protamines, which ultimately substitute histones during spermiogenesis, thereby synthesizing rigid disulphide bonds between cysteine residues. As a matter of fact, ROS produced by redox reactions are involved in the appropriate chromatin packaging of humans and various other mammalian species [158–160]. ROS act as oxidants delivering oxidative capacity, thereby contributing to the generation of disulphide bonds, too [161]. Moreover, peroxides are involved in the development of a defensive cover around mitochondria called 'mitochondrial capsule', which if not formed properly may result in cellular impairment and problems related to power generation in the spermatozoa [162].

2.7.2 Capacitation

Capacitation is an important maturational event of the spermatozoa which occurs in the female genital tract, during which they acquire a state of hyperactive motility along with an increase in cellular sensitivity to chemotactic signalling [163]. The molecular mechanism of capacitation includes efflux of cholesterol, influx of bicarbonate (HCO_3^-) and Ca^{2+}. The oxidative stress drives various physiological changes related to capacitation of spermatozoa through the activation of a cAMP/protein kinase A phosphorylation pathway, which includes the stimulation of kinase-like proteins, upregulation of tyrosine phosphorylation in the spermatozoa tail and the initiation of sterol oxidation [164]. Capacitation has also been associated with increased thiol abundance on certain proteins on the surface of spermatozoa, which triggers the release of fully capacitated cells from their resting place on the surface of oviductal epithelium [165, 166]. Increased profusion of thiol too has been described to be an outcome of oxidative stress [167].

It has been confirmed that $O_2^{\bullet-}$ triggers a cascade of events during capacitation. The production of $O_2^{\bullet-}$ is initiated immediately following the induction of capacitation which gradually declines in due course of time. However, increased amount of $O_2^{\bullet-}$ is required to hyperactivate spermatozoa and basal concentration of the ion is necessary to maintain the cell in that condition [168]. In fact, $O_2^{\bullet-}$ has been found to carry out the most important function in hyperactive motility [169].

It is often speculated that H_2O_2 occupies the centre-stage in the progression of capacitation, whereas $O_2^{\bullet-}$ acts as a regulatory molecule for hyperactivation [170]. Furthermore, $NO^{\bullet}$ has been reported to increase capacitation explicitly in the female reproductive tract. The physiological effects elicited by NO$^\bullet$ are due to its intricate involvement with H_2O_2. This is established by the fact that cAMP, a complicatedly involved molecule in the biochemical framework of capacitation gets exhausted in the presence of ROS foragers and NOS inhibitors [171].

Nitric oxide regulates capacitation by functioning in conjunction with $O_2^{\bullet-}$. A hypothesis of ROS-induced ROS generation has also been proposed based on the fact that NO$^\bullet$ activates $O_2^{\bullet-}$ production and vice versa in the growth of spermatozoa. Additionally, these free radicals may react to form another ROS, ONOO$^-$. The generation of ONOO$^-$ is significant because it acts as an effector of capacitation; however, it is devoid of any stimulating influence on NO$^\bullet$ and $O_2^{\bullet-}$ synthesis [172].

2.7.3 Acrosome Reaction

The last maturational stage of a spermatozoon is the acrosome reaction, which involves the modification of the frontal region of the head and the release of acrosome enzymes succeeding interaction with the zona pellucida of the oocyte [173]. The primary proteolytic enzymes released during the process are acrosin and hyaluronidase, which facilitate the movement of spermatozoa through the cumulus oophorus and consequent union of spermatozoon and oocyte [3].

In vivo, the acrosome reaction gets initiated when the zona pellucida glycoprotein 3 (ZP3) ligands on the surface of the oocyte unite with the membrane of spermatozoa [174]. ROS facilitate this process by increasing the affinity of spermatozoa to the ZP3 ligand, due to the upregulation of corresponding receptors on the spermatozoa membrane [175]. In contrast, in vitro studies have suggested that acrosome reaction can be originated by the growth of spermatozoa with recombinant ZP3, progesterone or ROS. However, conflicting reports exist regarding the in vitro involvement of ROS in acrosome reaction. Some investigators denied the existence of any distinct association between the concentration of free radicals and the action of acrosome enzymes [176]. Although, other reports state that the addition of ROS to incubation media activates acrosome reaction [177, 178]. Free radicals such as $O_2^{\bullet-}$, H_2O_2 and even NO$^\bullet$ are found to influence acrosome reaction when present in low concentrations [179, 180].

However, there is a lack of absolute accord as to which free radical is completely involved in the regulation the acrosome reaction. H_2O_2 has also been shown to play an essential role in spermatozoa entrance of zona-free hamster eggs. [181]. On the other hand, positive association was found between $O_2^{\bullet-}$ amplification and induction of acrosome reaction. Moreover, no change in the level of these free radicals or acrosome reaction could be observed in the presence of SOD [182]. Later, it has been clarified that both $O_2^{\bullet-}$ and H_2O_2 are necessary to induce acrosome reaction [179].

From the mechanistic perspective, acrosome reaction is thought to be initiated by an influx of Ca^{2+} from the accumulated calcium stock of acrosome and the extracellular compartments into the cytosol [183]. It has been considered that the induction of Ca^{2+} influx is due to the production of ROS by enzymatic pathways [184].

2.7.4 Sperm–Oocyte Fusion

Immediately following acrosome reaction, the fusion of spermatozoa and oocyte takes place. The fluidity of plasma membrane is an essential criterion for the accomplishment of this process and is reliant on polyunsaturated fatty acid along with phospholipase A2 (PLA2) activity. It has been proposed that $O_2^{\bullet-}$ produced by spermatozoa function as membrane phospholipase, thus facilitating in membrane fluidity [185]. ROS are proposed to enhance the PLA2 activity by inhibiting protein tyrosine phosphatase, thereby reducing the chances of dephosphorylation-mediated PLA2 reduction [182]. ROS may also optimize PLA2 activity by triggering protein kinase C (PKC) through the action of protein tyrosine kinase. PKC phosphorylates phospholipases, followed by subsequent initiation of an augmented PLA2 activity [186, 187].

The association of ROS and spermatozoon–oocyte union is further established. Researchers have observed a marked rise in the spermatozoon–oocyte union in spermatozoa possessing major tyrosine phosphorylation properties, which is suggestive of the fact that the actions which are associated with capacitation and acrosome reaction are related to the ones involved in sperm–oocyte fusion. In this regard, $O_2^{\bullet-}$ and H_2O_2 have been found to upregulate fertilization rates [183, 188].

Additionally, ROS have been found to facilitate chemotactic movements of the spermatozoa towards the oocyte in the female reproductive tract as a result of the secretion of progesterone by the cumulus oophorus cells. This has further been confirmed in vitro where catalase treatment has been shown to lead to a substantial reduction in the number of chemotactic spermatozoa. These point towards the existence of some sort of functional gateway of ROS at which they aid in chemotactic movements of spermatozoa. Beyond that point, ROS induce oxidative stress, which may turn detrimental for chemotactic spermatozoa. In such situations, catalase may have its own functional significance by helping in the reduction of cellular oxidative damage [189].

2.8 Conclusions

ROS are not totally harmful to the reproductive system. Preserving a physiological concentration of ROS in the reproductive tract is of paramount importance. Indeed, an optimum balance of ROS is a prerequisite to ensure efficacious lifespan of spermatozoa i.e., from synthesis to fertilization of the oocyte. ROS are critical in confirming structural remodelling of spermatozoa by triggering intracellular pathways, thereby assisting in the chromatin condensation, spermatozoa motility, capacitation, acrosome reaction and chemotactic movements. Similarly, the role of ROS in female reproduction cannot be underestimated as well. ROS function in numerous biological processes from the maturation of oocyte to fertilization and development of the embryo. Conversely, a disbalance in the concentration of ROS and antioxidants confers pathological consequences on spermatozoa, which range from reduced concentration, declined motility and reduction in fertilization ability to even apoptosis. Furthermore, overproduction of ROS can lead to myriad infertility problems and a spectrum of female reproductive disorders [190].

A physiological reference value of ROS is yet to be defined. Due to their transient nature of action, experimental tests remain complex and perplexing. Due to the lack of an accurate cut-off value of ROS, clinicians face difficulties in quantitatively differentiating between the physiological and pathological levels of ROS. Moreover, measurement of oxidative stress in vivo is both challenging and controversial, due to the prevalence of interlaboratory and interobserver differences. The sensitivity and specificity of various oxidative stress biomarkers being unknown adds up to the challenge of unifying a comprehensive method of assessing oxidative stress. Due to the fact that most of the studies on oxidative stress are either observational, case control or in vitro in nature, it creates an abstinence in the way of getting a comparative aspect of these studies. Newer studies should focus on uniform populations and similar outcome measures so that the results can be perceived more easily by establishing a comprehensive comparison.

Despite extensive studies being conducted in this field, further research is required to lucidly understand the mechanisms involved in ROS-mediated activation of intracellular machinery encompassing various reproductive processes. An improved knowledge of the subtle oxidative balance influencing diverse reproductive functions and regulations could be of utmost significance in the diagnosis, prevention and treatment of both female and male reproductive disorders.

References

1. Ray PD, Huang B-W, Tsuji Y. Reactive oxygen species (ROS) homeostasis and redox regulation in cellular signalling. Cell Signal. 2012;24:981–90. https://doi.org/10.1016/j.cellsig.2012.01.008.
2. Snezkina AV, Kudryavtseva AV, Kardymon OL, Savvateeva MV, Melnikova NV, Krasnov GS, Dmitriev AA. ROS generation and antioxidant defence systems in normal and malignant cells. Oxidative Med Cell Longev. 2019;2019:6175804. https://doi.org/10.1155/2019/6175804.
3. Agarwal A, Gupta S, Sharma RK. Role of oxidative stress in female reproduction. Reprod Biol Endocrinol. 2005;3:28. https://doi.org/10.1186/1477-7827-3-28.
4. Al-Gubory KH, Fowler PA, Garrel C. The roles of cellular reactive oxygen species, oxidative stress and antioxidants in pregnancy outcomes. Int J Biochem Cell Biol. 2010;42:1634–50. https://doi.org/10.1016/j.biocel.2010.06.001.
5. Finkel T, Holbrook NJ. Oxidants, oxidative stress and the biology of ageing. Nature. 2000;408:239–47. https://doi.org/10.1038/35041687.

6. Schieber M, Chandel NS. ROS function in redox signalling and oxidative stress. Curr Biol. 2014;24:R453–62. https://doi.org/10.1016/j.cub.2014.03.034.
7. Ahmad G, Almasry M, Dhillon AS, Abuayyash MM, Kothandaraman N, Cakar Z. Overview and sources of reactive oxygen species (ROS) in the reproductive system. In: Agarwal A, Sharma RK, Gupta S, Harlev A, Ahmad G, du Plessis SS, Esteves SC, Wang SM, Durairajanayagam D, editors. Oxidative stress in human reproduction shedding light on a complicated phenomenon. Springer; 2017. p. 1–16. https://doi.org/10.1007/978-3-319-48427-3_1.
8. Finkel T. Signal transduction by reactive oxygen species. J Cell Biol. 2011;194:7–15. https://doi.org/10.1083/jcb.201102095.
9. Roychoudhury S, Agarwal A, Virk G, Cho C-L. Potential role of green tea catechins in the management of oxidative stress-associated infertility. Reprod Biomed Online. 2017;34:487–98. https://doi.org/10.1016/j.rbmo.2017.02.006.
10. Wood ZA, Poole LB, Karplus PA. Peroxiredoxin evolution and the regulation of hydrogen peroxide signalling. Science. 2003;300:650–3. https://doi.org/10.1126/science.1080405.
11. du Plessis SS, Harlev A, Mohamed MI, Habib E, Kothandaraman N, Cakar Z. Physiological roles of reactive oxygen species (ROS) in the reproductive system. In: Agarwal A, Sharma RK, Gupta S, Harlev A, Ahmad G, du Plessis SS, Esteves SC, Wang SM, Durairajanayagam D, editors. Oxidative stress in human reproduction shedding light on a complicated phenomenon. Springer; 2017. p. 47–64. https://doi.org/10.1007/978-3-319-48427-3_3.
12. Aitken RJ, Clarkson JS, Fishel S. Generation of reactive oxygen species, lipid peroxidation and human sperm function. Biol Reprod. 1989;41:183–97. https://doi.org/10.1095/biolreprod41.1.183.
13. Martindale JL, Holbrook NJ. Cellular response to oxidative stress: signalling for suicide and survival. J Cell Physiol. 2002;192:1–15. https://doi.org/10.1002/jcp.10119.
14. Li R, Jia Z, Trush MA. Defining ROS in biology and medicine. React Oxyg Species. 2016;1:9–21. https://doi.org/10.20455/ros.2016.803.
15. Di Meo S, Reed TT, Venditti P, Victor VM. Role of ROS and RNS sources in physiological and pathological conditions. Oxidative Med Cell Longev. 2016;2016:1245049. https://doi.org/10.1155/2016/1245049.
16. Phaniendra A, Jestadi DB, Periyasamy L. Free radicals: properties, sources, targets and their implication in various diseases. Indian J Clin Biochem. 2015;30:11–26. https://doi.org/10.1007/s12291-014-0446-0.
17. Nunes-Silva A, Freitas-Lima L. The association between physical exercise and reactive oxygen species (ROS) production. J Sports Med Doping Stud. 2014;4:2161-0673.1000152. https://doi.org/10.4172/2161-0673.1000152.
18. Buettner GR. Superoxide dismutase in redox biology: the roles of superoxide and hydrogen peroxide. Anti Cancer Agents Med Chem. 2011;11:341–6. https://doi.org/10.2174/187152011795677544.
19. Droge W. Free radicals in the physiological control of cell function. Physiol Rev. 2002;82:47–95. https://doi.org/10.1152/physrev.00018.2001.
20. Kehrer JP, Robertson JD, Smith CV. Free radicals and reactive oxygen species. In: McQueen CA, editor. Comprehensive toxicology, vol. 1. 2nd ed. London: Elsevier Inc; 2010. p. 277–307.
21. Fridovich I. Superoxide radical: an endogenous toxicant. Annu Rev Pharmacol Toxicol. 1983;23:239–57. https://doi.org/10.1146/annurev.pa.23.040183.001323.
22. Fridovich I. The biology of oxygen radicals. Science. 1978;201:875–80. https://doi.org/10.1126/science.210504.
23. Tchobanoglous G, Burton FL, Stensel HD. Wasterwater engineering: treatment and reuse. 4th ed. New York: Mc Graw Hill; 2003.
24. Munter R. Advanced oxidation processes: current status and prospects. Proc Estonian Acad Sci Chem. 2001;50:59–80.
25. Underbakke ES, Surmeli NB, Smith BC, Wynia-Smith SL, Marletta MA. Nitric oxide signalling. In: Reedijk J, Poeppelmeier K, editors. Comprehensive inorganic chemistry II:

from elements to applications. 2nd ed. London: Elsevier Inc; 2013. p. 241–62. https://doi.org/10.1016/B978-0-08-097774-4.00320-X.
26. Radi R. Oxygen radicals, nitric oxide and peroxynitrite: redox pathways in molecular medicine. Proc Natl Acad Sci U S A. 2018;115:5839–48. https://doi.org/10.1073/pnas.1804932115.
27. Bruckdorfer R. The basics about nitric oxide. Mol Asp Med. 2005;26:3–31. https://doi.org/10.1016/j.mam.2004.09.002.
28. Genestra M. Oxyl radicals, redox-sensitive signalling cascades and antioxidants. Cell Signal. 2007;19:1807–19. https://doi.org/10.1016/j.cellsig.2007.04.009.
29. Wong H-S, Dighe PA, Mezera V, Monternier P-A, Brand MD. Production of superoxide and hydrogen peroxide from specific mitochondrial sites under different bioenergetic conditions. J Biol Chem. 2017;292:16804–9. https://doi.org/10.1074/jbc.R117.789271.
30. Giorgio M, Trinei M, Migliaccio E, Pelicci PG. Hydrogen peroxide: a metabolic by-product or a common mediator of ageing signals? Nat Rev Mol Cell Biol. 2007;8:722–8. https://doi.org/10.1038/nrm2240.
31. Dringen R, Pawlowski PG, Hirrlinger J. Peroxide detoxification by brain cells. J Neurosci Res. 2005;79:157–65. https://doi.org/10.1002/jnr.20280.
32. Halliwell B, Clement MV, Long LH. Hydrogen peroxide in the human body. FEBS Lett. 2000;486:10–3. https://doi.org/10.1016/s0014-5793(00)02197-9.
33. Armogida M, Nistico R, Mercury NB. Therapeutic potential of targeting hydrogen peroxide metabolism in the treatment of brain ischemia. Br J Pharmacol. 2012;166:1211–24. https://doi.org/10.1111/j.1476-5381.2012.01912.x.
34. Aldred EM, Buck C, Vall K. Free radicals. In: Pharmacology: A handbook for Complementary Healthcare Professionals. London: Elsevier Inc; 2009. p. 1–3.
35. Hrycay EG, Bandiera SM. Involvement of cytochrome P450 in reactive oxygen species formation and cancer. Adv Pharmacol. 2015;74:35–84. https://doi.org/10.1016/bs.apha.2015.03.003.
36. Manda G, Nechifor M, Neagu M. Reactive oxygen species, cancer and anti-cancer therapies. Curr Chem Biol. 2009;3:22–46.
37. Stief TW. The physiology and pharmacology of singlet oxygen. Med Hypotheses. 2003;60:567–72. https://doi.org/10.1016/s0306-9877(03)00026-4.
38. Schoonbroodt S, Legrand-Poels S, Best-Belpomme M, Piette J. Activation of the NF-kappaB transcription factor in a T-lymphocytic cell line by hypochlorous acid. Biochem J. 1997;321:777–85. https://doi.org/10.1042/bj3210777.
39. Wentworth P Jr, McDunn JE, Wentworth AD, Takeuchi C, Nieva J, Jones T, Bautista C, Ruedi JM, Gutierrez A, Janda KD, Babior BM, Eschenmoser A, Lerner RA. Evidence for antibody-catalyzed ozone formation in bacterial killing and inflammation. Science. 2002;298:2195–9. https://doi.org/10.1126/science.1077642.
40. Zhu X, Wentworth P Jr, Wentworth AD, Eschenmoser A, Lerner RA, Wilson IA. Probing the antibody-catalyzed water-oxidation pathway at atomic resolution. Proc Natl Acad Sci U S A. 2004;101:2247–52. https://doi.org/10.1073/pnas.0307311101.
41. Lerner RA, Eschenmoser A. Ozone in biology. Proc Natl Acad Sci U S A. 2003;100:3013–5. https://doi.org/10.1073/pnas.0730791100.
42. Bocci V. Ozone as a bioregulator: pharmacology and toxicology of ozonetherapy today. J Biol Regul Homeost Agents. 1996;10:31–53.
43. Agarwal A, Allamaneni SS. Role of free radicals in female reproductive diseases and assisted reproduction. Reprod Biomed Online. 2004;9:338–47. https://doi.org/10.1016/s1472-6483(10)62151-7.
44. Li W, Young JF, Sun J. NADPH oxidase-generated reactive oxygen species in mature follicles are essential for *Drosophila* ovulation. Proc Natl Acad Sci U S A. 2018;115:7765–70. https://doi.org/10.1073/pnas.1800115115.
45. Kampfer C, Saller S, Windschuttl S, Berg D, Berg U, Mayerhofer A. Pigment-epithelium derived factor (PEDF) and the human ovary: a role in the generation of ROS in granulosa cells. Life Sci. 2014;97:129–36. https://doi.org/10.1016/j.lfs.2013.12.007.

46. Fujii J, Iuchi Y, Okada F. Fundamental roles of reactive oxygen species and protective mechanisms in the female reproductive system. Reprod Biol Endocrinol. 2005;3:43. https://doi.org/10.1186/1477-7827-3-43.
47. Agarwal A, Aponte-Mellado A, Premkumar BJ, Shaman A, Gupta S. The effects of oxidative stress on female reproduction: a review. Reprod Biol Endocrinol. 2012;10:49. https://doi.org/10.1186/1477-7827-10-49.
48. Ruder EH, Hartman TJ, Goldman MB. Impact of oxidative stress on female fertility. Curr Opin Obstet Gynecol. 2009;21:219. https://doi.org/10.1097/gco.0b013e32832924ba.
49. Behrman HR, Kodaman PH, Preston SL, Gao S. Oxidative stress and the ovary. J Soc Gynecol Investig. 2001;8:S40–2. https://doi.org/10.1016/s1071-5576(00)00106-4.
50. Freitas C, Neto AC, Matos L, Silva E, Ribeiro A, Silva-Carvalho J, Almeida H. Follicular fluid redox involvement for ovarian follicle growth. J Ovarian Res. 2017;10:44.
51. Luddi A, Governini L, Capaldo A, Campanella G, De Leo V, Piomboni P, Morgante G. Characterization of the age-dependent changes in antioxidant defences and protein's sulfhydryl/carbonyl stress in human follicular fluid. Antioxidants. 2020;9:927. https://doi.org/10.3390/antiox9100927.
52. Nelson SM, Telfer EE, Anderson RA. The ageing ovary and uterus: new biological insights. Hum Reprod Update. 2013;19:67–83. https://doi.org/10.1093/humupd/dms043.
53. Roselli M, Dubey RK, Imthurn B, Macas E, Keller PJ. Effects of nitric oxide on human spermatozoa: evidence that nitric oxide decreases sperm motility and induces sperm toxicity. Hum Reprod. 1995;10:1786–90. https://doi.org/10.1093/oxfordjournals.humrep.a136174.
54. El Mouatassim S, Guerin P, Menezo Y. Expression of genes encoding antioxidant enzymes in human and mouse oocytes during the final stages of maturation. Mol Hum Reprod. 1999;5:720–5. https://doi.org/10.1093/molehr/5.8.720.
55. Ekerhovd E, Norstrom A. Involvement of a nitric oxide-cyclic guanosine monophosphate pathway in control of fallopian tube contractility. Gynecol Endocrinol. 2004;19:239–46. https://doi.org/10.1080/09513590400019296.
56. Al-Azemi M, Refaat B, Amer S, Ola B, Chapman N, Ledger W. The expression of inducible nitric oxide synthase in the human fallopian tube during the menstrual cycle and in ectopic pregnancy. Fertil Steril. 2010;94:833–40. https://doi.org/10.1016/j.fertnstert.2009.04.020.
57. Shao R, Zhang SX, Weijdegard B, Shein Z, Egecioglu E, Norstrom A, Brannstrom M, Billig H. Nitric oxide synthases and tubal ectopic pregnancies induced by Chlamydia infection: basic and clinical insights. Mol Hum Reprod. 2010;16:907–15.
58. Griffith OW, Stuehr DJ. Nitric oxide synthases: properties and catalytic mechanisms. Annu Rev Physiol. 1995;57:707–36. https://doi.org/10.1146/annurev.ph.57.030195.003423.
59. Morris SM Jr, Billiar TR. New insights into the regulation of inducible nitric oxide synthesis. Am J Phys. 1994;266:E829–39. https://doi.org/10.1152/ajpendo.1994.266.6.E829.
60. Tranguch S, Huet-Hudson Y. Decreased viability of nitric oxide synthase double knockout mice. Mol Reprod Dev. 2003;65:175–9. https://doi.org/10.1002/mrd.10274.
61. Zhan X, Li D, Johns RA. Expression of endothelial nitric oxide synthase in ciliated epithelia of rats. J Histochem Cytochem. 2003;51:81–7. https://doi.org/10.1177/002215540305100110.
62. Kalra A, Wehrle CJ, Tuma F. Anatomy, abdomen and pelvis, peritoneum. In: StatPearls. StatPearls Publishing; 2020.
63. Hoare BS, Khan YS. Anatomy, Abdomen and pelvis, female internal genitals. In: StatPearls. StatPearls Publishing; 2020.
64. Van Langerdonckt A, Casanas-Roux F, Donnez J. Oxidative stress and peritoneal endometriosis. Fertil Steril. 2002;77:861–70. https://doi.org/10.1016/s0015-0282(02)02959-x.
65. Das S, Chattopadhyay R, Ghosh S, Ghosh S, Goswami SK, Chakravarty BN, Chaudhury K. Reactive oxygen species level in follicular fluid: embryo quality marker in IVF? Hum Reprod. 2006;21:2403–7. https://doi.org/10.1093/humrep/del156.
66. Donnez J, Binda MM, Donnez O, Dolmans M-M. Oxidative stress in the pelvic cavity and its role in the pathogenesis of endometriosis. Fertil Steril. 2016;106:1011–7. https://doi.org/10.1016/j.fertnstert.2016.07.1075.

67. Agarwal A, Durairajanayagam D, du Plessis SS. Utility of antioxidants during assisted reproductive techniques: an evidence-based review. Reprod Biol Endocrinol. 2014;12:112. https://doi.org/10.1186/1477-7827-12-112.
68. DeCherney A, Hill MJ. The future of imaging and assisted reproduction. In: Rizk B, editor. Ultrasonography in reproductive medicine and infertility. Cambridge University Press; 2010. p. 1–10.
69. Rizk B, Badr M, Talerico C. Oxidative stress and the endometrium. In: Agarwal A, Aziz N, Rizk B, editors. Studies on women's health. Springer; 2013. p. 61–74. https://doi.org/10.1007/978-1-62703-041-0_3.
70. Ngo C, Chereaau C, Nicco C, Weill B, Chapron C, Batteux F. Reactive oxygen species controls endometriosis progression. Am J Pathol. 2009;175:225–34. https://doi.org/10.2353/ajpath.2009.080804.
71. Sikka SC. Relative impact of oxidative stress on male reproductive function. Curr Med Chem. 2001;8:851–62. https://doi.org/10.2174/0929867013373039.
72. Chen S-J, Allam J-P, Duan Y-G, Haidi G. Influence of reactive oxygen species on human sperm functions and fertilizing capacity including therapeutical approaches. Arch Gynecol Obstet. 2013;288:191–9. https://doi.org/10.1007/s00404-013-2801-4.
73. Haber F, Weiss J. Uber die katalyse des hydroperoxydes. Naturwissenschaften. 1932;20:948–50.
74. Sen CK. Antioxidants and redox regulation of cellular signalling: introduction. Med Sci Sports Exerc. 2001;33:368–70. https://doi.org/10.1097/00005768-200103000-00005.
75. Henkel RR. Leukocytes and oxidative stress: dilemma for sperm function and male fertility. Asian J Androl. 2011;13:43–52. https://doi.org/10.1038/aja.2010.76.
76. Agarwal A, Virk G, Ong C, du Plessis SS. Effect of oxidative stress on male reproduction. World J Mens Health. 2014;32:1–17. https://doi.org/10.5534/wjmh.2014.32.1.1.
77. Sabeur K, Ball BA. Characterization of NADPH oxidase 5 in equine testis and spermatozoa. Reproduction. 2007;134:263–70. https://doi.org/10.1530/REP-06-0120.
78. Turrens JF. Mitochondrial formation of reactive oxygen species. J Physiol. 2003;552:335–44. https://doi.org/10.1113/jphysiol.2003.049478.
79. Attaran M, Pasqualotto E, Falcone T, Goldberg JM, Miller KF, Agarwal A, Sharma RK. The effect of follicular fluid reactive oxygen species on the outcome of in vitro fertilization. Int J Fertil Womens Med. 2000;45:314–20.
80. Gharagozloo P, Aitken RJ. The role of sperm oxidative stress in male infertility and the significance of oral antioxidant therapy. Hum Reprod. 2011;26:1628–40. https://doi.org/10.1093/humrep/der132.
81. Agarwal A, Roychoudhury S, Bjugstad KB, Cho C-L. Oxidation-reduction potential of semen: what is its role in the treatment of male infertility? Ther Adv Urol. 2016;8:302–18. https://doi.org/10.1177/1756287216652779.
82. Saleh RA, Agarwal A, Nada EA, El-Tonsy MH, Sharma RK, Meyer A, Nelson DR, Thomas AJ. Negative effects of increased sperm DNA damages in relation to seminal oxidative stress in men with idiopathic and male factor infertility. Fertil Steril. 2003;79:1597–605. https://doi.org/10.1016/s0015-0282(03)00337-6.
83. Gil-Guzman E, Ollero M, Lopez MC, Shamra RK, Alvarez JG, Thomas AJ Jr, Agarwal A. Differential production of reactive oxygen species by subsets of human spermatozoa at different stages of maturation. Hum Reprod. 2001;16:1922–30. https://doi.org/10.1093/humrep/16.9.1922.
84. Agarwal A, Saleh RA, Bedaiwy MA. Role of reactive oxygen species in the pathophysiology of human reproduction. Fertil Steril. 2003;79:829–43. https://doi.org/10.1016/s0015-0282(02)04948-8.
85. Lavranos G, Balla M, Tzortzopoulou A, Syriou V, Angelopoulou R. Investigating ROS sources in male infertility: a common end for numerous pathways. Reprod Toxicol. 2012;34:298–307. https://doi.org/10.1016/j.reprotox.2012.06.007.

86. Shang Y, Liu C, Cui D, Han G, Yi S. The effect of chronic bacterial prostatitis on semen quality in adult men: a meta-analysis of case-control studies. Sci Rep. 2014;4:7233. https://doi.org/10.1038/srep07233.
87. Lu J-C, Huang Y-F, Lu N-Q. WHO laboratory manual for the examination and processing of human semen: its applicability to andrology laboratories in China. Zhonghua Nan Ke Xue. 2010;16:867–71.
88. Rengan AK, Agarwal A, van der Linde M, du Plessis SS. An investigation of excess residual cytoplasm in human spermatozoa and its distinction from the cytoplasmic droplet. Reprod Biol Endocrinol. 2012;10:92. https://doi.org/10.1186/1477-7827-10-92.
89. Aziz N, Saleh RA, Sharma RK, Lewis-Jones I, Esfandiari N, Thomas AJ Jr, Agarwal A. Novel association between sperm reactive oxygen species production, sperm morphological defects and the sperm deformity index. Fertil Steril. 2004;81:349–54. https://doi.org/10.1016/j.fertnstert.2003.06.026.
90. Will MA, Swain J, Fode M, Sonksen J, Christman GM, Ohl D. The great debate: varicocele treatment and impact on fertility. Fertil Steril. 2011;95:841–52. https://doi.org/10.1016/j.fertnstert.2011.01.002.
91. Cho C-L, Esteves SC, Agarwal A. Novel insight into the pathophysiology of varicocele and its association with reactive oxygen species and sperm DNA fragmentation. Asian J Androl. 2016;18:186–93. https://doi.org/10.4103/1008-682X.170441.
92. Shiraishi K, Matsuyama H, Takihara H. Pathophysiology of varicocele in male infertility in the era of assisted reproductive technology. Int J Urol. 2012;19:538–50. https://doi.org/10.1111/j.1442-2042.2012.02982.x.
93. Voglmayr JK, Setchell BP, White IG. The effects of heat on the metabolism and ultrastructure of ram testicular spermatozoa. J Reprod Fertil. 1971;24:71–80. https://doi.org/10.1530/jrf.0.0240071.
94. Morgan D, Cherny VV, Murphy R, Xu W, Thomas LL, DeCoursey TE. Temperature dependence of NADPH oxidase in human eosinophils. J Physiol. 2003;550:447–58. https://doi.org/10.1113/jphysiol.2003.041525.
95. Sakamoto Y, Ishikawa T, Kondo Y, Yamaguchi K, Fujisawa M. The assessment of oxidative stress in infertile patients with varicocele. BJU Int. 2008;101:1547–52. https://doi.org/10.1111/j.1464-410X.2008.07517.x.
96. Mostafa T, Anis T, Imam H, El-Nashar AR, Osman IA. Seminal reaxctive oxygen species-antioxidant relationship in fertile males with and without varicocele. Andrologia. 2009;41:125–9. https://doi.org/10.1111/j.1439-0272.2008.00900.x.
97. Monniaux D, Cadoret V, Clement F, Dalbies-Tran R, Elis S, Fabre S, Maillard V, Monget P, Uzbekova S. Folliculogenesis. In: Huhtaniemi I, Martini L, editors. Encyclopedia of endocrine diseases, vol. 2. 2nd ed; 2018. p. 377–98.
98. Tropea A, Miceli F, Minici F, Tiberi F, Orlando M, Gangale MF, Romani F, Catino S, Mancuso S, Navarra P, Lanzone A, Apa R. Regulation of vascular endothelial growth factor synthesis and release by human luteal cells in vitro. J Clin Endocrinol Metab. 2006;91:2303–9. https://doi.org/10.1210/jc.2005-2457.
99. Basini G, Grasselli F, Bianco F, Tirelli M, Tamanini C. Effects of reduced oxygen tension on reactive oxygen species production and activity of antioxidant enzymes in swine granulosa cells. Biofactors. 2004;20:61–9. https://doi.org/10.1002/biof.5520200201.
100. Laloraya M, Pradeep KG, Laloraya MM. Changes in the levels of superoxide anion radical and superoxide dismutase during the estrous cycle of *Rattus norvegicus* and induction of superoxide dismutase in rat ovary by lutropin. Biochem Biophys Res Commun. 1988;157:146–53. https://doi.org/10.1016/s0006-291x(88)80025-1.
101. Devine PJ, Perreault SD, Luderer U. Roles of reactive oxygen species and antioxidants in ovarian toxicity. Biol Reprod. 2012;86:27. https://doi.org/10.1095/biolreprod.111.095224.
102. Yu YS, Sui HS, Han ZB, Li W, Luo MJ, Tan JH. Apoptosis in granulosa cells during follicular atresia: relationship with steroids and insulin-like growth factors. Cell Res. 2004;14:341–6. https://doi.org/10.1038/sj.cr.7290234.

103. Kaipia A, Hsueh AJ. Regulation of ovarian follicle atresia. Annu Rev Physiol. 1997;59:349–63. https://doi.org/10.1146/annurev.physiol.59.1.349.
104. Sugino N. Reactive oxygen species in ovarian physiology. Reprod Med Biol. 2005;4:31–44. https://doi.org/10.1007/BF03016135.
105. Forman HJ, Torres M. Reactive oxygen species and cell signalling: respiratory burst in macrophage signalling. Am J Respir Crit Care Med. 2002;166:S4–8. https://doi.org/10.1164/rccm.2206007.
106. Rizzo A, Roscino MT, Binetti F, Sciorsci RL. Roles of reactive oxygen species in female reproduction. Reprod Domest Anim. 2012;47:344–52. https://doi.org/10.1111/j.1439-0531.2011.01891.x.
107. Ishikawa Y, Hirai K, Ogawa K. Cytochemical localization of hydrogen peroxide production in the rat uterus. J Histochem Cytochem. 1984;32:674–6. https://doi.org/10.1177/32.6.6725936.
108. Hemler ME, Cook HW, Lands WE. Prostaglandin biosynthesis can be triggered by lipid peroxidase. Arch Biochem Biophys. 1979;193:340–5. https://doi.org/10.1016/0003-9861(79)90038-9.
109. Morrow JD, Hill KE, Burk RF, Nammour TM, Badr KF, Roberts LJ 2nd. A series of prostaglandin F2-like compounds are produced in vivo in humans by a non-cyclooxygenase, free radical-catalyzed mechanism. Proc Natl Acad Sci U S A. 1990;87:9383–7. https://doi.org/10.1073/pnas.87.23.9383.
110. Riley JC, Behrman HR. Oxygen radicals and reactive oxygen species in reproduction. Proc Soc Exp Biol Med. 1991;198:781–91. https://doi.org/10.3181/00379727-198-43321c.
111. Russel DL, Robker RL. Molecular mechanism of ovulation: co-ordination through the cumulus complex. Hum Reprod Update. 2007;13:289–312. https://doi.org/10.1093/humupd/dml062.
112. Richards JS, Russell DL, Ochsner S, Espey LL. Ovulation: new dimensions and new regulators of the inflammatory-like response. Annu Rev Physiol. 2002;64:69–92. https://doi.org/10.1146/annurev.physiol.64.081501.131029.
113. Shkolnik K, Tadmor A, Ben-Dor S, Nevo N, Galiani D, Dekel N. Reactive oxygen species are indispensable in ovulation. Proc Natl Acad Sci U S A. 2011;108:1462–7. https://doi.org/10.1073/pnas.1017213108.
114. Brannstrom M, Mayrhofer G, Robertson SA. Localization of leukocyte subsets in the rat ovary during the pre-ovulatory period. Biol Reprod. 1993;48:277–86. https://doi.org/10.1095/biolreprod48.2.277.
115. Van der Hoek KH, Maddocks S, Woodhouse CM, van Rooijen N, Robertson SA, Norman RJ. Intrabursal injection of clodronate liposomes cause macrophage depletion and inhibits ovulation in the mouse ovary. Biol Reprod. 2000;62:1059–66. https://doi.org/10.1095/biolreprod62.4.1059.
116. DeYulia GJ Jr, Carcamo JM. EFG receptor-ligand interaction generates extracellular hydrogen peroxide that inhibits EGFR-associated protein tyrosine phosphatases. Biochem Biophys Res Commun. 2005;334:38–42. https://doi.org/10.1016/j.bbrc.2005.06.056.
117. Hsieh M, Lee D, Panigone S, Horner K, Chen R, Theologis A, Lee DC, Threadgill DW, Conti M. Luteinizing hormone-dependent activation of the epidermal growth factor network is essential for ovulation. Mol Cell Biol. 2007;27:1914–24. https://doi.org/10.1128/MCB.01919-06.
118. Miyazaki T, Sueoka K, Dharmarajan AM, Atlas SJ, Bulkley GB, Wallach EE. Effect of inhibition of oxygen free radical on ovulation and progesterone production by the in-vitro perfused rabbit ovary. J Reprod Fertil. 1991;91:207–12. https://doi.org/10.1530/jrf.0.0910207.
119. Chen L, Russell PT, Larsen WJ. Functional significance of cumulus expansion in the mouse: roles for the preovulatory synthesis of hyaluronic acid within the cumulus mass. Mol Reprod Dev. 1993;34:87–93. https://doi.org/10.1002/mrd.1080340114.

120. Kodaman PH, Behrman HR. Endocrine-regulated and protein kinase C-dependent generation of superoxide by rat preovulatory follicles. Endocrinology. 2001;142:687–93. https://doi.org/10.1210/endo.142.2.7961.
121. de Lamirande E, Gagnon C. Reactive oxygen species (ROS) and reproduction. In: Armstrong D, editor. Free radicals in diagnostic medicine. Springer; 1994. p. 185–97. https://doi.org/10.1007/978-1-4615-1833-4_14.
122. Al-Gubory KH, Garrel C, Faure P, Sugino N. Roles of antioxidant enzymes in corpus luteum rescue from reactive oxygen species-induced oxidative stress. Reprod Biomed Online. 2012;25:551–60. https://doi.org/10.1016/j.rbmo.2012.08.004.
123. Sawada M, Carlson JC. Studies on the mechanism controlling generation of superoxide radical in luteinized rat ovaries during regression. Endocrinology. 1994;135:1645–50. https://doi.org/10.1210/endo.135.4.7925128.
124. Tanaka M, Miyazaki T, Tanigaki S, Kasai K, Minegishi K, Miyakoshi K, Ishimoto H, Yoshimura Y. Participation of reactive oxygen species in PGF2alpha-induced apoptosis in rat luteal cells. J Reprod Fertil. 2000;120:239–45. https://doi.org/10.1530/jrf.0.1200239.
125. Noda Y, Ota K, Shirasawa T, Shimizu T. Copper/zinc superoxide dismutase insufficiency impairs progesterone secretion and fertility in female mice. Biol Reprod. 2012;86:1–8. https://doi.org/10.1095/biolreprod.111.092999.
126. Shimamura K, Sugino N, Yoshida Y, Nakamura Y, Ogino K, Kato H. Changes in lipid peroxide and antioxidant enzyme activities in corpora lutea during pseudopregnancy in rats. J Reprod Fertil. 1995;105:253–7. https://doi.org/10.1530/jrf.0.1050253.
127. Sawada M, Carlson JC. Superoxide radical production in plasma membrane samples from regressing rat corpora lutea. Can J Physiol Pharmacol. 1989;67:465–71. https://doi.org/10.1139/y89-074.
128. Behrman HR, Preston SL. Luteolytic actions of peroxide in rat ovarian cells. Endocrinology. 1989;124:2895–900. https://doi.org/10.1210/endo-124-6-2895.
129. Behrman HR, Aten RF. Evidence that hydrogen peroxide blocks hormone-sensitive cholesterol transport into mitochondria of rat luteal cells. Endocrinology. 1991;128:2958–66. https://doi.org/10.1210/endo-128-6-2958.
130. Gagioti S, Colepicolo P, Bevilacqua E. Post-implantation mouse embryos have the capability to generate and release reactive oxygen species. Reprod Fertil Dev. 1995;7:1111–6. https://doi.org/10.1071/rd9951111.
131. Nasr-Esfahani MH, Winston NJ, Johnson MH. Effect of glucose, glutamine, ethylenediaminetetraacetic acid and oxygen tension on the concentration of reactive oxygen species and on development of the mouse preimplantation embryo in vitro. J Reprod Fertil. 1992;96:219–31. https://doi.org/10.1530/jrf.0.0960219.
132. Yamashita T, Yamazaki H, Kon Y, Watanabe T, Arikawa J, Miyoshi I, Kasai N, Kuwabara M. Progressive effect of alpha-phenyl-N-tert-butyl nitrone (PBN) on rat embryo development in vitro. Free Radic Biol Med. 1997;23:1073–7. https://doi.org/10.1016/s0891-5849(97)00139-1.
133. Guerin P, El Mouatassim S, Menezo Y. Oxidative stress and protection against reactive oxygen species in the pre-implantation embryo and its surroundings. Hum Reprod Update. 2001;7:175–89. https://doi.org/10.1093/humupd/7.2.175.
134. Mannaerts D, Faes E, Cos P, Briede JJ, Gyselaers W, Cornette J, Gorbanev Y, Bogaerts A, Spaanderman M, Van Craenenbroek E, Jacquemyn Y. Oxidative stress in healthy pregnancy and preeclampsia in linked to chronic inflammation, iron status and vascular function. PLoS One. 2018;13:e0202919. https://doi.org/10.1371/journal.pone.0202919.
135. Myatt L, Cui X. Oxidative stress in the placenta. Histochem Cell Biol. 2004;122:369–82. https://doi.org/10.1007/s00418-004-0677-x.
136. Abrahams VM, Kim YM, Straszewski SL, Romero R, Mor G. Macrophage and apoptotic cell clearance during pregnancy. Am J Reprod Immunol. 2004;51:275–82. https://doi.org/10.1111/j.1600-0897.2004.00156.x.

137. Khalil A, Jauniaux E, Cooper D, Harrington K. Pulse wave analysis in normal pregnancy: a prospective longitudinal study. PLoS One. 2009;4:e6134. https://doi.org/10.1371/journal. pone.0006134.
138. Wu F, Tian F-J, Lin Y, Xu W-M. Oxidative stress: placenta function and dysfunction. Am J Reprod Immunol. 2016;76:258–71. https://doi.org/10.1111/aji.12454.
139. Krause BJ, Hanson MA, Casanello P. Role of nitric oxide in placental vascular development and function. Placenta. 2011;32:797–805. https://doi.org/10.1016/j.placenta.2011.06.025.
140. Wu F, Tian F-J, Lin Y. Oxidative stress in placenta: health and diseases. Biomed Res Int. 2015;2015:293271. https://doi.org/10.1155/2015/293271.
141. Liu A-X, Jin F, Zhang W-W, Zhou T-H, Zhou C-Y, Yao W-M, Qian Y-L, Huang H-F. Proteomic analysis on the alteration of protein expression in the placental villous tissue of early pregnancy loss. Biol Reprod. 2006;75:414–20. https://doi.org/10.1095/biolreprod.105.049379.
142. Conner EM, Grisham MB. Inflammation, free radicals and antioxidants. Nutrition. 1996;12:274–7. https://doi.org/10.1016/s0899-9007(96)00000-8.
143. Jenkin G, Young IR. Mechanism responsible for parturition: the use of experimental models. Anim Reprod Sci. 2004;82–83:567–81. https://doi.org/10.1016/j.anireprosci.2004.05.010.
144. Golightly E, Jabbour HN, Norman JE. Endocrine immune interactions in human parturition. Mol Cell Endocrinol. 2011;335:52–9. https://doi.org/10.1016/j.mce.2010.08.005.
145. Masumoto N, Tasaka K, Miyake A, Tanizawa O. Superoxide anion increases intracellular free calcium in human myometrial cells. J Biol Chem. 1990;265:22533–6.
146. Appiah I, Milovanovic S, Radojicic R, Nikolic-Kokic A, Orescanin-Dusic Z, Slavic M, Trbojevic S, Skrbic R, Spasic M, Blagojevic D. Hydrogen peroxide affects contractile activity and antioxidant enzymes in rat uterus. Br J Pharmacol. 2009;158:1932–41. https://doi.org/10.1111/j.1476-5381.2009.00490.x.
147. O'Donovan DJ, Fernandes CJ. Free radicals and diseases in premature infants. Antioxid Redox Signal. 2004;6:169–76. https://doi.org/10.1089/152308604771978471.
148. Kovalski N, de Lamirande E, Gagnon C. Reactive oxygen species generated by human neutrophils inhibit sperm motility: protective effect of seminal plasma and scavengers. Fertil Steril. 1992;58:809–16. https://doi.org/10.1016/S0015-0282(16)55332-1.
149. Guerriero G, Trocchia S, Abdel-Gawad FK, Ciarcia G. Roles of reactive oxygen species in the spermatogenesis regulation. Front Endocrinol (Lausanne). 2014;5:56. https://doi.org/10.3389/fendo.2014.00056.
150. Shi Y, Buffenstein R, Pulliam DA, Van Remmen H. Comparative study of oxidative stress and mitochondrial function in aging. Integr Comp Biol. 2010;50:869–79. https://doi.org/10.1093/icb/icq079.
151. Fujii J, Imai H. Redox reactions in mammalian spermatogenesis and the potential targets of reactive oxygen species under oxidative stress. Spermatogenesis. 2014;4:e979108. https://doi.org/10.4161/21565562.2014.979108.
152. Lui W-Y, Cheng CY. Transcription regulation in spermatogenesis. Adv Exp Med Biol. 2008;636:115–32. https://doi.org/10.1007/978-0-387-09597-4_7.
153. Chen C, Ouyang W, Grigura V, Zhou Q, Carnes K, Lim H, Zhao G-Q, Arber S, Kurpios N, Murphy TL, Cheng AM, Hassell JA, Chandrashekar V, Hofmann M-C, Hess RA, Murphy KM. ERM is required for transcriptional control of the spermatogonial stem cell niche. Nature. 2005;436:1030–4. https://doi.org/10.1038/nature03894.
154. Grimes SR. Testis-specific transcriptional control. Gene. 2004;343:11–22. https://doi.org/10.1016/j.gene.2004.08.021.
155. Aitken RJ, Roman SD. Antioxidant system and oxidative stress in the stress. Oxidative Med Cell Longev. 2008;1:15–24. https://doi.org/10.4161/oxim.1.1.6843.
156. Montano MM, Deng H, Liu M, Sun X, Singal R. Transcriptional regulation by the estrogen receptor of antioxidative stress enzymes and its functional implications. Oncogene. 2004;23:2442–53. https://doi.org/10.1038/sj.onc.1207358.

157. Aitken RJ, Gordon E, Harkiss D, Twigg JP, Milne P, Jennings Z, Irvine DS. Relative impact of oxidative stress on the functional competence and genomic integrity of human spermatozoa. Biol Reprod. 1998;59:1037–46. https://doi.org/10.1095/biolreprod59.5.1037.
158. Saowaros W, Panyim S. The formation of disulphide bonds in human protamines during sperm maturation. Experientia. 1979;35:191–2. https://doi.org/10.1007/BF01920608.
159. Cheng W-M, An L, Wu Z-H, Zhu Y-B, Liu J-H, Gao H-M, Li X-H, Zheng S-J, Chen D-B, Tian J-H. Effects of disulphide bond reducing agents on sperm chromatin structural integrity and developmental competence of in vitro matured oocytes after intracytoplasmic sperm injection in pigs. Reproduction. 2009;137:633–43. https://doi.org/10.1530/REP-08-0143.
160. Hutchinson JM, Rau DC, DeRouchey JE. Role of disulphide bonds on DNA packaging forces in bull sperm chromatin. Biophys J. 2017;113:1925–33. https://doi.org/10.1016/j.bpj.2017.08.050.
161. Rousseaux J, Rousseaux-Prevost R. Molecular localization of free thiols in human sperm chromatin. Biol Reprod. 1995;52:1066–72. https://doi.org/10.1095/biolreprod52.5.1066.
162. Roveri A, Ursini F, Flohe L, Maiorino M. PHGPx and spermatogenesis. Biofactors. 2001;14:213–22. https://doi.org/10.1002/biof.5520140127.
163. Eisenbach M. Mammalian sperm chemotaxis and its association with capacitation. Dev Genet. 1999;25:87–94. https://doi.org/10.1002/(SICI)1520-6408(1999)25:2<87::AID-DVG2>3.0.CO;2-4.
164. Aitken RJ. Reactive oxygen species as mediators of sperm capacitation and pathological damage. Mol Reprod Dev. 2017;84:1039–52. https://doi.org/10.1002/mrd.22871.
165. Gualtieri R, Mollo V, Duma G, Talevi R. Redox control of surface protein sulphhydryls in bovine spermatozoa reversibly modulates sperm adhesion to the oviductal epithelium and capacitation. Reproduction. 2009;138:33–43. https://doi.org/10.1530/REP-08-0514.
166. O'Flaherty C. Redox regulation of mammalian sperm capacitation. Asian J Urol. 2015;17:583–90. https://doi.org/10.4103/1008-682X.153303.
167. Kralikova M, Crha I, Huser M, Melounova J, Zakova J, Matejovicova M, Ventruba P. The intracellular concentration of homocysteine and related thiols is negatively correlated to sperm quality after highly effective method of sperm lysis. Andrologia. 2017;49:e12702. https://doi.org/10.1111/and.12702.
168. de Lamirande E, Harakat A, Gagnon C. Human sperm capacitation induced by biological fluids and progesterone, but not by NADH or NADPH, is associated with the production of superoxide anion. J Androl. 1998;19:215–25. https://doi.org/10.1002/j.1939-4640.1998.tb01991.x.
169. Suarez SS. Control of hyperactivation in sperm. Hum Reprod Update. 2008;14:647–57. https://doi.org/10.1093/humupd/dmn029.
170. de Lamirande E, Jiang H, Zini A, Kodama H, Gagnon C. Reactive oxygen species and sperm physiology. Rev Reprod. 1997;2:48–54. https://doi.org/10.1530/ror.0.0020048.
171. Zini A, de Lamirande E, Gagnon C. Low levels of nitric oxide promote human sperm capacitation in vitro. J Androl. 1995;16:424–31. https://doi.org/10.1002/j.1939-4640.1995.tb00558.x.
172. de Lamirande E, Lamothe G. Reactive oxygen-induces reactive oxygen formation during human sperm capacitation. Free Radic Biol Med. 2009;46:502–10. https://doi.org/10.1016/j.freeradbiomed.2008.11.004.
173. du Plessis SS, Agarwal A, Halabi J, Tvrda E. Contemporary evidence on the physiological role of reactive oxygen species in human sperm function. J Assist Reprod Genet. 2015;32:509–20. https://doi.org/10.1007/s10815-014-0425-7.
174. Breitbart H, Spungin B. The biochemistry of the acrosome reaction. Mol Hum Reprod. 1997;3:195–202. https://doi.org/10.1093/molehr/3.3.195.
175. Breitbart H, Naor Z. Protein kinase in mammalian sperm capacitation and the acrosome reaction. Rev Reprod. 1999;4:151–9. https://doi.org/10.1530/ror.0.0040151.

176. Ichikawa T, Oeda T, Ohmori H, Schill WB. Reactive oxygen species influence the acrosome reaction but not acrosin activity in human spermatozoa. Int J Androl. 1999;22:37–42. https://doi.org/10.1046/j.1365-2605.1999.00145.x.
177. de Lamirande E, Gagnon C. A positive role for the superoxide anion in triggering hyperactivation and capacitation of human spermatozoa. Int J Androl. 1993;16:21–5. https://doi.org/10.1111/j.1365-2605.1993.tb01148.x.
178. Boerke A, Tsai PS, Garcia-Gill N, Brewis IA, Gadella BM. Capacitation-dependent reorganization of microdomains in the apical sperm head plasma membrane: functional relationship with zona binding and the zona-induced acrosome reaction. Theriogenology. 2008;70:1188–96. https://doi.org/10.1016/j.theriogenology.2008.06.021.
179. de Lamirande E, Tsai C, Harakat A, Gagnon C. Involvement of reactive oxygen species in human sperm acrosome reaction induced by A23187, lysophosphatidylcholine and biological fluid ultrafiltrates. J Androl. 1998;19:585–94. https://doi.org/10.1002/j.1939-4640.1998.tb02061.x.
180. Herrero MB, de Lamirande E, Gagnon C. Nitric oxide is a signalling molecule in spermatozoa. Curr Pharm Des. 2003;9:419–25. https://doi.org/10.2174/1381612033391720.
181. Aitken RJ, Paterson M, Fisher H, Buckingham DW, van Duin M. Redox regulation of tyrosine phosphorylation in human spermatozoa and its role in the control of human sperm function. J Cell Sci. 1995;108:2017–25. https://doi.org/10.1242/jcs.108.5.2017.
182. Griveau JF, Renard P, Le Lannou D. Superoxide anion production by human spermatozoa as a part of the ionophore-induced acrosome reaction process. Int J Androl. 1995;18:67–74. https://doi.org/10.1111/j.1365-2605.1995.tb00388.x.
183. Aitken RJ, Buckingham DW, Harkiss D, Paterson M, Fisher H, Irvine DS. The extragenomic action of progesterone on human spermatozoa is influenced by redox regulated changes in tyrosine phosphorylation during capacitation. Mol Cell Endocrinol. 1996;117:83–93. https://doi.org/10.1016/0303-7207(95)03733-0.
184. Aitken RJ, Buckingham DW, West KM. Reactive oxygen species and human spermatozoa: analysis of the cellular mechanism involved in luminol- and lucigenin-dependent chemiluminescence. J Cell Physiol. 1992;151:466–77. https://doi.org/10.1002/jcp.1041510305.
185. Ohzu E, Yanagimachi R. Acceleration of acrosome reaction in hamster spermatozoa by lysolecithin. J Exp Zool. 1982;224:259–63. https://doi.org/10.1002/jez.1402240216.
186. Goldman R, Ferber E, Zort U. Reactive oxygen species are involved in the activation of cellular phospholipase A2. FEBS Lett. 1992;309:190–2. https://doi.org/10.1016/0014-5793(92)81092-z.
187. Zor U, Ferber E, Gergely P, Szucs K, Dombradi V, Goldman R. Reactive oxygen species mediate phorbol ester-regulated tyrosine phosphorylation and phospholipase A2 activation: potentiation by vanadate. Biochem J. 1993;295:879–88. https://doi.org/10.1042/bj2950879.
188. Aitken RJ. Molecular mechanisms regulating human sperm function. Mol Hum Reprod. 1997;3:169–73. https://doi.org/10.1093/molehr/3.3.169.
189. Sanchez R, Sepulveda C, Risopatron J, Villegas J, Giojalas LC. Human sperm chemotaxis depends on critical levels of reactive oxygen species. Fertil Steril. 2010;93:150–3. https://doi.org/10.1016/j.fertnstert.2008.09.049.
190. Sharma R, Roychoudhury S, Alsaad R, Bamajbuor F. Negative effects of oxidative stress (OS) on reproductive system at cellular level. In: Agarwal A, Sharma RK, Gupta S, Harlev A, Ahmad G, du Plessis SS, Esteves SC, Wang SM, Durairajanayagam D, editors. Oxidative stress in human reproduction shedding light on a complicated phenomenon. Springer; 2017. p. 65–88. https://doi.org/10.1007/978-3-319-48427-3_4.

Chapter 3
Pathological Roles of Reactive Oxygen Species in Male Reproduction

Saptaparna Chakraborty and Shubhadeep Roychoudhury

Abstract Reactive oxygen species (ROS) are free radicals that have at least one unpaired electron and play specific roles in the human body. An imbalance of ROS and antioxidant levels gives rise to a condition called oxidative stress. High levels of ROS in the male reproductive tract can interfere with its normal functioning and can even pose as toxic to the sperm, inhibiting sperm functioning (including motility) and metabolism. Oxidative stress resulting from ROS and lipid peroxidation is one of the major causes of male infertility including infertility in varicocele patients. These may cause DNA and peroxidative damage and apoptosis. Production of ROS in excess also leads to erectile dysfunction (ED). In recent years, studies have also linked oxidative stress with the development, progress, and therapy response of prostate cancer patients. The present study summarizes the pathological roles of ROS in male reproductive problems such as infertility, ED, and prostate cancer and also provide an insight into the probable mechanism through which ROS exert their pathological impact.

Keywords Oxidative stress · Male infertility · Erectile dysfunction · Prostate cancer

3.1 Introduction

Reactive oxygen species (ROS) are extremely reactive oxidizing agents that play specific pathological and physiological roles in the human body [1]. ROS are free radicals that are tagged as molecular entities having a minimum of one unpaired electron that creates a group of highly reactive compounds [2]. However, there are several ROS that do not contain any unpaired electron [3]. ROS is a mutual term that includes superoxide ($O_2^{\bullet-}$), singlet oxygen ($^1O_2$), hydroxyl radical ($OH^\bullet$), hydrogen

---

S. Chakraborty · S. Roychoudhury (✉)
Department of Life Science and Bioinformatics, Assam University, Silchar, India

© The Author(s), under exclusive license to Springer Nature Switzerland AG 2022
K. K. Kesari, S. Roychoudhury (eds.), *Oxidative Stress and Toxicity in Reproductive Biology and Medicine*, Advances in Experimental Medicine and Biology 1358, https://doi.org/10.1007/978-3-030-89340-8_3

peroxide (H_2O_2), peroxyl radical (LOO•), alkoxyl radical (LO•), lipid hydroperoxide (LOOH), peroxynitrite ($ONOO^-$), ozone (O_3), and hypochlorous acid (HOCl) [4]. Reactive nitrogen species (RNS) are nitrogen containing reactive species such as nitric oxide (NO•), nitrogen dioxide radical (NO_2•), and peroxynitrite (NO^-_3), and RNS is also classified as ROS as they are also oxygen containing species [3]. In fact, ROS are the by-products of oxygen metabolism [5] and are formed during oxidative metabolism in mitochondria and also in cellular response to invasion of bacteria, cytokines, and xenobiotics [6]. Oxidative stress is a condition that is caused from an imbalance between accumulation and production of ROS in cells and tissues and the capability of a living body to detoxify them [7].

ROS production in the male reproductive tract is concerning as these reactive species are potentially toxic to metabolism, motility, and other functions of spermatozoa [5, 8]. Studies show that to attain fertilizing ability, spermatozoa require low levels of ROS, and epididymal and testicular functions are sensitive to ROS levels, which is why it is important for the male reproductive system to have a balanced redox state [9].

Although levels of ROS are necessary for different physiological aspects of spermatozoa such as capacitation, hyperactivation, acrosomal reaction, and signalling process for fertilization, however, ROS-induced oxidative stress is one of the major causes of male infertility via damage to sperm DNA, apoptosis and peroxidative damage [10]. Men with varicocele reportedly have increased levels of chemo-attractive ROS. Oxidative stress resulting from lipid peroxidation and high levels of ROS like NO is one of the most important elements in the pathophysiology of varicocele-related male infertility [11]. Overproduction of ROS such as H_2O_2 and $O_2^{•-}$ is a common pathological attribute of erectile dysfunction (ED), which has been associated with both neurogenic and vasculogenic ED [12]. In recent years oxidative stress, resulting from ROS, has also been identified as one of the characteristics for aggressive phenotype of prostate cancer and it is associated with the development, progress, and therapy response of the disease [13].

In this study, the pathological roles of ROS in different male reproductive problems such as male infertility, ED, and prostate cancer have been reviewed. It also attempts to provide an insight into the probable oxidative stress-dependent pathway through which ROS exert their pathological impact.

3.2 Male Infertility

Infertility is the condition where a couple is unable to conceive a baby even after one year of having frequent unprotected sexual intercourse, and this condition affects about 180 million people around the world [14]. Male infertility is the condition where a man is unable to make a fertile female pregnant after having unprotected sexual intercourse for at least a year, and according to a World Health Organization (WHO) report, in developing countries, one in every four couple suffers from infertility [15, 16]. In about 50% of the cases, infertility is due to the male factor [17]. Male infertility is mainly of three types, namely untreatable sterility including primary seminiferous tubule failure; treatable conditions including obstructive azoospermia, sperm autoimmunity, deficiency of gonadotropin, sexual function disorders, reversible toxin effects; and untreatable subfertility including oligozoospermia, asthenozoospermia, tetratozoospermia, and normozoospermia with functional defects [18].

Male infertility is a complex pathological condition and is associated with several causes and risk factors, which can be classified into three broad categories – congenital, acquired, and idiopathic risk factors. Conditions like anorchia, cryptorchidism, congenital absence of vas deferens, microdeletion of Y chromosome, genetic or chromosomal anomalies like Klinefelter syndrome or Robertsonian translocation, congenital obstruction, and genetic endocrinopathy fall under congenital factors. Varicocele, tumours of germ cells, testicular trauma and torsion, sexual dysfunctions like ED, acquired hypogonadotropic hypogonadism, infection, and obstruction of urinogenital tract, anti-sperm antibodies, systematic disease like liver cirrhosis, and exogenous factors such as radiotherapy or heat constitute the acquired factors. Idiopathic risk factors include smoking, alcohol consumption, obesity, advancing age, drugs, and occupational or environmental exposure to toxins [19, 20]. Genetic factors account for about 15% of male infertility [21].

3.2.1 Effect of ROS on Sperm Structural and Functional Integrity

ROS-induced oxidative stress primarily contributes to the aetiology of male infertility, as it damages both the structural and functional integrity of spermatozoa [22]. The membrane fluidity of spermatozoa comes from the polyunsaturated fatty acids (PUFA) present in the sperm plasma membrane, which also makes them more susceptible to oxidative damage [23]. When PUFAs are attacked by ROS, a propagating cycle called 'radical chain reaction' is triggered which results in the conversion of PUFA to malondialdehyde (MDA), 4-hydroxynonenal (4-HNE), and acrolein [24]. MDA, 4-HNE, and acrolein are the biomarkers of oxidative stress that cause damage, mitochondrial dysregulation, and increase the yield of ROS along with leakage of ROS from inner mitochondrial membrane. When the level of ROS surpasses the physiological need of a body, the spermatozoa experiences oxidative stress which ultimately results in infertility [25]. When ROS is generated in excess in semen by potential ROS sources such as leukocytes and abnormal spermatozoa, some spermatozoa experience oxidative damage along with a loss in their function which may ultimately lead to male infertility. Apart from that, deficiency in antioxidant glutathione level in the midpiece of spermatozoa may also cause male infertility [26]. ROS-induced oxidative stress can cause defect in the head and midpiece of spermatozoa such as damaged acrosome, amorphous heads, and thickened midpiece, too [27].

3.2.2 Effect of ROS on Sperm DNA

For fertilization to occur, the DNA of the spermatozoa must be healthy, as any change in the structure of DNA can cause alteration or loss of function of any protein [28]. ROS have the ability to cause base modifications and hence fragmentation of sperm DNA is one of the key outcomes of oxidative stress, either directly or indirectly [29]. Oxidative stress, induced by ROS or RNS, impairs the function and integrity of DNA via a homeostatic imbalance of antioxidant enzyme levels in the spermatozoa and seminal plasma [30, 31]. During the transfer of spermatozoa to the epididymis from the seminiferous tubule, immature spermatozoa produce ROS, which further leads to DNA damage in mature spermatozoa, and this causes an increase in apoptosis of germ cell and decrease in sperm count, which ultimately leads to infertility [32, 33]. On the basis of the type, ROS can lead to protein modification via oxidation of thiol, tyrosine nitration, sulfonation, or glutathionylation, and among these the most common is the oxidation of thiol, which affects chromatin remodelling during spermiogenesis [29]. Oxidative stress causes lipid peroxidation that disrupts cell membrane fluidity, increases cell permeability for non-specific ions, inactivates the membrane enzymes and the electrophilic aldehydes damage the nuclear DNA [34, 35].

3.2.3 Effect of ROS on Sperm Motility

Motility is an important aspect of sperm function since they have to swim through the female reproductive tract post ejaculation in order to fertilize the oocyte [36]. Spermatozoa are sensitive to ROS-induced damage due to the lack of cytoplasm and low level of oxygen radical-scavenging enzymes in them. ROS-induced oxidative stress in one of the common factors which affect sperm motility. Motility declines when ROS, induced in mitochondrial complex I, cause peroxidative damage to the mid-piece of spermatozoa [37, 38]. NO reacts with O_2- to produce peroxynitrite that further generates OH- which leads to lipid peroxidation and DNA damage thus ultimately leading to sperm dysfunction and low sperm motility [39]. In oligoasthenozoospermic and asthenozoospermic men, high levels of MDA are reported and changes in the membrane fluidity, due to oxidation, affects sperm motility [40]. ROS causes a rapid loss of ATP that affects sperm motility, viability, acrosome reaction, and capacitation ultimately leading to infertility [38].

3.2.4 Effect of Varicocele-Induced ROS on Infertility

Varicocele is one of the endogenous sources of ROS in seminal plasma [41]. The word 'varicocele' has its origins from a Latin word 'varix' meaning 'dilated vein' and a German word 'kele' that means 'tumour or swelling' [42]. The condition in which the pampiniform plexus of scrotal veins dilates and enlarges abnormally is called varicocele. Pampiniform plexus is the vein that drains blood from both the testicles [43]. Although the condition of varicocele is generally painless, it is clinically very significant. It is because varicocele is the most commonly identified cause of abnormal semen analysis including decreased sperm count, motility, and abnormal sperm morphology [44]. Based on appearance, varicoceles are categorized as small, medium, or large, and 80–90% of the clinically detectable varicoceles occur in the left side of the body due to the anatomy of the testicular veins. The internal spermatic vein on the right side directly terminates into the inferior vena cava having low pressure, whereas the internal spermatic veins of the left side joins with the left renal vein which has relatively high pressure [45]. Also, the level of ROS is linked with the grade of the varicocele [46].

Several theories suggest that varicocele-induced male infertility is characterized by apoptosis, increase in oxidative stress, DNA damage of sperm, tissue hypoxia, deterioration of seminiferous tubules, and immunological infertility caused by a decline in the production of Fas protein [47]. Oxidative stress, resulting from elevated levels of ROS, is associated with varicocele and other male subfertility issues such as testicular torsion, cryptorchidism, genitourinary tract infection, or inflammation [47, 48]. Studies showed that reduced antioxidant capacity or oxidative stress associated with increase in ROS is associated varicocele-induced male infertility. Patients with varicocele also report significant increase of chemo-attractive ROS [49].

Yoon and co-workers studied infertile patients having varicocele and stated that internal spermatic vein (ISV) of such patients had ROS in higher levels. ROS are produced in midpiece of sperm and thus the spermatozoa are susceptible to ROS and associated oxidative stress [47]. Oxidative stress resulting from ROS shortens the telomere of the chromosome that directly causes genomic instability [50]. Production of ROS damages the Leydig cells, which reduces the secretion of testosterone [51]. Oxidative stress also effects the epididymis and accessory glands [52]. Yoon and co-workers also reported that there is a relation between testicular hypotrophy and increase in ROS associated with varicocele [47]. Another study on infertile men with and without varicocele also suggested that testicular dysfunction associated with varicocele might be due to ROS [53]. Lipid peroxidation of sperm membrane also occurs due to damage by ROS which is also considered to be the key mechanism of infertility [54]. In a meta-analysis on oxidative stress, varicocele, and infertility, infertile men with varicocele had high levels of ROS-induced oxidative stress, and increase in ROS production might have led to an increase in the utilization of antioxidants and thus lowered their concentrations [55]. This imbalance between antioxidant levels and ROS concentration in the body causes oxidative stress [56]. An increased oxidative stress hampers the process of spermatogenesis and leads to poor semen parameters [57]. Varicocele is also reported to induce scrotal hyperthermia by increasing the temperature by refluxing of abdominal warm blood via the incompetent valves of internal spermatic veins and cremasteric veins into the pampiniform plexus [48]. Several studies have demonstrated the relationship between ROS generation and heat exposure [58]. In presence of heat, production of mitochondrial ROS increases that results in the transfer of electrons to molecular oxygen, and hence formation of ROS and inhibition of adenosine

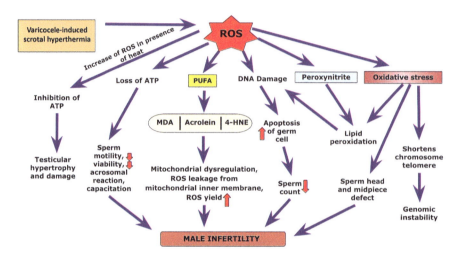

Fig. 3.1 Possible mechanism of ROS-induced infertility in the male. ROS attack PUFAs present in the plasma membrane of sperm, which leads to the formation of acrolein, MDA, and 4-HNE, which cause mitochondrial dysfunction and leakage of ROS from mitochondrial inner membrane, which in turn increases ROS production. This may lead to male infertility. ROS present in semen cause oxidative damage to sperm, which may also result in infertility. ROS also decrease ATP which affects sperm motility, viability, acrosomal reaction, and capacitation and ultimately leads to infertility. ROS produced from immature spermatozoa damage the DNA of mature spermatozoa, which causes germ cell apoptosis, which in turn decreases sperm count, all these resulting in infertility. ROS cause oxidative stress, which damages the head and midpiece of sperm, which leads to male infertility. Upon reaction with NO, O^{2+} forms peroxinitrite which causes lipid peroxidation, damages sperm DNA, and decreases sperm motility leading to sperm dysfunction. Oxidative stress leads to shortening of chromosome telomere, too, which results in genomic instability. ROS cause lipid peroxidation, which may also result in varicocele-induced infertility in affected men. Varicocele induces scrotal hyperthermia, which leads to an increase in ROS. Increase of ROS in presence of heat leads to inhibition in ATP generation, which further results in testicular hypertrophy and dysfunction. Red arrows indicate the decrease of the respective substance, which has a negative impact and leads to male infertility. ROS: reactive oxygen species, PUFA: polyunsaturated fatty acid, MDA: malondialdehyde, 4-HNE: 4-hydroxy-2-nonenal, ATP: adenosine triphosphate, DNA: deoxyribonucleic acid.

triphosphate (ATP) synthesis [59]. Heat-induced upregulation of inducible NOS increases NO production, which may contribute to testicular damage associated with varicocele [60].

Potential mechanisms through which ROS cause male infertility are presented in Fig. 3.1.

3.3 Erectile Dysfunction (ED)

The inability of a man to achieve and maintain a rigid penile erection, appropriate for a satisfactory coitus is termed as ED or impotence [61]. In men above the age of 40 years, it is a common condition, and its predominance is rising sharply with age and co-morbidities [62]. The Massachusetts Male Ageing Study (MMAS) reported a prevalence of 52% men with some degree of ED while 10% of men had complete ED in an age group of 40–70 years [63]. However, a study showed that in real life situation, 1 out of 4 men seeking medical help for ED was <40 years of age [64].

The causes of ED are often multifactorial. Diseases that affect the hormone levels, nerves, arteries, smooth muscle tissue, corporal endothelium, or tunica albuginea of penis can cause ED [65]. Although the process of ageing is an essential factor for the occurrence of ED, conditions like depression, performance anxiety along with other sexual diseases and disorders are also considered as important contributing factors. Apart from these, other causes of ED include neurological and hormonal disorders, diabetes mellitus, stroke, glaucoma, constructive obstructive pulmonary disease (COPD), multiple sclerosis, sequela of priapism, and prostatic hyperplasia [66].

In the absence of any stimulation the penis remains in a flaccid state, which means the smooth muscles remain contracted. The contraction of the smooth muscle of penis is controlled by noradrenaline, and intrinsic and endothelium-derived contracting factors [67]. In the presence of a sexual stimulation, non-adrenergic non-cholinergic (NANC) nerve fibres release NO and parasympathetic cholinergic nerve fibres release acetylcholine (Ach) [68]. Production of cyclic guanosine monophosphate (cGMP) is stimulated by NO when it enters the smooth muscle and the levels of Ca^{2+} also decrease [66, 68]. cGMP activates protein kinase G and closes Ca^{2+} channels while opens the K^+ channels. A decline in the level of intercellular Ca^{2+} causes penile intracavernosal smooth muscles to relax. As the smooth muscles relax, there is an increase in the atrial flow and the lacunar spaces in the corpora cavernosa fill with blood, which leads the subtunical veins to compress thereby blocking the venu-occlusive activity or the venous outflow. All these activities ultimately result in a firm erection with minimal inflow or outflow of blood in the corpora cavernosa once the erection is established. When penile phosphodiesterase 5 (PDE 5) is degraded by cGMP, the penile smooth muscle again contracts and the process is reversed. Interruption in any of these processes can give rise to ED [69, 70].

In penile tissues, ROS are produced from mitochondrial electron transport chain, NADPH oxidase, NOS, xanthine oxidase, lipoxygenase, cyclooxygenase, haemoxygenases, and haemoproteins along with redox cycling of small molecules, autooxidation, hypoxia-reoxygenation, hyperoxia, and oxygenation of haemoglobin [71]. Roumeguere and co-workers also showed that oxidative stress, inflammation, and apoptosis of human penile endothelial cells decrease the levels of NO which contributes to the vascular aetiology of ED [72]. An important mechanism in the pathophysiology of ED is the interaction between ROS and penile NO [73]. When $O_2^{•-}$ reacts with NO, peroxynitrite is formed that in turn forms fatty deposits in arteries called atherogenesis [74]. When peroxynitrite reacts with tyrosyl residue of proteins, it inactivates SOD and decreases the removal of $O_2^{•-}$ which in turn increases peroxynitrite production and diminishes the available concentration of NO. Peroxynitrite, which causes smooth-muscle relaxation, is less potent than NO [75]. An experimental study on rabbits reported that compared to short-lived and immediate relaxation induced by NO, relaxation induced by peroxynitrite is slow and prolonged. With NO, penile tissues return to their original tension immediately, but with peroxynitrite, they are unable to do so. Hence these mechanisms produce ED due to an ineffective relaxation in the penile cavernosal tissues [76]. Peroxynitrite and $O_2^{•-}$ may also upsurge the endothelial apoptosis leading to its denudation and decrease in the bioavailability of NO. $O_2^{•-}$ mobilizes Ca^{2+} ions that may have direct vasoconstriction effect. This may also lead to ED [71, 76, 77]. A decreased concentration of NO worsens the adhesion of leukocytes and platelets to the vascular endothelial cells and releases leukotrienes and thromboxane A2 that causes vasoconstriction. These substances in turn intensify ED [78]. Apart from this, production of NO in excessive amounts also directly damages the penile cavernosa. In inflammatory conditions, high levels of NO are generated in corpora cavernosa, which elevates the formation of peroxynitrite, which has cytotoxic effects on the muscles of corpora cavernosa [79]. Ageing is an important factor associated with ED as oxidative damage caused by $O_2^{•-}$ to the vasculature plays a crucial role in course of natural ageing. Ageing also diminishes the bioavailability of NO and reduces the activity of NOS [80].

Potential mechanisms through which ROS cause ED are presented in Fig. 3.2.

3.4 Prostate Cancer

In men, prostate is a small walnut-shaped gland located at the base of the urinary bladder, and it produces seminal fluid that nourishes and transports sperm [81, 82]. Prostate cancer is one of the most frequently diagnosed cancers and is also a prime reason behind the cancer-related deaths in men [83]. Generally, majority of the prostate cancers grow slowly, they are low grade, and are also not very destructive [84]. Due to multiple genetic alterations, prostate cancer is considered as a highly heterogeneous cancer [83]. Though the risk of prostate cancer increases with age,

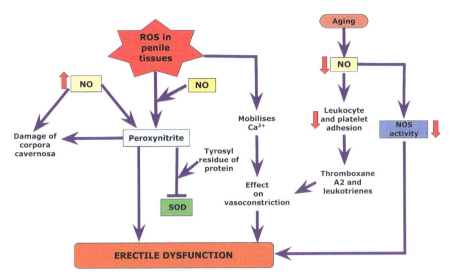

Fig. 3.2 Possible mechanism of ROS-induced erectile dysfunction. ROS is formed from the midpiece of sperm, NADPH oxidase, mitochondrial electron transport chain, xanthine oxidase, cyclooxygenase, NOS, lipooxygenase, haemooxygenases, and haemoproteins along with redox cycling of small molecules, hypoxia-reoxygenation, auto-oxidation, hyperoxia, and oxygenation of haemoglobin. ROS induces oxidative stress which along with inflammation and apoptosis of endothelial cells decreases the level of NO in penis. ROS such as O^{2-} mobilize Ca^{2+}, which affects vasoconstriction and results in ED. O^{2+} reacts with NO in penis and forms peroxynitrite. Peroxynitrite reacts with tyrosyl residue, which inactivates SOD, which in turn decreases the removal of O^{2+}. This leads to decrease in NO level in penis and increase in peroxynitrite formation. Peroxynitrite slows down and prolongs the relaxation, and the penis is unable to return to its original state, which leads to ED. O^{2+} and peroxynitrite also cause apoptosis in penile endothelium, which decreases the NO in penis, which further decreases the adhesion of leukocytes and platelets to vascular endothelium; this releases thromboxane A2 and leukotrienes, which causes vasoconstriction and leads to ED. When the level of NO increases in penis, it damages corpora cavernosa either directly or by increasing peroxynitrite, finally resulting in ED. Moreover, natural ageing also decreases NO and NOS activities, which may lead to ED. Red arrows indicate decrease or increase of the respective substances, which has a negative effect and results in ED. ROS: reactive oxygen species, ED: erectile dysfunction, NADPH: reduced form of nicotinamide adenine dinucleotide phosphate, NO: nitric oxide, NOS: nitric oxide synthase.

mainly after the age of 50, the prevalence of prostate cancer appears to be increasing in old adolescents and young adults [85, 86].

Age, ethnicity, genetic factors, family history, diet, alcohol consumption, cigarette smoking, obesity, sexually transmitted diseases, environmental carcinogens, vasectomy, exposure to X rays, insulin and insulin like growth factors, chronic inflammation and prostatitis, and vitamin and mineral supplements are some of the risk factors associated with prostate cancer [87]. In white men after the age of 50 years the risk of prostate cancer increases even without having any family history of the disease, and in the case of black men or men having a family history of prostate cancer, the risk increases after 40 years of age [88]. Prevalence of prostate cancer varies highly with different ethnic groups although the highest incidence is seen among African-American men which might be due to the presence of chromosome 8q24 variants, which have been associated with a risk of increased prostate cancer [89, 90]. A 'Hereditary Prostate Cancer' susceptibility gene or *HPC* gene (*HPC/ELAC2*), identified in chromosome 17p11, encodes a protein *ELAC2* that is associated with prostate cancer development [91]. Due to the presence of androgen receptor, X chromosome is also believed to be associated with prostate cancer inheritance [92]. Apart from genetic factors, following a westernized lifestyle and diet also increases the risk of prostate cancer [87]. A study reported a relationship between risk of prostate cancer with high intake of alcohol [93]. Obesity, combined with physical inactivity, causes insulin resistance with decreased uptake of glucose that leads to high levels of blood insulin and thus it might initiate cancer progression in the prostate as insulin promotes growth and proliferation [94]. Male smokers also have high levels of circulating sex hormones that might contribute to prostate cancer progression [95]. Furthermore, a Finnish cohort study linked the seropositivity for human papilloma virus (HPV-16 and HPV-18) with prostate cancer [96]. Carcinogens such as bisphenol-A (BPA), chlordecone, herbicides like agent orange also pose as risk factors for prostate cancer [87]. Vasectomy may also be associated with an increased risk of prostate cancer [97]. Myles and co-workers showed that exposure to hip or pelvic X-ray increases the risk of prostate cancer [98]. Inflammation of prostate gland or prostatitis has a strong link with prostate cancer, too [99].

Development and growth of prostate cancer is associated with elevated ROS levels [100]. Androgen and androgenic hormones play a vital role in normal development, differentiation, and function of the prostate gland apart from maintaining ROS balance. Fluctuations in the fraction or ratios of androgens, androgenic factors and other growth stimulatory factors are associated with abnormal prostate growth [101]. Fat-rich diet has been shown to promote ROS production accompanied by a rise in the secretion of tumor necrosis factor-alpha (TNF-α) and adipokine, thus causing chronic inflammation [102]. Hence, excessive generation of ROS or insufficiency in the antioxidant defence system of a normal cell (or both) may cause oxidative stress in the cell and the increased level of ROS may be linked with initiation and development of prostate cancers [103] A study also showed that fat-rich diet results in a rapid growth of prostate tumors and increases the ratio of M2/M1 macrophage and fraction of myeloid-derived suppressor cells through interleukin (IL)-6/phospho-signal transducer and activator of transcription 3 (pSTAT3) signalling [104]. Mitochondria are the major source of generation and mutations in mitochondrial DNA as well as changes in the mitochondrial bioenergetics may be the underlying cause of prostate cancer development. Studies showed that in malignant prostate cancer cells, there are changes in nuclear-encoded subunit IV of mitochondria [105, 106]. According to the free radical theory of ageing, cells age due to the accumulation of oxidative stress in due course of time and the age-related changes in the prostate might be linked with an upsurge in oxidative stress. Thus, a hypothesis emerged that all men would develop microscopic prostate cancer in their old age regardless of diet, lifestyle, occupation, or other factors [107]. Bacterial or non-bacterial chronic prostatitis damages prostate epithelial cells and in most cases leads to inflammation [108]. Continuous exposure of prostate cells to inflammation can

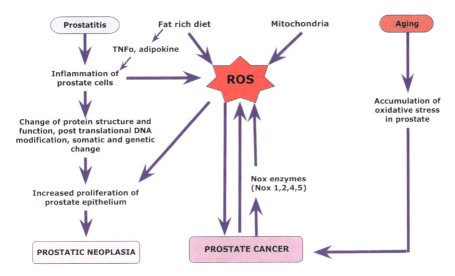

Fig. 3.3 Possible mechanism of ROS-induced prostate cancer. Fat rich diet promotes ROS formation. Mitochondria are the major sources of ROS generation that may lead to prostate cancer. ROS are also generated from the inflammation of prostate cell, caused due to bacterial or non-bacterial prostatitis. Fat-rich diet also produces TNF-α and adipokine, which lead to chronic inflammation of the prostate. Inflammation of prostate cells also results in structural and functional alterations in proteins, post-translational DNA modifications, and some somatic and genetic changes. All these may lead to increased proliferation of the prostate epithelium. ROS along with increased prostate epithelial growth result in prostatic neoplasia. Apart from that, ageing leads to accumulation of oxidative stress in prostate cells which may also cause prostate cancer. Furthermore, prostate cancer cells produce ROS and NAD(P)H oxidase or Nox enzymes such as Nox 1, Nox 2, Nox 4, and Nox 5, which generate ROS that are helpful in maintaining the malignant phenotype of the prostate cancer cells. ROS: reactive oxygen species, TNF-α: tumor necrosis factor alpha, Nox: NAD(P)H oxidase.

dramatically increase the levels of ROS, alter structure and function of protein, induce somatic and genetic changes, and also cause post-translational DNA modification [109]. Such changes can cause damage to prostate tissues, and to compensate this damage, prostate epithelial cells undergo enhanced proliferation, which ultimately leads to prostatic tumour or neoplasia [110]. Malignant prostate cells produce ROS and NAD(P)H oxidase or Nox enzymes which are also an important source of ROS. These are very crucial for the growth and maintenance of malignant phenotype of prostate cells [111, 112]. Nox 1, Nox 5, and various isoforms of Nox enzymes like Nox 4, Nox 2, and also Nox 5 are found in prostate cancer cells as reported from a study. Abnormal expression of Nox1 intensifies growth, tumorigenicity, and angiogenicity in prostate cancer cells [113].

Potential mechanisms through which ROS cause prostate cancer are presented in Fig. 3.3.

Table 3.1 summarizes the effects ROS on male infertility, ED, and prostate cancer.

Table 3.1 Effects of ROS along with possible mechanism(s) of action in male reproductive problems: infertility, erectile dysfunction (ED), and prostate cancer

Andrological problem	Possible mechanism of ROS action	Reference(s)
Male infertility	ROS attacks PUFA of sperm plasma membrane, which forms acrolein, MDA, and 4-HNE. MDA and 4-HNE cause mitochondrial dysfunction, leakage of ROS from mitochondrial inner membrane, which in turn increases ROS production, which then leads to infertility	[24, 25]
	ROS, when generated in excess in semen by leukocytes and abnormal sperm, some sperm experience oxidative damage along with a loss in their function which may lead to infertility	[26]
	Deficiency in antioxidant glutathione level in the midpiece of sperm may also cause infertility	[26]
	ROS produced from immature sperm damage DNA of mature spermatozoa that increases apoptosis of germ cell and decreases sperm count that may lead to infertility	[32, 33]
	Sperm motility declines when ROS, induced in mitochondrial complex I, cause peroxidative damage to midpiece of sperm	[37, 38]
	High levels of MDA change the membrane fluidity of sperm and hence affect motility	[40]
	ROS decreases ATP that affects sperm motility, viability, acrosome reaction, capacitation and ultimately leads to infertility	[38]
	Reduced antioxidant capacity or oxidative stress, associated with an increase in ROS, is linked to varicocele-induced infertility	[49].
	Increased ROS causes testicular hypertrophy associated with varicocele	[47]
	ROS also causes testicular dysfunction associated with varicocele	[53]
	Lipid peroxidation of sperm membrane caused due to ROS damage results in varicocele-induced infertility	[54]
	Oxidative stress caused due to imbalance between antioxidant level and ROS concentration hampers spermatogenesis and leads to poor sperm parameters	[55, 56]
	Heat increases mitochondrial ROS production that transfers electrons to molecular oxygen, and hence formation of ROS and inhibition of ATP synthesis. Heat-induced upregulation of inducible NOS increases NO production that may contribute to varicocele-associated testicular damage.	[59, 60]

(continued)

Table 3.1 (continued)

Andrological problem	Possible mechanism of ROS action	Reference(s)
Erectile dysfunction (ED)	Oxidative stress, inflammation, and apoptosis of penile endothelial cells decrease NO and contribute to ED	[72]
	Peroxynitrite reacts with tyrosyl residue of proteins, inactivates SOD, leads to decreased removal of $O^{2-\cdot}$ that increases peroxynitrite formation, and reduces the available NO concentration. Peroxynitrite slows down and prolongs the relaxation, and the penis is unable to return to the original state. All these lead to ED.	[75, 76]
	Peroxynitrite and $O^{2-\cdot}$ increase endothelial apoptosis, and the bioavailability of NO. $O^{2-\cdot}$ mobilizes Ca^{2+} ions that may have direct effect on vasoconstriction which leads to ED	[71, 76, 77]
	Low NO concentration worsens leukocytes and platelets adhesion to vascular endothelial cells, releases thromboxane A2 and leukotrienes that causes vasoconstriction. All these lead to ED.	[71]
	Excessive NO production in inflammatory conditions directly damages corpora cavernosa, which elevates peroxynitrite formation which has cytotoxic effects on corpora cavernosa muscles.	[79]
	Oxidative damage caused by $O^{2-\cdot}$ to the vasculature plays an important role in the process of natural ageing which is an important factor associated with ED; ageing also reduces NOS activity and bioavailability of NO.	[80]
Prostate cancer	Fat-rich diet promotes ROS, increases TNF-α and adipokine secretion causing chronic inflammation. Fat-rich diet also increases prostate tumour proliferation, M2/M1 macrophage ratio and myeloid-derived suppressor cells fraction via IL-6/pSTAT3 signalling.	[102, 104]
	Mitochondria are the major sources that generate ROS and mutations in mitochondrial DNA, and changes in mitochondrial bioenergetics may be the underlying cause for the development of prostate cancer	[105]
	According to the free radical theory of ageing, cells accumulate oxidative stress in due course of time and the age-related changes in the prostate might be linked with an increase in oxidative stress. Thus, a hypothesis emerged that all men would develop microscopic prostate cancer in their old age regardless of diet, lifestyle, occupation, or other factors	[107]
	Prostatitis leads to prostate inflammation and continuous exposure to inflammation increases ROS levels that leads to several changes in the prostate tissue; the prostate epithelium undergoes rapid proliferation that ultimately results in prostatic neoplasia	[108–110]

ROS reactive oxygen species, *NO* nitric oxide, *SOD* superoxide dismutase, *NOS* nitric oxide synthase, *ATP* adenosine triphosphate, *PUFA* polyunsaturated fatty acid, *MDA* malondialdehyde, *4-HNE* 4-hydroxy-2-nonenal, *DNA* deoxyribonucleic acid, *TNF-α* tumour necrosis factor-alpha, *IL* interleukin, *pSTAT3* phospho-signal transducer and activator of transcription 3

3.5 Future Perspectives

For normal reproductive functions to take place in the male, maintaining a balanced redox state is important [9]. A class of endogenous ROS scavengers called 'peroxiredoxins' or PRDX, which are mainly associated with sperm capacitation and oocyte binding remodelling, also act as an ROS control system [114]. However, dysfunction in PRDX and associated enzymes needed for their reactivation might hamper sperm capacitation and motility as well as damage sperm DNA [115].

Drugs like tadalafil, sildenafil, and vardenafil are effective in improving sexual function in men having ED and work well in younger men as well as those above 50 years of age. Apart from androgen therapy, other treatment strategies for ED include X-ray visualization and penile implant surgery [116]. Microsurgical subinguinal varicocelectomy, inguinal varicocelectomy, microscopic inguinal varicocelectomy, laproscopic varicocelectomy, and percutaneous varicocele embolization are some of the current treatment procedures of varicocele [117]. Problems related to male infertility is treated with either medicines or surgeries. Hormones such as clomiphene citrate and human chorionic gonadotropin are used for the treatment of infertility [118]. Infertility from varicocele is surgically treated by varicocelectomy, and infertility resulting from vasectomy is treated with reversal vasectomy [119]. Antioxidant therapy is also used to improve semen quality including sperm motility as antioxidants can decrease ROS-induced oxidative stress [120]. Gene therapy and gene modifications of germ line and Sertoli cells are available for treating male infertility although there are ethical questions [121]. A powerful therapeutic approach for treating male infertility is stem cell therapy [122]. Pluripotent stem cells are used, and male germ cells are developed either via in vitro differentiation into an advanced haploid product or a combination of in vitro differentiation and in vivo transplantation [123–126]. Therapies for prostate cancer include docetaxel, cabazitaxel, enzalutamide, abiraterone acetate, radium-223, sipuleucel-T, olaparib, and pembrolizumab [127]. Other therapeutic techniques for the management of prostate cancer include contemporary imaging technique like Tc-based and F-NaF positron emission tomography or PET, locoregional therapy like radical prostatectomy, and stereotactic body radiotherapy [128–130].

Apart from the above-mentioned treatment procedures, herbal medications are also useful in managing male infertility, ED, and prostate cancer [131, 132]. Roots of the Ayurvedic herb "Ashwagandha" or *Withania somnifera* are used against low sperm count, semen volume, and also low levels of testosterone and LH [133]. Other herbs such as *Orchis latifolia*, *Mucuna pruriens*, *Asparagus racemosus* and *Tinospora cardifolia* are used in the Unani and Ayurvedic systems of medicine, which may also help in increasing semen quantity, work as aphrodisiac, and in the management of disorders of male reproductive tract and combat ageing [134]. Other Ayurvedic herbs including "Ashvatta" or *Ficus relegiosa* and Fenugreek or *Trigonella foenun-graecum* are used in the management of ED [135, 136] Polysaccharide from a traditional Chinese herb *Morinda officinalis* can reduce the varicocele-induced damage to the seminiferous tubules, promote spermatogenesis, and also regulate abnormal levels of sex hormones [137]. Studies also report that Chinese herbal medicine "Bushen Shengjing Decoction" can increase motility, viability, and density of sperm along with decreasing sperm deformity rate and ROS levels in men with oligospermia [138]. Ethanol extract of a traditional Chinese medicine *Ganoderma tsugae* have shown to significantly suppress metastatic prostate cancer in cell lines PC-3 and DU145 [139]. A traditional Japanese medicine "Saikokaryukotsuboreito" showed effectiveness in the treatment of sexual dysfunction in schizophrenia [140]. A Korean herbal formula KH-204 also called "Ojayeonjonghwan" have shown to treat diabetes mellitus-induced ED in experimental rats [141]. Herbs like *Curcuma longa, Ginkgo biloba*, and *Camellia sinensis* might also be helpful in the management of male infertility, ED, and prostate cancer [131]. *Pausinystalia yohimbe* and *Panax ginseng* are useful against ED, too [142]. Considering the effectiveness of these herbal medications, the clinicians may combine these with the current treatment procedures which might prove helpful in managing the problems of male infertility, ED, and prostate cancer, thereby presenting a potentially wide range of curative measures.

3.6 Conclusions

In conclusion, the present study identifies various pathways through which ROS may hamper the normal functioning of the male reproductive system, induce oxidative stress, and cause men's health issues including infertility, ED, and prostate cancer. This chapter also presents a potentially wide range of curative measures ranging from modern medicine to traditional herbal alternatives, although more dose-dependent clinical trials are needed to validate and incorporate their use as potential alternatives for the clinical management of ROS-associated reproductive health problems in the male.

References

1. Rizzo A, Roscino MT, Binetti F, Sciorsci RL. Roles of reactive oxygen species in reactive. Reprod Domest Anim. 2012;42(2):344–52.
2. Riley JC, Behrman HR. Oxygen radicals and reactive oxygen species in reproduction. Proc Soc Exp Biol Med. 1991;198(3):781–91.
3. Li R, Jia Z, Trush MA. Defining ROS in biology and medicine. React Oxygen Species (Apex). 2016;1(1):9–21.
4. Helliwell B. Antioxidants in human diseases. Annu Rev Nutr. 1996;16:33–50.
5. Sharma R, Roychoudhury R, Alsaad R, Bamajbuor F. Negative effects of oxidative stress (OS) on reproductive system at cellular level. In: Agarwal A, Sharma R, Gupta S, Harlev A, Ahmed G, du Plessis SS, Esteves SC, Wang SM, Durairajanayagam D, editors. Oxidative stress in human reproduction. 1st ed. Berlin/Heidelberg/New York: Springer; 2017. p. 65–88.
6. Ray PD, Huang BW, Tsuji Y. Reactive oxygen species (ROS) homeostasis and redox regulation in cellular signaling. Cell Signal. 2012;24(5):981–90.

7. Pizzino G, Irrera N, Cucinotto M, Pallio G, Mannino F, Arcoraci V, Squadrito F, Altavilla D, Bitto A. Oxidative stress: harms and benefits for human health. Oxidative Med Cell Longev. 2017;2017:8416763.
8. Wagner H, Cheng JW, Ko EY. Role of reactive oxygen species in male infertility: an updated review of literature. Arab J Urol. 2018;16(1):35–43.
9. Baskaran S, Finelli R, Agarwal A, Henkel R. Reactive oxygen species in male reproduction: a boon or a bane? Andrologia. 2020;00:e13577.
10. Tafuri S, Ciani F, Iorio EL, Esposito L, Cocchia N. Reactive oxygen species (ROS) and male fertility. In new discoveries in embryology, Bin Wu. IntechOpen; 2015. https://doi.org/10.5772/60632. Accessed from: https://www.intechopen.com/books/newdiscoveries-in-embryology/reactive-oxygen-species-ros-and-male-fertility
11. Cho C, Esteves SC, Agarwal A. Novel insights into the pathophysiology of varicocele and its association with reactive oxygen species and sperm DNA fragmentation. Asian J Androl. 2016;18(2):186–93.
12. Jeremy JY, Yim AP, Wan S, Angelini GD. Oxidative stress, nitric oxide and vascular disease. J Cardiovasc Surg. 2002;17:324–7.
13. Valko M, Izakovic M, Mazur M, Rhodes CJ, Telser J. Role of oxygen radicals in DNA damage and cancer incidence. Mol Cell Biochem. 2004;266:37–56.
14. Leslie SW, Siref LE, Soon-Sutton TL, Khan MAB. Male Infertility. [Updated 2021 Feb 12]. In: StatPearls [Internet]. Treasure Island (FL): StatPearls Publishing. (2021); Available from: https://www.ncbi.nlm.nih.gov/books/NBK562258/. Accessed 21 Apr 2021.
15. Kumar N, Singh AK. Trends of male factor infertility, an important cause of infertility: a review of literature. J Hum Reprod Sci. 2015;8(4):191–6.
16. Mascarenhas MN, Flaxman SR, Boerma T, Vanderpoel S, Stevens GA. National, regional, and global trends in infertility prevalence since 1990: a systematic analysis of 277 health surveys. PLoS Med. 2012;9:e1001356.
17. Aston KI, Krausz C, Laface I, Ruiz-Castané E, Carrell DT. Evaluation of 172 candidate polymorphisms for association with oligozoospermia or azoospermia in a large cohort of men of European descent. Hum Reprod. 2010;25(6):1383–97.
18. Barak S, Baker HWG. Clinical Management of Male Infertility. [Updated 2016 Feb 5]. In: Feingold KR, Anawalt B, Boyce A, et al., editors. Endotext [Internet]. South Dartmouth (MA): MDText.com, Inc.; 2000-. Available from: https://www.ncbi.nlm.nih.gov/books/NBK279160/. Accessed 22 Apr 2021.
19. Tournaye H, Krausz C, Oates RD. Novel concepts in the aetiology of male reproductive impairment. Lancet Diabetes Endocrinol. 2017;5(7):544–53.
20. Agarwal A, Baskaran S, Parekh N, Cho CL, Henkel R, Vij S, Arafa M, Selvam MKP, Shah R. Male infertility. Lancet. 2019;397(10271):319–33.
21. Krausz C, Riera-Escamilla A. Genetics of male infertility. Nat Rev Urol. 2018;15(6):369–84.
22. Bisht S, Faiq M, Tolahunase M, Dada R. Oxidative stress and male infertility. Nat Rev Urol. 2017;14(8):470–85. https://doi.org/10.1038/nrurol.2017.69.
23. Aitken R, Harkiss D, Buckingham D. Relationship between ironcatalysed lipid peroxidation potential and human sperm function. J Reprod Fertil. 1993;98(1):257–65. https://doi.org/10.1530/jrf.0.0980257.
24. Aitken RJ. Reactive oxygen species as mediators of sperm capacitation and pathological damage. Mol Reprod Dev. 2017;84(10):1039–52.
25. Agarwal A, Rana M, Qiu E, AlBunni H, Bui AD, Henkel R. Role of oxidative stress, infection and inflammation in male infertility. Andrologia. 2018;50(11):e13126. https://doi.org/10.1111/and.13126.
26. Adeoye O, Olawumi J, Opeyemi A, Christiania O. Review on the role of glutathione on oxidative stress and infertility. JBRA Assist Reprod. 2018;22(1):61–6. https://doi.org/10.5935/1518-0557.20180003.
27. Aziz N, Saleh RA, Sharma RK, Lewis-Jones I, Esfandiari N, Thomas AJ Jr, Agarwal A. Novel association between sperm reactive oxygen species production, sperm morpho-

logical defects, and the sperm deformity index. Fertil Steril. 2004;81(2):34954. https://doi.org/10.1016/j.fertnstert.2003.06.026.
28. Talwar P, Hayatnagarkar S. Sperm function test. J Hum Reprod Sci. 2015;8(2):61–9.
29. Ribas-Maynou J, Yeste M. Oxidative stress in male infertility: causes, effects in assisted reproductive techniques, and protective support of antioxidants. Biology (Basel). 2020;9(4):77. https://doi.org/10.3390/biology9040077.
30. Aitken RJ, Curry BJ. Redox regulation of human sperm function: from the physiological control of sperm capacitation to the etiology of infertility and DNA damage in the germ line. Antioxid Redox Signal. 2011;14(3):367–81. https://doi.org/10.1089/ars.2010.3186.
31. Gong S, Gabriel MCS, Zini A, Chan P, O'Flaherty C. Low amounts and high thiol oxidation of peroxiredoxins in spermatozoa of infertile men. J Androl. 2013;33(6):1342–51.
32. Kullisaar T, Turk S, Punab M, Korrovits P, Kisand K, Rehema A, Zilmer K, Zilmer M, Mandar R. Oxidative stress in leucocytospermic prostatitis patients: preliminary results. Andrologia. 2008;40(3):161–72.
33. Agarwal A, Saleh RA, Bedaiwy MA. Role of reactive oxygen species in the pathophysiology of human reproduction. Fertil Steril. 2003;79(4):829–43.
34. Wong-Ekkabut J, Xu Z, Triampo W, Tang IM, Tieleman DP, Monticelli L. Effect of lipid peroxidation on the properties of lipid bilayers: a molecular dynamics study. Biophys J. 2007;93(12):4225–36. https://doi.org/10.1529/biophysj.107.112565. Epub 2007 Aug 31. PMID: 17766354; PMCID: PMC2098729
35. Aitken RJ, Baker MA. Causes and consequences of apoptosis in spermatozoa; contributions to infertility and impacts on development. Int J Dev Biol. 2013;57(2–4):26572. https://doi.org/10.1387/ijdb.130146ja.
36. Inaba K. Molecular architecture of the sperm flagella: molecules for motility and signaling. Zool Sci. 2003;20(9):1043–56. https://doi.org/10.2108/zsj.20.1043.
37. Amaral A, Ramalho-Santos J, St John JC. The expression of polymerase gamma and mitochondrial transcription factor A and the regulation of mitochondrial DNA content in mature human sperm. Hum Reprod. 2007;22(6):1585–96. https://doi.org/10.1093/humrep/dem030.
38. Ihsan AU, Khan FU, Khongorzul P, Ahmad KA, Naveed M, Yasmeen S, Cao Y, Taleb A, Maiti R, Akhter F, Liao X, Li X, Cheng Y, Khan HU, Alam K, Zhou X. Role of oxidative stress in pathology of chronic prostatitis/chronic pelvic pain syndrome and male infertility and antioxidants function in ameliorating oxidative stress. Biomed Pharmacother. 2018;106:714–23. https://doi.org/10.1016/j.biopha.2018.06.139.
39. Wink DA, Mitchell JB. Chemical biology of nitric oxide: insights into regulatory, cytotoxic, and cytoprotective mechanisms of nitric oxide. Free Radic Biol Med. 1998;25(4–5):434–4356.
40. Potts RJ, Notarianni LJ, Jefferies TM. Seminal plasma reduces exogenous oxidative damage to human sperm, determined by the measurement of DNA strand breaks and lipid peroxidation. Mutat Res. 2000;447(2):249–56. https://doi.org/10.1016/s00275107(99)00215-8.
41. Dutta S, Majzoub A, Agarwal A. Oxidative stress and sperm function: a systematic review on evaluation and management. Arab J Urol. 2019;2019.17(2):87–97. https://doi.org/10.1080/2090598X.2019.1599624.
42. Varicocele. Lexico UK Dictionary. https://www.lexico.com/definition/varicocele. Accessed 17 Apr 2021.
43. Yetkin E, Ozturk S. Dilating vascular diseases: pathophysiology and clinical aspects. Int J Vasc Med. 2018;2018:9024278.
44. Paick S, Choi WS. Varicocele and testicular pain: a review. World J Mens Health. 2019;37(1):4–11.
45. Leslie SW, Sajjad H, Siref LE. [Updated 2021 Feb 14]. Varicocele. In: StatPearls [Internet]. Treasure Island (FL): StatPearls Publishing; 2021 Jan-. Available from: https://www.ncbi.nlm.nih.gov/books/NBK448113/. Accessed 17 Apr 2021.
46. Will MA, Swain J, Fode M, Sonksen J, Christman GM, Ohl D. The great debate: varicocele treatment and impact on fertility. Fertil Steril. 2011;95(3):841–52. https://doi.org/10.1016/j.fertnstert.2011.01.002.

47. Yoon CJ, Park HJ, Park NC. Reactive oxygen species in the internal spermatic and brachialveins of patients with varicocele-induced infertility. Korean J Urol. 2010;51:348–53.
48. Cho CL, Esteves SC, Agarwal A. Novel insights into the pathophysiology of varicocele and its association with reactive oxygen species and sperm DNA fragmentation. Asian J Androl. 2016;18(2):186–93.
49. Weese DL, Peaster ML, Himsl KK, Leach GE, Lad PM, Zimmern PE. Stimulated reactive oxygen species generation in the spermatozoa of infertile men. J Urol. 1993;149:64–7.
50. Santos JH, Meyer JN, Van Houten B. Mitochondrial localization of telomerase as a determinant for hydrogen peroxide-induced mitochondiral DNA damage and apoptosis. Hum Mol Genet. 2006;15:1757–68.
51. Zirkin BR, Chen H. Regulation of Leydig cell steroidogenic function during aging. Biol Reprod. 2000;63:977–81.
52. El-Taieb MA, Herwig R, Nada EA, Greilberger J, Marberger M. Oxidative stress and epididymal sperm transport, motility and morphological defects. Eur J Obstet Gynecol Reprod Biol. 2009;144(Suppl 1):S199–203.
53. Koksal IT, Tefekli A, Usta M, Erol H, Abbasoglu S, Kadioglu A. The role of reactive oxygen species in testicular dysfunction associated with varicocele. BJU Int. 2000;86:549–52.
54. Alvarez JG, Touchstone JC, Blasco L, Storey BT. Spontaneous lipid peroxidation and production of hydrogen peroxide ansuperoxide in human spermatozoa: superoxide dismutase as major enzyme protectant against oxygen toxicity. J Androl. 1987;8(5):338–48.
55. Agarwal A. Relationship between oxidative stress, varicocele and infertility: a meta-analysis. Reprod Biomed Online. 2006;12(5):630–3.
56. Agarwal A, Gupta S, Sharma RK. Role of oxidative stress in female reproduction. Reprod Biol Endocrinol. 2005;3:28.
57. Zini A, Blumenfeld A, Libman J, Willis J. Beneficial effect of microsurgical varicocelectomy on human sperm DNA integrity. Hum Reprod. 2005;20(4):1018–21.
58. Allamaneni SS, Naughton CK, Sharma RK, Thomas AJ Jr, Agarwal A. Increased seminal reactive oxygen species levels in patients with varicoceles correlate with varicocele grade but not with testis size. Fertil Steril. 2004;82:1684–6.
59. Voglmayr JK, Setchell BP, White IG. The effects of heat on the metabolism and ultrastructure of ram testicular spermatozoa. J Reprod Fertil. 1971;24:71–80.
60. Rosselli M, Dubey RK, Imthurn B, Macas E, Keller PJ. Effects of nitric oxide on human spermatozoa: evidence that nitric oxide decreases sperm motility and induces sperm toxicity. Hum Reprod. 1995;10:1786–90.
61. Muneer A, Kalsi J, Nazareth I, Arya M. Erectile dysfunction. BMJ. 2014;348:g129.
62. Shamloul R, Ghanem H. Erectile dysfunction. Lancet. 2013;381(9861):153–65.
63. Feldman HA, Goldstein I, Hatzichristou DG, Krane RJ, McKinlay JB. Impotence and its medical and psychosocial correlates: results of the Massachusetts male aging study. J Urol. 1994;151(1):54–61.
64. Capogrosso P, Colicchia M, Ventimiglia E, Castagna G, Clementi MC, Suardi N, Castiglione F, Briganti A, Cantiello F, Damiano R, Montorsi F, Salonia A. One patient out of four with newly diagnosed erectile dysfunction is a young man-worrisome picture from the everyday clinical practice. J Sex Med. 2013;10(7):1833–41.
65. Matsui H, Sopko NA, Hannan JL, Bivalacqua TJ. Pathophysiology of erectile dysfunction. Curr Drug Targets. 2015;16(5):411–9.
66. Sooriyamoorthy T, Leslie SW. [Updated 2021 Feb 22]. Erectile Dysfunction. In: StatPearls [Internet]. Treasure Island (FL): StatPearls Publishing; 2021 Jan-. Available from: https://www.ncbi.nlm.nih.gov/books/NBK562253/. Accessed 13 Aug 2021.
67. Lue TF. Erectile dysfunction. N Engl J Med. 2000;342(24):1802–13.
68. Yafi FA, Jenkins L, Albersen M, Corona G, Isidori AM, Goldfarb S, Maggi M, Nelson CJ, Parish S, Salonia A, Tan R, Mulhall JP, Hellstrom WJ. Erectile dysfunction. Nat Rev Dis Primers. 2016;2016(2):16003.

69. Dean RC, Lue TF. Physiology of penile erection and pathophysiology of erectile dysfunction. Urol Clin North Am. 2005;32(4):379–95.
70. Giuliano F. Neurophysiology of erection and ejaculation. J Sex Med. 2011;8(Suppl 4):310–5.
71. Jeremy JY, Jones RA, Koupparis AJ, Hotston M, Persad R, Angelini GD, Shukla N. Reactive oxygen species and erectile dysfunction: possible role of NADPH oxidase. Int J Impot Res. 2006;19(3):265–80.
72. Roumeguère T, Antwerpen PV, Fathi H, Rousseau A, Vanhamme L, Franck T, Costa C, Morelli A, Lelubre C, Hauzeur C, Raes M, Serteyn E, Wespes E, Vanhaeverbeek M, Boudjeltia Z. Relationship between oxidative stress and erectile function. Free Radic Res. 2017; ISSN: 1071-5762 (Print) 1029–2470.
73. Jones RW, Rees RW, Minhas S, Ralph D, Persad RA, Jeremy JY. Oxygen free radicals and the penis. Expert Opin Pharm. 2002;3:889–97.
74. Beckman JS, Koppenol WH. Nitric oxide, superoxide, and peroxynitrite: the good, the bad, and ugly. Am J Phys. 1996;271:1424–37.
75. Zou M, Martin C, Ullrich V. Tyrosine nitration as a mechanism of selective inactivation of prostacyclin synthase by peroxynitrite. Biol Chem. 1997;378:707–13.
76. Khan MA, Thompson CS, Mumtaz FH, Mikhailidis DP, Morgan RJ, Bruckdorfer RK, Naseem KM. The effect of nitric oxide and peroxynitrite on rabbit cavernosal smooth muscle relaxation. World J Urol. 2001;19:220–4.
77. Katusic ZS, Vanhoutte PM. Superoxide anion is an endothelium-derived contracting factor. Am J Phys. 1989;257:33–7.
78. Jeremy JY, Angelini GD, Khan M, Mikhailidis DP, Morgan RJ, Thompson CS, Bruckdorfer KR, Naseem KM. Platelets, oxidant stress and erectile dysfunction: a hypothesis. Cardiovasc Res. 2002;46:50–4.
79. Wink DA, Feelisch M, Fukuto J, Chistodoulou D, Jourd'heuil D, Grisham MB, Vodovotz Y, Cook JA, Krishna M, DeGraff WG, Kim S, Gamson J, Mitchell JB. The cytotoxicity of nitroxyl: possible implications for the pathophysiological role of NO. Arch Biochem Biophys. 1998;351:66–74.
80. Agarwal A, Nandipati KC, Sharma RK, Zippe CD, Raina R. Role of oxidative stress in the pathophysiological mechanism of erectile dysfunction. J Androl. 2006;27(3):335–47.
81. Aron LT, Franco O, Hayward SW. Review of prostate anatomy and embryology and the etiology of BPH. Urol Clin North Am. 2016;43(3):279–88. https://doi.org/10.1016/j.ucl.2016.04.012.
82. Prostate cancer. Mayo Clinic. Available from: mayoclinic.org/diseasesconditions/prostate-cancer/symptoms-causes/syc-20353087. Accessed on 5 May 2021.
83. Sugiura M, Sato H, Kanesaka M, Imamura Y, Sakamoto S, Ichikawa T, Kaneda A. Epigenetic modifications in prostate cancer. Int J Urol. 2020;28(2):140–9. https://doi.org/10.1111/iju.14406.
84. Leslie SW, Soon-Sutton TL, Sajjad H, Siref LE. Prostate Cancer. In: StatPearls [Internet]. Treasure Island (FL): StatPearls Publishing. 2021; Available from: https://www.ncbi.nlm.nih.gov/books/NBK470550/. Accessed on 5 May 2021.
85. Prostate cancer: risk factors and prevention. Cancer.Net. 2021; Available from: https://www.cancer.net/cancer-types/prostate-cancer/risk-factors-and-prevention. Accessed on 5 May 2021.
86. Bleyer A, Spreafico F, Barr R. Prostate cancer in young men: an emerging young adult and older adolescent challenge. Cancer. 2020;126(1):46–57. https://doi.org/10.1002/cncr.32498.
87. Rawla P, Epidermiology of prostate cancer. World J. Oncologia. 2019;10(2):63–89. https://doi.org/10.14740/wjon1191.
88. Cancer Stat Facts: Prostate Cancer [Internet]. SEER, 2018. Available from: https://seer.cancer.gov/statfacts/html/prost.html. Accessed on 16 May 2021.
89. SEER Cancer Statistics Review, 1975–2013 [Internet]. National Cancer Institue, Bethesda, MD. 2016. Available from: https://seer.cancer.gov/csr/1975_2015/. Accessed 16 May 2021. [Internet]. SEER, 2018. Available from: https://seer.cancer.gov/explorer/application.php

90. Okobia MN, Zmuda JM, Ferrell RE, Patrick AL, Bunker CH. Chromosome 8q24 variants are associated with prostate cancer risk in a high risk population of African ancestry. Prostate. 2011;71(10):1054–63.
91. Noda D, Itoh S, Watanabe Y, Inamitsu M, Dennler S, Itoh F, Koike S, Danielpour D, ten Dijke P, Kato M. ELAC2, a putative prostate cancer susceptibility gene product, potentiates TGF-beta/Smad-induced growth arrest of prostate cells. Oncogene. 2006;25(41):5591–600.
92. Xu J, Meyers D, Freije D, Isaacs S, Wiley K, Nusskern D, Ewing C, Wilkens E, Bujnovszky P, Bova GS, Walsh P, Isaacs W, Schleutker J, Matikainen M, Tammela T, Visakorpi T, Kallioniemi OP, Berry R, Schaid D, French A, McDonnell S, Schroeder J, Blute M, Thibodeau S, Grönberg H, Emanuelsson M, Damber JE, Bergh A, Jonsson BA, Smith J, Bailey-Wilson J, Carpten J, Stephan D, Gillanders E, Amundson I, Kainu T, Freas-Lutz D, Baffoe-Bonnie A, Van Aucken A, Sood R, Collins F, Brownstein M, Trent J. Evidence for a prostate cancer susceptibility locus on the X chromosome. Nat Genet. 1998;20(2):175–9.
93. Rota M, Scotti L, Turati F, Tramacere I, Islami F, Bellocco R, Negri E, Corrao G, Boffetta P, La Vecchia C, Bagnardi V. Alcohol consumption and prostate cancer risk: a meta-analysis of the dose-risk relation. Eur J Cancer Prev. 2012;21(4):350–9.
94. Kaaks R, Stattin P. Obesity, endogenous hormone metabolism, and prostate cancer risk: a conundrum of "highs" and "lows". Cancer Prev Res (Phila). 2010;3(3):25962.
95. Huncharek M, Haddock KS, Reid R, Kupelnick B. Smoking as a risk factor for prostate cancer: a meta-analysis of 24 prospective cohort studies. Am J Public Health. 2010;100(4):693–701.
96. Dillner J, Knekt P, Boman J, Lehtinen M, Af Geijersstam V, Sapp M, Schiller J, Maatela J, Aromaa A. Sero-epidemiological association between humanpapillomavirus infection and risk of prostate cancer. Int J Cancer. 1998;75(4):5647.
97. Cox B, Sneyd MJ, Paul C, Delahunt B, Skegg DC. Vasectomy and risk of prostate cancer. JAMA. 2002;287(23):3110–5.
98. Myles P, Evans S, Lophatananon A, Dimitropoulou P, Easton D, Key T, Pocock R, Dearnaley D, Guy M, Edwards S, O'Brien L, Gehr-Swain B, Hall A, Wilkinson R, Eeles R, Muir K. Diagnostic radiation procedures and risk of prostate cancer. Br J Cancer. 2008;98(11):1852–6.
99. Jiang J, Li J, Yunxia Z, Zhu H, Liu J, Pumill C. The role of prostatitis in prostate cancer: meta-analysis. PLoS One. 2013;8(12):e85179.
100. Tan BL, Norhaizan ME. Oxidative stress, diet and prostate cancer. World J Mens Health. 2021;39(2):195–207. https://doi.org/10.5534/wjmh.200014.
101. Gonthier K, Poluri RTK, Audet-Walsh É. Functional genomic studies reveal the androgen receptor as a master regulator of cellular energy metabolism in prostate cancer. J Steroid Biochem Mol Biol. 2019;191:105367.
102. Maurizi G, Della Guardia L, Maurizi A, Poloni A. Adipocytes properties and crosstalk with immune system in obesityrelated inflammation. J Cell Physiol. 2018;233:88–97.
103. Khandrika L, Kumar B, Koul S, Maroni P, Koul HK. Oxidative stress in prostate cancer. Cancer Lett. 2009;282(2):125–36. https://doi.org/10.1016/j.canlet.2008.12.011.
104. Hayashi T, Fujita K, Nojima S, Hayashi Y, Nakano K, Ishizuya Y, et al. High-fat diet-induced inflammation accelerates prostate cancer growth via IL6 signaling. Clin Cancer Res. 2018;24:4309–18.
105. Petros JA, Baumann AK, Ruiz-Pesini E, Amin MB, Sun CQ, Hall J, Lim S, Issa MM, Flanders WD, Hosseini SH, Marshall FF, Wallace DC. mtDNA mutations increase tumorigenicity in prostate cancer. Proc Natl Acad Sci U S A. 2005;102:719–24.
106. Chen JZ, Gokden N, Greene GF, Green B, Kadlubar FF. Simultaneous generation of multiple mitochondrial DNA mutations in human prostate tumors suggests mitochondrial hypermutagenesis. Carcinogenesis. 2003;24:1481–7.
107. Paschos A, Pandya R, Duivenvoorden WCM, Pinthus JH. Oxidative stress in prostate cancer: changing research concepts towards a novel paradigm for prevention and therapeutics. Prostate Cancer and Prostatic Dis. 2013;16(3):217–25.

108. de Marzo AM, Platz EA, Sutcliffe S, Xu J, Grönberg H, Drake CG, Nakai Y, Isaacs WB, Nelson WG. Inflammation in prostate carcinogenesis. Nat Rev Cancer. 2007;7:256–69.
109. Olinski R, Gackowski D, Foksinski M, Rozalski R, Roszkowski K, Jaruga P. Oxidative DNA damage: assessment of the role in carcinogenesis, atherosclerosis, and acquired immunodeficiency syndrome. Free Radic Biol Med. 2002;33:192–200.
110. Naber KG, Weidner W. Chronic prostatitis-an infectious disease? J Antimicrob Chemother. 2000;46:157–61.
111. Lim SD, Sun C, Lambeth JD, Marshall F, Amin M, Chung L, Petros JA, Arnold RS. Increased Nox1 and hydrogen peroxide in prostate cancer. Prostate. 2005;62:200–7.
112. Brar SS, Corbin Z, Kennedy TP, Hemendinger R, Thornton L, Bommarius B, Arnold RS, Whorton AR, Sturrock AB, Huecksteadt TP, Quinn MT, Krenitsky K, Ardie KG, Lambeth JD, Hoidal JR. NOX5 NAD(P)H oxidase regulates growth and apoptosis in DU 145 prostate cancer cells. Am J Physiol Cell Physiol. 2003;285:C353–69.
113. Arbiser JL, Petros J, Klafter R, Govindajaran B, McLaughlin ER, Brown LF, Cohen C, Moses M, Kilroy S, Arnold RS, Lambeth JD. Reactive oxygen generated by Nox1 triggers the angiogenic switch. Proc Natl Acad Sci (USA). 2002;99:715–20.
114. Lee D, Moawad AR, Morielli T, Fernandez MC, O'Flaherty C. Peroxiredoxins prevent oxidative stress during human sperm capacitation. Mol Hum Reprod. 2016;23(2):106–15. https://doi.org/10.1093/molehr/gaw081.
115. O'Flaherty C. Redox regulation of mammalian sperm capacitation. Asian J Androl. 2015;17:583–90.
116. Roychoudhury S, Chakraborty S, Choudhury AP, Das A, Jha NK, Slama P, Nath M, Massanyi P, Ruokolainen J, Kesari KK. Environmental factors-induced oxidative stress: hormonal and molecular pathway disruptions in hypogonadism and erectile dysfunction. Antioxidants. 2021;10:837. https://doi.org/10.3390/antiox10060837.
117. Johnson D, Sandlow J. Treatment of varicoceles: techniques and outcomes. Fertil Steril. 2017;108(3):378–84. https://doi.org/10.1016/j.fertnstert.2017.07.020.
118. Pan MM, Hockenberry MS, Kirby EW, Lipshultz LI. Male infertility diagnosis and treatment in the era of in vitro fertilization and intracytoplasmic sperm injection. Med Clin North Am. 2018;102(2):337–47. https://doi.org/10.1016/j.mcna.2017.10.008.
119. Pourmoghadam Z, Aghebati-Maleki L, Motalebnezhad M, Yousefi B, Yousefi M. Current approaches for the treatment of male infertility with stem cell therapy. J Cellular Physiol. 2018;233(10):6455–69. https://doi.org/10.1002/jcp.26577.
120. Cocuzza M, Sikka SC, Athayde KS, Agarwal A. Clinical relevance of oxidative stress and sperm chromatin damage in male infertility: an evidence based analysis. Int Braz J Urol. 2007;33:603–21.
121. Gassei K, Orwig KE. Experimental methods to preserve male fertility and treat male factor infertility. Fertil Steril. 2016;105:256–66.
122. Beckmann J, Scheitza S, Wernet P, Fischer JC, Giebel B. Asymmetric cell division within the human hematopoietic stem and progenitor cell compartment: identification of asymmetrically segregating proteins. Blood. 2007;109:5494–501.
123. Malik N, Rao MS. A review of the methods for human iPSC derivation. Pluripotent stem cells. Methods Mol Biol. 2013;997:23–33.
124. Xu XL, Yi F, Pan HZ, Duan SL, Ding ZC, Yuan GH, Qu J, Zhang HC, Liu GH. Progress and prospects in stem cell therapy. Acta Pharmacol Sin. 2013;34:741–6.
125. Cai H, Xia X, Wang L, Liu Y, He Z, Guo Q, Xu C. In vitro and in vivo differentiation of induced pluripotent stem cells into male germ cells. Biochem Biophys Res Commun. 2013;433(3):286–91.
126. Zhu Y, Hu HL, Li P, Yang S, Zhang W, Ding H, Tian RH, Ning Y, Zhang LL, Guo XZ, Shi ZP, Li Z, He Z. Generation of male germ cells from induced pluripotent stem cells (iPS cells): an in vitro and in vivo study. Asian J Androl. 2012;14:574.
127. Teo MY, Rathkopf DE, Kantoff P. Treatment of advanced prostate cancer. Annu Rev Med. 2019;70:479–99. https://doi.org/10.1146/annurev-med-051517-011947.

128. Choline C 11 injection. Rochester, MN: Mayo Clinic; 2012. https://www.accessdata.fda.gov/drugsatfda_docs/label/2012/203155s000lbl.pdf
129. Sooriakumaran P, Karnes J, Stief C, Copsey B, Montorsi F, Hammerer P, Beyer B, Moschini M, Gratzke C, Steuber T, Suardi N, Briganti A, Manka L, Nyberg T, Dutton SJ, Wiklund P, Graefen M. A multi-institutional analysis of perioperative outcomes in 106 men who underwent radical prostatectomy for distant metastatic prostate cancer at presentation. Eur Urol. 2016;69:788–94.
130. Ost P, Jereczek-Fossa BA, As NV, Zili T, Muacevic A, Oliver K, Henderson D, Casamassima F, Orecchia R, Surgo A, Brown L, Tree A, Miralbell R, Meerleer GD. Progression-free survival following stereotactic body radiotherapy for oligometastatic prostate cancer treatment-naive recurrence: a multi-institutional analysis. Eur Urol. 2016;69(1):9–12.
131. Roychoudhury S, Chakraborty S, Das A, Guha P, Agarwal A, Henkel R. Herbal medicine used to treat andrological problems: Asian and Indian subcontinent: *Ginkgo biloba*, *Curcuma longa*, and *Camellia sinensis*. In: Henkel R, Agarwal A, editors. Herbal medicine in andrology an evidence-based update. 1st ed. Cambridge, MA: Academic Press; 2021. p. 129–46.
132. Kim SJ, Jeon SH, Kwon EB, Jeong HC, Choi SW, Bae WJ, Cho HJ, Ha US, Hong SH, Lee JY, Hwang SY, Kim SW. Early and synergistic recovery effect of herbal combination on surgically corrected varicocele. Altern Ther Health Med. 2020;26(3):24–31.
133. Sengupta P, Agarwal A, Pogrebetskaya M, Roychoudhury S, Durairajanayagam D, Henkel R. Role of *Withania somnifera* (Ashwagandha) in the management of male infertility. Reprod Biomed. 2017;36:311–26.
134. Lohiya NK, Balasubramanian K, Ansari AS. Indian folklore medicine in managing men's health and wellness. Andrologia. 2016;48:894–907.
135. Virani NV, Chandola HM, Vyas SN, Jadeja DB. Clinical study on erectile dysfunction in diabetic and non-diabetic subjects and its management with *Ficus relegiosa* Linn. Ayu. 2010;31(3):272–9. https://doi.org/10.4103/0974-8520.77148.
136. George A, Liske E. Acceptance of herbal medicine in andrology. In: Henkel R, Agarwal A, editors. Herbal medicine in andrology an evidence-based update. 1st ed. Cambridge, MA: Academic Press; 2021. p. 215–55.
137. Zhang L, Zhao X, Wang F, Lin Q, Wang W. Effects of *Morinda officinalis* polysaccharide on experimental varicocele rats. Evidence based complementary and alternative medicine. 2016; 2016, Article ID 5365291. https://doi.org/10.1155/2016/5365291.
138. Zhang HQ, Zhao HX, Zhang AJ. Male infertility with severe oligospermatism and azoospermia treated by Bhusen Shengjing decoction combined with intracytoplasmic sperm injection. Chin J Integr Med. 2007;27:972–5.
139. Huang WC, Chang MS, Huang SY, Tsai CJ, Kuo PH, Chang HW, Huang ST, Kuo CL, Lee SL, Kao MC. Chinese herbal medicine *Ganoderma tsugae* displays potential anti-cancer efficacy on metastatic prostate cancer cells. Int J Mol Sc. 2019;20(18):4418. https://doi.org/10.3390/ijms20184418.
140. Takashi T, Uchida H, Suzuki T, Mimura M. Effectiveness of Saikokaryukotsuboreito (herbal medicine) for antipsychotic-induced sexual dysfunction in male patients with schizophrenia: a description of two cases. Rep Psychiatry. 2014;2014(2014):784671. https://doi.org/10.1155/2014/784671.
141. Jeong HC, Bae WJ, Zhu GQ, Jeon SH, Choi SW, Kim SJ, Cho HJ, Hong SH, Lee JY, Hwang SY, Kim SW. Synergistic effects of extracorporeal shockwave therapy and modified Ojayeonjonghwan on erectile dysfunction in an animal model of diabetes. Investig Clin Urol. 2019;60(4):285–94. https://doi.org/10.4111/icu.2019.60.4.285.
142. Lee JKC, Tan RBW, Chung E. Erectile dysfunction treatment and traditional medicine—can east and west medicine coexist? Transl Androl Urol. 2017;6:91–100.

Chapter 4
Molecular Interactions Associated with Oxidative Stress-Mediated Male Infertility: Sperm and Seminal Plasma Proteomics

Manesh Kumar Panner Selvam, Damayanthi Durairajanayagam, and Suresh C. Sikka

Abstract Male factor issues are responsible for 50% of couples infertility. Seminal oxidative stress is one of the major factors that affect the normal physiological aspects of sperm function such as motility and progression, hyperactivation, capacitation, acrosome reaction and zona-pellucida penetration prior to fertilization. In recent times, high-throughput proteomic platforms are used to identify the proteins associated with these aspects of sperm function as associated with oxidative stress. In this review, we have provided a workflow that includes an overview of advanced proteomic techniques and bioinformatic tools used to interpret proteomic results. Furthermore, we have highlighted proteins associated with dysregulated molecular pathways in sperm and seminal plasma due to oxidative stress. We have also described the molecular interactions between proteins associated with oxidative stress and their potential role in male infertility.

Keywords Male Infertility · Sperm · Seminal Plasma · Oxidative Stress · Proteomics

---

M. K. Panner Selvam (✉) · S. C. Sikka (✉)
Department of Urology, Tulane University School of Medicine, New Orleans, LA, USA
e-mail: mpannerselvam@tulane.edu; ssikka@tulane.edu

D. Durairajanayagam
Department of Physiology, Faculty of Medicine, Universiti Teknologi MARA, Sungai Buloh Campus, Selangor, Malaysia

© The Author(s), under exclusive license to Springer Nature Switzerland AG 2022
K. K. Kesari, S. Roychoudhury (eds.), *Oxidative Stress and Toxicity in Reproductive Biology and Medicine*, Advances in Experimental Medicine and Biology 1358, https://doi.org/10.1007/978-3-030-89340-8_4

4.1 Introduction

Infertility is classified as a condition associated with a couple's inability to achieve a clinical pregnancy for more than a year despite having regular, well-timed, unprotected sexual intercourse [1]. The male partner contributes to approximately half of the total infertility cases and is solely responsible for 20–30% of these [2]. Infertility rates are estimated to be similar among males in North America (4.5–6%), Europe (7.5%), and Australia (8–9%), and highest in Africa and Eastern Europe (8–12%) [3]. On average, about 15% of male infertility cases have no identifiable cause and are classified as unexplained or idiopathic male infertility [4]. Men with unexplained infertility are unable to reproduce despite having normal sexual history, physical examination, endocrine assessment, semen parameters, and no female factor involvement [5].

Among the contributory factors implicated in cases of unexplained male infertility is oxidative stress. Oxidative stress represents a state of imbalance between the generation of reactive oxygen species (ROS) and the antioxidant defense mechanisms within the body [4–6]. ROS are generated as byproducts of normal cellular metabolism and are essential at physiological levels for normal reproduction [7]. The major sources of seminal ROS are leukocytes and spermatozoa. When present at elevated levels in semen, ROS become harmful to the sperm and can negatively impact fertility [8, 9]. The adverse effects of oxidative stress on sperm quality and function makes it a major factor in the etiology of male infertility. Studies have demonstrated the role of oxidative stress in several other male infertility diagnoses including varicocele [10], leukocytospermia [11], prostatitis [12], and idiopathic infertility [13].

Proteomics is the systematic analysis and characterization of the entire set of proteins present within the cell / tissue under certain conditions. Proteomics studies not only do help identify the proteins present but can also provide valuable information in terms of protein abundance, localization, posttranslational modifications, isoforms, and molecular interactions [14]. While the genome is a constant, the proteome is variable and dynamic. Thus, in proteomics studies related to infertility, the proteome can be examined to elicit an overall picture (at a protein level) of the underlying cellular processes in fertile and infertile males. Moreover, proteomics analysis can contribute to the understanding of critical biological pathways underlying the development of male infertility [15].

Mature spermatozoa are transcriptionally and translationally quiescent and thereby depend on their proteins to perform their biological functions [16]. Proteins are abundant in semen and obtainable with relative ease, making spermatozoa and seminal plasma suitable samples for proteomic analysis [17]. By characterizing the proteins in spermatozoa and seminal plasma, the function of specific fertility-related proteins can be determined [18]. Through the application of advancing proteomics tools and techniques, researchers could potentially identify critical changes in key proteins in spermatozoa and seminal plasma of male infertile patients [19]. Oxidative stress has been reported to impact metabolic processes in spermatozoa, regulatory pathways in seminal plasma, and stress responses in both sperm and seminal plasma [20].

In this review, we provide an overview of the workflow involved when applying advanced proteomics and bioinformatics approaches to sperm and seminal plasma samples. We have highlighted the spermatozoa and seminal plasma proteins related to dysregulated molecular pathways in males with oxidative stress-induced infertility. We also describe the interactions between the proteins associated with oxidative stress and examine their potential role as an underlying cause of male infertility.

4.2 Proteomic Evaluation in Sperm and Seminal Plasma

Most proteomics studies in male infertility aim to compare, identify, and subsequently quantify the proteins that are differentially expressed between samples (seminal plasma or spermatozoa) from fertile males and infertile patients [10, 21, 22]. Detection of these proteins are commonly performed using both the conventional proteomic techniques, e.g., gel electrophoresis; and the advanced techniques, e.g., matrix-assisted laser desorption/ionization time-of-flight (MALDI-TOF) and liquid chromatography-tandem mass spectrometry (LC-MS/MS) [17]. The methodological workflow commonly used for the processing of semen samples with oxidative stress in proteomics studies of spermatozoa and seminal plasma are shown in Fig. 4.1 and briefly described as follows.

The initial step is to extract the proteins from spermatozoa and seminal plasma. Once the samples are ready for analysis, the complex mixture of proteins within the samples are separated using gel-based methods [20]. In one-dimensional sodium dodecyl sulphate-polyacrylamide gel electrophoresis (SDS-PAGE), proteins are separated based only on the differences in their molecular weight [23]. The more commonly used technique is the two-dimensional gel electrophoresis (2D-GE), where proteins are separated based on their charge (isoelectric point, first dimension) and mass (relative molecular weight, second dimension) [24].

A modified form of 2D-GE is the difference gel electrophoresis (DIGE). Here, proteins within each sample are covalently tagged with fluorescent dyes of different colors to facilitate simultaneous comparison of two to three protein samples on the same gel. Common proteins between the samples appear as "spots" that have a fixed ratio of fluorescent signals, while proteins that are differentially expressed show different fluorescence ratios [25]. However, these gel-based methods lack the sensitivity to detect proteins present in low abundance [26]. The integration of gel-based methods and advanced techniques (such as MALDI-TOF MS and LC-MS/MS) overcomes this technical limitation to provide enhanced protein separation and identification [27, 28].

After separation, protein spots in the gels are cut out and then digested by an endopeptidase (e.g., trypsin) into small peptide fragments. Next, the digested peptides will either be separated by LC or directly analyzed using MS. In LC-MS, the digested peptides are passed through a high-performance liquid chromatography

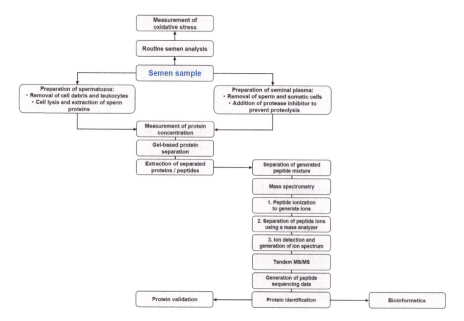

Fig. 4.1 Workflow for the processing of semen samples with oxidative stress in proteomics studies of spermatozoa and seminal plasma. Following sample collection and liquefaction, routine semen analysis and measurement of oxidative stress are performed. Examples of oxidative stress measurement testing include oxidative-reductive potential (ORP), ROS, malonaldehyde (MDA), and protein carbonyl. Following this, seminal plasma and spermatozoa respectively from the semen samples are prepared for proteomic analysis. Spermatozoa and seminal plasma are separated. For the spermatozoa sample, cell debris and leukocytes are removed, while for seminal plasma samples, spermatozoa and other somatic cells are removed. Sperm cells are lysed, the proteins extracted, and its concentration is determined. For seminal plasma, protease inhibitors are added to prevent proteolysis. Protein concentration in both types of samples is then measured. Next, the proteins are separated usually using gel-based methods such as SDS-PAGE, 2D-GE, and DIGE. Thereafter, the separated proteins/peptides are extracted using in-gel digestion. The generated peptide mixture is then separated into individual or homogenous fragments using liquid chromatography, e.g., HPLC. Next, the peptides are subjected to mass spectrometry, which consists of three parts: peptide ionization that generates ion in a gaseous phase using MALDI or ESI; separation of peptide ions based on mass-to-charge ratio using a mass analyzer (e.g., TOF, FTIC, ion trap); and lastly ion detection and generation of ion spectrum (intensity vs. m/z). The process of fragmentation and mass analysis is repeated several times in tandem MS/MS and will eventually generate peptide sequencing data or MS/MS spectra. Next, protein identification is carried out based on the peptide sequences using available protein database search engines such as SEQUEST, Mascot, and X!Tandem. Finally, the identified proteins are subjected to bioinformatics where functional annotation and enrichment analysis, protein–protein interactions, pathway network, and posttranslational modifications are analyzed using the appropriate databases, such as STRING, IPA, MetaCore, UniPROT, Reactome, and DAVID. The identified proteins of interest are also subjected to protein validation such as Western blotting, immunohistochemistry, and selected / multiple or parallel reaction monitoring

(HPLC) for separation of the fragments [10, 21]. Smaller homogenous fragments of peptides eluting from the LC column are introduced into the mass spectrometer. The individual peptide fragments are then ionized by MALDI or electrospray ionization, ESI, and subsequently the ions will be analyzed by MS [29].

MS is a high-throughput technique that provides high sensitivity and specificity. It is used for protein identification and for the determination of protein sequences. There are three main components to a mass spectrometer: an ion source (e.g., MALDI or ESI), a mass analyzer (e.g. TOF, Quadrupole Fourier-transform ion cyclotron (FTIC), or ion trap), and an ion detection system [27]. Thus, in protein analysis using the MS technique, proteins are initially ionized by the ion source of the mass spectrometer where ions are generated in a gas phase. The resulting ions are then separated by their mass-to-charge ratio (m/z) using a mass analyzer, before being detected [30].

In tandem MS (MS/MS), peptides are subjected to several rounds of fragmentation and mass analysis to obtain peptide sequences. The resulting large number of peptide sequences (MS/MS spectra) are then compared against the predicted spectra of proteins in existing databases. These databases are generally protein databases translated from genomic data [31], but spectral libraries have also been utilized [32].

The comprehensive list of proteins can be determined using suitable algorithms in computational software / protein database search engines such as SEQUEST, Mascot, X!-Tandem or X!!Tandem, and MaxQuant-Andromeda. The database search approach is normally followed up with further analysis to reduce false discovery rates (FDRs) in peptide identification [10, 33]. To ensure that comparable and reliable results are generated for downstream analysis of proteins, raw quantitative data is first subjected to preprocessing and normalization before any statistical analyses are carried out [33].

Functional annotation is an analytical technique used to annotate and classify proteins based on their biological functions. It is based on a statistical assessment termed enrichment. Enrichment analysis can be performed using the Gene Ontology (GO) library where proteins are annotated with GO terms provide information indicating biological processes, molecular functions, and cellular components [34]. Enrichment analysis can also be performed using pathway annotation (e.g., Kyoto Encyclopedia of Genes and Genomes (KEGG), Reactome) to elicit previously known regulatory pathway networks [35]. Enrichment protocols also help identify posttranslational modifications (e.g., phosphorylation, glycosylation, acetylation, methylation, ubiquitination) [36].

Proteomics data could also be used to decipher protein–protein interaction (PPI) networks (via online databases such as Search Tool for the Retrieval of Interacting Genes/Proteins (STRING)) and signaling networks [37]. Software programs such as Ingenuity Pathway Analysis (IPA) and MetaCore™ provide a comprehensive picture of the interactions between the proteins and the transcriptional factors that regulate their expression [20]. Finally, systems or network biology can be applied to examine PPIs and the functional relationship between genes and proteins within a large network [33].

Differentially expressed proteins (DEPs) identified between study groups could also be subjected to functional annotation and enrichment analysis using bioinformatics tools and databases (e.g., UniProt, Reactome, and Database for Annotation Visualization and Integrated Discovery (DAVID)) along with IPA analysis to ascertain the role of DEPs in biological processes and pathways, PPIs, and network regulation [38, 39].

Western blotting is used for detection of low abundance proteins and for validation of selected proteins of interest [10, 21, 22, 40]. This technique involves the separation of proteins using electrophoresis, its transfer onto a nitrocellulose membrane, and accurate detection of a target protein using enzyme-conjugated antibodies [41]. Other validation techniques to confirm protein identifications include microscopic imaging such as immunohistochemical staining for localization and relative abundance of protein expression. Further MS analysis such as selected/multiple reaction monitoring or the more advanced parallel reaction monitoring can be performed for validation of protein abundance differences between samples [42].

4.3 Oxidative Stress-Mediated Changes in Semen Proteins

Seminal oxidative stress is prevalent in 80% of infertile men [43]. Excess amounts of ROS have deleterious effects on the physiological function of spermatozoa [9, 44]. Several sperm and seminal plasma studies have been carried out to understand the impact of oxidative stress on cellular pathways and sperm proteins associated with the reproductive process (reviewed in [45, 46]). This section covers the global proteomic studies conducted on sperm and seminal plasma of patients with seminal oxidative stress.

4.3.1 Molecular Changes in Sperm

A continuous state of oxidative stress in semen can alter the expression of proteins in sperm. Using the LC-MS/MS approach, Sharma et al. observed that a total of 74 proteins were differentially expressed due to oxidative stress in sperm [47]. Majority of these DEPs were involved in the CREM signaling pathway that regulates spermatid differentiation. Key proteins such as HIST1H2BA, MDH2, TGM4, GPX4, GLUL, HSP90B1, and HSPA5 were proposed as biomarkers of oxidative stress [47]. In addition to these proteins, computational and data mining approaches have identified ACE, HSPA2, RPS27A, MAP3K3, and APP proteins associated with sperm function to be altered in patients with high levels of seminal ROS [48]. Another global sperm proteomic study revealed a new set of six proteins (CLGN, TPPII, DNAI2, EEA1, HSPA4L, and SERPINA5) common to patients with low, medium, and high levels of seminal oxidative stress [49]. These six DEPs were associated with reproductive function affecting the normal physiological function of sperm [49]. Recently, Dias et al. observed alterations in the sperm and seminal plasma proteome of fertile men with high levels of ROS. A total of 371 and 41 DEPs were reported in sperm and seminal plasma of fertile men with oxidative stress.

Bioinformatic analysis revealed activation of proteasomal system and antioxidant defense mechanisms in fertile men with high ROS levels [22].

4.3.2 Molecular Changes in Seminal Plasma

To date, there are only very few proteomic studies that have been conducted using seminal plasma of infertile men exhibiting high levels of seminal oxidative stress (Table 4.1). Sharma et al. compared the seminal plasma proteome profile of men (infertile and healthy donors) with high reactive species levels vs. men with low levels of ROS. Only 14 proteins were differentially expressed in the seminal plasma of ROS positive samples [50]. Among them, ACPP, AZGP1, CLU, KLK3, and PIP were regulated by the androgen receptor [50]. In another proteomic study, MME and FAM3D proteins were proposed as biomarkers in evaluating the fertility status of men with low, medium and high levels of seminal ROS [51]. Intasqui et al. identified 94 DEPs in the seminal plasma of normozoospermic men exhibiting high levels of oxidative stress (lipid peroxidation). Furthermore, mucin-5B was suggested as a marker of oxidative stress in semen [52].

4.4 Enriched Molecular Pathways Associated with Oxidative Stress in Male Infertility-Related Conditions

4.4.1 Sperm Abnormalities

Sperm and seminal plasma proteins involved in molecular pathways associated with oxidative stress mechanisms were dysregulated in men with sperm abnormalities. Saraswat et al. reported that the eNOS signaling pathway linked with regulation of oxidative stress was enriched in the sperm of asthenozoospermic men [53]. Whereas, the phosphoproteome profile of spermatozoa showed that the cAMP-mediated PKA signaling pathway associated with the oxidative stress mechanism was dysregulated in asthenozoospermic men [54]. Pathway analysis indicated that the oxidation-reduction process was upregulated in the oligoasthenozoospermic condition [55]. Whereas in seminal plasma, molecular functions such as oxidoreductase activity and antioxidant activity were dysregulated in the oligoasthenozoospermic condition [56]. Samanta et al. reported that cell redox homeostasis and oxidoreductase activity were enriched in immature sperm fractions of infertile men [39].

Table 4.1 Proteomic studies on sperm and seminal plasma related to oxidative stress

Study reference	Sample type	Subjects	Key DEPs	Results
Hamada et al. [74]	Sperm	20 donors and 32 infertile men's semen samples divided into two ROS +ve and ROS −ve groups	lactotransferrin-2, peroxiredoxin-1	1265 and 1343 proteins identified in ROS +ve and ROS −ve groups, respectively.
Sharma et al. [47]	Sperm	20 normal donors and 32 infertile men's semen samples divided into two ROS +ve and ROS −ve groups	HIST1H2BA, MDH2, TGM4, GPX4, GLUL, HSP90B1 and HSPA5	74 DEPs. Energy metabolism and regulation, gluconeogenesis and glycolysis, protein modifications, and oxidative stress regulation were affected in the ROS+ group.
Ayaz et al. [49]	Sperm	42 semen samples of infertile men were divided into low, medium, and high ROS groups and 17 samples from fertile men as the control group	CLGN, TPPII, DNAI2, EEA1, HSPA4L, and SERPINA5	305 DEPs reported in low, medium, and ROS groups. 51 DEPs unique to the low ROS group. 47 DEPs unique to medium ROS group. 104 DEPs unique to the high ROS group.
Dias et al. [22]	Sperm and seminal plasma	20 fertile men's semen samples divided into ROS +ve and ROS −ve groups	NDUFS1, SOD1 and PRDX4	371 DEPs in sperm. 44 DEPs in seminal plasma. Activation of antioxidant defense system and proteasomal system.
Sharma et al. [50]	Seminal plasma	20 donors and 32 infertile men's semen samples divided into two ROS +ve and ROS −ve groups	FN1, PIP, KLK3, SEMG2, CLU, MIF, and LTF	14 DEPs were identified and associated with antioxidative activity and regulatory processes.
Agarwal et al. [51]	Seminal plasma	42 semen samples from infertile men were divided into low, medium, and high ROS groups and 17 samples from fertile men as the control group	MME and FAM3D	Proteins associated with posttranslational modifications of proteins, protein folding, and developmental disorder were overexpressed in the high ROS group.

(continued)

Table 4.1 (continued)

Study reference	Sample type	Subjects	Key DEPs	Results
Intasqui et al. [52]	Seminal plasma	46 normozoospermic samples divided into low (control) and high lipid peroxidation level groups	mucin-5B	94 DEPs. Enriched pathways: unsaturated fatty acids biosynthesis, oxidants and antioxidants activity, cellular response to heat stress, and immune response.

DEPs Differentially expressed proteins, *ROS* Reactive oxygen species

4.4.2 Varicocele

Semen parameters are compromised in infertile men with varicocele [57]. Several proteomic studies have reported dysregulation of proteins associated with reproductive function in sperm of varicocele patients (reviewed in [28]). Varicocele induces oxidative stress in semen [57], hence it is imperative to understand the functional status of molecular pathways linked to seminal oxidative stress in the varicocele condition. Agarwal et al. reported that 13.7% of DEPs detected in unilateral varicocele condition were involved in oxidoreductase activity of sperm [58]. Furthermore, Swain et al. demonstrated that an altered redox molecular mechanism was responsible for sperm dysfunction in unilateral varicocele [59]. In cases of bilateral varicocele, molecular pathways such as cell redox homeostasis and oxidoreductase were dysregulated due to altered expression of sperm proteins [60]. Moreover, sperm proteins associated with oxidative stress mechanism were acetylated in varicocele condition [61]. Aside from sperm proteins, seminal plasma proteins were also involved in regulation of oxidative stress related pathways. Panner Selvam et al. reported that response to ROS and oxidative stress networks were affected in seminal plasma of males with bilateral varicocele condition [10]. In unilateral varicocele condition, transcription factor NRF2 responsible for prevention of oxidative damage to sperm was altered in seminal plasma [62].

4.4.3 Testicular Cancer

Both testicular germ cell seminomas and non-seminomas affect semen parameters and the fertility potential of sperm [63, 64]. Recent proteomic studies have reported the involvement of DEPs in oxidative stress-related pathways. Panner Selvam et al. reported that molecular pathways such as NRF2-mediated oxidative stress response and oxidative phosphorylation process were affected in testicular cancer patients with normal and abnormal semen parameters [65]. Proteins linked with the oxidative phosphorylation pathway were altered in the sperm of testicular cancer seminoma patients [66]. Free radical scavenging pathway was enriched along with activation of production and synthesis of ROS in spermatozoa of testicular cancer patients [67]. Overall, it is clear that the molecular mechanisms related to oxidative stress are either directly or indirectly dysregulated in the semen of men with different infertility issues, suggesting that redox-homeostasis is disturbed in the sperm and seminal plasma of infertile men. Hence, maintenance of antioxidant levels in semen is pivotal to counteract such seminal oxidative stress.

4.5 Antioxidant Treatment and Its Effect on Sperm Proteins

Oral antioxidant supplementation is widely used in the management of male infertility. They counterbalance the seminal oxidative stress developed due to overproduction of ROS in semen. A recent global survey conducted among reproductive specialists revealed that 85.6% of clinicians prescribe antioxidants for the treatment of male infertility [68]. Antioxidant treatment has proven to be effective in improving semen parameters in men with idiopathic and unexplained infertility [69, 70]. Apart from semen parameters, antioxidant supplementation was found to be beneficial to fertility-associated sperm proteins in idiopathic infertile men [71]. Post-antioxidant treatment, transcription factors such as PPARGC1A and NFE2L2 were predicted to be activated in spermatozoa. Furthermore, antioxidant supplementation also modulated the expression of proteins associated with the CREM signaling pathway, which is essential for proper differentiation of spermatids in testis [71]. In addition, molecular chaperon complex (TRiC) was activated in spermatozoa of idiopathic infertile men post-antioxidant treatment. TRiC complex regulates the sperm telomere length [72] and expression of sperm–oocyte interacting proteins required for successful fertilization [73]. Finally, sperm proteins such as NDUFS1, PRKAR1A, CCT3, and SPA17 were proposed as potential biomarkers that can help with evaluating the effect of antioxidant therapy on fertility-associated molecular functions in sperm [71].

4.6 Conclusions

Oxidative stress is known to play a significant role in the etiology of male infertility. Modifications in the sperm and seminal plasma proteome are associated with poor sperm function and semen quality. Proteomics studies have demonstrated significant changes in both the sperm and seminal plasma protein profiles of male infertility patients based on etiologies linked to oxidative stress. Utilization and the advancement of proteomics-based approaches provide vast opportunities to gain deeper insights into the key underlying factors and cellular pathways leading to the development of infertility associated with the loss of redox homeostasis in the male. Discovery of proteins of interest that could serve as reliable biomarkers in various fertility disorders would greatly contribute to the development of novel techniques for the clinical diagnosis and potential treatment of oxidative stress-related male infertility, especially those that are idiopathic in nature.

References

1. Definitions of infertility and recurrent pregnancy loss: a committee opinion. Fertil Steril. 2020;113(3):533–5.
2. Winters BR, Walsh TJ. The epidemiology of male infertility. Urol Clin North Am. 2014;41(1):195–204.
3. Agarwal A, Mulgund A, Hamada A, Chyatte MR. A unique view on male infertility around the globe. Reprod Biol Endocrinol. 2015;13:37.
4. Hamada A, Esteves SC, Nizza M, Agarwal A. Unexplained male infertility: diagnosis and management. Int Braz J Urol. 2012;38(5):576–94.
5. Doshi S, Sharma R, Agarwal A, editors. Oxidative Stress in Unexplained Male Infertility. 2015.
6. Sikka SC. Role of oxidative stress and antioxidants in andrology and assisted reproductive technology. J Androl. 2004;25(1):5–18.
7. Du Plessis SS, Agarwal A, Halabi J, Tvrda E. Contemporary evidence on the physiological role of reactive oxygen species in human sperm function. J Assist Reprod Genet. 2015;32(4):509–20.
8. Tremellen K. Oxidative stress and male infertility–clinical perspective. Hum Reprod Update. 2008;14(3):243–58.
9. Sikka SC. Relative impact of oxidative stress on male reproductive function. Curr Med Chem. 2001;8(7):851–62.
10. Panner Selvam MK, Agarwal A, Baskaran S. Proteomic analysis of seminal plasma from bilateral varicocele patients indicates an oxidative state and increased inflammatory response. Asian J Androl. 2019;21(6):544–50.
11. Agarwal A, Mulgund A, Alshahrani S, Assidi M, Abuzenadah AM, Sharma R, et al. Reactive oxygen species and sperm DNA damage in infertile men presenting with low level leukocytospermia. Reprod Biol Endocrinol. 2014;12:126.
12. Kagedan D, Lecker I, Batruch I, Smith C, Kaploun I, Lo K, et al. Characterization of the seminal plasma proteome in men with prostatitis by mass spectrometry. Clin Proteomics. 2012;9(1):2.
13. Mayorga-Torres BJM, Camargo M, Cadavid ÁP, du Plessis SS, Cardona Maya WD. Are oxidative stress markers associated with unexplained male infertility? Andrologia. 2017;49(5)
14. Graves PR, Haystead TA. Molecular biologist's guide to proteomics. Microbiol Mol Biol Rev. 2002;66(1):39–63. Table of contents.
15. du Plessis SS, Kashou AH, Benjamin DJ, Yadav SP, Agarwal A. Proteomics: a subcellular look at spermatozoa. Reprod Biol Endocrinol. 2011;9:36.
16. Baker MA, Nixon B, Naumovski N, Aitken RJ. Proteomic insights into the maturation and capacitation of mammalian spermatozoa. Syst Biol Reprod Med. 2012;58(4):211–7.
17. Panner Selvam MK, Agarwal A. Update on the proteomics of male infertility: a systematic review. Arab J Urol. 2018;16(1):103–12.
18. Duncan MW, Thompson HS. Proteomics of semen and its constituents. Proteomics Clin Appl. 2007;1(8):861–75.
19. Johnston DS, Wooters J, Kopf GS, Qiu Y, Roberts KP. Analysis of the human sperm proteome. Ann NY Acad Sci. 2005;1061:190–202.
20. Agarwal A, Durairajanayagam D, Halabi J, Peng J, Vazquez-Levin M. Proteomics, oxidative stress and male infertility. Reprod Biomed Online. 2014;29(1):32–58.

21. Panner Selvam MK, Agarwal A. Proteomic profiling of seminal plasma proteins in varicocele patients. World J Mens Health. 2021;39(1):90–8.
22. Dias TR, Samanta L, Agarwal A, Pushparaj PN, Panner Selvam MK, Sharma R. proteomic signatures reveal differences in stress response, antioxidant defense and proteasomal activity in fertile men with high seminal ROS levels. Int J Mol Sci. 2019;20(1)
23. Pilch B, Mann M. Large-scale and high-confidence proteomic analysis of human seminal plasma. Genome Biol. 2006;7(5):R40.
24. Magdeldin S, Enany S, Yoshida Y, Xu B, Zhang Y, Zureena Z, et al. Basics and recent advances of two dimensional- polyacrylamide gel electrophoresis. Clin Proteomics. 2014;11(1):16.
25. Viswanathan S, Unlü M, Minden JS. Two-dimensional difference gel electrophoresis. Nat Protoc. 2006;1(3):1351–8.
26. Vazquez-Levin MH. Proteomic analysis and sperm physiopathology: the two-dimensional difference in gel electrophoresis approach. Fertil Steril. 2013;99(5):1199–200.
27. Chandramouli K, Qian PY. Proteomics: challenges, techniques and possibilities to overcome biological sample complexity. Hum Genomics Proteomics. 2009;2009
28. Panner Selvam MK, Baskaran S, Agarwal A, Henkel R. Protein profiling in unlocking the basis of varicocele-associated infertility. Andrologia. 2021;53(1):e13645.
29. Steen H, Mann M. The ABC's (and XYZ's) of peptide sequencing. Nat Rev Mol Cell Biol. 2004;5(9):699–711.
30. Siuzdak G. An introduction to mass spectrometry ionization: An excerpt from the expanding role of mass spectrometry in biotechnology, 2nd ed.; MCC Press: San Diego, 2005. J Assoc Lab Autom. 2004;9(2):50–63.
31. Kumar C, Mann M. Bioinformatics analysis of mass spectrometry-based proteomics data sets. FEBS Lett. 2009;583(11):1703–12.
32. Lam H. Building and searching tandem mass spectral libraries for peptide identification. Mol Cell Proteomics. 2011;10(12):R111.008565.
33. Chen C, Hou J, Tanner JJ, Cheng J. Bioinformatics methods for mass spectrometry-based proteomics data analysis. Int J Mol Sci. 2020;21(8)
34. Manzoni C, Kia DA, Vandrovcova J, Hardy J, Wood NW, Lewis PA, et al. Genome, transcriptome and proteome: the rise of omics data and their integration in biomedical sciences. Brief Bioinform. 2018;19(2):286–302.
35. Reimand J, Isserlin R, Voisin V, Kucera M, Tannus-Lopes C, Rostamianfar A, et al. Pathway enrichment analysis and visualization of omics data using g:Profiler, GSEA, cytoscape and enrichmentmap. Nat Protoc. 2019;14(2):482–517.
36. Samanta L, Swain N, Ayaz A, Venugopal V, Agarwal A. Post-translational modifications in sperm proteome: the chemistry of proteome diversifications in the pathophysiology of male factor infertility. Biochim Biophys Acta. 2016;1860(7):1450–65.
37. Szklarczyk D, Franceschini A, Wyder S, Forslund K, Heller D, Huerta-Cepas J, et al. STRING v10: protein-protein interaction networks, integrated over the tree of life. Nucleic Acids Res. 2015;43(Database issue):D447–52.
38. Panner Selvam MK, Agarwal A, Dias TR, Martins AD, Baskaran S, Samanta L. Molecular pathways associated with sperm biofunction are not affected by the presence of round cell and leukocyte proteins in human sperm proteome. J Proteome Res. 2019;18(3):1191–7.
39. Samanta L, Sharma R, Cui Z, Agarwal A. Proteomic analysis reveals dysregulated cell signaling in ejaculated spermatozoa from infertile men. Asian J Androl. 2019;21(2):121–30.
40. Wang J, Wang J, Zhang HR, Shi HJ, Ma D, Zhao HX, et al. Proteomic analysis of seminal plasma from asthenozoospermia patients reveals proteins that affect oxidative stress responses and semen quality. Asian J Androl. 2009;11(4):484–91.
41. Aslam B, Basit M, Nisar MA, Khurshid M, Rasool MH. Proteomics: technologies and their applications. J Chromatogr Sci. 2017;55(2):182–96.
42. Handler DC, Pascovici D, Mirzaei M, Gupta V, Salekdeh GH, Haynes PA. The art of validating quantitative proteomics data. Proteomics. 2018;18(23):e1800222.

43. Agarwal A, Parekh N, Panner Selvam MK, Henkel R, Shah R, Homa ST, et al. Male Oxidative Stress Infertility (MOSI): proposed terminology and clinical practice guidelines for management of idiopathic male infertility. World J Mens Health. 2019;37(3):296–312.
44. Baskaran S, Finelli R, Agarwal A, Henkel R. Reactive oxygen species in male reproduction: a boon or a bane? Andrologia. 2021;53(1):e13577.
45. Agarwal A, Panner Selvam MK, Baskaran S. Proteomic analyses of human sperm cells: understanding the role of proteins and molecular pathways affecting male reproductive health. Int J Mol Sci. 2020;21(5)
46. Cannarella R, Crafa A, Barbagallo F, Mongioì LM, Condorelli RA, Aversa A, et al. Seminal plasma proteomic biomarkers of oxidative stress. Int J Mol Sci. 2020;21(23)
47. Sharma R, Agarwal A, Mohanty G, Hamada AJ, Gopalan B, Willard B, et al. Proteomic analysis of human spermatozoa proteins with oxidative stress. Reprod Biol Endocrinol. 2013;11:48.
48. Ayaz A, Agarwal A, Sharma R, Kothandaraman N, Cakar Z, Sikka S. Proteomic analysis of sperm proteins in infertile men with high levels of reactive oxygen species. Andrologia. 2018;50(6):e13015.
49. Ayaz A, Agarwal A, Sharma R, Arafa M, Elbardisi H, Cui Z. Impact of precise modulation of reactive oxygen species levels on spermatozoa proteins in infertile men. Clin Proteomics. 2015;12(1):4.
50. Sharma R, Agarwal A, Mohanty G, Du Plessis SS, Gopalan B, Willard B, et al. Proteomic analysis of seminal fluid from men exhibiting oxidative stress. Reprod Biol Endocrinol. 2013;11:85.
51. Agarwal A, Ayaz A, Samanta L, Sharma R, Assidi M, Abuzenadah AM, et al. Comparative proteomic network signatures in seminal plasma of infertile men as a function of reactive oxygen species. Clin Proteomics. 2015;12(1):23.
52. Intasqui P, Antoniassi MP, Camargo M, Nichi M, Carvalho VM, Cardozo KH, et al. Differences in the seminal plasma proteome are associated with oxidative stress levels in men with normal semen parameters. Fertil Steril. 2015;104(2):292–301.
53. Saraswat M, Joenväärä S, Jain T, Tomar AK, Sinha A, Singh S, et al. Human spermatozoa quantitative proteomic signature classifies normo- and asthenozoospermia. Mol Cell Proteomics. 2017;16(1):57–72.
54. Parte PP, Rao P, Redij S, Lobo V, D'Souza SJ, Gajbhiye R, et al. Sperm phosphoproteome profiling by ultra performance liquid chromatography followed by data independent analysis (LC–MSE) reveals altered proteomic signatures in asthenozoospermia. J Proteomics. 2012;75(18):5861–71.
55. Herwig R, Knoll C, Planyavsky M, Pourbiabany A, Greilberger J, Bennett KL. Proteomic analysis of seminal plasma from infertile patients with oligoasthenoteratozoospermia due to oxidative stress and comparison with fertile volunteers. Fertil Steril. 2013;100(2):355–66.e2.
56. Liu X, Wang W, Zhu P, Wang J, Wang Y, Wang X, et al. In-depth quantitative proteome analysis of seminal plasma from men with oligoasthenozoospermia and normozoospermia. Reprod BioMed Online. 2018;37(4):467–79.
57. Gill K, Kups M, Harasny P, Machalowski T, Grabowska M, Lukaszuk M, et al. The negative impact of varicocele on basic semen parameters, sperm nuclear DNA dispersion and oxidation-reduction potential in semen. Int J Environ Res Public Health. 2021;18(11)
58. Agarwal A, Sharma R, Durairajanayagam D, Ayaz A, Cui Z, Willard B, et al. Major protein alterations in spermatozoa from infertile men with unilateral varicocele. Reprod Biol Endocrinol. 2015;13:8.
59. Swain N, Samanta L, Agarwal A, Kumar S, Dixit A, Gopalan B, et al. Aberrant upregulation of compensatory redox molecular machines may contribute to sperm dysfunction in infertile men with unilateral varicocele: a proteomic insight. Antioxid Redox Signal. 2020;32(8):504–21.
60. Agarwal A, Sharma R, Durairajanayagam D, Cui Z, Ayaz A, Gupta S, et al. Spermatozoa protein alterations in infertile men with bilateral varicocele. Asian J Androl. 2016;18(1):43–53.

61. Panner Selvam MK, Samanta L, Agarwal A. Functional analysis of differentially expressed acetylated spermatozoal proteins in infertile men with unilateral and bilateral varicocele. Int J Mol Sci. 2020;21(9)
62. Panner Selvam MK, Agarwal A, Sharma R, Samanta L, Gupta S, Dias TR, et al. Protein fingerprinting of seminal plasma reveals dysregulation of exosome-associated proteins in infertile men with unilateral varicocele. World J Mens Health. 2021;39(2):324–37.
63. Ping P, Gu BH, Li P, Huang YR, Li Z. Fertility outcome of patients with testicular tumor: before and after treatment. Asian J Androl. 2014;16(1):107–11.
64. Bahadur G, Ozturk O, Muneer A, Wafa R, Ashraf A, Jaman N, et al. Semen quality before and after gonadotoxic treatment. Hum Reprod. 2005;20(3):774–81.
65. Panner Selvam MK, Agarwal A, Pushparaj PN. Altered molecular pathways in the proteome of cryopreserved sperm in testicular cancer patients before treatment. Int J Mol Sci. 2019;20(3)
66. Dias TR, Agarwal A, Pushparaj PN, Ahmad G, Sharma R. Reduced semen quality in patients with testicular cancer seminoma is associated with alterations in the expression of sperm proteins. Asian J Androl. 2020;22(1):88–93.
67. Panner Selvam MK, Agarwal A, Pushparaj PN. A quantitative global proteomics approach to understanding the functional pathways dysregulated in the spermatozoa of asthenozoospermic testicular cancer patients. Andrology. 2019;7(4):454–62.
68. Agarwal A, Finelli R, Selvam MKP, Leisegang K, Majzoub A, Tadros N, et al. A global survey of reproductive specialists to determine the clinical utility of oxidative stress testing and antioxidant use in male infertility. World J Mens Health. 2021;39(3):470–88.
69. Arafa M, Agarwal A, Majzoub A, Panner Selvam MK, Baskaran S, Henkel R, et al. Efficacy of antioxidant supplementation on conventional and advanced sperm function tests in patients with idiopathic male infertility. Antioxidants (Basel). 2020;9(3)
70. Agarwal A, Leisegang K, Majzoub A, Henkel R, Finelli R, Panner Selvam MK, et al. Utility of antioxidants in the treatment of male infertility: clinical guidelines based on a systematic review and analysis of evidence. World J Mens Health. 2021;39(2):233–90.
71. Agarwal A, Panner Selvam MK, Samanta L, Vij SC, Parekh N, Sabanegh E, et al. Effect of antioxidant supplementation on the sperm proteome of idiopathic infertile men. Antioxidants (Basel). 2019;8(10)
72. Freund A, Zhong FL, Venteicher AS, Meng Z, Veenstra TD, Frydman J, et al. Proteostatic control of telomerase function through TRiC-mediated folding of TCAB1. Cell. 2014;159(6):1389–403.
73. Dun MD, Smith ND, Baker MA, Lin M, Aitken RJ, Nixon B. The chaperonin containing TCP1 complex (CCT/TRiC) is involved in mediating sperm-oocyte interaction. J Biol Chem. 2011;286(42):36875–87.
74. Hamada A, Sharma R, du Plessis SS, Willard B, Yadav SP, Sabanegh E, et al. Two-dimensional differential in-gel electrophoresis-based proteomics of male gametes in relation to oxidative stress. Fertil Steril. 2013;99(5):1216–26.e2.

Chapter 5
Unraveling the Molecular Impact of Sperm DNA Damage on Human Reproduction

Renata Finelli, Bruno P. Moreira, Marco G. Alves, and Ashok Agarwal

Abstract Semen analysis is the cornerstone in the investigation of fertility status of male partner. However, more advanced tests have emerged including the analysis of sperm chromatin integrity and DNA damage as markers of semen quality. This is of particular interest, as preserving the genetic information is essential to achieve a successful reproductive event. Moreover, the presence of unrepaired DNA lesions can affect cellular functions, resulting in the onset of pathological conditions associated with male infertility, and the transmission of diseases to the offspring. Hence, in this chapter, we aim to review the main factors leading to sperm DNA damage, along with the different types of damage which can occur. Furthermore, molecular mechanisms involved in DNA repair during spermatogenesis or after fertilization of the oocyte are described, and the laboratory techniques currently used in diagnostics and research, for the analysis of sperm DNA damage are also presented. Finally, the impact of sperm DNA damage on reproductive outcomes such as fertilization and pregnancy rates will be discussed with a focus on animal and human studies, along with the identification of new markers of sperm chromatin integrity.

Keywords Chromatin · DNA damage · DNA repair · Reproduction · Spermatozoa

Abbreviations

8-OHdG	8-hydroxyguanosine
APE1	AP endonuclease1
AZF	Azoospermia factor
BER	Base excision repair
DAPI	4′,6-diamidino-2-phenylindole
DDR	DNA damage response
DFI	DNA fragmentation index
DSB	Double-strand break

R. Finelli · A. Agarwal (✉)
American Center for Reproductive Medicine, Cleveland Clinic, Cleveland, OH, USA
e-mail: agarwaa@ccf.org

B. P. Moreira · M. G. Alves
Department of Anatomy, Unit for Multidisciplinary Research in Biomedicine (UMIB), Institute of Biomedical Sciences Abel Salazar (ICBAS), University of Porto, Porto, Portugal

dUTP	Deoxyuridine triphosphate
FEN-1	Flap endonuclease-1
GG-NER	Global genome NER
HSPA2	Heat shock protein
HR	Homologous recombination
IUI	Intrauterine insemination
miRNAs	microRNAs
MMEJ	Microhomology-mediated end joining pathway
mtDNA	Mitochondrial DNA
NAHR	Non-allelic homologous recombination
NHEJ	Nonhomologous end-joining
NER	Nucleotide excision repair
NGS	Next generation sequencing
ORP	Oxidation-reduction potential
PCNA	Proliferating cell nuclear antigen
PARP	Poly (ADP-ribose) polymerase
Q-FISH	Quantitative fluorescent in situ hybridization
RPL	Recurrent pregnancy loss
RR	Relative risk
RPA	Replication protein-A
RNA pol II	RNA polymerase II
rNTPs	Ribose nucleoside triphosphate
ROS	Reactive oxygen species
SCD	Sperm chromatin dispersion
SCSA	Sperm chromatin structure assay
STS-PCR	Sequence-tagged site polymerase chain reaction
SSB	Single-strand break
ssDNA	Single-stranded DNA
TC-NER	Transcription-coupled NER
TERC	Telomerase RNA component
TERT	Telomerase reverse transcriptase
TUNEL	Terminal deoxynucleotidyl transferase-mediated deoxyuridine tri-phosphate nick end labeling
XRCC1	X-Ray cross-complementing protein 1.

5.1 Sperm DNA Damage

5.1.1 DNA Damage and Male Infertility: Brief Overview

Infertility is defined as the incapacity of a couple to achieve pregnancy even after 12 months of unprotected sexual intercourse. The male partner is responsible for up to 50% of these cases [1]. Semen analysis is done to determine the male reproductive potential; however, its significance in diagnosis of male infertility is limited, as a percentage of infertile patients reports normal semen parameters or altered semen quality with no identifiable reason [2, 3]. This has led to the development of more advanced molecular tests, which have shed light on the importance of sperm DNA integrity in reproduction and the role of sperm DNA damage in causing male infertility. In this chapter, we describe the causes of sperm DNA damage, along with the different types of damage which can occur, and the molecular mechanisms involved in its repair during spermatogenesis or after fertilization of the oocyte. Moreover, the techniques which are currently being used in the diagnostic and research of sperm DNA damage are briefly presented, along with the implication of cellular damage on fertilization and pregnancy. Finally, future areas of investigation are presented, where we report several new markers of sperm chromatin integrity which could be of scientific relevance in the years to come.

5.1.2 Brief Introduction to the Causes of DNA Damage

The preservation of genetic information is fundamental for every form of life to ensure the reproduction of the species. If not correctly repaired, lesions in DNA can lead to genome coding errors, causing diseases and compromising cell viability [4]. DNA damage can be categorized as endogenous and exogenous based on its origin [4] (Fig. 5.1). Endogenous DNA damage is caused by factors inherent to the cell, i.e., the machinery involved in DNA replication and repair (e.g. DNA polymerases) [5], or the reaction of DNA with oxidative compounds [6].

The cellular machinery responsible for DNA replication carries the burden of being responsible for errors in a range of 10^{-7} to 10^{-1} [7]. Most of these errors are repaired by the DNA damage response (DDR) system but, in some cases, they can be transmitted to future generations. One of the most common errors in replication by DNA polymerase is the failure to discriminate between ribose nucleoside triphosphate (rNTPs) versus deoxyribose nucleoside triphosphate (dNTPs) [8]. This leads to mis-incorporated nucleotides which are a major source of spontaneous mutagenesis [9]. Base substitutions and insertions/deletions are also common errors in replication in case of slipped strand mispairing events [10]. Topoisomerase enzymes are another potential source of DNA damage. These enzymes act on the topology of DNA, participating in the over/underwinding of DNA. During replication, at the replication fork stage, these enzymes transiently nick the supercoiled DNA and relieve the torsional strain, allowing DNA polymerases to continue working on the strand [11]. These breaks are realigned, resolving the complex. However, if one of the strands has a single-strand break (SSB), a missing base or a mispaired loop, these enzymes can generate double-strand breaks (DSB) or mismatches, resulting in the formation of suicide complexes between topoisomerases and the DNA [11]. Abasic sites and spontaneous base deamination are also a major source of DNA damage. The former is defined as the absence in the DNA of a purine or pyrimidine base [12], and can occur spontaneously or as an intermediate step in the one of the mechanisms of DDR (base excision repair) which will be

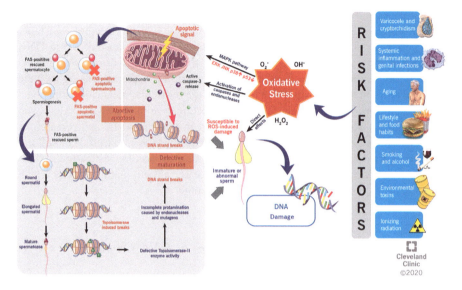

Fig. 5.1 Overview of the origins of DNA damage. DNA damage results from endogenous mechanisms such as defective maturation, abortive apoptosis, and oxidative stress. Moreover, exogenous (age, infection, cancer, hormonal imbalances, obesity, diabetes, and environmental) risk factors lead to DNA damage

described later in the chapter in more detail. Spontaneous base deamination is the removal of the amino group from DNA bases resulting in the formation of uracil, thymine, xanthine, and hypoxanthine from cytosine, 5-methylcytosine, guanine, and adenine respectively [13]. Reactive oxygen species (ROS) are an important player as promoters of endogenous DNA damage. These are byproducts of metabolic pathways, with important roles in cell signaling and homeostasis, and include superoxide radicals ($•O_2^-$), hydrogen peroxide (H_2O_2), and the hydroxyl radical ($•OH$) [6]. ROS damage includes abasic sites, base modifications, and induction of SSB and DSB [6].

Exogenous DNA damage can occur due to the action of several chemical and environmental agents (e.g. radiation, plant toxins, viruses, genotoxic agents). Ionizing radiation can directly or indirectly damage DNA, inducing both SSB and DSB through the generation of ROS and the formation of dimers [14, 15]. Particularly, they produce two types of dimers in the DNA: cyclobutane pyrimidine dimers and pyrimidine (6–4) pyrimidone photoproducts. Alkylating agents such as nitrogen mustards, aziridines and epoxides, alkyl sulfonates, and nitrosoureas are also capable of damaging DNA by reacting with the highly nucleophilic base ring nitrogens and causing abasic sites and mispair of DNA bases [16]. Bisphenol A is one of the most common examples [17], but several food additives have also been described [18, 19]. In a broad classification, these genotoxic agents can be grouped in two categories: agents capable of inducing chemical changes (e.g. base modifying agents) and agents capable of inducing physical changes in the DNA (e.g.

induction of strand breaks). Nevertheless, some agents can induce both types of damage [18, 19].

If not repaired, damaged DNA leads to genomic stress and apoptosis. Crucially, cells have built-in highly sophisticated mechanisms to mitigate the nefarious consequences of DNA damage, as explained in the subsection number 5.3 of the chapter.

5.2 Different Types of Sperm DNA Damage

5.2.1 Mismatched Bases

In DNA helix, bases are paired according to the general rule of thymine-adenine and guanine-cytosine, while the word "mismatch" indicates the presence of a wrong base-pairing. Crucially, DNA polymerases are able to repair such mistakes as they show an intrinsic proofreading activity as exonuclease [7]. This process is not unfailing, and a wrong base can be still introduced during the process of DNA synthesis and replication, increasing the rate of mutation. The frequency of mismatched bases increases in microsatellite sequences, characterized by the repetition of 1, 2, or 3 nucleotides, as these sequences are more susceptible to strand misalignments [20]. Misalignments can also occur during the physiological process of DNA transcription, when the DNA helices are open to allow the activity of the RNA polymerases [21].

5.2.2 Base modifications: Oxidation, Alkylation, Hydrolysis, Deamination, Pyrimidine Dimers, and Intrastrand Crosslinking

Single base in the DNA helix can be chemically modified resulting in higher mutation rate and altered protein expression. The most common types of base modifications include oxidation, alkylation, hydrolysis, and deamination (Fig. 5.2).

Oxidation

Oxidation of purine and pyrimidine bases is mainly mediated by ROS [22]. Pyrimidines (cytosine, methylcytosine, and thymine) are oxidized at the level of the double bond between the carbons in position 5 and 6; this generates hydrate and glycol derivative molecules, as well as urea. Moreover, when cytosine and methylcytosine bases are oxidized, they can spontaneously undergo further deamination, and converted to uracil and thymine [23]. Purine bases (adenine, guanine) can also be oxidized at the level of their imidazole ring: oxidation opens the ring, which can be further rearranged, generating several derivative molecules, with

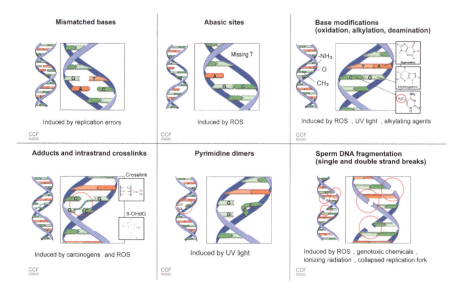

Fig. 5.2 Different types of DNA damage that can occur: mismatched bases, abasic sites, base modifications (oxidation, alkylation, deamination), adducts and intrastrand crosslinks, pyrimidine dimers, and single- and double-strand fragmentation

formamidopyrimidine, 8-oxo and 2-hydroxyadenine derivatives among others [24]. The hydrolysis of the bases results in the creation of an abasic site.

Alkylation

Alkylation is a chemical modification where an alkyl group is transferred from a molecule to another. In the DNA helix, all the atoms of oxygen and nitrogen present in the bases can be alkylated, resulting in relatively stable adducts which are high mutagenic and cytotoxic products (i.e. O^6-chloroethylguanine, 1-O^6-ethanoguanine) [25]. Of these, the atom of nitrogen in position 7 of the guanine base has been identified as the most reactive site for the alkylating reaction [25]. Alkylating agents can be endogenous metabolic intermediates, such as s-adenosylmethionine, nitrosated amines, and glycine derivatives. Furthermore, alkylating agents are also produced by smoking tobacco (i.e. 4-(methylnitrosamino)-1-(3-pyridyl)-1-butanone, 4-(methylnitrosamino)-1-(3-pyridyl)-1-butanol), and are used in the cancer therapy [25].

Hydrolysis

The hydrolysis of chemical bonds can occur between the bases and the deoxyribose sugar spontaneously, and results in the generation of an abasic site. Particularly, the spontaneous removal of purine occurs more frequently than that of pyrimidines

(~10,000 times per cell per day) [12]. Abasic sites can lead to single-strand breaks, as well as to the generation of interstrand crosslink through the reaction of the aldehyde residue with a guanine on the opposite strand [26]. Further, the hydrolysis of the sugar-phosphate bond can produce strand breaks in the DNA backbone helix [27].

Deamination

Purine and pyrimidine bases can also undergo spontaneous deamination, with the loss of their amino groups and conversion to uracil, hypoxanthine, xanthine, and thymine [13]. This is not an uncommon phenomenon, as the conversion of cytosine to uracil has been estimated to occur around 500 times per day in a single genome, and it can be responsible for transition mutations [13].

Dimers of Pyrimidines

Dimers of pyrimidines are UV-induced mutations generated through a photochemical reaction, which can lead to the generation of cyclobutane pyrimidine dimers and (6-4) pyrimidine-pyrimidone photoproducts [28]. These are unstable mutations which distort the DNA double helix and can easily undergo deamination and conversion into a uracil base. Pyrimidine dimers mainly occur at 5-methylcytosine residue present in a 5′-CpG-3′ island, hence DNA regions which are abundant in such residues are highly susceptible to UV-mediated DNA lesions [29].

Interstrand Crosslinks

Some agents are able to create adducts with DNA on both strands, generating interstrand crosslinks [30]. These have a clastogenic effect: whereas mutagenic agents increase the rate of DNA mutations, clastogenic ones affect the separation of DNA strands, causing damage at the chromosomal levels. In fact, intrastrand crosslinks are responsible for impaired chromatid exchanges, loss and rearrangements of chromosomal sections [30].

5.2.3 DNA Fragmentation

DNA fragmentation at one or both strands can occur due to a variety of factors, including pathological conditions, lifestyle risk factors, aging, infections, inflammatory conditions, or exposure to environmental toxicants and pollutants [31] (Fig. 5.2). The mechanisms by which such conditions lead to DNA fragmentation include, but are not limited to, the induction of apoptosis, or failure to complete

DNA repair process, which lead to the generation of nicks in the DNA helix [31]. A third mechanism involves the establishment of oxidative stress [31].

During spermatogenesis, immature or abnormal sperm are physiologically removed by triggering apoptosis; however, when this process fails (abortive apoptosis), apoptotic sperm are released in the ejaculate. Apoptotic markers have been observed in ejaculated sperm cells, such as the expression of FAS receptor in the sperm membrane and fragmented DNA due to the action of endonucleases (caspases 1, 3, 8, and 9) [32]. During their maturation, elongated spermatids undergo protamination, i.e. the process where histones are gradually substituted by alkaline protamines in order to tightly compact the chromatin. During this process, the DNA helix must be open, and the enzyme topoisomerase II contributes to reduce the torsional stress by creating nicks in the DNA helix. If not repaired, nicks can result in DNA fragmentation. Finally, oxidative stress plays a significant contributory role in the generation of fragmented DNA [31]. Oxidative stress is characterized by a disequilibrium between oxidant and antioxidant powers, with a shift towards increased concentration of ROS. Oxidative stress can increase DNA fragmentation rate in testis, along the transit in the male reproductive tract, or post-ejaculation. This is mediated indirectly by the products of lipid peroxidation, which are mutagenic, or directly by the action of ROS on DNA bases [33, 34]. The oxidation of DNA guanosine results in the generation of 8-hydroxyguanosine (8-OHdG), which is therefore considered a marker of oxidative stress-related DNA damage [31].

Although presented individually in this chapter, these mechanisms are not mutually exclusive, but concur simultaneously to the generation of fragmented DNA. In fact, failure in the sperm maturation can trigger the apoptosis, or result in less compact DNA which is more susceptible to oxidative attacks. Similarly, oxidative stress-related DNA damage can be responsible for the induction of the apoptosis.

5.2.4 Telomere Shortening

Telomeres are terminal chromosomal DNA regions which play a role in protecting the DNA ends from degradation. These regions are rich in TTAGGG repetitions (between 10 and 15 kb), associated in loops with several proteins, including telomeric repeat-binding factor 1 (TRF1), telomeric repeat-binding factor 2 (TRF2), ras-proximate-1 (RAP1), tripeptidyl peptidase 1 (TPP1) among others, which contribute to the formation of the so-called shelterin complex [35]. The shelterin complex creates a protein cap which functionally regulates the length of telomeric ends and protects them from being recognized as damaged DNA and then being further degraded. During DNA replication, a segment of telomeric DNA is gradually lost due to the nature itself of the replicative mechanism. In fact, whereas the leading DNA strand can be replicated continuously in 5'-3' direction, this does not occur for the lagging strand, whose replication relies on the generation of Okazaki fragments (around 10 nucleotides length). The replication of each fragment is triggered by small RNA primers, which are finally replaced by DNA [36]. However, DNA synthases cannot replace RNA at the very last telomeric fragment, resulting in the loss of that DNA section. Although minimal, the repetition of this process consistently leads to the shortening of telomeres after several rounds of cell division in somatic cells. Hence, a cell can undergo a maximum number of divisions (the so-called Hayflick limit) before the excessive shortening of telomeres induces cellular senescence [37].

Some human cells (i.e. male germ, cancer, and stem cells) express a holoenzyme called telomerase, which can prevent the shortening of telomeric ends [35]. This includes a subunit called telomerase reverse transcriptase (TERT), with RNA-dependent DNA polymerase activity, which can synthetize DNA by using RNA as template. Moreover, telomerase includes an RNA component, a subunit named telomerase RNA component (TERC), which is complementary to telomeric DNA sequence, and it is used as template to elongate the telomeres by adding DNA repeats. Once that DNA strand is elongated, the replicative machine synthesizes a new primer and replicates the double DNA strand. Telomerase enzyme is highly expressed in type A spermatogonia, which show highly replicative rate [38], while it is gradually reduced during spermatogenesis, until being inactive in mature sperm. Interestingly, a correlation between increased leukocyte and sperm telomere length and advanced male age has been observed, along with increased leukocyte telomere length in the offspring [38]. Although it is prevented by the expression of telomerase, the shortening of telomeres reportedly correlates with the establishment of oxidative stress, particularly in those conditions characterized by chronic inflammation and high levels of ROS [39]. Oxidative stress is supposed to damage the telomeres by inducing SSB or by affecting the activity of those proteins involved in repairing and elongation [39]. Several studies have reported that sperm telomere length is significantly lower in infertile men compared to fertile men [40, 41]. Sperm telomere length was also positively correlated with several sperm parameters including sperm motility, sperm count, and sperm viability [40, 42].

5.2.5 Y Chromosome Microdeletions

A microdeletion is defined as the loss of a small part of the gene sequence, undetectable by the karyotype analysis, but visible by molecular biology techniques [43]. The microdeletions detected on Y chromosome are variable in size (0.8–7.7 Mb) and occur in approximately 10–15% of azoospermic patients and 5–10% of severe oligospermic patients [43]. Conversely, subjects with sperm count greater than 5 million sperm/ml show infrequent microdeletions, while they are never reported in normozoospermic patients [43, 44]. Y microdeletions are commonly found in the azoospermia factor (AZF) locus. This locus contains several genes that are transcribed in the testis and have a well defined role in spermatogenesis. Conventionally, it is subdivided into three subregions: AZFa, in the proximal part of Yq (Yq11.21), AZFb, and AZFc (partially overlapping) in the most distal part of Yq (Yq11.23) [45]. The molecular mechanism that is responsible for complete AZF deletion is named non-allelic homologous recombination (NAHR), as it takes place between highly homologous palindromic sequences with the same orientation, and leads to the loss of genetic material between them [46]. The AZFa region is the only one which does not contain palindromes, and the mechanism underlying the formation of deletions is represented by the recombination between two retroviral elements (HERV15yq1 and HERV15yq2) present at the extremities of the region itself [47].

5.2.6 Epigenetic Abnormalities

All those mechanisms which regulate gene expression in an organism without altering DNA sequence are classified as epigenetics mechanisms. These include post-translational modifications such as methylation, acetylation, phosphorylation, ubiquitination, and sumoylation of both DNA and proteins responsible for chromatin compaction in a dynamic and plastic process. DNA methylation is mediated by methyltransferase enzymes and occurs at the cytosine residues which are mainly located in CpG islands, and exerts an inhibitory effect on gene expression [48]. Histones can be methylated as well, as the methylation of histone H3 on lysine residues in position 9 and 27 (H3K9 and H3K27, respectively) are reportedly associated with compaction of the chromatin [48]. Conversely, the acetylation of histone proteins (i.e. H4K5, H3K9K14) facilitates the opening of the chromatin, the binding of transcriptional factors and promotes gene expression [49]. When DNA is damaged, chromatin is highly acetylated in order to be decompacted, and made accessible for proteins involved in DNA repair [49]. In this context, some chromatin remodeling factors have been identified, such as activity-regulated cytoskeleton associated protein 1 (Arc1), chromodomain-helicase-DNA-binding protein 2 (CHD2), 3 (CHD3) 4 (CHD4), and SWI/SNF-related matrix-associated actin-dependent regulator of chromatin subfamily A member 5 (SMARCA5), which promote chromatin decondensation by adding several polyADP-ribose units to the chromatin and the histone proteins [50]. Furthermore, in case of DNA damage, the histone H2AX is phosphorylated at the serine residue in position 139 and contributes to the recruitment of factors involved in DNA DSB repair [51].

During the spermatogenesis, around 90%-95% of the histones are substituted by protamines 1 and 2, which are basic proteins responsible for compacting sperm chromatin over ten times stricter than histones [52]. This results in the inhibition of protein transcription in mature spermatozoa, along with increased protection of sperm DNA from external attacks. The substitution of histones with protamines occurs in a dynamic process where the core histones are first highly acetylated to allow chromatin to open, and then ubiquitinated to be degraded by the proteasomal machinery [52]. In fact, murine model showed impaired spermatogenesis and higher apoptosis rate when the testis-specific ubiquitin conjugating enzyme HR6B was mutated [53]. Furthermore, global DNA methylation was found significantly higher in oligoasthenozoospermic patients compared to normozoospermic men [54], while differences in DNA methylation levels and in gene expression were found when comparing infertile with fertile men [55]. An alteration of such epigenetics mechanisms can lead to an incorrect protamination, resulting in the random chromatin packaging and male infertility.

5.3 Molecular Mechanisms of Sperm DNA Repair

5.3.1 Excision Repair

Base Excision Repair

Base excision repair (BER) mechanism is responsible for the editing of those DNA modifications which do not significantly distort the helix structure [56]. This is a step-by-step process, which starts with the removal of the altered base by means of DNA glycosylase and the generation of an apurinic/apyrimidinic (AP) site. In mammals, 11 different DNA glycosylases have been characterized, which are able to identify specific DNA lesions, including mismatch and base loss, among other bases modifications [57]. Glycosylases can be classified as mono- or bifunctional, depending on whether they are able to cleave only the glycosidic bond or they also have β-/β-δ lyase activity, respectively [57]. When a glycosylases creates an AP site, AP endonuclease 1 (APE1) further processes the DNA backbone to generate a single nucleotide gap in the DNA helix. The free 3′-OH end is then recognized by DNA polymerase β, which fills the gap, while the nick in the DNA backbone is closed by the simultaneous action of DNA ligase IIIα and x-ray cross-complementing protein 1 (XRCC1) [56, 58]. The above-mentioned mechanism is responsible for repairing 80% of the lesions, and it is termed as "short-patch" BER. In a small percentage of cases, DNA Pol β is substituted by DNA Pol δ or ε (Pol δ/ε), which generate a 5′-DNA flap structure by mediating the inclusion of 2–8 nucleotides into the gap [56]. In this "long-patch" BER mechanism, the flap structure is removed by the combined action of flap endonuclease-1 (FEN-1) and proliferating cell nuclear antigen (PCNA), and then the gap is filled. The nick is finally closed by DNA ligase I [56]. The choice between short- or long-patch pathways depends on the status of the cells, with the latter being more common in proliferative cells during the S-phase of the cell cycle [59].

Nucleotide Excision Repair

Nucleotide excision repair (NER) mechanism is responsible for repairing those lesions that significantly distort the DNA helix, which may occur in the entire genome or specifically in those tracts of DNA which are actively transcribed. This is mediated by two distinct subpathways: the global genome NER (GG-NER) and transcription-coupled NER (TC-NER) [60, 61]. The former is responsible for the detection of a large variety of DNA lesions that thermodynamically alter the DNA structure along the entire genome, and relies on multiple molecular actors including XPC, RAD23B, CETN2, and UV-DDB for the identification of such lesions [60, 61]. The TC-NER pathway is activated when RNA polymerase II (RNA pol II), the enzyme responsible for RNA transcription, is blocked by the presence of a lesion in the DNA template [60, 61]. Here, a different set of molecular mediators (e.g. UVSSA, USP7, CSB) are recruited [60, 61]. Although it is still unclear, RNA Pol II is suggested to degrade the nascent transcript, in a process termed as "backtracking," to expose the DNA damage to the repair machinery complex.

In both pathways, the presence of these large multiprotein complexes further recruits transcription factor II H (TFIIH), a transcriptional factor with helicase-like activity, that opens the DNA double helix [60]. If damage is not detected, the entire process is aborted, while the presence of damage induces incision of the DNA by means of endonucleases XPF-ERCC1 and XPG and the removal of 22–30 nucleotides [62]. The first nick in 5' triggers the DNA synthesis for filling the gap, which is then closed by a ligase enzyme. In replicating cells, the enzymes which are mainly involved are DNA Pol ε and DNA ligase 1, while DNA Pol δ, DNA Pol κ, and DNA ligase III mainly act in noncycling cells.

5.3.2 Mismatch Repair

Mechanisms for repairing DNA mismatch recognize and repair base–base mismatches along with DNA misalignments [63]. Moreover, the recognition of mismatches also inhibits the homologous recombination, hence limiting the generation of lethal DNA rearrangements [63]. In eukaryotic cells, the mismatches are identified by heterodimeric complexes, MSH2-MSH6 (or MutSα) and MSH2-MSH3 (or MutSβ) [64]. MutSα selectively recognizes mismatches along the DNA helix and DNA extra-loops of 1–4 bases, while MutSβ recognizes DNA extra-loops of 2–10 bases. Both MutSα and MutSβ show DNA binding function along with ATPase activity, which is triggered by the presence of heteroduplex DNA. When the error is recognized, the hydrolysis of ATP activates EXOI, a 5'-3' exonuclease which removes a DNA segment of few nucleotides, starting from a break in the DNA single helix [65]. The DNA break acts as an initiation site for the removal of the incorrect base. When a DNA break is present at the 5' direction of the mismatch, EXOI is able to directly process the DNA [65]. However, when the break is at 3' direction of the mismatch, the heterodimeric MutLα complex (MLH1-PMS2) is recruited by both MutSα and MutSβ to generate a break at the 5'. This is ensured by the recruitment of additional factors, RFC proteins and PCNA, which are also responsible for the identification of the correct DNA helix to repair [66]. The gap is then protected by replication protein-A (RPA) until the DNA is resynthesized by DNA polymerase δ. The DNA strands can be finally linked together by DNA ligase I [67].

Epigenetic factors play a role on the efficiency of the mismatch repair mechanisms. In fact, in vitro studies have shown that MutSα is recruited on the replicating chromatin when the histone H3 is trimethylated [68], whereas H3 mutations which impair the methylation result in reduced chromatin stability and increased rate of mutations [68].

5.3.3 DNA Double-Strand Break Repair/ Recombinational Repair

DSBs are among the most deleterious lesions occurring in the DNA [69]. Nevertheless, eukaryotic cells are well equipped to deal with these lesions, with apoptosis being triggered when the DSB cannot be repaired. The mechanisms for DSBs repair are distinct and were initially thought of consisting two main pathways called homologous end-joining (also known as homologous recombination – HR) and nonhomologous end-joining (NHEJ) [70]. However, several studies identified a third pathway in NHEJ-deficient yeast cells, which was named alternative NHEJ pathway or microhomology-mediated end-joining pathway (MMEJ) [71].

The choice between these pathways depends on the cell cycle status [72], posttranslational modifications [73], and the structure of DNA breaks [74]. HR is predominantly present in the S and G2 phases of cell cycle, where the DSBs are repaired using a template sister chromatid [72]. On the other hand, NHEJ does not require a sister chromatid as template, thus occurring in any phase of the cell cycle. Several posttranslational modifications, such as phosphorylation, help regulate these pathways and the activity of associated proteins [73]. Also, HR is generally considered an error-proof pathway, while NHEJ is more suitable to loss of genetic information or chromosomal rearrangements [74].

HR requires a homologous sequence as template for repair. In this pathway, DSB is resected 5–3′ on one strand of the DSB ends, producing terminal 3′-OH single-stranded DNA (ssDNA) tail which is then compared with other sequences to find a template, generally the sister chromatid [75]. The presence of DSB recruits the MRE11/RAD50/NBS1 complex [76], which in turn recruits phosphorylated CtIP and activates MRE11 endonuclease. This process results in the generation of a short 3′-ssDNA overhang [75]. The single strand that is being cleaved in this process is further cleaved by either of the two nucleases, EXO1 or DNA2, which generates a long 3′-ssDNA overhang. The tail of the ssDNA is coated by RPA, to avoid self-annealing of ssDNA and the action of nucleases. RPA is then substituted by RAD51, when the homologous strand is located, and strand invasion occurs. A D-loop is formed within the displaced ssDNA which is stabilized by RPA. RAD51 is removed from the double-stranded DNA and DNA synthesis occurs [75].

Unlike HR, NHEJ does not require nucleolytic resection of DNA ends. In fact, the occurrence of nucleolytic resection of DNA ends typically inhibits canonical NHEJ [77]. In this pathway, the Ku70 and Ku80 heterodimers recognize and bind DSBs in the DNA, encapsulating them to prevent degradation [78]. The MRE11/

RAD50/NBS1 complex is then recruited to promote binding of the compatible DNA ends [79]. If the ends are not compatible due to DNA overhangs and/or deletions of nucleotides, nucleases and/or DNA polymerases are recruited to remove nucleotides on DNA ends and/or insert nucleotides on the DNA gaps [80]. Afterwards, DNA ligation occurs between the fixed strands.

MMEJ pathway is slightly different than NHEJ canonical pathway. First, NEHJ does not require homology between the DNA ends of the DSBs (4 nucleotides or less), while MMEJ requires homology that varies between 2 and 20 nucleotides [81]. Additionally, MMEJ is not regulated by NHEJ regulating factors [81]. Interestingly, NHEJ factors like Ku70 and Ku80 heterodimers and RPA inhibit resection of DNA and the MMEJ [82]. Additionally, several factors involved in DNA resection like CtIP and DNA2 have been described as crucial to MMEJ pathway [83]. Interestingly, DNA polymerase θ (POLQ) has shown to facilitate the removal of RPA from resected DSBs to promote their annealing by MMEJ pathway [84].

5.4 DNA Repair Mechanisms During Spermatogenesis

Spermatogenesis is a dynamic process which starts with spermatogonia and results in a fully mature spermatozoon. These proliferative cells undergo mitotic clonal expansion in order to maintain a pool of progenitor cells, as well as meiosis, to halve the chromosome content and become primary spermatocyte. During meiosis I, primary spermatocyte matures into two secondary spermatocytes, which in turn produce four spermatids during the meiosis II. Finally, spermatids undergo a series of morphological and maturation changes, resulting in mature spermatozoa [85].

Sperm DNA can be repaired until the third week of spermatogenesis, while there are no repair mechanisms active during the transit and storage in the epididymis or post-ejaculation. In spermatogonia, DNA damage is repaired by the proofreading feature of the DNA polymerase along with the mismatch repair mechanism. Furthermore, these diploid cells rely on HR mechanism for the repair of DSB. At the spermatid stage, the repair of DSB relies on MMEJ pathways, as these are haploid cells and do not have a chromatin sister to be used as template during HR [86]. Furthermore, BER mechanism is also active and involved in the removal of oxidative stress-modified DNA bases [87]. Finally, spermatozoa show a truncated BER pathway, due to the expression of OGG1 protein [88]. This is involved in the detection and removal of oxidized bases, generating an abasic site which is later repaired by the oocyte.

5.5 Role of Oocyte in Repairing Sperm DNA Damage

The first study investigating the oocyte capability to repair sperm DNA damage dates back to the end of 1980s, with Matsuda and Tobari in 1989 using a murine in vitro system [89]. In this study, sperm were treated with increased doses of X-ray to induce DNA damage; by performing in vitro fertilization, authors demonstrated that fertilization can still occur if the sperm DNA is damaged, as the oocyte possesses the ability to repair the DNA damage in sperm. Furthermore, the different rate of chromosome aberrations observed after treating the oocyte with diverse DNA repair inhibitors suggested that multiple mechanisms take place in the oocyte for sperm DNA repair [89]. This has been demonstrated later in 2007 and 2009, by analyzing the transcription of DNA repair genes in oocyte at the germinal vesicle stage and in human in vitro matured oocytes [90, 91]. Although not all the mRNAs present in the oocyte are translated into proteins, authors detected the expression of genes involved in all the repair mechanisms (BER, NER, DSB, mismatch repair), highlighting the importance of oocyte in repairing paternal DNA damage.

However, the repair capabilities of oocyte are limited, as reported by Ahmadi and Ng in 1999, who investigated the effects of sperm DNA damage on reproductive outcomes by using an in vitro model [92]. Here, increased doses of radiations were used to induce SDF, analyzed by terminal deoxynucleotidyl transferase-mediated deoxyuridine triphosphate nick end labeling (TUNEL) assay. Hamster oocytes were used to analyze the reproductive potential of the damaged sperm in terms of penetration and fertilization rates, along with blastocysts development. Results showed that increased rate of SDF did not affect the sperm ability to penetrate the zona pellucida or to fertilize the oocyte. Conversely, blastocysts development was significantly impaired, with a reduction of 50% when SDF was equal to 8.1%. This suggested that oocyte might be able to repair paternal DNA only when the rate of damage is low (<8%).

Also, the oocyte age seems to significantly influence its capability to repair sperm DNA damage. In 2020, Horta et al. performed an in vitro fertilization by using oocytes retrieved from young (5–8 weeks) and old (42–45 weeks) mice, and sperm irradiated to obtain three distinct levels of DNA damage [93]. The fertilization rate varied according to the rate of sperm DNA damage. However, no difference was observed between the oocyte age groups, whereas a different blastocyst formation rate was reported. In humans, the transcriptomic profile of oocytes isolated from patients older than 37 years significantly differed in comparison with oocytes from younger patients (<36 years), with differential expression of genes involved in DNA repair and response to DNA damage [94], in agreement with previous results, showing increased number of chromosomal alterations in the paternal DNA when the DSB repair mechanisms in the oocyte were disrupted [95].

5.6 Molecular Tests for the Analysis of DNA Damage in Sperm

5.6.1 Brief Overview of SDF Testing

The most common tests for SDF testing are the sperm chromatin dispersion (SCD) assay, the sperm chromatin structure assay (SCSA), the TUNEL assay and the single cell gel electrophoresis (Comet) assay.

Unlike other SDF tests, the SCD test measures the absence of damage instead of the damaged DNA in sperm. By using lysis solution that induces acid denaturation and removal of nuclear proteins, sperm with nonfragmented DNA exhibit a characteristic halo of dispersed DNA [96]. In sperm with fragmented DNA, this characteristic halo is absent or exists as a minimal halo around the sperm nucleoid, observed under bright field microscopy. Staining with 4′,6-diamidino-2-phenylindole (DAPI) can be also performed, and sperm analyzed by fluorescence microscopy. Conveniently, commercial kits such as the Halosperm ® kit (Halotech DNA SL, Spain, Madrid) are available in the market [97]. Several advantages make this an appealing test: it is easy to perform, produces reproducible results, and the overall protocol is not time-consuming. However, despite SDF detected by SCD test being negatively correlated with the embryo quality and fertilization rates in ARTs, it is not correlated with clinical pregnancy rates or births [98].

SCSA is a flow cytometric test based on the principle that abnormal sperm chromatin has a greater susceptibility to physical induction of partial DNA denaturation in situ [99]. This test can be divided into two different protocols, SCSA-acid, and SCSA-heat according to the physical way of inducing DNA denaturation in situ, although the first one is much easier to use and has become the standard protocol [100]. The extent of DNA denaturation is assessed by flow cytometry, using the metachromatic properties of a fluorescent stain (acridine orange). This test measures the shift from green fluorescence, corresponding to native, double-stranded DNA, to red fluorescence, corresponding to denatured, single-stranded DNA [99]. Due to the use of a flow cytometer, a large number of spermatozoa can be analyzed. Moreover, this is the only SDF test that has clearly established cut-off values to identify samples that are not compatible with in vivo pregnancy [101–103]. Additionally, this test can accurately predict ARTs outcomes such as fertilization rates [103]. However, the need for costly instruments such as the flow cytometer can limit its use.

TUNEL assay uses a template-independent DNA polymerase that adds deoxyuridine triphosphate (dUTP) tagged with a fluorochrome to the 3′-hydroxyl termini of SSB and DSB, detected by flow cytometry [104, 105]. Several studies have tried to identify a cut-off value to differentiate between fertile and infertile men; however, this value is still under debate. Nevertheless, this test is a good indicator for fertilization and pregnancy rates using ART [106, 107]. Fluorescence microscopy can also be used; however, to achieve accurate and reliable results a flow cytometer is recommended.

The Comet assay is based on the lysis of sperm cells and their electrophoretic migration under neutral or alkaline conditions, leaving a trail that resembles a comet when observed by fluorescence microscopy [108]. The tail length and intensity are then considered to determine DNA damage. Several studies have tried to determine cut-off levels for infertility diagnosis [109, 110]. Overall, it is an easy and sensitive assay to determine DNA damage, which correlates significantly with data from other SDF assays.

5.6.2 Real-Time PCR for mtDNA Copy Number and Long PCR for mtDNA Integrity

Due to their role as the powerhouse of the cell, mitochondrial physiology in sperm cells is widely considered as a crucial parameter of sperm quality. In fact, several studies have shown that mitochondrial DNA (mtDNA) copy number, which reflects the number of mtDNA copies per nuclear DNA copy, can be used as a biomarker of sperm quality [111, 112]. Traditional tests use real-time PCR to amplify the target gene (usually MT-ND1) and then compare its expression to the amplification of a single copy nuclear gene used as a reference (usually hemoglobulin or beta-actin). This results in a relative quantification [113]. More recently, a new method known as droplet digital mitochondrial DNA measurement has been developed to determine the absolute mtDNA copy number, not only from cell populations but also from single cells, with several advantages comparing to real-time PCR [114]. MtDNA copy number in sperm has been determined in several species, including stallions and humans, where higher mtDNA copy number is associated with worse sperm quality [111, 112]. Studies reported a significant increase in the mtDNA copy number of men with abnormal semen parameters compared with men with normal semen parameters, along with a negative correlation between mtDNA copy number and sperm count [115, 116]. Interestingly, the authors also report a significant increase in mtDNA copy number on the fraction corresponding to the sperm with lower fertilizing capability compared with the fraction with the best fertilizing capability, after separating sperm samples through density gradient [116]. Higher mtDNA copy number is also associated with lower odds of fertilization and embryo quality in ART [117].

Long range PCR is a technique that allows the amplification of DNA fragments whose length cannot be typically amplified using real-time PCR (over 30 kilobases) by modifying DNA polymerases [118]. This is of special utility for sperm samples since it can detect mtDNA deletions and/or fragmentations of mtDNA. In fact, studies have shown that sperm samples from patients with abnormal sperm parameters were associated with lower mtDNA integrity compared with normozoospermic men [115, 119]. Additionally, men with obstructive azoospermia present a higher prevalence of large mtDNA deletions when compared with fertile men [119]. One of the most common mtDNA deletions covers 4977 base pair, and men with abnormal sperm parameters have a higher incidence of this mutation [120]. Additionally, higher mtDNA deletion frequency is also associated with lower odds of fertilization and embryo quality in ART [117]. Altogether, a clear relation between mtDNA integrity and sperm quality exists and it deserves a closer look in years to come.

5.6.3 Telomere Length Measurement

Real-time PCR and quantitative fluorescent in situ hybridization (Q-FISH) are mostly used to determine telomere length [121]. Both methods have pros and cons. Real-time PCR is simple and fast compared to Q-FISH. On the other hand, Q-FISH is more precise and is more accurate for measuring telomere length. However, it is a more complex method, with a lengthy protocol [121]. Interestingly, a negative correlation between sperm telomere length and SDF was reported [40]. Sperm telomere length also positively correlated with good quality embryos [122]. Additionally, pregnancy rates were null when using sperm with abnormal sperm telomere length, suggesting the existence of threshold values for sperm telomere length concerning a correct telomere function [123]. However, an agreement concerning the correlation between sperm telomere length, sperm parameters and clinical outcomes is still under debate, and more studies are needed.

5.6.4 PCR for Y Microdeletions

The gold standard method to detect microdeletions of chromosome Y is the sequence-tagged site polymerase chain reaction (STS-PCR). This technique is fast, simple, and reproducible but is highly dependent on the selection of the STS markers, which should be highly specific [124]. A wide array of STS markers has been described as specific of AZFa region. However, multiplex STS-based PCR can be performed using multiple STS markers for each AZF region, which can currently detect over 95% of existing deletions [125]. Guidelines have been published for the molecular diagnosis of Y microdeletions [124], supporting the optimization of the panel of selected primers and detection methods. Furthermore, a multiplex PCR protocol has been proposed, which eliminates the need for downstream amplicon identification [125]. Overall, STS-PCR is a crucial molecular test in the diagnosis of male infertility due to sperm DNA damage.

5.6.5 DNA Methylation Profiling Using Pyrosequencing and Next Generation Sequencing

Pyrosequencing has been used to detect the DNA methylation profile. This DNA sequencing technique relies on the sequencing by synthesis principle [126]. Studies have also established protocols for genome-wide DNA methylation and for specific genome-wide sperm DNA methylation [127, 128]. Next generation sequencing (NGS) has allowed for high-throughput functional genomic research with its ability to sequence thousands of genes or gene regions simultaneously. Indeed, recent studies using NGS have shown that the DNA regions of imprinted genes have different methylation levels between asthenozoospermic, oligozoospermic, and oligoasthenozoospermic individuals [129]. In ART setting, sperm DNA methylation positively correlates with fertilization rate and embryo quality [130]. Furthermore, studies have suggested that men with aberrant epigenetic markers, particularly in imprinted genes, can transmit them to their offspring, highlighting the importance of DNA methylation profiling during ARTs [131]. A study comparing men undergoing IVF treatment and fertile men observed different DNA methylation patterns between both groups and, more importantly, that DNA methylation status can be used to predict IVF success, since abnormal sperm DNA methylation patterns were associated with poor embryo quality [132]. Urdinguio et al. reported that idiopathic infertile men had abnormal DNA methylation patterns in 2752 CpG islands compared to fertile men [133]. Furthermore, abnormal sperm DNA methylation status in several imprinted genes has been found to be associated with recurrent pregnancy loss [134].

5.6.6 Detection of DNA Repair Proteins as Molecular Markers: γH2AX and XRCC1

Cells have developed a complex machinery to repair DNA damage. γH2AX and XRCC1 are two proteins involved in DNA repair pathways, thus their detection can be highly valuable as molecular markers of DNA damage. γH2AX is the phosphorylated form of H2A histone family member X (H2AX) at mammalian Ser-139 [51]. This phosphorylation is an early response to DSBs induced by several cytotoxic agents and, thus can be used as an indirect biomarker for diagnosis using immunofluorescence-based assays. Phosphorylation of H2AX can be observed microscopically using γH2AX specific fluorescent antibodies. Manually counting the number of γH2AX foci directly relates with the number of DNA DSBs as there is a one-to-one relation between the number of γH2AX foci per nucleus observed and DSBs in the sample [135]. Computational approaches to automatically analyze and count γH2AX foci have also been developed as alternatives to reduce the time for the analysis [136]. As γH2AX is formed every time that a DSBs occurs, it also allows the study of kinetics of DNA DSB repair [137]. Garolla et al. demonstrated that sperm from pregnant couples undergoing ICSI had higher γH2AX percentage when compared with sperm from couples which had a negative ICSI outcome [138]. Therefore, γH2AX assay is likely to be a good predictive test for ICSI outcome, with better predictive values than the TUNEL assay [138]. Additionally, another study with patients undergoing IVF found that those cycles which resulted in live birth exhibited lower levels of γH2AX than those with no birth [139]. Using flow cytometry, Zhong et al., described that γH2AX levels are negatively correlated with sperm parameters and overall higher in the sperm of infertile men [140]. This study also indicated a threshold of 18.55% as a cutoff value to differentiate fertile from infertile men. Overall, γH2AX levels are a good biomarker for DNA damage and several studies heavily suggest that it can be used as a predictive marker for successful ARTs.

XRCC1 gene encodes a 70 kDa multi-domain protein involved in the SSB repair, and NER and BER pathways, acting as a scaffolding protein which interacts with multiple repair enzymes [141]. Several studies, using real time PCR, have shown that individuals exposed to different environmental threats associated with DNA damage (e.g. radiation, heavy metals) have lower XRCC1 mRNA levels, highlighting the potential of this gene as an indirect DNA damage biomarker [142, 143]. Singh et al. reported that individuals with azoospermia had lower expression of XRCC1 mRNA than individuals with normal spermatogenesis in testicular biopsies [144]. Additionally, using age-matched controls, transcript levels of XRCC1 were lower in severe oligozoospermic individuals compared to controls and even lower in severe oligoasthenozoospermic individuals comparing to controls and to oligozoospermic individuals [144]. XRCC1 protein levels, identified by Western blot, were also decreased in both conditions when compared with healthy, fertile patients. Although more studies are needed, XRCC1 is suggested as a potential biomarker of DNA damage in spermatozoa.

5.7 Sperm DNA Damage and Fertilization of the Oocyte

5.7.1 In Vitro *and Animal Studies on the Role of Sperm DNA in Supporting Sperm Movement and Oocyte Fertilization*

There is a considerable lack of animal studies correlating sperm motility with sperm DNA damage. Cryopreserved sperm is used for the preservation of animal models. A study by Li & Lloyd has shown that cryopreservation of mouse sperm increases DNA fragmentation index (DFI) by TUNEL both in wildtype C57BL/6N males and in mutant males proven fertile by natural mating [145]. Additionally, total sperm motility and progressive motility were decreased post-thaw in both groups. Interestingly, the authors found that post-thaw DFI was negatively correlated with total sperm motility and progressive sperm motility assessed before freezing. Frozen-thawed sperm from fertile mutant males also had a significantly lower DFI when compared with frozen-thawed sperm from mice with abnormal sperm motility

[145]. Fertilization rate was also assessed using sperm from mice that failed to produce offspring after IVF and embryo transfer, and sperm from mice that failed to produce offspring after ICSI and embryo transfer. In comparison with a fertile murine control, frozen-thawed sperm from the groups that failed to produce pups showed higher DFI levels in both IVF and ICSI groups [145]. Furthermore, the fertilization rate after IVF, assessed by 2-cell rate, was also lower in the groups with higher DFI compared with the control groups [145]. Nevertheless, no significant changes were found in the fertilization rates of the group undergoing ICSI. In agreement with reports from human studies, this suggests that the fertilization event could not be heavily impacted by sperm DNA damage, at least as much as the other developmental stages such as blastocyst formation and fetal development. Another study on bulls reported that DFI can be used to discriminate between below average, average, and above average fertility bulls, with the former having the highest values of DFI [146]. A study by Waterhouse et al., also revealed using flow cytometry that sperm DNA damage is tightly related to bulls fertility [147]. Gliozzi et al., using the estimated relative conception rate, classified bulls as high and low fertility and analyzed a wide array of sperm parameters through CASA and flow cytometry. Higher sperm motility and lower levels of SDF were found in the high fertility group, among other differences [148]. Applying stepwise multiple regression analysis, the authors established a predictive model with nine parameters to be evaluated resulting in a positive relationship ($R^2 = 0.84$, $P < 0.05$) when using this model to determine between field and predicted fertility, which could help significantly in the selection of fertile animals in artificial insemination centers [148].

A case-control study performed in dogs showed a higher level of DFI in the animals with poor-quality ejaculates. The DFI also negatively correlated with several sperm parameters, including total and progressive sperm motility, reinforcing the link between DNA damage in sperm and the loss of sperm motility [149]. Sperm motility is logically a crucial sperm characteristic for a successful fertilization, i.e., if the spermatozoa does not reach the oocyte, fertilization does not occur. Sperm DNA damage and sperm motility are heavily related with age and seasonality in Chilean Purebred Stallions, a horse breed. Rodrigo et al., reported that ejaculates collected from stallions in the oldest group had lower levels of sperm progressive motility and higher levels of SDF when compared with ejaculates from younger stallions [150].

5.7.2 Clinical Studies in ART

Although the oocyte has the potential to repair sperm DNA damage, this depends on the extension of such damage. Hence, some studies have reported a negative association between SDF and fertilization rate. A study conducted by Tang et al., in 2020, analyzed the fertilization rate in 116 men with mild-to-moderate asthenozoospermia compared with 407 normozoospermic men [151]. When SFD was analyzed by SCD, increased levels were reportedly associated with higher incidence of cycles with total fertilization failure and/or low fertilization rate after IVF. By performing IVF, Simon et al. also reported a decline in fertilization rate based on the percentage of sperm with SSBs, analyzed by alkaline comet [152]. However, the physiological step of fertilization can be bypassed by performing ICSI: SDF was not associated with impaired fertilization rate after ICSI and euploid blastocyst transfer [153], regardless of the high percentage of sperm with SSBs or DSBs [152, 154]. A recently published meta-analysis by Ribas-Maynou et al. also highlighted a significant difference in the number of studies reporting a negative association between sperm DNA damage and fertilization rate when IVF or ICSI was used (76% vs 52%, respectively) [155], regardless of the type of technique used to investigate DNA damage.

Besides SDF, the clinical association of other types of sperm DNA damage with fertilization outcome is overall poorly investigated. In case of Y microdeletions, phenotype may vary from azoospermia to oligozoospermia, so testicular sperm retrieval is an option to consider in case of ART. In a study investigating reproductive outcomes of patients with Y chromosome microdeletions compared to oligo-/azoospermic men without microdeletion, fertilization rate was reportedly lower in the first group, when either ejaculated or testicular extracted sperm were used [156]. These observations were further confirmed by a systematic review published on the topic in 2021 [157]. However, these studies did not perform any sub-analysis based on the type of microdeletion. While no clear data are available for AZFa and AZFb deletions, patients with AZFc deletion in Y chromosome showed lower fertilization rate after ICSI with both testicular retrieved or ejaculated sperm [157, 158]. This is due to the presence of genes in AZFc region which play an important role in spermatogenesis, and whose deletion is associated with spermatogenetic failure [46]. However, data published on this topic is still controversial and under debate, with other studies reporting lower fertilization rate [159, 160].

The influence of epigenetics regulation on fertilization outcome is also under investigation. Lower levels of protamination have been associated with reduce fertilization rate [161], suggesting an impairment of the chromatin remodeling mechanisms. However, global sperm methylation status did not correlate with the fertilization rate, when IVF was performed [162]. After fertilization, sperm DNA is activated and is involved in the embryo development at the stage of 4–8 cells [163]. Hence, the degree of sperm compaction may not affect the oocyte fertilization and reduce fertilization rate.

5.8 Sperm DNA Damage and Pregnancy

5.8.1 *In Vitro and Animal Studies on the Role of Sperm DNA in Supporting Embryo Implantation and Pregnancy*

Several studies have studied the role of sperm DNA damage in embryo implantation and pregnancy by using animal models, further cementing the importance of this parameter in the characterization of sperm quality and fertility potential of mammalian species. A study by Li et al. in mice has shown that embryos derived by sperm with high DFI failed to litter after embryo transfer when compared with embryos derived by sperm with low DFI [145]. A study by Seifi-Jamadi et al., in Belgian Blue bulls also reported interesting observations. Semen collected from bulls which were exposed to high temperature and humidity index showed lower sperm parameters including sperm motility and higher levels of hydrogen peroxide and aberrant chromatin condensation, when compared with the control group [164]. They also reported lower blastocyst rates and higher apoptotic rates, reflecting a poorer embryo development [164]. Interestingly, several studies have associated protamine deficiency with sperm DNA damage in animals [165, 166]. However, some studies in bulls reported that protamine deficiency and abnormal chromatin packaging cannot directly explain the differences in fertility and embryonic development between different bulls, highlighting the need for more studies in this particular topic [167, 168]. A study in boars has shown that fertility, defined by farrow rate and average number of pigs born, negatively correlated with DFI [169]. Interestingly, a study exploring the relationship between different methods of cryopreserving mouse sperm and the resulting SDF reported that freeze-dried mouse sperm with lower DFI values significantly correlated with the embryonic development rates after ICSI [170], highlighting the importance of DNA damage for embryo implantation and development. A study in rams using epidydimal sperm also showed that sperm with higher rate of intact spermatozoa and lower rate of SSB and DSB had better results concerning embryo development, with higher rates of development to the blastocyst stage than sperm with higher rates of DNA damage [171], further confirming the results of the previous study in another species. A negative correlation between pregnancy levels and sperm with SDF was also reported in mares [172].

DFI values of 25% for stallions and around 10–20% for bulls have been suggested as thresholds representing the maximum value of DFI from which fertility is compromised for these species [173]. Nevertheless, this value heavily dependent on the methodology used (e.g., type of SDF test used, method for sample collection, processing procedure for fertilization) which limits the conclusions that can be drawn from these studies.

A study by Ahmadi et al., reported that human spermatozoa in which DNA damage was induced by radiation still retained the ability to fertilize oocytes. However, sperm which were exposed to higher intensities of radiation-induced DNA damage had lower level of embryonic development, lower rates of embryo implantation and lower numbers of live fetuses when compared with groups exposed to lower radiation doses and the control group [92]. Similar results were reported by Upadhya et al., in which mouse sperm exposed to higher radiation levels had lower number of embryos reaching blastocyst stage and a lower total cell number in preimplantation stage embryos [174].

The idea of inheritability of negative traits caused by the level of sperm DNA damage was further explored in a study where mouse sperm with DNA damage was used through ICSI to generate embryos. As expected, ICSI resulted in lower rates of embryo development, implantation, and number of offspring when sperm with damaged DNA were used [175]. Interestingly, mouse generated by DNA fragmented sperm had worse results in behavior tests when compared with in-vivo fertilized controls and a higher tendency to develop mesenchymal tumors, premature aging and lower life expectancy [175]. This clearly indicates that DNA integrity is key for a correct embryo development and to avoid further health complications down the line.

5.8.2 Clinical Studies in Natural Pregnancies and ART

Studies investigating the impact of SDF on natural pregnancy were analyzed by a meta-analysis conducted by Zini in 2011 (Table 5.1). This included three studies with a total of 616 couples, showing that high levels of SDF, measured by SCSA, were associated with a failure to achieve natural pregnancy, with a cumulative odds-ratio of 7.01 [176]. However, small number of studies included and their large heterogeneity made the findings inconclusive. When intrauterine insemination (IUI) was used, high levels of SDF were associated with lower pregnancy (relative risk – RR: 0.34, $P < 0.001$) and delivery rates (RR: 0.14, $P < 0.001$), by analyzing the results from 10 studies and a total of 2,839 IUI cycles [177] (Table 5.1). Conversely, couples with low levels of SDF were reportedly three times more likely to conceive [178]. When IVF was performed, four different meta-analyses published between 2006 and 2017 reported an association between high levels of SDF and lower pregnancy rates [176, 179–181], while higher risk of miscarriage rate [176, 180, 182, 183] and lower live birth rate was also reported after both IVF and ICSI [184] (Table 5.1). In 2019, a meta-analysis investigating the role of SDF in recurrent pregnancy loss (RPL) analyzed 13 studies for a total of 530 couples [185] (Table 5.1). Authors reported that couples with idiopathic RPL had higher levels of SDF than fertile couples ($P < 0.001$), in agreement with another meta-analysis published in the same year [186]. In summary, high SDF levels seem to be associated with decreased pregnancy rate after IVF and increased miscarriage rate after both IVF and ICSI. This might be explained by the technique used for ART: in fact, the natural fertilization process is bypassed when ICSI is used, so that a clinical pregnancy can be achieved easier with ICSI than IVF. Also, gametes are not cultured for prolonged time before ICSI and are less exposed to oxidative stress and laboratory-induced damage. Finally, DNA damage may significantly affect genes responsible for embryo development and implantation. Although these studies seems to suggest an impact of sperm DNA damage on clinical reproductive outcomes, we should not forget that the studies performed so far are highly heterogeneous in terms of population recruited, techniques applied for SDF analysis and thresholds used to define high and low SDF levels, along with the presence of possible bias and confounding factors inherent in the original data. Also, SDF is always measured in semen before the initiation of the cycle, hence this value may not truly reflect the status of the sperm collected for insemination.

Table 5.1 Clinical studies investigating the impact of high sperm DNA fragmentation (SDF) in natural pregnancies and assisted reproduction.

Study	Main findings
Li et al. [179]	Association between high SDF and low pregnancy rates after IVF
Zini et al. [182]	Association between high SDF and high risk of miscarriage rate
Zini [176]	Association between high SDF and failure to achieve natural pregnancy (OR: 7.01)
	Association between high SDF and low pregnancy rates after IVF
	Association between high SDF and high risk of miscarriage rate
Ribas-Maynou et al. [183]	Association between high SDF and high risk of miscarriage rate
Zhao et al. [180]	Association between high SDF and low pregnancy rates after IVF
	Association between high SDF and high risk of miscarriage rate
Osman et al. [184]	Association between high SDF and low live birth rate after IVF and ICSI
Simon et al. [181]	Association between high SDF and low pregnancy rates after IVF
Chen et al. [177]	Association between high SDF and low pregnancy and delivery rates after IUI
Sugihara et al. [178]	Couples with low SDF were reportedly three times more likely to conceive
	Association between high SDF and low pregnancy rates after IVF
Tan et al. [185]	Idiopathic RPL patients had higher SDF than fertile couples
McQueen et al. [186]	Idiopathic RPL patients had higher SDF than fertile couples

ICSI intracytoplasmic sperm injection, *IVF* in vitro fertilization, *OR* odds-ratio, *RPL* recurrent pregnancy loss, *SDF* sperm DNA fragmentation

Whether Y microdeletions have an impact on pregnancy and miscarriage rates has been investigated in several studies, although data for AZFa and AZFb are still poor and inconclusive. When patients with different Y microdeletions were analyzed, no statistical difference was observed in these reproductive outcomes, either when ejaculated or testicular retrieved sperm were used for ICSI [156]. In azoo- and oligozoospermic patients with AZFc microdeletion, pregnancy (45.4% and 67.5%, respectively) and live birth rates (35.1% and 53.4%, respectively) were reportedly lower than in control group even when ICSI was performed by using either ejaculated or testicular retrieved sperm [158], with no difference reported for miscarriage rate. However, data published on this topic is controversial, as other studies reported no difference in these reproductive outcomes in case of AZFc microdeletions [156, 159, 187]. ICSI has been successfully used so far in clinics for those patients who are severely oligozoospermic; however, it still must be clarified if the presence of Y microdeletions can adversely affect the achievement of a clinical pregnancy. Similarly, the association between Y microdeletions and pregnancy loss needs to be further investigated and elucidated, as the reports are still very controversial [157].

An important area of investigation focuses on the impact of epigenetics alteration on reproductive outcomes. As the inactive sperm DNA is highly methylated, a variation in global methylation pattern has been reported to be associated with reduced pregnancy rate after IVF [162]. This may be explained considering that reduced methylation rate results in less compacted chromatin and higher DNA susceptibility to DNA damage [188]. Also, sperm proteomics analysis of couples who did not achieve a pregnancy showed an overexpression of proteins involved in chromatin assembly (i.e. histones variants, and SRSF protein kinase 1) in comparison with those couples who achieved pregnancy [189]. However, further studies are needed to clarify any association between sperm chromatin status and pregnancy rate.

5.9 Molecular Markers of Sperm DNA Damage and Future Areas of Investigation

Classic semen analysis remains the method of choice worldwide for the assessment and prediction of male fertility potential. Nevertheless, the last decades of clinical reproductive medicine have shown that routine semen analysis cannot accurately detect all the underlying causes of male infertility [190]. Hence, identification of novel molecular markers of sperm quality has been highly desired and pursued by the reproductive scientific community. As discussed earlier, ROS are a major endogenous source of sperm DNA damage. Consequently, several methods have been developed which are based on the ability to, directly or indirectly, measure oxidative radicals. Using appropriate probes, intra- and extracellular ROS levels (e.g., OH^-, H_2O_2) can be detected by chemiluminescence outputting a signal which represents the ROS levels [191]. Determination of total antioxidant capacity using seminal plasma also gives a direct measurement of the antioxidant potential of the sample [192]. Indirect measurements rely on the detection of malondialdehyde (a lipid peroxidation product), carbonyl groups (resulting of protein oxidation), and 8-OHdG. Oxidation-reduction potential (ORP) assessment allows for an accurate measurement of the balance between reductants and oxidants in semen samples [193].

A study by Heidari et al., aimed to assess the potential of heat shock protein (HSPA2) and metallopeptidase domain2 (ADAM2) as biomarkers for sperm quality, including sperm DNA damage. These two glycoproteins are involved in the fertilization of the oocyte, with important roles in preserving DNA integrity and cell adhesion and signaling [194, 195]. Importantly, this study revealed that sperm from asthenoteratozoospermic men had lower levels of HSPA2 and ADAM2 when compared with normozoospermic men [196]. In fact, a significant negative correlation was also identified between HSPA2 and ADAM2 levels, and SDF which could be justified due to their role in the maintenance of sperm DNA integrity [196]. The poly (ADP-ribose) polymerase (PARP) has also been hypothesized as a possible biomarker of sperm DNA damage. PARP is a nuclear enzyme specialized in the repairing of DNA lesions, described as having important functions in spermatogenesis [197]. During apoptosis, PARP is cleaved by caspase-3 which inhibits PARP's DNA repairing abilities [198]. Crucially, the presence of the cleaved form of PARP in ejaculated sperm samples was recently described, which further underlines a possible use of PARP levels as a biomarker of sperm DNA integrity [199]. Nevertheless, there is still a need for functional studies establishing this link. γH2AX, the phosphorylated form of H2A histone family member X (H2AX), was also suggested as a possible biomarker of SDF. H2AX is a type of histone protein, whose phosphorylated form at Ser-139 has a crucial role in an earlier response to the appearance of DSBs [200]. In a study by Zhong et al., γH2AX levels were found to be positively correlated with DSBs in semen samples, highlighting a possible use of γH2AX as a biomarker for sperm DNA damage [140]. Nevertheless, more studies are needed in sperm, as the presence of γH2AX by itself is not a specific marker of DSBs [201]. Interestingly, a study by Behrouzi et al., highlights the difficult in finding suitable biomarkers for male infertility due to the heterogeneity between semen samples for different patients. In this work, the authors established four experimental groups, group 1 – normozoospermic patients with DFI <15%, group 2 – samples with low motility but DFI <15%, group 3 – samples with DFI >30% and low motility, and group 4 – samples with DFI >30% but normal motility levels [202]. When analyzing the sperm protein profile of these groups, a profile from the samples with normal sperm parameters and low DFI could not be established due to a great level of heterogeneity between samples from the same group [202]. Furthermore, the same occurred in samples from groups 2–4. However, the authors still identified, through liquid chromatography tandem mass spectrometry, 128 proteins in the different experimental groups, which could be divided in four main protein groups: (a) mitochondria-related proteins, oxidative stress, and energy pathway-related proteins, (b) DNA binding and histone proteins, (c) protein processing, ubiquitination, proteasome, and (d) proteins involved in sperm motility [202]. Further studies are necessary to validate some of these candidate proteins as potential biomarkers of SDF and male infertility.

A recent study by Li et al., investigated the microRNAs (miRNAs) expression profile in seminal plasma of normozoospermic individuals with different SDF levels. Using NGS, the authors identified 431 differentially expressed miRNAs in the seminal plasma of normozoospermic individuals [203]. The authors validated the differential expression of two out of seven miRNAs (miR-374b-5p and miR-26b-5p) by qPCR among the different experimental groups [203]. The miR-374b-5p and miR-26b-5p expression levels were significantly lower in the higher SDF group when compared with the group with SDF below 15%. Overall, the expression levels of these two miRNAs could be used as a biomarker of SDF and improve the diagnosis of idiopathic male infertility as this study clearly highlights the existence of distinct miRNA profiles in the seminal plasma of normozoospermic individuals with different levels of SDF.

In an effort to better understand idiopathic male infertility and to identify novel markers that could help clinical diagnosis of this condition, the OMICS sciences (i.e. genomics, transcriptomics, proteomics, metabolomics) have been looked with particular interest. Studies have investigated the differences in the proteomic profiles between sperm samples with high and low SDF with interesting results. Although 163 proteins were conserved between these groups, 23 proteins were unique or overexpressed in the group with high DFI while 71 proteins were unique or overexpressed in the group with low DFI [204]. Functional assessment of these proteins found that processes such as energy metabolism, protein folding and acrosome assembly were affected [204]. Further studies are required to understand if any of these proteins could be a good biomarker for SDF and, consequently, for clinical diagnosis of male infertility.

5.10 Conclusions

Infertility is one of the greatest health and societal challenges of the twenty-first century. Despite being initially overlooked, the male factor is responsible for up to 50% of overall infertility cases. Due to the inability of the standard semen analysis to predict male factor as a major cause in couple infertility, there is a need for advanced techniques to assess sperm quality, e.g., sperm chromatin integrity and DNA damage. Endogenous and exogenous factors can damage DNA by (a) inducing mismatched bases on the DNA helix, (b) chemically altering the single bases structure, (c) inducing DNA fragmentation and microdeletions, (d) augmenting the rate at which telomeres are shortened, or (e) through epigenetic abnormalities. Molecular mechanisms involved in sperm DNA damage repair are essential during spermatogenesis, and after oocyte fertilization for a successful birth following natural pregnancy or assisted reproduction. Currently, new molecular markers of sperm DNA damage are being investigated. In the coming years there will be a need for further studies to develop reliable and well-tested biomarkers capable of giving accurate results when used in clinical diagnosis, improving our current diagnostic methods. This will also help unveil the mystery behind millions of cases around the world that are currently described as with idiopathic male infertility.

References

1. Zegers-Hochschild F, Adamson GD, Dyer S, et al. The international glossary on infertility and fertility care, 2017. Hum Reprod. 2017;32(9):1786–801.
2. Hamada A, Esteves SC, Nizza M, et al. Unexplained male infertility: diagnosis and management. Hum Androl. 2011;1(1):2–16.
3. Agarwal A, Parekh N, Panner Selvam M, et al. Male oxidative stress infertility (MOSI): proposed terminology and clinical practice guidelines for management of idiopathic male infertility. World J Mens Health. 2019;37(3):296–312.
4. Baldi E. Genetic damage in human spermatozoa. 2nd ed. Baldi E, Muratori M, editors. Vol. 1166. Springer; 2014. XIII, 210.
5. Lindahl T, Barnes DE. Repair of endogenous DNA damage. Cold Spring Harb Symp Quant Biol. 2000;65:127–33.
6. Poetsch AR. The genomics of oxidative DNA damage, repair, and resulting mutagenesis. Comput Struct Biotechnol J. 2020;18:207–19.

7. Kunkel T. Evolving views of DNA replication (in)fidelity. Cold Spring Harb Symp Quant Biol. 2009;74:91–101.
8. Potenski CJ, Klein HL. How the misincorporation of ribonucleotides into genomic DNA can be both harmful and helpful to cells. Nucleic Acids Res. 2014;42(16):10226–34.
9. McElhinny SAN, Kumar D, Clark AB, et al. Genome instability due to ribonucleotide incorporation into DNA. Nat Chem Biol. 2010;6(10):774–81.
10. Lujan S, Williams J, Clausen A, et al. Evidence that ribonucleotides are signals for mismatch repair of leading strand replication errors. Mol Cell. 2013;50(3):437–43.
11. Bush NG, Evans-roberts K, Maxwell A. DNA Topoisomerases. EcoSal Plus. 2015;6(2):1–34.
12. Wilson DM, Barsky D. The major human abasic endonuclease: formation, consequences and repair of abasic lesions in DNA. Mutat Res – DNA Repair. 2001;485(4):283–307.
13. Lindahl T. Instability and decay of the primary structure of DNA. Nature. 1993;362(6422):709–15.
14. Vignard J, Mirey G, Salles B. Ionizing-radiation induced DNA double-strand breaks: a direct and indirect lighting up. Radiother Oncol. 2013;108(3):362–9.
15. Mullenders LHF. Solar UV damage to cellular DNA: from mechanisms to biological effects. Photochem Photobiol Sci. 2018;17(12):1842–52.
16. Fu D, Calvo J, a, Samson LD. SERIES: genomic instability in cancer balancing repair and tolerance of DNA damage caused by alkylating agents. Nat Rev Cancer. 2013;12(2):104–20.
17. Jalal N, Surendranath AR, Pathak JL, et al. Bisphenol A (BPA) the mighty and the mutagenic. Toxicol Rep. June 2017;2018(5):76–84.
18. Zengin N, Yüzbaşioğlu D, Ünal F, et al. The evaluation of the genotoxicity of two food preservatives: sodium benzoate and potassium benzoate. Food Chem Toxicol. 2011;49(4):763–9.
19. Pandir D. DNA damage in human germ cell exposed to the some food additives in vitro. Cytotechnology. 2016;68(4):725–33.
20. Levinson G, Gutman GA. Slipped-strand mispairing: a major mechanism for DNA sequence evolution. Mol Biol Evol. 1987;4(3):203–21.
21. Carey LB. RNA polymerase errors cause splicing defects and can be regulated by differential expression of RNA polymerase subunits. Elife. 2015;4:e09945.
22. Baskaran S, Finelli R, Agarwal A. Reactive oxygen species in male reproduction: a boon or a bane? Andrologia. 2020;00:e13577.
23. Tremblay S, Wagner JR. Dehydration, deamination and enzymatic repair of cytosine glycols from oxidized poly(dG-dC) and poly(dI-dC). Nucleic Acids Res. 2008;36(1):284–93.
24. Jena NR, Mishra PC. Formation of ring-opened and rearranged products of guanine: mechanisms and biological significance. Free Radic Biol Med. 2012 Jul 1;53(1):81–94.
25. Drabløs F, Feyzi E, Aas PA, et al. Alkylation damage in DNA and RNA – repair mechanisms and medical significance. DNA Repair (Amst). 2004 Nov 2;3(11):1389–407.
26. Dutta S, Chowdhury G, Gates KS. Interstrand cross-links generated by abasic sites in duplex DNA. J Am Chem Soc. 2007;129(7):1852–3.
27. Gates K. An overview of chemical processes that damage cellular DNA: spontaneous hydrolysis, alkylation, and reactions with radicals. Chem Res Toxicol. 2009;22(11):1747–60.
28. Banaś AK, Zgłobicki P, Kowalska E, et al. All you need is light. Photorepair of uv-induced pyrimidine dimers. Genes (Basel). 2020;11(11):1–17.
29. Tommasi S, Denissenko MF, Pfeifer GP. Sunlight induces pyrimidine dimers preferentially at 5-methylcytosine bases. Cancer Res. 1997;57(21):4727–30.
30. Noll DM, McGregor Mason T, Miller PS. Formation and repair of interstrand cross-links in DNA. Chem Rev. 2006;106(2):277–301.
31. Muratori M, Tamburrino L, Marchiani S, et al. Investigation on the origin of sperm DNA fragmentation: role of apoptosis, immaturity and oxidative stress. Mol Med. 2015 Jan 30;21(1):109–22.
32. Sakkas D, Mariethoz E, St. John JC. Abnormal sperm parameters in humans are indicative of an abortive apoptotic mechanism linked to the fas-mediated pathway. Exp Cell Res. 1999;251(2):350–5.

33. Esterbauer H, Muskiet F, Horrobin DF. Cytotoxicity and genotoxicity of lipid-oxidation products. Am J Clin Nutr. 1993;57(5 Suppl)
34. Łuczaj W, Skrzydlewska E. DNA damage caused by lipid peroxidation products. Cell Mol Biol Lett. 2003;8(2):391–413.
35. Turner K, Vasu V, Griffin D. Telomere biology and human phenotype. Cells. 2019;8(73):1–19.
36. Burgers PMJ. Polymerase dynamics at the eukaryotic DNA replication fork. J Biol Chem. 2009;284(7):4041–5.
37. Shay JW, Wright WE. Hayflick, his limit, and cellular ageing. Nat Rev Mol Cell Biol. 2000;1(1):72–6.
38. Kimura M, Cherkas LF, Kato BS, et al. Offspring's leukocyte telomere length, paternal age, and telomere elongation in sperm. PLoS Genet. 2008;4(2):e37.
39. Barnes R, Fouquerel E, Opresko P. The impact of oxidative DNA damage and stress on telomere homeostasis. Mech Ageing Dev. 2019;177:37–45.
40. Rocca MS, Speltra E, Menegazzo M, et al. Sperm telomere length as a parameter of sperm quality in normozoospermic men. Hum Reprod. 2016;31(6):1158–63.
41. Thilagavathi J, Kumar M, Mishra SS, et al. Analysis of sperm telomere length in men with idiopathic infertility. Arch Gynecol Obstet. 2013;287(4):803–7.
42. Yang Q, Zhang N, Zhao F, et al. Processing of semen by density gradient centrifugation selects spermatozoa with longer telomeres for assisted reproduction techniques. Reprod Biomed Online. 2015;31(1):44–50.
43. Sadeghi-Nejad H, Oates RD. The Y chromosome and male infertility. Curr Opin Urol. 2008 Nov;18(6):628–32.
44. Ferlin A, Arredi B, Speltra E, et al. Molecular and clinical characterization of Y chromosome microdeletions in infertile men: a 10-year experience in Italy. J Clin Endocrinol Metab. 2007;92(3):762–70.
45. Vogt PH, Edelmann A, Kirsch S, et al. Human Y chromosome azoospermia factors (AZF) mapped to different subregions in Yq11. Hum Mol Genet. 1996;5(7):933–43.
46. Yu XW, Wei ZT, Jiang YT, et al. Y chromosome azoospermia factor region microdeletions and transmission characteristics in azoospermic and severe oligozoospermic patients. Int J Clin Exp Med. 2015;8(9):14634–46.
47. Kamp C, Hirschmann P, Voss H, et al. Two long homologous retroviral sequence blocks in proximal Yq11 cause AZFa microdeletions as a result of intrachromosomal recombination events. Hum Mol Genet. 2000;9(17):2563–72.
48. Gunes S, Esteves SC. Role of genetics and epigenetics in male infertility. Andrologia. 2021 Feb 1;53(1):e13586.
49. Liu Y, Lu C, Yang Y, et al. Influence of histone tails and H4 tail acetylations on nucleosome-nucleosome interactions. J Mol Biol. 2011 Dec 16;414(5):749–64.
50. Smeenk G, Wiegant WW, Marteijn JA, et al. Poly(ADP-ribosyl)ation links the chromatin remodeler SMARCA5/SNF2H to RNF168-dependent DNA damage signaling. J Cell Sci. 2013;126(4):889–903.
51. Scully R, Xie A. Double strand break repair functions of histone H2AX. Mutat Res – Fundam Mol Mech Mutagen. 2013;750(1–2):5–14.
52. Steger K, Balhorn R. Sperm nuclear protamines: a checkpoint to control sperm chromatin quality. J Vet Med Ser C Anat Histol Embryol. 2018;47(4):273–9.
53. Roest HP, Van Klaveren J, De Wit J, et al. Inactivation of the HR6B ubiquitin-conjugating DNA repair enzyme in mice causes male sterility associated with chromatin modification. Cell. 1996 Sep 6;86(5):799–810.
54. Rahiminia T, Yazd EF, Fesahat F, et al. Sperm chromatin and DNA integrity, methyltransferase mRNA levels, and global DNA methylation in oligoasthenoteratozoospermia. Clin Exp Reprod Med. 2018;45(1):17–24.
55. Laqqan M, Hammadeh ME. Alterations in DNA methylation patterns and gene expression in spermatozoa of subfertile males. Andrologia. 2018;50(3):1–9.

56. Krokan HE, Bjoras M. Base excision repair. In: Friedberg E, Elledge S, Lehmann A, Lindahl T, Muzi-Falconi M, editors. Additional Perspectives on DNA Repair, Mutagenesis, and Other Responses to DNA Damage. Cold Spring Harbor Laboratory Press; 2013. p. a012583.
57. Stivers JT, Jiang YL. A mechanistic perspective on the chemistry of DNA repair glycosylases. Chem Rev. 2003 Jul;103(7):2729–59.
58. Cappelli E, Taylor R, Cevasco M, et al. Involvement of XRCC1 and DNA ligase III gene products in DNA base excision repair. J Biol Chem. 1997 Sep 19;272(38):23970–5.
59. Sung JS, Demple B. Roles of base excision repair subpathways in correcting oxidized abasic sites in DNA. FEBS J. 2006;273(8):1620–9.
60. Scharer O. Nucleotide excision repair in eukaryotes. In: Friedberg E, Elledge S, Lehmann A, Lindahl T, Muzi-Falconi M, editors. Additional Perspectives on DNA Repair, Mutagenesis, and Other Responses to DNA Damage. Cold Spring Harbor Laboratory Press; 2013. p. 341–4.
61. Spivak G. Nucleotide excision repair in humans. DNA Repair. 2015;36:13–8.
62. Fagbemi AF, Orelli B, Schärer OD. Regulation of endonuclease activity in human nucleotide excision repair. DNA Repair (Amst). 2011 Jul 15;10(7):722–9.
63. Jiricny J. Postreplicative mismatch repair. Cold Spring Harb Perspect Biol. 2013 Apr;5(4):1–23.
64. Weßbecher IM, Brieger A. Phosphorylation meets DNA mismatch repair. DNA Repair (Amst). 2018 Dec 1;72:107–14.
65. Ijsselsteijn R, Jansen JG, de Wind N. DNA mismatch repair-dependent DNA damage responses and cancer. DNA Repair (Amst). 2020 Sep;1(93):102923.
66. Paul Solomon Devakumar LJ, Gaubitz C, Lundblad V, et al. Effective mismatch repair depends on timely control of PCNA retention on DNA by the Elg1 complex. Nucleic Acids Res. 2019 Jul 26;47(13):6826–41.
67. Bateman AC. DNA mismatch repair proteins: scientific update and practical guide. J Clin Pathol. 2021;74:264–8.
68. Huang Y, Li GM. DNA mismatch repair in the context of chromatin. Cell Biosci. 2020;10(10):1–8.
69. Jackson SP, Bartek J. The DNA-damage response in human biology and disease. Nature. 2009;461(7267):1071–8.
70. Her J, Bunting SF. How cells ensure correct repair of DNA double-strand breaks. J Biol Chem. 2018;293(27):10502–11.
71. Boulton SJ, Jackson SP. Saccharomyces cerevisiae Ku70 potentiates illegitimate DNA double-strand break repair and serves as a barrier to error-prone DNA repair pathways. EMBO J. 1996;15(18):5093–103.
72. Huertas P, Jackson SP. Human CtIP mediates cell cycle control of DNA end resection and double strand break repair. J Biol Chem. 2009;284(14):9558–65.
73. Lee KJ, Saha J, Sun J, et al. Phosphorylation of Ku dictates DNA double-strand break (DSB) repair pathway choice in S phase. Nucleic Acids Res. 2015;44(4):1732–45.
74. Reynolds P, Anderson JA, Harper JV, et al. The dynamics of Ku70/80 and DNA-PKcs at DSBs induced by ionizing radiation is dependent on the complexity of damage. Nucleic Acids Res. 2012;40(21):10821–31.
75. Ranjha L, Howard SM, Cejka P. Main steps in DNA double-strand break repair: an introduction to homologous recombination and related processes. Chromosoma. 2018;127(2):187–214.
76. Johzuka K, Ogawa H. Interaction of Mre11 and Rad50: two proteins required for DNA repair and meiosis-specific double-strand break formation in Saccharomyces cerevisiae. Genetics. 1995;139(4):1521–32.
77. Huertas P. DNA resection in eukaryotes: deciding how to fix the break. Nat Struct Mol Biol. 2010;17(1):11–6.
78. Ramsden DA, Geliert M. Ku protein stimulates DNA end joining by mammalian DNA ligases: a direct role for Ku in repair of DNA double-strand breaks. EMBO J. 1998;17(2):609–14.

79. Huang J, Dynan WS. Reconstitution of the mammalian DNA double-strand break end-joining reaction reveals a requirement for an Mre11/Rad50/NBS1-containing fraction. Nucleic Acids Res. 2002;30(3):667–74.
80. Ma Y, Pannicke U, Schwarz K, et al. Hairpin opening and overhang processing by an Artemis/DNA-dependent protein kinase complex in nonhomologous end joining and V(D)J recombination. Cell. 2002;108(6):781–94.
81. Chang HHY, Pannunzio NR, Adachi N, et al. Non-homologous DNA end joining and alternative pathways to double-strand break repair. Nat Rev Mol Cell Biol. 2017;18(8):495–506.
82. Deng SK, Gibb B, De Almeida MJ, et al. RPA antagonizes microhomology-mediated repair of DNA double-strand breaks. Nat Struct Mol Biol. 2014;21(4):405–12.
83. Howard SM, Yanez DA, Stark JM. DNA damage response factors from diverse pathways, including DNA crosslink repair, mediate alternative end joining. PLoS Genet. 2015;11(1):1–25.
84. Mateos-Gomez P, Kent T, Deng S, et al. The helicase domain of Polθ counteracts RPA to promote alt- NHEJ. Nat Struct Mol Biol. 2017;24(12):1116–23.
85. Gunes S, Al-Sadaan M, Agarwal A. Spermatogenesis, DNA damage and DNA repair mechanisms in male infertility. Reprod Biomed Online. 2015;31(3):309–19.
86. Ahmed EA, Scherthan H, de Rooij DG. DNA double strand break response and limited repair capacity in mouse elongated spermatids. Int J Mol Sci. 2015 Dec 16;16(12):29923–35.
87. Boiteux S, Radicella JP. Base excision repair of 8-hydroxyguanine protects DNA from endogenous oxidative stress. Biochimie. 1999;81(1–2):59–67.
88. Smith TB, Dun MD, Smith ND, et al. The presence of a truncated base excision repair pathway in human spermatozoa that is mediated by OGG1. J Cell Sci. 2013 Mar 15;126(6):1488–97.
89. Matsuda Y, Tobari I. Repair capacity of fertilized mouse eggs for X-ray damage induced in sperm and mature oocytes. Mutat Res. 1989;210(1):35–47.
90. Jaroudi S, Kakourou G, Cawood S, et al. Expression profiling of DNA repair genes in human oocytes and blastocysts using microarrays. Hum Reprod. 2009;24(10):2649–55.
91. Menezo Y, Russo GL, Tosti E, et al. Expression profile of genes coding for DNA repair in human oocytes using pangenomic microarrays, with a special focus on ROS linked decays. J Assist Reprod Genet. 2007;24(11):513–20.
92. Ahmadi A, Ng SC. Fertilizing ability of DNA-damaged spermatozoa. J Exp Zool. 1999 Nov;284(6):696–704.
93. Horta F, Catt S, Ramachandran P, et al. Female ageing affects the DNA repair capacity of oocytes in IVF using a controlled model of sperm DNA damage in mice. Hum Reprod. 2020;35(3):529–44.
94. Grøndahl ML, Yding Andersen C, Bogstad J, et al. Gene expression profiles of single human mature oocytes in relation to age. Hum Reprod. 2010;25(4):957–68.
95. Marchetti F, Essers J, Kanaar R, et al. Disruption of maternal DNA repair increases sperm-derived chromosomal aberrations. Proc Natl Acad Sci USA. 2007;104(45):17725–9.
96. Fernández JL, Muriel L, Rivero MT, et al. The sperm chromatin dispersion test: a simple method for the determination of sperm DNA fragmentation. J Androl. 2003 Jan;24(1):59–66.
97. Fernández JL, Muriel L, Goyanes V, et al. Halosperm® is an easy, available, and cost-effective alternative for determining sperm DNA fragmentation. Fertil Steril. 2005;84(4):860.
98. Velez de la Calle JF, Muller A, Walschaerts M, et al. Sperm deoxyribonucleic acid fragmentation as assessed by the sperm chromatin dispersion test in assisted reproductive technology programs: results of a large prospective multicenter study. Fertil Steril. 2008;90(5):1792–9.
99. Evenson DP. Sperm Chromatin Structure Assay (SCSA®). Methods Mol Biol. 2013;927:147–64.
100. Evenson DP, Jost LK, Marshall D, et al. Utility of the sperm chromatin structure assay as a diagnostic and prognostic tool in the human fertility clinic. Hum Reprod. 1999;14(4):1039–49.
101. Bungum M, Bungum L, Giwercman A. Sperm chromatin structure assay (SCSA): a tool in diagnosis and treatment of infertility. Asian J Androl. 2011;13(1):69–75.

102. Evenson DP, Kasperson K, Wixon RL. Analysis of sperm DNA fragmentation using flow cytometry and other techniques. Soc Reprod Fertil Suppl. 2007;65(August):93–113.
103. Miciński P, Pawlicki K, Wielgus E, et al. The sperm chromatin structure assay (SCSA) as prognostic factor in IVF/ICSI program. Reprod Biol. 2009;9(1):65–70.
104. Sharma R, Ahmad G, Esteves SC, et al. Terminal deoxynucleotidyl transferase dUTP nick end labeling (TUNEL) assay using bench top flow cytometer for evaluation of sperm DNA fragmentation in fertility laboratories: protocol, reference values, and quality control. J Assist Reprod Genet. 2016;33:291–300.
105. Sharma RK, Sabanegh E, Mahfouz R, et al. TUNEL as a test for sperm DNA damage in the evaluation of male infertility. Urology. 2010 Dec;76(6):1380–6.
106. Sergerie M, Laforest G, Bujan L, et al. Sperm DNA fragmentation: threshold value in male fertility. Hum Reprod. 2005;20(12):3446–51.
107. Benchaib M, Braun V, Lornage J, et al. Sperm DNA fragmentation decreases the pregnancy rate in an assisted reproductive technique. Hum Reprod. 2003;18(5):1023–8.
108. Simon L, Carrell DT. Sperm DNA damage measured by comet assay. In: Carrell D, Aston K, editors. Spermatogenesis: Methods and Protocols. Springer; 2013. p. 137–46.
109. Simon L, Lutton D, McManus J, et al. Sperm DNA damage measured by the alkaline comet assay as an independent predictor of male infertility and in vitro fertilization success. Fertil Steril. 2011;95(2):652–7.
110. Lewis SEM, Agbaje IM. Using the alkaline comet assay in prognostic tests for male infertility and assisted reproductive technology outcomes. Mutagenesis. 2008;23(3):163–70.
111. Faja F, Carlini T, Coltrinari G, et al. Human sperm motility: a molecular study of mitochondrial DNA, mitochondrial transcription factor A gene and DNA fragmentation. Mol Biol Rep. 2019;46(4):4113–21.
112. Darr CR, Moraes LE, Connon RE, et al. The relationship between mitochondrial DNA copy number and stallion sperm function. Theriogenology. 2017;94:94–9.
113. Grady JP, Murphy JL, Blakely EL, et al. Accurate measurement of mitochondrial DNA deletion level and copy number differences in human skeletal muscle. PLoS One. 2014;9(12):e114462.
114. O'Hara R, Tedone E, Ludlow A, et al. Quantitative mitochondrial DNA copy number determination using droplet digital PCR with single cell resolution: a focus on aging and cancer. Genome Res. 2019;29(11):1878–88.
115. Song GJ, Lewis V. Mitochondrial DNA integrity and copy number in sperm from infertile men. Fertil Steril. 2008;90(6):2238–44.
116. May-Panloup P, Chrétien MF, Savagner F, et al. Increased sperm mitochondrial DNA content in male infertility. Hum Reprod. 2003;18(3):550–6.
117. Wu H, Whitcomb BW, Huffman A, et al. Associations of sperm mitochondrial DNA copy number and deletion rate with fertilization and embryo development in a clinical setting. Hum Reprod. 2019;34(1):163–70.
118. Nelson WS, Prodöhl PA, Avise JC. Development and application of long-PCR for the assay of full-length animal mitochondrial DNA. Mol Ecol. 1996;5(6):807–10.
119. O'Connell M, McClure N, Lewis SEM. A comparison of mitochondrial and nuclear DNA status in testicular sperm from fertile men and those with obstructive azoospermia. Hum Reprod. 2002;17(6):1571–7.
120. Kao SH, Chao HT, Wei YH. Mitochondrial deoxyribonucleic acid 4977-bp deletion is associated with diminished fertility and motility of human sperm. Biol Reprod. 1995;52(4):729–36.
121. Amir S, Vakonaki E, Tsiminikaki K, et al. Sperm telomere length: diagnostic and prognostic biomarker in male infertility (Review). World Acad Sci J. 2020:259–63.
122. Yang Q, Zhao F, Dai S, et al. Sperm telomere length is positively associated with the quality of early embryonic development. Hum Reprod. 2015;30(8):1876–81.
123. Cariati F, Jaroudi S, Alfarawati S, et al. Investigation of sperm telomere length as a potential marker of paternal genome integrity and semen quality. Reprod Biomed Online. 2016;33(3):404–11.

124. Simoni M, Bakker E, Krausz C. EAA/EMQN best practice guidelines for molecular diagnosis of y-chromosomal microdeletions. State of the art 2004. Int J Androl. 2004;27(4):240–9.
125. Kozina V, Cappallo-Obermann H, Gromoll J, et al. A one-step real-time multiplex PCR for screening Y-chromosomal microdeletions without downstream amplicon size analysis. PLoS One. 2011;6(8):e23174.
126. Harrington CT, Lin EI, Olson MT, et al. Fundamentals of pyrosequencing. Arch Pathol Lab Med. 2013;137(9):1296–303.
127. Masser DR, Stanford DR, Freeman WM. Targeted DNA methylation analysis by next-generation sequencing. J Vis Exp. 2015;96:1–11.
128. Hammoud SS, Cairns BR, Carrell DT. Analysis of gene-specific and genome-wide sperm DNA methylation. In: Carrell D, Aston K, editors. Spermatogenesis: Methods and Protocols. Springer; 2013. p. 451–8.
129. He W, Sun Υ, Zhang S, et al. Profiling the DNA methylation patterns of imprinted genes in abnormal semen samples by next-generation bisulfite sequencing. J Assist Reprod Genet. 2020;37(9):2211–21.
130. Benchaib M, Ajina M, Lornage J, et al. Quantitation by image analysis of global DNA methylation in human spermatozoa and its prognostic value in in vitro fertilization: A preliminary study. Fertil Steril. 2003;80(4):947–53.
131. Marques PI, Fernandes S, Carvalho F, et al. DNA methylation imprinting errors in spermatogenic cells from maturation arrest azoospermic patients. Andrology. 2017;5(3):451–9.
132. Aston KI, Uren PJ, Jenkins TG, et al. Aberrant sperm DNA methylation predicts male fertility status and embryo quality. Fertil Steril. 2015;104(6):1388–97. e1–5
133. Urdinguio RG, Bayón GF, Dmitrijeva M, et al. Aberrant DNA methylation patterns of spermatozoa in men with unexplained infertility. Hum Reprod. 2015;30(5):1014–28.
134. Khambata K, Raut S, Deshpande S, et al. DNA methylation defects in spermatozoa of male partners from couples experiencing recurrent pregnancy loss. Hum Reprod. 2021;36(1):48–60.
135. Rothkamm K, Löbrich M. Evidence for a lack of DNA double-strand break repair in human cells exposed to very low x-ray doses. Proc Natl Acad Sci USA. 2003;100(9):5057–62.
136. Ivashkevich AN, Martin OA, Smith AJ, et al. γH2AX foci as a measure of DNA damage: a computational approach to automatic analysis. Mutat Res. 2011;711(1–2):49–60.
137. Sharma A, Singh K, Almasan A. Histone H2AX phosphorylation: a marker for DNA damage. In: Bjergbæk L, editor. DNA Repair Protocols, Methods in Molecular Biology. New York: Springer; 2012. p. 613–26.
138. Garolla A, Cosci I, Bertoldo A, et al. DNA double strand breaks in human spermatozoa can be predictive for assisted reproductive outcome. Reprod Biomed Online. 2015;31(1):100–7.
139. Coban O, Serdarogullari M, Yarkiner Z, et al. Investigating the level of DNA double-strand break in human spermatozoa and its relation to semen characteristics and IVF outcome using phospho-histone H2AX antibody as a biomarker. Andrology. 2019:1–6.
140. Zhong HZ, Lv FT, Deng XL, et al. Evaluating γH2AX in spermatozoa from male infertility patients. Fertil Steril. 2015;104(3):574–81.
141. London R. The structural basis of XRCC1-mediated DNA repair. DNA Repair. 2015;30:90–103.
142. Abbas Kadhum R, Abbas AW. Detection of XRCC1 expression and (8-OHdG) levels as a marker of oxidative DNA damage in individuals exposed to low dose of gamma rays. IOP Conf Ser Mater Sci Eng. 2019;557(1):012082.
143. Singh P, Mitra P, Goyal T, et al. Evaluation of DNA damage and expressions of DNA repair gene in occupationally lead exposed workers (Jodhpur, India). Biol Trace Elem Res. 2021;199(5):1707–14.
144. Singh V, Kumar Mohanty S, Verma P, et al. XRCC1 deficiency correlates with increased DNA damage and male infertility. Mutat Res – Genet Toxicol Environ Mutagen. 2019;839(January):1–8.
145. Li MW, Lloyd KCK. DNA fragmentation index (DFI) as a measure of sperm quality and fertility in mice. Sci Rep. 2020;10(1):1–11.

146. Kumaresan A, Johannisson A, Al-Essawe EM, et al. Sperm viability, reactive oxygen species, and DNA fragmentation index combined can discriminate between above- and below-average fertility bulls. J Dairy Sci. 2017 Jul 1;100(7):5824–36.
147. Waterhouse KE, Haugan T, Kommisrud E, et al. Sperm DNA damage is related to field fertility of semen from young Norwegian Red bulls. Reprod Fertil Dev. 2006;18(7):781–8.
148. Gliozzi TM, Turri F, Manes S, et al. The combination of kinetic and flow cytometric semen parameters as a tool to predict fertility in cryopreserved bull semen. Animal. 2017 Apr 11;11(11):1975–82.
149. Prinosilova P, Rybar R, Zajicova A, et al. DNA integrity in fresh, chilled and frozen-thawed canine spermatozoa. Vet Med (Praha). 2012;57(3):133–42.
150. Castro R, Morales P, Parraguez VH. Post-thawing sperm quality in Chilean purebred stallions: effect of age and seasonality. J Equine Vet Sci. 2020 Sep;1(92):103170.
151. Tang L, Rao M, Yang W, et al. Predictive value of the sperm DNA fragmentation index for low or failed IVF fertilization in men with mild-to-moderate asthenozoospermia. J Gynecol Obstet Hum Reprod. 2020;101868
152. Simon L, Brunborg G, Stevenson M, et al. Clinical significance of sperm DNA damage in assisted reproduction outcome. Hum Reprod. 2010;25(7):1594–608.
153. Green KA, Patounakis G, Dougherty MP, et al. Sperm DNA fragmentation on the day of fertilization is not associated with embryologic or clinical outcomes after IVF/ICSI. J Assist Reprod Genet. 2020;37(1):71–6.
154. Casanovas A, Ribas-Maynou J, Lara-Cerrillo S, et al. Double-stranded sperm DNA damage is a cause of delay in embryo development and can impair implantation rates. Fertil Steril. 2019;111(4):699–707.e1.
155. Ribas-Maynou J, Yeste M, Becerra-Tomás N, et al. Clinical implications of sperm DNA damage in IVF and ICSI: updated systematic review and meta-analysis. Biol Rev. 2021;1–17
156. Zhu YC, Wu TH, Li GG, et al. Decrease in fertilization and cleavage rates, but not in clinical outcomes for infertile men with AZF microdeletion of the y chromosome. Zygote. 2014;23:771–7.
157. Golin AP, Yuen W, Flannigan R. The effects of y chromosome microdeletions on in vitro fertilization outcomes, health abnormalities in offspring and recurrent pregnancy loss. Transl Androl Urol. 2021;10(3):1457–66.
158. Zhang L, Ming MJ, Li M, et al. Poor intracytoplasmic sperm injection outcome in infertile males with azoospermia factor c microdeletions. Fertil Steril. 2021;2:1–8.
159. Zhu Y, Wu T, Li G, et al. The sperm quality and clinical outcomes were not affected by sY152 deletion in Y chromosome for oligozoospermia or azoospermia men after ICSI treatment. Gene. 2015;573(2):233–8.
160. Liu XY, Wang RX, Fu Y, et al. Outcomes of intracytoplasmic sperm injection in oligozoospermic men with Y chromosome AZFb or AZFc microdeletions. Andrologia. 2017 Feb 1;49(1)
161. Nasr-Esfahani MH, Razavi S, Tavalaee M. Failed fertilization after ICSI and spermiogenic defects. Fertil Steril. 2008 Apr;89(4):892–8.
162. Benchaib M, Braun V, Ressnikof D, et al. Influence of global sperm DNA methylation on IVF results. Hum Reprod. 2005;20(3):768–73.
163. Braude P, Bolton V, Moore S. Human gene expression first occurs between the four- and eight-cell stages of preimplantation development. Nature. 1988;332:459–61.
164. Seifi-Jamadi A, Zhandi M, Kohram H, et al. Influence of seasonal differences on semen quality and subsequent embryo development of Belgian Blue bulls. Theriogenology. 2020;158:8–17.
165. Fortes MRS, Satake N, Corbet DH, et al. Sperm protamine deficiency correlates with sperm DNA damage in Bos indicus bulls. Andrology. 2014;2(3):370–8.
166. Dogan S, Vargovic P, Oliveira R, et al. Sperm protamine-status correlates to the fertility of breeding bulls. Biol Reprod. 2015;92(4): 92, 1–9.
167. Castro LS, Siqueira AFP, Hamilton TRS, et al. Effect of bovine sperm chromatin integrity evaluated using three different methods on in vitro fertility. Theriogenology. 2018;107:142–8.

168. Kipper BH, Trevizan JT, Carreira JT, et al. Sperm morphometry and chromatin condensation in Nelore bulls of different ages and their effects on IVF. Theriogenology. 2017;87:154–60.
169. Didion BA, Kasperson KM, Wixon RL, et al. Boar fertility and sperm chromatin structure status: a retrospective report. J Androl. 2009;30(6):655–60.
170. Kawase Y, Wada NA, Jishage K. Evaluation of DNA fragmentation of freeze-dried mouse sperm using a modified sperm chromatin structure assay. Theriogenology. 2009;72(8):1047–53.
171. Palazzese L, Gosálvez J, Anzalone DA, et al. DNA fragmentation in epididymal freeze-dried ram spermatozoa impairs embryo development. J Reprod Dev. 2018;64(5):393–400.
172. Crespo F, Quiñones-Pérez C, Ortiz I, et al. Seasonal variations in sperm DNA fragmentation and pregnancy rates obtained after artificial insemination with cooled-stored stallion sperm throughout the breeding season (spring and summer). Theriogenology. 2020;148:89–94.
173. Kumaresan A, Das Gupta M, Datta TK, et al. Sperm DNA integrity and male fertility in farm animals: a review. Front Vet Sci. 2020;7(321):1–15.
174. Upadhya D, Kalthur G, Kumar P, et al. Association between the extent of DNA damage in the spermatozoa, fertilization and developmental competence in preimplantation stage embryos. J Turkish Ger Gynecol Assoc. 2010;11(4):182–6.
175. Fernández-Gonzalez R, Moreira PN, Pérez-Crespo M, et al. Long-term effects of mouse intracytoplasmic sperm injection with DNA-fragmented sperm on health and behavior of adult offspring. Biol Reprod. 2008;78(4):761–72.
176. Zini A. Are sperm chromatin and DNA defects relevant in the clinic? Syst Biol Reprod Med. 2011;57(1–2):78–85.
177. Chen Q, Zhao JY, Xue X, et al. The association between sperm DNA fragmentation and reproductive outcomes following intrauterine insemination, a meta analysis. Reprod Toxicol. 2019;86:50–5.
178. Sugihara A, Van Avermaete F, Roelant E, et al. The role of sperm DNA fragmentation testing in predicting intra-uterine insemination outcome: a systematic review and meta-analysis. Eur J Obstet Gynecol Reprod Biol. 2020;244:8–15.
179. Li Z, Wang L, Cai J, et al. Correlation of sperm DNA damage with IVF and ICSI outcomes: a systematic review and meta-analysis. J Assist Reprod Genet. 2006;23(9–10):367–76.
180. Zhao J, Zhang Q, Wang Y, et al. Whether sperm deoxyribonucleic acid fragmentation has an effect on pregnancy and miscarriage after in vitro fertilization/intracytoplasmic sperm injection: a systematic review and meta-analysis. Fertil Steril. 2014;102(4):998–1005. e8
181. Simon L, Zini A, Dyachenko A, et al. A systematic review and meta-analysis to determine the effect of sperm DNA damage on in vitro fertilization and intracytoplasmic sperm injection outcome. Asian J Androl. 2017;19(1):80–90.
182. Zini A, Boman JM, Belzile E, et al. Sperm DNA damage is associated with an increased risk of pregnancy loss after IVF and ICSI: systematic review and meta-analysis. Hum Reprod. 2008;23(12):2663–8.
183. Ribas-Maynou J, García-Peiró A, Fernandez-Encinas A, et al. Double stranded sperm DNA breaks, measured by comet assay, are associated with unexplained recurrent miscarriage in couples without a female factor. PLoS One. 2012;7(9)
184. Osman A, Alsomait H, Seshadri S, et al. The effect of sperm DNA fragmentation on live birth rate after IVF or ICSI: a systematic review and meta-analysis. Reprod Biomed Online. 2015;30(2):120–7.
185. Tan J, Taskin O, Albert A, et al. Association between sperm DNA fragmentation and idiopathic recurrent pregnancy loss: a systematic review and meta-analysis. Reprod Biomed Online. 2019;38(6):951–60.
186. McQueen DB, Zhang J, Robins JC. Sperm DNA fragmentation and recurrent pregnancy loss: a systematic review and meta-analysis. Fertil Steril. 2019 Jul;112(1):54–60.e3.
187. Gonçalves C, Cunha M, Rocha E, et al. Y-chromosome microdeletions in nonobstructive azoospermia and severe oligozoospermia. Asian J Androl. 2017;19(3):338–45.
188. Aitken RJ, De Iuliis GN. Origins and consequences of DNA damage in male germ cells. Reprod Biomed Online. 2007 Jan 1;14(6):727–33.

189. Azpiazu R, Amaral A, Castillo J, et al. High-throughput sperm differential proteomics suggests that epigenetic alterations contribute to failed assisted reproduction. Hum Reprod. 2014;29(6):1225–37.
190. Douglas C, Parekh N, Kahn L, et al. A novel approach to improving the reliability of manual semen analysis: a paradigm shift in the workup of infertile men. World J Mens Health. 2021;39(2):172–85.
191. Agarwal A, Allamaneni SSR, Said TM. Chemiluminescence technique for measuring reactive oxygen species. Reprod Biomed Online. 2004;9(4):466–8.
192. Mahfouz R, Sharma R, Sharma D, et al. Diagnostic value of the total antioxidant capacity (TAC) in human seminal plasma. Fertil Steril. 2009;91(3):805–11.
193. Agarwal A, Bui AD. Oxidation-reduction potential as a new marker for oxidative stress: correlation to male infertility. Investig Clin Urol. 2017;58(6):385.
194. Yuan R, Primakoff P, Myles DG. A role for the disintegrin domain of cyritestin, a sperm surface protein belonging to the ADAM family, in mouse sperm-egg plasma membrane adhesion and fusion. J Cell Biol. 1997;137(1):105–12.
195. Nixon B, Bromfield EG, Cui J, et al. Heat shock protein A2 (HSPA2): regulatory roles in germ cell development and sperm function. In: MacPhee D, editor. Advances in Anatomy Embryology and Cell Biology. Cham: Springer; 2017. p. 67–93.
196. Heidari M, Darbani S, Darbandi M, et al. Assessing the potential of HSPA2 and ADAM2 as two biomarkers for human sperm selection. Hum Fertil. 2020;23(2):123–33.
197. Gagné JP, Moreel X, Gagné P, et al. Proteomic investigation of phosphorylation sites in poly(ADP-ribose) polymerase-1 and poly(ADP-ribose) glycohydrolase. J Proteome Res. 2009;8(2):1014–29.
198. Amours DD, Sallmann FR, Dixit VM, et al. Gain-of-function of poly(ADP-ribose) polymerase-1 upon cleavage by apoptotic proteases: implications for apoptosis. J Cell Sci. 2001;114(20):3771–8.
199. Mahfouz RZ, Sharma RK, Poenicke K, et al. Evaluation of poly(ADP-ribose) polymerase cleavage (cPARP) in ejaculated human sperm fractions after induction of apoptosis. Fertil Steril. 2009;91(5 Suppl):2210–20.
200. Kuo LJ, Yang LX. γ-H2AX- a novel biomaker for DNA double-strand breaks. In Vivo (Brooklyn). 2008;22(3):305–10.
201. Cleaver JE, Feeney L, Revet I. Phosphorylated H2Ax is not an unambiguous marker for DNA double strand breaks. Cell Cycle. 2011;10(19):3223–4.
202. Behrouzi B, Kenigsberg S, Alladin N, et al. Evaluation of potential protein biomarkers in patients with high sperm DNA damage. Syst Biol Reprod Med. 2013;59(3):153–63.
203. Li L, Li H, Tian Y, et al. Differential microRNAs expression in seminal plasma of normospermic patients with different sperm DNA fragmentation indexes. Reprod Toxicol. 2020;94:8–12.
204. Intasqui P, Camargo M, Del Giudice PT, et al. Unraveling the sperm proteome and postgenomic pathways associated with sperm nuclear DNA fragmentation. J Assist Reprod Genet. 2013 Sep 27;30(9):1187–202.

Chapter 6
Role of Infection and Leukocytes in Male Infertility

Sandipan Das, Shubhadeep Roychoudhury, Shatabhisha Roychoudhury, Ashok Agarwal, and Ralf Henkel

Abstract Male infertility is considered as a multifactorial complex reproductive illness, and male urogenital infection and inflammation are crucial etiologies contributing up to 35% of all cases. Mostly triggered by sexually transmitted diseases and uropathogens, chronic manifestation of such infection may cause irreversible infertility in the male. Male urogenital infection involves bacterial, viral, protozoal, and fungal infections many of which remain asymptomatic most of the time and are passed to the sexual partner leading to fertilization failure, pregnancy loss, and even development of illness in the offspring. The abundance of leukocytes in semen can be used as an indicator of urogenital infection. Its contribution in male infertility can be as high as 30% and the clinical condition is referred to as leukocytospermia. Seminal bacterial load together with increased leukocytes contribute to the impairment of male fertility parameters such as, sperm motility, DNA integrity, acrosome reaction, and damage sperm molecular structure. Pathophysiology of bacteriospermia-induced impairment of male infertility is probably mediated by the involvement of bacterial pathogens in the intrinsic apoptotic pathway resulting in

---

S. Das · S. Roychoudhury
Department of Life Science and Bioinformatics, Assam University, Silchar, India

S. Roychoudhury
Department of Microbiology, R. G. Kar Medical College and Hospital, Kolkata, India

Health Centre, Assam University, Silchar, India

A. Agarwal
American Center for Reproductive Medicine, Cleveland Clinic, Cleveland, OH, USA

R. Henkel (✉)
Department of Metabolism, Digestion and Reproduction, Imperial College London, London, UK

Department of Medical Bioscience, University of the Western Cape, Bellville, South Africa

American Center for Reproductive Medicine, Cleveland Clinic, Cleveland, OH, USA

Logix Pharma, Theale, Reading, UK
e-mail: r.henkel@imperial.ac.uk

© The Author(s), under exclusive license to Springer Nature Switzerland AG 2022
K. K. Kesari, S. Roychoudhury (eds.), *Oxidative Stress and Toxicity in Reproductive Biology and Medicine*, Advances in Experimental Medicine and Biology 1358, https://doi.org/10.1007/978-3-030-89340-8_6

sperm death, whereas that of seminal leukocytes operates through excessive generation of ROS. Although the application of antibiotics forms the frontline therapeutic approach, the growing resistance to antibiotics poses a concern in the management of microbes-induced male urogenital infection. Complementary and alternative medicine may offer additional management options in combating such infections. On the other hand, both broad spectrum antibiotics and antioxidant therapy have showed promising results in the management of infertile men with leukocytospermia. Use of herbal medicine may also play a promising role in the management of such patients. However, recent molecular biology techniques have noted the association of elevated levels of IL-8 with both the Chlamydial infection of the male urogenital tract as well as the clinical condition of leukocytospermia. On the basis of such common pathogenesis, further research involving advanced molecular techniques may pave the way towards the development of better diagnostic tools in the clinical management of male urogenital infection and leukocytospermia.

Keywords Male urogenital infection · Inflammation · Uropathogens · Leukocytospermia · ROS · Infertility

6.1 Introduction: Infection and Male Infertility

Infertility can be defined as the failure to attain clinical pregnancy after having unprotected sexual intercourse for one year or more [189]. Globally, approximately 15% couples of reproductive age suffer from infertility, and male factor infertility contributes to 50% of the cases [4]. Male infertility is a multifactorial, complex reproductive disease, and urogenital infections are believed to be a contributory factor in 12-35% of the cases [163], thereby affecting several parts of the urogenital system such as testis, epididymis, prostate, and accessory sex glands [105]. In men, infection and inflammatory response of the urogenital tract is considered to be a crucial etiology of infertility [155].

In 2001, the European Association of Urology (EAU) for the first time classified male urinary and genital tract infections, a way which assisted clinicians in their workup algorithm of diagnosis and management of the disease. The classes include: (i) uncomplicated cystitis, (ii) uncomplicated pyelonephritis, (iii) complicated urinary tract infection (UTI) (with or without pyelonephritis), (iv) urosepsis, (v) urethritis, and (vi) prostatitis, epididymitis, and orchitis. In their latest guideline in 2018, the EAU has further specified these categories into two broad classes: (i) complicated and (ii) uncomplicated UTIs [22, 117].

The presence of pathogens in semen [34, 88] can alter sperm quality and function both directly and indirectly [111, 140]. Bacteria can directly interact with spermatozoa [13, 41, 174, 191] or indirectly via modification of the microenvironment thereby consuming the available energy and consequently resulting in loss of sperm motility [19], by triggering pro-inflammatory cytokines such as interleukin (IL)-1β, IL-6, IL-8, and tumor necrosis factor (TNF)α and the release of high levels of reactive oxygen species (ROS) by neutrophils and macrophages [11, 38, 194] resulting in oxidative stress and infertility [2], as well as by scarring with subsequent anatomical obstructions [14, 138].

Infectious inflammatory diseases of the male genital tract are mostly triggered by ascending sexually transmitted diseases (STDs) and uropathogens. High manifestations are associated with irreversible infertility. Chronic inflammation has also been associated with reduced integrity of sperm flagellar membrane, concentration, motility, increased sperm DNA fragmentation, and apoptosis [156]. Infections can be mediated by the involvement of various microbes/microbial agents including bacteria, virus, protozoa, and fungi [26, 124]. Depending on the pathogen, male genital tract infections remain asymptomatic in 10% to 50% of the cases and are then passed to the sexual partner through intercourse resulting in fertilization failure, pregnancy loss, and in some cases even development of illness in offspring [30, 32, 49, 163] or resulting in long-term consequences for the affected men [72].

6.1.1 Cystitis and Pyelonephritis

UTIs involve both the lower (characterized by cystitis) and upper (characterized by pyelonephritis) parts of the urinary tract. Typical features of cystitis comprise the presence of dysuria (painful urination), pollakisuria (frequent urination), incomplete bladder emptying sensation, and cloudy urine [17, 78]. Pyelonephritis in the male is considered as an irreversible complicated infection often associated with underlying pathologies such as enlargement of prostate/hypertrophy, neurogenic bladder, urinary tract stones, catheterization, tumors, or immune suppression due to steroid intake or diabetes [76, 78, 100]. Most common pathogens involved in these two diseases are *Staphylococcus* sp. *Pseudomonas* sp., *Candida* sp., and some species of Enterobacteriaceae such as *Escherichia coli*, *Klebsiella* sp., and *Proteus* sp. [78].

6.1.2 Urosepsis

Urosepsis is referred to as a sepsis of the urinary tract which arises as a result of an acute infection in the genital tract. The frequency of an urosepsis is much higher in males than the females. Generally, it amounts only to approximately 9% to 31% of all sepsis cases [43] with reported mortality rates between 25% and 60% [21]. Generally, the prevalence is among aged individuals and people having history of diabetes, immunosuppression (due to steroid intake or chemotherapy), obstruction, and neurogenic bladder [22, 143]. Pathogens causing this condition include *E. coli* followed by *Klebsiella* sp., *Proteus* sp., *Serratia* sp., *Enterococcus* sp., and *Staphylococcus* sp. [78]. The intensity of the disease largely depends on the immune response of the host.

6.1.3 Urethritis

An inflammation of the urethra is called a urethritis. It can be divided into two types: i) gonococcal urethritis (GU) and ii) non-gonococcal urethritis (NGU). Symptoms include irritation at the urethral orifice, urethral discomfort with or without itching, dysuria, and urethral discharge. Sometimes, symptoms may even be absent [75]. Commonly found microbes causing urethritis are *Chlamydia trachomatis*, *Neisseria gonorrhoeae*, *Mycoplasma genitalium*, *Ureaplasma* sp., and less commonly adenovirus, herpes simplex virus (HSV), and *Trichomonas* sp. [24].

6.1.4 Prostatitis, Epididymitis, and Orchitis

Prostatitis is considered as one of the most commonly occurring disorders of the genital tract affecting males of all age groups, but middle-aged men have been reported with maximum frequency [119]. Treatment of prostatitis is very difficult as common antibiotics do not properly penetrate into the prostate and its secretions [116]. Therefore, modern fluoroquinolones are considered best for its treatment and management [9].

It is characterized by irritation, obstruction, and urogenital pain, including sexual dysfunctions such as hematospermia and ejaculatory discomfort in some cases [92, 93]. The National Institute of Health (NIH) categorized the disease into four types [85]: i) acute bacterial prostatitis (ABP), ii) chronic bacterial prostatitis (CBP), iii) chronic prostatitis/chronic pelvic pain syndrome (CPPS) with or without inflammation, and iv) asymptomatic inflammatory prostatitis (AIP). Common bacterial pathogens involved in prostatitis include *E. coli* (50% to 80%), *Klebsiella* sp. and *Proteus* sp. (10% to 30%), *Enterococcus* sp. (5% to 10%), *Pseudomonas* sp. (more than 5%), *Staphylococcus* sp., and *Streptococcus* sp. [86, 93, 115, 118].

Epididymitis is referred to as the inflammation of the epididymis, whereas testicular inflammation in known as orchitis which normally develops when inflammation in the epididymis ascends to the testicle(s) [172]. Sometimes epididymitis can also involve an orchitis and this clinical condition is called epididymo-orchitis, which is often characterized by painful swelling of the epididymis and the testes resulting from an ascending infection by sexually transmitted microorganisms or by non-sexually transmitted microbes from the urinary tract [155, 167]. Various types

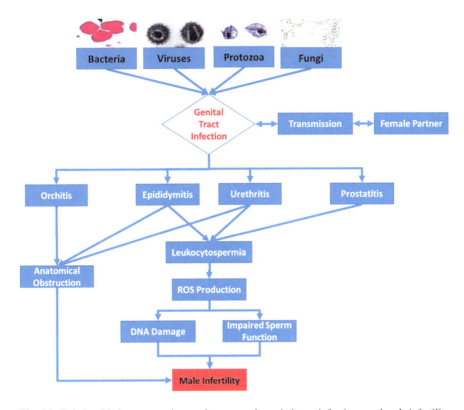

Fig. 6.1 Relationship between various pathogens, male genital tract infections, and male infertility

of male genital tract infections associated with anatomical obstructions, sperm quality alterations, and impairment of sperm function ultimately lead to male infertility [71, 72] (Fig. 6.1).

6.2 Types of Infection

6.2.1 Bacterial Infection

Although controversially reported [175], bacteriospermia may alter the semen quality as evident from its association with reduction of total sperm count, concentration, normal morphology, total motility, progressive motility, and sperm vitality [131] as well as sperm DNA fragmentation and fertilization rates in vitro [192]. Different types of bacterial species such as *Escherichia coli, Chlamydia trachomatis, Staphylococcus aureus, Staphylococcus haemolyticus, Neisseria gonorrhea,*

Ureaplasma urealyticum, or *Mycoplasma spec.* can be present in the semen of both fertile and infertile men. Infection by some of the important bacterial species and their role in male infertility is discussed below.

Escherichia coli

E. coli is mostly responsible for asexually transmitted epididymo-orchitis which is one of the possible causes of male infertility. It also affects the testes, prostate, and accessory sex glands and is possibly involved in altering the quality of semen including sperm damage [128]. In fact, it is the most frequently isolated microorganism in urogenital infections [94, 135]. In a study by Mehta et al. [103] aerobic cocci could be isolated in 49% of the analyzed ejaculates with *Streptococcus faecalis* in 53% of the positive samples. Moretti et al. [111] conducted semen culture in 246 patients and identified *Enterococcus faecalis* (32.1%) as the most common pathogen followed by E. *coli* (20.3%) as the second. The interaction between *E. coli* and spermatozoa has been shown to be mediated by mannose-receptors [181] and specific attachment organelles such as pili or type-1 fimbriae [1]. Binding of the bacteria to sperm heads and tails has been shown to result in sperm agglutination, thus impairing sperm fertilizing potential [41]. Sperm immobilization factor produced by *E.coli* is believed to be involved in alteration of morphological structure of sperm particularly curling of the tails which induces sperm immobilization rather than agglutination [135]. Incubation of sperm with *E. coli* has also been found to reduce their capability to induce acrosome reaction that may ultimately compromise the fertilization capacity [50]. In addition, soluble factors from *E. coli* affect sperm functions including mitochondrial membrane potential [154].

Chlamydia trachomatis

Chlamydia trachomatis is an obligatory intracellular Gram-negative bacterium and with an estimated 127.2 million new infections annually recognized globally as the most prevalent cause of sexually transmitted diseases (STDs) [144]. It is characterized by a broad range of clinical prevalence from sub-clinical infection to inflammatory response [137, 148]. In about 75% of the women and about 50% of the men, infections with *C. trachomatis* remain asymptomatic [32] and may consequently cause acute epididymitis, urinary tract and pelvic inflammation, ectopic pregnancy, and infant pneumonia apart from association with low progressive motility and sperm DNA fragmentation thereby leading to male infertility [68, 83, 107]. *C. trachomatis* infection has also been linked with negative reproductive outcome, particularly with an impairment of embryo implantation and/or immune rejection after uterine transfer of in vitro fertilized embryos [179]. Furthermore, seminal IL-8 levels have been associated with *C. trachomatis* infection, which may potentially be used as a biomarker for such urogenital infection in the male [83].

Staphylococcus sp.

Staphylococcus sp. are one of the most commonly isolated bacteria from the male genital tract, and *S. aureus* infection has been associated with reduced semen volume, sperm motility, morphology, vitality, and concentration [19, 126, 160, 192]. In a recent study involving 100 patients, *S. aureus* has been found to colonize 16% infertile males and their infection affected sperm quality possibly contributing to infertility in such men [53]. These authors also recommended the inclusion of *S. aureus* examination into the clinical workup algorithm of male infertility patients. Zayed et al. [192] reported on 84 bacteriospermic males with *Staphylococcus* sp. (*S. aureus*, *S. epidermidis*, *S. haemolyticus*), *E. coli*, *E. faecalis*, and *S. agalactiae* infections. This study shows that these microorganisms exert significant negative influence on total sperm motility, progressive motility, concentration, and chromatin condensation, with poor fertilization success in intracytoplasmic sperm injection (ICSI) patients.

Neisseria gonorrhoeae

This Gram-negative diplococci bacterium causes the urogenital tract infection called gonorrhea, which is with 86.9 million cases annually considered the second most prevalent sexually transmitted infection globally [144, 184]. *N. gonorrhoeae* commonly causes urethritis and epididymitis in men and has also been associated with acute epididymitis [65]. If this condition remains untreated it may lead to infertility [28]. A study on asymptomatic patients confirmed that gonorrhea infection does not change sperm quality parameters such as semen volume, sperm count, motility, and morphology, but a significant reduction in the citric acid level has been observed which may result in prostatitis in affected men [130].

Ureaplasma sp.

There are two species (*Ureaplasma urealyticum* and *Ureaplasma parvum*) that normally colonize the male reproductive tract and can then be found in the ejaculate [190]. Even though both species are pathogenic, it is mainly *U. urealyticum* that is causing urethritis, pelvic inflammatory disease, or infertility [64, 153, 177]. The pathophysiology of *Ureaplasma* sp. infection in male infertility is not completely known yet [195]. So far, *Ureaplasma urealyticum* infection has been found to exert negative impact on semen quality such as, higher seminal viscosity, lower pH, reduced sperm concentration and sperm count [176]. Another in vitro study also confirmed the potential of *Ureplasma* sp. in deteriorating sperm quality including decline in sperm motility and alteration of membrane integrity [123]. In addition, Potts et al. [134] reported significantly increased seminal reactive oxygen species (ROS) levels that can cause more sperm DNA fragmentation [139].

***Mycoplasma* sp.**

In the genus of Mycoplasma also two pathogens, *Mycoplasma hominis* and *Mycoplasma genitalium*, are associated with genital tract infections and infertility [10, 36]. Although the presence of *Mycoplasma* sp. in male urogenital tract is considered to be commensal, it may colonize the urinary tract without any visible symptom and accelerate infection possibly impairing the fertility status of affected men [6, 7]. Infection incidences for *M. hominis* and *M. genitalium* have been reported with 10.8% and 5%, respectively [64]. On the other hand, incidences between 19% and 41% have been reported for patients with recurrent urethritis for *M. genitalium* [178]. *Mycoplasma hominis* has been shown to influence semen quality negatively including sperm count, total motility, progressive motility, and morphology in infected males. After antimicrobial treatment these parameters showed improvement suggesting the potential involvement of *Mycoplasma hominis* in male infertility [7]. *Mycoplasma genitalium* is also capable of binding with sperm head, mid-piece, and tail thereby making it immobile upon binding of large numbers of pathogens. When bound, this bacterium may even travel along with motile sperm in some cases. This may lead to transmission of infection to the female partner during sexual intercourse and possibly impairs the fertilization capacity mainly by sperm immobilization [168].

6.3 Viral Infections

Viral infections in the male urogenital tract have been associated with the potential capacity of transmission to the female partner and even impairment of fertility by affecting several parts of the male reproductive system. Longer persistence of viruses in semen in comparison to other body fluids probably indicates their presence in male genital organs. This needs adequate research focus for designing proper diagnostic and management approaches of infected men [51, 95].

6.3.1 Mumps Virus (MuV)

The Mumps virus is an RNA virus belonging to the Paramyxoviridae family and often associated with parotid gland inflammation along with additional complications such as orchitis and epididymo-orchitis [33, 77]. Mumps-orchitis is capable of inducing testicular cancer, too [63, 132]. Although the exact pathophysiology is not understood, Sertoli cells appear as the prominent targets of MuV infection. A mouse model shows that the infection disrupts the blood–testis barrier by induction of TNF-α in the Sertoli cells, which is believed to be one of the possible mechanisms of MuV-induced male infertility [185]. Reduction in testis size has also been associated with impairment of spermatogenesis [16, 33]. After 3 months of infection, mumps-orchitis was found to negatively influence sperm motility and morphology although the effect on sperm count was negligible [16]. In post-pubescent men, MuV infection is frequently observed together with epididymo-orchitis which is also associated with a reduction in sperm count and increased morphological abnormalities of spermatozoa. Furthermore, the presence of anti-sperm antibodies in semen after 84 days of onset of the infection has been associated with male infertility [77].

6.3.2 Human Papilloma Virus (HPV)

Although HPV infections are mainly known to cause pregnancy complications and cervical cancer [66], in men, the virus is evidently decreasing sperm motility and increasing sperm DNA damage [57, 142]. A proposed mechanism of HPV-induced male infertility includes compromised semen quality parameters, production of anti-sperm antibody leading to reduction in sperm motility, thwarting of sperm–oocyte interaction, and hampering the sperm DNA fidelity [95]. HPV is one of the most common causes of sexually transmitted infections and has been associated with abnormal semen quality parameters [79]. The prevalence of HPV-positive semen as a risk factor for infertility has been suggested in the male in a meta-analysis [97]. Recently, men infected with HPV have been shown to possess low number of sperm with normal morphology and reduced motility which are important parameters for successful fertilization [108]. Furthermore, a number of complications including sperm apoptosis, reduced count, increased anti-sperm antibody production, and miscarriage have been associated with HPV infection, which assumes greater significance from the perspective of male fertility evaluation [164].

6.3.3 Human Immunodeficiency Virus (HIV)

While HIV infection can affect people belonging to any age group, it has been estimated to be as high as 86% among the age group between 15 to 44 years. Although HIV infection is not curable, proper treatment strategies can help to some extent [54]. HIV type-1 is the most common infection in men and women of reproductive age group; however, the accurate origin of HIV-1 in the male genital tract has not been confirmed yet [173]. It has been identified with altered semen characteristics including notable reduction in semen volume, sperm progressive motility, total count [45], concentration, and normal morphology in comparison to unaffected healthy men [121].

6.3.4 Zika Virus (ZIKV)

Zika virus is a virus of the Flaviviridae family and its infection spreads commonly through the *Aedes* mosquito vector. However, it may also be transmitted through sexual intercourse [110] thereby posing a new threat to male fertility [95]. Persistence of Zika virus in the male urogenital tract is not completely understood yet, but infection is likely to alter semen quality parameters and sperm function, in some cases leading to symptoms such as hematospermia, uncomfortable ejaculation, prostatitis, and penile discharge [87]. In a murine model, it has also been shown that the virus can actively reproduce in the male reproductive organs and can cause orchitis [51]. Here, it appears that ZIKV specifically infects early spermatogenic stages and Sertoli cells [98]. Moreover, the virus is persistent in semen with intermittent shedding [62] also in patients with non-obstructive azoospermia [61]. A recent ex vivo study demonstrated the replication of Zika virus in human testicular tissues including germ cells apart from a vast range of somatic cell types. Moreover, propagation of viral shedding in human semen indicated towards its replication in the genital tract of the male [101], which is perceived as a possible emerging threat to male fertility. Another study on 15 Zika virus-infected men revealed deterioration of semen characteristics including decreased sperm count, and the presence of virus in motile sperm has been found to increase the chance of viral transmission during sexual intercourse or in assisted reproductive processes [80].

6.3.5 SARS-CoV-2

The Severe acute respiratory syndrome coronavirus 2 (SARS-CoV-2) is an RNA virus of the Coronaviridae family that causes the corona virus disease-2019 (COVID-19). SARS-CoV-2 enters human cells via the angiotensin converting enzyme-2 (ACE2). Since the male reproductive system, specifically Leydig cells, Sertoli cells, and early spermatogenic cells, highly express this receptor protein, the male reproductive system is affected by this infection and sexual transmission might be possible [91] and negative long-term effects on male reproductive health is not excluded. However, negative short-term effects of COVID-19 on spermatogenesis have been shown in a multicenter study involving 69 patients [52], and more controlled studies are necessary in order to come forward with solid conclusions.

6.3.6 Protozoan Infection

Among protozoal infections affecting male reproductive health, *Trichomonas vaginalis* is with annual 248 million new infections [183] the most prevalent protozoa found in urethra, epididymis, and prostate of males, and it has also been associated with subsequent infection [74, 161]. This anaerobic parasite can impair male fertility potential by hindering sperm function and quality [106]. In patients with non-obstructive azoospermia, *Trichomonas vaginalis* has been shown to infect the testes thereby impairing the process of spermatogenesis and ultimately leading to infertility in such men [67]. An early study by Gopalkrishnan et al. [69] showed that the infection leads to significant impairments of sperm motility, viability, and morphology. The infection could possibly have even negative effects on the success of reproductive outcome in assisted reproduction as has been indicated in a case study by Lucena et al. [96] involving an asthenozoospermic patient.

6.3.7 Fungal Infection

In recent times, *Candida* sp. has been associated with male infertility. The infection mediated by *Candida* sp. is known as semen candidiasis, and it can alter semen quality parameters leading to male infertility [27]. *Candida albicans* possesses the ability to reduce sperm motility by binding with the sperm cell and lower the mitochondrial membrane potential and induce sperm cell apoptosis [26]. Another in vitro study confirmed the association of yeast with reduction in sperm motility and normal morphology [151].

6.4 Management of Bacterial Infection

Generally, the treatment of any infection depends on the type of pathogen. However, since male genital tract infections often remain dormant and asymptomatic, this is making their diagnosis quite difficult. Furthermore, the growing antibiotic resistance poses a concern in the management of microorganism-induced male genital infection [163].

For treatment of bacterial infections such as cystitis, trimethoprim-sulphamethoxazole is recommended, sometimes fluoroquinolones are also used to cure the disease. A broad range of antibiotics can be used to treat urosepsis such as cefotaxime, ceftazidime, piperacillin/tazobactam, ceftolozane/tazobactam, ceftazidime/avibactam, imipenem/cilastatin, and meropenem [22]. Intramuscular ceftriaxone along with oral administration of doxycycline is generally prescribed to treat epididymo-orchitis. If the infection is caused by *Chlamydia* sp. or by pathogens other than gonococci, oral administration of either ofloxacin or doxycycline can be taken for the management of the condition [166].

Complementary and alternative medicine also offer additional management options to control such infections. The ethanolic essence of *Zingiber officinale* and *Punica granatum* has shown to possess strong antibacterial effects against *E. coli*. Similarly, *Ocimum sanctum* also showed strong preventive action against *Klebsiella pneumoniae* and *Enterococcus faecalis*. *Terminalia chebula* and *Azadirachta indica* can also inhibit *Klebsiella pneumoniae* and *Enterococcus faecalis*, respectively [159]. Further validation of herbal products holds promise towards standardization of herbal management approaches of such urogenital infections.

6.5 Leukocytes and Male Infertility

Leukocytes are the primary components of the defense mechanism of the body which impart protection against foreign invaders. Leukocytes include granulocytes, lymphocytes, and monocytes, and the granulocytes are further divided into basophils, eosinophils, and neutrophils. In fact, they accumulate at the sites of infection and eradicate pathogens. In particular, neutrophils can attack the lipid membrane of pathogens through the production ROS [129].

Leukocytes occur throughout the male urogenital system and are commonly found in the seminal ejaculates of both fertile and infertile men irrespective of the history of infection [47, 147, 149, 171]. The abundance of leukocytes in semen can be used as an indicator of urogenital infection and inflammation (Fig. 6.2).

Following idiopathic infertility and varicocele, male genital tract infections are with 11.6% the third most common cause, but potentially correctable [177] of male

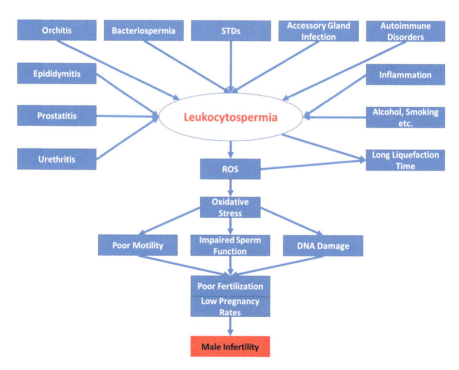

Fig. 6.2 Leukocytospermia as central point of reactive oxygen species (ROS) and oxidative stress-induced sperm dysfunction, poor fertilization, and male infertility

infertility [122]. As a possible reason, excessive amounts of seminal leukocytes, which physiologically defend the infection and have been reported to produce high amounts of reactive oxygen species (ROS) [3, 8, 73, 158], have been shown to decrease sperm motility and increase sperm DNA fragmentation [56, 113]. Although the seminal concentration of leukocytes should not exceed 1×10^6 leukocytes/mL [182], the clinical value of leukocytospermia in predicting a man's fertility status is still debated [5, 39, 90]. A concentration of seminal polymorphonuclear neutrophils of more than 0.6×10^6/ml has been linked with deleterious effects on sperm quality including ROS-induced reduction in sperm motility [84]. Excessive seminal ROS levels induce damage to the sperm plasma membrane integrity, DNA fidelity, and apoptosis while reducing the sperm count [2]. On the other hand, seminal leukocyte concentrations of $<0.2\times10^6$ leukocytes/ml have been suggested to be sensitive for the clinical evaluation [136]. However, even concentrations as low as 0.1×10^6 leukocytes/ml semen may contribute to ROS-induced sperm DNA damage [3]. Henkel et al. [73] associated significantly lower sperm motility and increased seminal ROS production and sperm DNA fragmentation with leukocyte concentrations of $<0.1\times10^6$/ml.

Microbial infections in the urogenital tract pose a serious threat to a man's fertility potential through deterioration of sperm quality and function [18, 140, 165], alteration of the process of spermatogenesis [18], scarring and obstruction of seminal tract [18, 70, 72, 102], and interference with fertilization process [165]. Infections have been associated with high seminal pH and hyperviscosity [44], low semen volume [99], sperm motility, and vitality [150] as well as poor reproductive outcome after IVF [140].

6.6 Bacteriospermia

The presence of bacteria in semen is considered to be the most harmful among the infectious etiologies [145]. Infections mediated by bacteria affect the male urogenital system including the epididymitis, prostate, testis, and urethra [131, 146, 169]. Bacterial infections compromise sperm quality [111] and function via deterioration of motility and morphology, generation of anti-sperm antibodies, and triggering inflammatory response [191]. Most prominent bacterial species identified from semen include *E. coli, Staphylococcus aureus, Klebsiella* sp., *Chlamydia trachomatis, Mycoplasma hominis,* and *Enterococcus faecalis*. The presence of seminal bacterial species has been reported in 15% infertile men which is characterized by the presence of >1000 colony forming units of bacteria per ml semen. This condition is clinically termed as bacteriospermia [42, 146].

Infections mediated by bacteria have been associated with reduced sperm parameters, sperm DNA damage, and poor IVF outcomes [111, 191]. A bacterial load in semen is also linked with increased seminal leukocyte concentrations, and together they contribute to the impairment of male fertility by compromising sperm concentration, motility, DNA integrity and morphology [42]. The pathophysiology of bacteriospermia-induced impairment of sperm quality is probably mediated by the involvement of bacterial pathogens in the intrinsic apoptotic pathway that ultimately results in sperm death, while that of leukocytospermia operates through the induction of oxidative stress leading to altered semen characteristics [60]. Leukocytospermia-induced subfertility in the male is not only characterized by the production of ROS but also by pro-inflammatory cytokines and bacterial adhesion to sperm leading to sperm agglutination [41, 46, 59]. Bacteriospermia also reduces the binding ability of sperm to oocytes via promotion of peroxidative damage to sperm cell membrane which further accelerates leukocyte-induced oxidative stress [58] (Fig. 6.2).

6.7 Etiology

It is essential to understand the causes of leukocytospermia in order to prevent the adverse effects of excess leukocytes in semen. Leukocytes are usually present in the urogenital tract of the male and do not harm the semen quality as can be seen by the large overlap in leukocyte numbers between fertile and infertile men [180]. Only if a threshold of activated leukocytes is exceeded and/or the natural seminal defense mechanisms are exhausted, male fertility is impaired by the induction of various pathologies. Urogenital infection is the one of the most prominent causes of excessive level of leukocytes in semen and such increase is triggered by seminal infection and inflammation [29]. Leukocytes are major sources for generation of free radicals in the male reproductive system and are regarded as one of the major etiologies of male infertility. Other etiologies of leukocytospermia include lifestyle factors such as smoking, consumption of alcohol, and marijuana [31], use of lubricants during sexual intercourse, illicit drugs, exposure to toxins, autoimmune disorders, varicocele, history of vasovasostomy, urethroplasty, etc. [15, 42, 104] (Fig. 6.2).

6.8 Pathophysiology

During the 1980s and 1990s, leukocytes were believed to be involved only in the process of phagocytosis of deformed sperm. However, other reports have revealed the implications of leukocytospermia in male infertility in particular and male reproductive health in general [149]. In fact, leukocytospermic men have hyperviscous semen, longer liquefaction time [113], lower sperm concentration [112, 187], total motility [112, 113], progressive motility [187], viability, morphology [112, 170] as well as lower ICSI outcomes (including fertilization and embryo development rates) [187]. Leukocytospermia has also been associated with impairment of sperm acrosome, tail abnormalities, DNA fragmentation, [12], damage to primary transcripts and telomere of chromosomes, and may further result into recurrent pregnancy loss which may also be accompanied with congenital deformities, neuropsychiatric disorders, or even cancer in the offspring [20].

Impairment of fertility parameters in leukocytospermic men may be attributed to the levels of excessive ROS generated by the seminal leukocytes [42]. The sperm plasma membrane being extraordinarily rich in polyunsaturated fatty acids makes the male germ cells more susceptible towards ROS-mediated deprivation [127, 149]. During infection, inflammation, or autoimmune diseases the leukocyte concentration increases in the male urogenital tract and the generation of ROS increases, causing peroxidation of membrane lipids and further decrease of the production of ATP through disruption of mitochondrial membrane potential [133, 141]. Such decline in ATP production is directly correlated with reduction of sperm motility and other sperm functions [35]. Simultaneously, excessive seminal leukocytes stimulate the sperm to produce ROS through cell-to-cell contact or through secretory products that again impair sperm quality including reduced motility and higher DNA damage [147].

6.9 Management of Leukocytospermia

Seminal leukocytes play an important role in male infertility. Therefore, the presence of leukocytes and the underlying disease (infection or lifestyle condition) need to be addressed for proper management. For the infections, use of broad spectrum antibiotics together with antioxidant therapy has shown promising results in the management of infertile patients with leukocytospermia. Antibiotics, by treating an underlying cause of elevated leukocyte counts, are believed to improve the overall semen quality parameters whereas antioxidants accelerate sperm function [81]. Antibiotics such as doxycycline, erythromycin, and trimethoprim/sulfamethaxazole have demonstrated significant improvement of sperm quality in leukocytospermic men, including sperm morphology, motility, and concentration [149, 162]. Levofloxacin also enhances both the macroscopic and microscopic parameters of semen quality including sperm motility in leukocytospermic patients [55], whereas ketotifen has shown promising results in enhancing sperm motility and morphology [125]. Similarly, use of anti-inflammatory drugs, particularly the Cox-2 inhibitors, can reduce seminal leukocytes and improve the sperm count [89, 109]. Sometimes, leukocytospermic men are advised to ejaculate frequently, while antibiotics have been found to minimize the seminal leukocytes, too [23]. Furthermore, herbal medicines may also play a promising role in the management of leukocytospermic patients as the supplementation of curcumin extract in vitro has been reported to enhance sperm motility in such men [193]. Similarly, the use of the dietary bioflavonoid quercetin can alleviate leukocyte-mediated oxidative stress and hence poses an alternative therapeutic option in leukocytospermic patients [40].

6.10 Perspective

The EAU guidelines recommend the evaluation of urogenital infections in the male, which includes medical history, urine analysis (dipstick), physical test, enumeration of leukocytes and erythrocytes, and nitrite reaction test [22]. On the other hand, the Canadian Urology Association asks for medical history of the patient followed by physical examination including a digital rectal examination and urine analysis as a mandatory diagnosis for lower urinary tract symptoms in the male [120]. However, whereas in the case of a chronic bacterial prostatitis, a positive semen culture may be sufficient to initiate an antibiotic treatment, a negative culture does not exclude the disease. Urine cultures, on the other hand, are usually not enough for the detection of bacteria [188]. In such cases, molecular biology techniques, for instance polymerase chain reaction (PCR)-based amplified ribosomal RNA gene regions maybe used to detect and isolate bacteria from semen samples of infertile patients thereby paving the way for the incorporation of these advanced tests in workup algorithms of the evaluation and subsequent management of urogenital infections in the male [82]. The use of real-time PCR (qPCR)-based approaches to quantify bacteria in semen also yielded promising results [157]. These PCR-based assays are more sensitive to diagnose the presence of microbes in semen. Another advanced approach based on a diafiltration matrix-assisted laser desorption/ionization time-of-flight mass spectrometry (MALDI-TOF MS) for identification of uropathogens in urine samples of males not only minimizes the currently required time of 24-48 hours to identify the microbes to as low as 2-3 hours but also maintains high accuracy [37]. Recently, next generation sequencing of DNA has been used experimentally for better diagnosis and individualized prevention and treatment of such men by detecting pathogens with resistant genes associated with urogenital infections [114].

Male urogenital infections are also associated with excessive seminal leukocytes, which in turn impair male fertility [25, 147]. Presence of immature germ cells along with leukocytes makes the identification of leukocytospermia difficult by using routine staining methods, hence, monoclonal antibodies to identify specific leukocyte antigens may be used as advanced diagnostic tool in clinical cases of leukocytospermia [186]. Recent studies have linked interleukin-6 (IL-6), interleukin-8 (IL-8), and tumor necrosis factor-α (TNF-α) with poor semen quality in leukocytospermic patients thereby indicating the possibility of emergence of IL-6, IL-8, and TNF-α as alternative diagnostic markers in the diagnosis of leukocytospermia in infertile patients [5, 48, 152]. However, molecular biology techniques also showed an association of elevated levels of IL-8 with both Chlamydial infection of the male urogenital tract and the clinical condition of leukocytospermia [83, 152]. Based on such common pathogenesis, further research involving advanced molecular techniques may pave the way towards the development of better diagnostic tools in the clinical management of male urogenital infection and leukocytospermia.

References

1. Abul-Milh M, Wu Y, Lau B, Lingwood CA, Barnett Foster D. Induction of epithelial cell death including apoptosis by enteropathogenic Escherichia coli expressing bundle-forming pili. Infect Immun. 2001;69:7356–64.
2. Agarwal A, Saleh RA, Bedaiwy MA. Role of reactive oxygen species in the pathophysiology of human reproduction. Fertil Steril. 2003;79:829–43.
3. Agarwal A, Mulgund A, Alshahrani S, Assidi M, Abuzenadah AM, Sharma R, Sabanegh E. Reactive oxygen species and sperm DNA damage in infertile men presenting with low level leukocytospermia. Reprod Biol Endocrinol. 2014;12:126.
4. Agarwal A, Mulgund A, Hamada A, Chyatte MR. A unique view on male infertility around the globe. Reprod Biol Endocrinol. 2015;13:37.
5. Aghazarian A, Stancik I, Pflüger H, Lackner J. Influence of pathogens and moderate leukocytes on seminal interleukin (IL)-6, IL-8, and sperm parameters. Int Urol Nephrol. 2013;45:359–65.
6. Ahmadi MH, Mirsalehian A, Bahador A. Prevalence of urogenital Mycoplasmas in Iran and their effects on fertility potential: A systematic review and meta-analysis. Iran J Public Health. 2016;45:409–22.
7. Ahmadi MH, Mirsalehian A, Sadighi Gilani MA, Bahador A, Talebi M. Asymptomatic infection with Mycoplasma hominis negatively affects semen parameters and leads to male infertility as confirmed by improved semen parameters after antibiotic treatment. Urology. 2017;100:97–102.
8. Aitken RJ, West KM. Analysis of the relationship between reactive oxygen species production and leukocyte infiltration in fractions of human semen separated on Percoll gradients. Int J Androl. 1990;13:433–51.
9. Almugbel SK, Alanezi FKB, Alhoshan FM, Alkhalifa RO, Alkhzaim AH, Almohideb MA. Classification and treatment of prostatitis: a review of literature. Int J Commun Med Pub Health. 2018;5:4941–6.
10. Andrade-Rocha FT. Ureaplasma urealyticum and Mycoplasma hominis in men attending for routine semen analysis. Prevalence, incidence by age and clinical settings, influence on sperm characteristics, relationship with the leukocyte count and clinical value. Urol Int. 2003;71:377–81.
11. Azenabor A, Ekun AO, Akinloye O. Impact of inflammation on male reproductive tract. J Reprod Infertil. 2015;16:123–9.
12. Aziz N, Agarwal A, Lewis-Jones I, Sharma RK, Thomas AJ Jr. Novel associations between specific sperm morphological defects and leukocytospermia. Fertil Steril. 2004;82:621–7.
13. Barbonetti A, Vassallo MRC, Cinque B, Filipponi S, Mastromarino P, Cifone MG, Francavilla S, Francavilla F. Soluble products of Escherichia coli induce mitochondrial dysfunction-related sperm membrane lipid peroxidation which is prevented by lactobacilli. PLoS One. 2013;8:e83136.
14. Bar-Chama N, Fisch H. Infection and pyospermia in male infertility. World J Urol. 1993;11:76–81.
15. Barratt CLR, Bolton AE, Cooke ID. Functional significance of white blood cells in the male and female reproductive tract. Hum Reprod. 1990;5:639–48.
16. Bartak V. Sperm count, morphology and motility after unilateral mumps orchitis. Reprod. 1973;32:491–4.
17. Beahm NP, Nicolle LE, Bursey A, Smyth DJ, Tsuyuki RT. The assessment and management of urinary tract infections in adults: Guidelines for pharmacists. Can Pharm J (Ott). 2017;150:298–305.
18. Berjis K, Ghiasi M, Sangy S. Study of seminal infection among an infertile male population in Qom, Iran, and its effect on sperm quality. Iran J Microbiol. 2018;10:111–6.

19. Berktas M, Aydin S, Yilmaz Y, Cecen K, Bozkurt H. Sperm motility changes after coincubation with various uropathogenic microorganisms: an in vitro experimental study. Int Urol Nephrol. 2008;40:383–9.
20. Bisht S, Faiq M, Tolahunase M, Dada R. Oxidative stress and male infertility. Nat Rev Urol. 2017;14:470–85.
21. Bone RC, Bone-Larson C. Gram-negative urinary tract infections and the development of SIRS. J Crit Ill. 1996;11(suppl):S20–9.
22. Bonkat G, Bartoletti R, Bruyère F, Cai T, Geerlings SE, Köves B, Schubert S, Wagenlehner F, Mezei T, Pilatz A, Pradere B, Veeratterapillay R. EAU guidelines on urological infections. EAU Guidelines. Edn. presented at the EAU Annual Congress Amsterdam the Netherlands 2020; 2021. ISBN 978-94-92671-07-3
23. Branigan EF, Muller CH. Efficacy of treatment and recurrence rate of leukocytospermia in infertile men with prostatitis. Fertil Steril. 1994;62:580–4.
24. Brill JR. Diagnosis and treatment of urethritis in men. Am Fam Phys. 2010;81:873–8.
25. Brunner RJ, Demeter JH, Sindhwani P. Review of guidelines for the evaluation and treatment of leukocytospermia in male infertility. World J Mens Health. 2019;37:128–37.
26. Burrello N, Salmeri M, Perdichizzi A, Bellanca S, Pettinato G, D'Agata R, Vicari E, Calogero AE. Candida albicans experimental infection: effects on human sperm motility, mitochondrial membrane potential and apoptosis. Reprod Biomed Online. 2009;18:496–501.
27. Castrillón-Duque EX, Suárez JP, Cardona Maya WD. Yeast and fertility: effects of in vitro activity of Candida spp. on sperm quality. J Reprod Infertil. 2018;19:49–55.
28. Chemaitelly H, Harfouche M, Blondeel K, Matsaseng TC, Kiarie J, Toskin I, Abu-Raddad LJ. Global epidemiology of Neisseria gonorrhoeae in infertile populations: protocol for a systematic review. MJ Open. 2019;9:e025808.
29. Chen SJ, Haidl G. Male genital tract infections and Leukocytospermia. In: Henkel R, Samanta L, Agarwal A, editors. Oxidants, antioxidants and impact of the oxidative status in male reproduction. London/San Diego/Cambridge/Oxford: Academic Press; 2018. p. 101–4.
30. Chen MY, Rohrsheim R, Donovan B. The differing profiles of symptomatic and asymptomatic Chlamydia trachomatis-infected men in a clinical setting. Int J STD AIDS. 2007;18:384–8.
31. Close CE, Roberts PL, Berger RE. Cigarettes, alcohol and marijuana are related to pyospermia in infertile men. J Urol. 1990;144:900–3.
32. Cunningham KA, Beagley KW. Male genital tract chlamydial infection: implications for pathology and infertility. Biol Reprod. 2008;79:180–9.
33. Davis NF, McGuire BB, Mahon JA, Smyth AE, O'Malley KJ, Fitzpatrick JM. The increasing incidence of mumps orchitis: a comprehensive review. BJU Int. 2010;105:1060–5.
34. De Francesco MA, Negrini R, Ravizzola G, Galli P, Manca N. Bacterial species present in the lower male genital tract: a five-year retrospective study. Eur J Contracep Reprod Health Care. 2011;16:47–53.
35. De Lamirande E, Gagnon C. Reactive oxygen species and human spermatozoa. II. Depletion of adenosine triphosphate plays an important role in the inhibition of sperm motility. J Androl. 1992;13:379–86.
36. Deguchi T, Maeda S. Mycoplasma genitalium: another important pathogen of nongonococcal urethritis. J Urol. 2002;167:1210–7.
37. DeMarco ML, Burnham CAD. Diafiltration MALDI-TOF mass spectrometry method for culture-independent detection and identification of pathogens directly from urine specimens. Am J Clin Pathol. 2014;141:204–12.
38. Depuydt CE, Bosmans E, Zalata A, Schoonjans F, Comhaire FH. The relation between reactive oxygen species and cytokines in andrological patients with or without male accessory gland infection. J Androl. 1996;17:699–707.
39. Derbel R, Sellami H, Sakka R, Ben Slima A, Mkaddem I, Gdoura R, Mcelreavey E, Ammar-Keskes L. Relationship between nuclear DNA fragmentation, mitochondrial DNA damage and standard sperm parameters in spermatozoa of infertile patients with leukocytospermia. J Gynecol Obstet Hum Reprod. 2021:102101. https://doi.org/10.1016/j.jogoh.2021.102101. Online ahead of print

40. Diao R, Gan H, Tian F, Cai X, Zhen W, Song X, Duan YG. In vitro antioxidation effect of Quercetin on sperm function from the infertile patients with leukocytospermia. Am J Reprod Immunol. 2019;82:e13155.
41. Diemer T, Weidner W, Michelmann HW, Schiefer HG, Rovan E, Mayer F. Influence of Escherichia coli on motility parameters of human spermatozoa in vitro. Int J Androl. 1996;19:271–7.
42. Domes T, Lo KC, Grober ED, Mullen JBM, Mazzulli T, Jarvi K. The incidence and effect of bacteriospermia and elevated seminal leukocytes on seminal parameter. Fertil Steril. 2012;97:1050–5.
43. Dreger NM, Degener S, Ahmad-Nejad P, Wöbker G, Roth S. Urosepsis–etiology, diagnosis, and treatment. Dtsch Arztebl Int. 2015;112:837–48.
44. Du Plessis SS, Gokul S, Agarwal A. Semen hyperviscosity: causes, consequences, and cures. Front Biosci (Elite Ed). 2013;5:224–31.
45. Dulioust E, Le Du A, Costagliola D, Guibert J, Kunstmann JM, Heard I, Juillard JC, Salmon D, Leruez-Ville M, Mandelbrot L, Rouzioux C, Sicard D, Zorn JR, Jouannet P, De Almeida M. Semen alterations in HIV-1 infected men. Hum Reprod. 2002;17:2112–8.
46. Eggert-Kruse W, Boit R, Rohr G, Aufenanger J, Hund M, Strowitzki T. Relationship of seminal plasma interleukin (IL)-8 and IL-6 with semen quality. Hum Reprod. 2001;16:517–28.
47. El Demiry MI, Hargreave TB, Busuttil A, Elton RE, James K, Chisholm GD. Immunocompetent cells in human testis in health and disease. Fertil Steril. 1987;48:470–9.
48. Eldamnhoury EM, Elatrash GA, Rashwan HM, El-Sakka AI. Association between leukocytospermia and semen interleukin-6 and tumor necrosis factor-alpha in infertile men. Andrology. 2018;6:775–80.
49. Elias J, Frosch M, Vogel U. Neisseria. In: Versalovic J, Carroll KC, Funke G, Jorgensen JH, editors. Manual of clinical microbiology, vol. 1. 10th ed. Washington, DC: ASM Press; 2011. p. 559–69.
50. El-Mulla KF, Kohn FM, Dandal M, El Beheiry AH, Schiefer HG, Weidner W, Schill WB. In vitro effect of Escherichia coli on human sperm acrosome reaction. Arch Androl. 1996;37:73–8.
51. Epelboin S, Dulioust E, Epelboin L, Benachi A, Merlet F, Patrat C. Zika virus and reproduction: facts, questions and current management. Hum Reprod Update. 2017;23:629–45.
52. Erbay G, Şanlı A, Türel H, Yavuz U, Erdoğan A, Karabakan M, Yariş M, Gültekin MH. Short-term effects of COVID-19 on semen parameters: a multicenter study of 69 cases. Andrology. 2021;2021 https://doi.org/10.1111/andr.13019.
53. Esmailkhani Esmailkhani A, Akhi MT, Sadeghi J, Niknafs B, Bialvaei AZ, Farzadi L, Safadel N. Assessing the prevalence of Staphylococcus aureus in infertile male patients in Tabriz, northwest Iran. Int J Reprod Biomed. 2018;16:469–74.
54. Ethics Committee of the American Society for Reproductive Medicine. Human immunodeficiency virus (HIV) and infertility treatment: a committee opinion. Fertil Steril. 2015;104:e1–8.
55. Fakhrildin MBMR, Selman MO, Najim DA. Effects of different antibiotics administered to infertile men with leukocytospermia on the sperm parameters. World J Pharmacol Res. 2015;4:144–9.
56. Fariello RM, Del Giudice PT, Spaine DM, Fraietta R, Bertolla RP, Cedenho AP. Effect of leukocytospermia and processing by discontinuous density gradient on sperm nuclear DNA fragmentation and mitochondrial activity. J Assist Reprod Genet. 2009;26:151–7.
57. Foresta C, Garolla A, Zuccarello D, Pizzol D, Moretti A, Barzon L, Palù G. Human papillomavirus found in sperm head of young adult males affects the progressive motility. Fertil Steril. 2010;93:802–6.
58. Fraczek M, Kurpisz M. Mechanisms of the harmful effects of bacterial semen infection on ejaculated human spermatozoa: potential inflammatory markers in semen. Folia Histochem Cytobiol. 2015;53:201–17.
59. Fraczek M, Szumala-Kakol A, Dworacki G, Sanocka D, Kurpisz M. In vitro reconstruction of inflammatory reaction in human semen: effect on sperm DNA fragmentation. J Reprod Immunol. 2013;100:76–85.

60. Fraczek M, Hryhorowicz M, Gill K, Zarzycka M, Gaczarzewicz D, Jedrzejczak P, Bilinska B, Piasecka M, Kurpisz M. The effect of bacteriospermia and leukocytospermia on conventional and nonconventional semen parameters in healthy young normozoospermic males. J Reprod Immunol. 2016;118:18–27.
61. Fréour T, Mirallié S, Hubert B, Splingart C, Barrière P, Maquart M, Leparc-Goffart I. Sexual transmission of Zika virus in an entirely asymptomatic couple returning from a Zika epidemic area, France, April 2016. Euro Surveill. 2016;21(23)
62. Froeschl G, Huber K, von Sonnenburg F, Nothdurft HD, Bretzel G, Hoelscher M, Zoeller L, Trottmann M, Pan-Montojo F, Dobler G, Woelfel S. Long-term kinetics of Zika virus RNA and antibodies in body fluids of a vasectomized traveller returning from Martinique: a case report. BMC Infect Dis. 2017;17:55.
63. Galazka AM, Robertson SE, Kraigher A. Mumps and mumps vaccine: a global review. Bull World Health Organ. 1999;77:3–14.
64. Gdoura R, Kchaou W, Chaari C, Znazen A, Keskes L, Rebai T, Hammami A. Ureaplasma urealyticum, Ureaplasma parvum, Mycoplasma hominis and Mycoplasma genitalium infections and semen quality of infertile men. BMC Infect Dis. 2007;7:129.
65. Gimenes F, Medina FS, Abreu AL, Irie MM, Esquicati IB, Malagutti N, Vasconcellos VR, Discacciati MG, Bonini MG, Maria-Engler SS, Consolaro ME. Sensitive simultaneous detection of seven sexually transmitted agents in semen by multiplex-PCR and of HPV by single PCR. PLoS One. 2014;9:e98862.
66. Gizzo S, Ferrari B, Noventa M, Ferrari E, Patrelli TS, Gangemi M, Nardelli GB. Male and couple fertility impairment due to HPV-DNA sperm infection: update on molecular mechanism and clinical impact–systematic review. Biomed Res Int. 2014;2014:230263.
67. Gong YH, Liu Y, Li P, Zhu ZJ, Hong Y, Fu GH, Xue YJ, Xu C, Li Z. A nonobstructive azoospermic patient with Trichomonas vaginalis infection in testes. Asian J Androl. 2018;20:97–8.
68. Gonzales GF, Munoz G, Sanchez R, Henkel R, Gallegos-Avila G, Díaz-Gutierrez O, Vigil P, Vasquez F, Kortebani G, Mazzolli A, Bustos-Obregón E. Update on the impact of Chlamydia trachomatis infection on male fertility. Andrologia. 2004;36:1–23.
69. Gopalkrishnan K, Hinduja IN, Kumar TC. Semen characteristics of asymptomatic males affected by Trichomonas vaginalis. J In Vitro Fert Embryo Transf. 1990;7:165–7.
70. Harkness AH. The pathology of gonorrhoea. Br J Vener Dis. 1948;24:137–47.
71. Henkel R. Infection in infertility. In: Parekattil SJ, Esteves SC, Agarwal A, editors. Male infertility: contemporary clinical approaches, andrology, ART and antioxidants. 2nd ed. Cham: Springer; 2020. p. 409–24.
72. Henkel R. Long-term consequences of sexually transmitted infections on men's sexual function: a systematic review. Arab J Urol. 2021; (in press)
73. Henkel R, Kierspel E, Stalf T, Mehnert C, Menkveld R, Tinneberg HR, Schill WB, Kruger TF. Effect of reactive oxygen species produced by spermatozoa and leukocytes on sperm functions in non-leukocytospermic patients. Fertil Steril. 2005;83:635–42.
74. Hezarjaribi HZ, Fakhar M, Shokri A, Teshnizi SH, Sadough A, Taghavi M. Trichomonas vaginalis infection among Iranian general population of women: a systematic review and meta-analysis. Parasitol Res. 2015;114:1291–300.
75. Horner PJ, Blee K, Falk L, van der Meijden W, Moi H. 2016 European guideline on the management of non-gonococcal urethritis. Int J STD AIDS. 2016;27:928–37.
76. Ishikawa K, Matsumoto T, Yasuda M, Uehara S, Muratani T, Yagisawa M, Sato J, Niki Y, Totsuka K, Sunakawa K, Hanaki H. The nationwide study of bacterial pathogens associated with urinary tract infections conducted by the Japanese Society of Chemotherapy. J Infect Chemother. 2011;17:126–38.
77. Jalal H, Bahadur G, Knowles W, Jin L, Brink N. Mumps epididymo-orchitis with prolonged detection of virus in semen and the development of anti-sperm antibodies. J Med Virol. 2004;73:147–50.

78. Japanese Association for Infectious Disease/Japanese Society of Chemotherapy; JAID/JSC Guide/Guidelines to Clinical Management of Infectious Disease Preparing Committee; Urinary tract infection/male genital infection working group, Yamamoto S, Ishikawa K, Hayami H, Nakamura T, Miyairi I, Hoshino T, Hasui M, Tanaka S, Kiyota H, Arakawa S. JAID/JSC guidelines for clinical management of infectious disease 2015 – Urinary tract infection/male genital infection. J Infect Chemother. 2017;23:733–51.
79. Jeršovienė V, Gudlevičienė Z, Rimienė J, Butkauskas D. Human papillomavirus and infertility. Medicina (Kaunas). 2019;55:377.
80. Joguet G, Mansuy JM, Matusali G, Hamdi S, Walschaerts M, Pavili L, Guyomard S, Prisant N, Lamarre P, Dejucq-Rainsford N, Pasquier C. Effect of acute Zika virus infection on sperm and virus clearance in body fluids: a prospective observational study. Lancet Infect Dis. 2017;17:1200–8.
81. Jung JH, Kim MH, Kim J, Baik SK, Koh SB, Park HJ, Seo JT. Treatment of leukocytospermia in male infertility: a systematic review. World J Men's Health. 2016;34:165–72.
82. Kiessling AA, Desmarais BM, Yin HZ, Loverde J, Eyre RC. Detection and identification of bacterial DNA in semen. Fertil Steril. 2008;90:1744–56.
83. Kokab A, Akhondi MM, Sadeghi MR, Modarresi MH, Aarabi M, Jennings R, Pacey AA, Eley A. Raised inflammatory markers in semen from men with asymptomatic chlamydial infection. J Androl. 2010;31:114–20.
84. Kovalski NN, de Lamirande E, Gagnon C. Reactive oxygen species generated by human neutrophiles inhibit sperm motility: Protective effect of seminal plasma and scavengers. Fertil Steril. 1992;58:809–16.
85. Krieger JN, Nyberg L Jr, Nickel JC. NIH consensus definition and classification of prostatitis. Jama. 1999;282:236–7.
86. Krieger JN, Ross SO, Limaye AP, Riley DE. Inconsistent localization of gram-positive bacteria to prostate-specific specimens from patients with chronic prostatitis. Urol. 2005;66:721–5.
87. Kurscheidt FA, Mesquita CS, Damke GM, Damke E, Analine RDA, Suehiro TT, Teixeira JJ, da Silva VR, Souza RP, Consolaro ME. Persistence and clinical relevance of Zika virus in the male genital tract. Nature Rev Urol. 2019;16:211–30.
88. La Vignera S, Condorelli RA, Vicari E, Salmeri M, Morgia G, Favilla V, Cimino S, Calogero AE. Microbiological investigation in male infertility: a practical overview. J Med Microbiol. 2014;63:1–14.
89. Lackner JE, Herwig R, Schmidbauer J, Schatzl G, Kratzik C, Marberger M. Correlation of leukocytospermia with clinical infection and the positive effect of anti-inflammatory treatment on semen quality. Fertil Steril. 2006;86:601–5.
90. Lackner JE, Märk I, Sator K, Huber J, Sator M. Effect of leukocytospermia on fertilization and pregnancy rates of artificial reproductive technologies. Fertil Steril. 2008;90:869–71.
91. Li D, Jin M, Bao P, Zhao W, Zhang S. Clinical characteristics and results of semen tests among men with coronavirus disease 2019. JAMA Netw Open. 2020;3:e208292.
92. Liang CZ, Li HJ, Wang ZP, Xing JP, Hu WL, Zhang TF, Ge WW, Hao ZY, Zhang XS, Zhou J, Li Y, Zhou ZX, Tang ZG. Treatment of chronic prostatitis in Chinese men. Asian J Androl. 2009;11:153–6.
93. Lipsky BA, Byren I, Hoey CT. Treatment of bacterial prostatitis. Clin Inf Dis. 2010;50:1641–52.
94. Liu JH, Li HY, Cao ZG, Duan YF, Li Y, Ye ZQ. Influence of several uropathogenic microorganisms on human sperm motility parameters in vitro. Asian J Androl. 2002;4:179–82.
95. Liu W, Han R, Wu H, Han D. Viral threat to male fertility. Andrologia. 2018;50:e13140.
96. Lucena E, Moreno-Ortiz H, Coral L, Lombana O, Moran A, Esteban-Pérez CI. Unexplained infertility caused by a latent but serious intruder: Trichomonas vaginalis? JFIV Reprod Med Genet. 2014;3:1.
97. Lyu Z, Feng X, Li N, Zhao W, Wei L, Chen Y, Yang W, Ma H, Yao B, Zhang K, Hu Z. Human papillomavirus in semen and the risk for male infertility: a systematic review and meta-analysis. BMC Inf Dis. 2017;17:714.

98. Ma W, Li S, Ma S, Jia L, Zhang F, Zhang Y, Zhang J, Wong G, Zhang S, Lu X, Liu M, Yan J, Li W, Qin C, Han D, Qin C, Wang N, Li X, Gao GF. Zika virus causes testis damage and leads to male infertility in mice. Cell. 2016;167:1511–24.
99. Marconi M, Pilatz A, Wagenlehner F, Diemer T, Weidner W. Impact of infection on the secretory capacity of the male accessory glands. Int Braz J Urol. 2009;35:299–309.
100. Matsumoto T, Hamasuna R, Ishikawa K, Takahashi S, Yasuda M, Hayami H, Tanaka K, Kiyota H, Muratani T, Monden K, Arakawa S, Yamamoto S. Nationwide survey of antibacterial activity against clinical isolates from urinary tract infections in Japan (2008). Int J Antimicrob Agents. 2011;37:210–8.
101. Matusali G, Houzet L, Satie AP, Mahé D, Aubry F, Couderc T, Frouard J, Bourgeau S, Bensalah K, Lavoué S, Joguet G. Zika virus infects human testicular tissue and germ cells. J Clin Invest. 2018;128:4697–710.
102. McMillan A, Pakianathan M, Mao JH, Macintyre CC. Urethral stricture and urethritis in men in Scotland. Genitourin Med. 1994;70:403–5.
103. Mehta RH, Sridhar H, Vijay Kumar BR, Anand Kumar TC. High incidence of oligozoospermia and teratozoospermia in human semen infected with the aerobic bacterium Streptococcus faecalis. Reprod Biomed Online. 2002;5:17–21.
104. Menkveld R. Leukocytospermia. In: Dya S, Harrison R, Kempers R, editors. Advances in fertility and reproductive medicine, International congress series No. 1266. Amsterdam: Elsevier; 2004. p. 218–24.
105. Mesbah N, Salem HK. Genital tract infection as a cause of male infertility; genital infections and infertility. IntechOpen; 2016. p. 63–8.
106. Mielczarek E, Blaszkowska J. Trichomonas vaginalis: pathogenicity and potential role in human reproductive failure. Infect. 2016;44:447–58.
107. Moazenchi M, Totonchi M, Salman Yazdi R, Hratian K, Mohseni Meybodi MA, Ahmadi Panah M, Chehrazi M, Mohseni Meybodi A. The impact of Chlamydia trachomatis infection on sperm parameters and male fertility: a comprehensive study. Int J STD AIDS. 2017;29:466–73.
108. Moghimi M, Zabihi-Mahmoodabadi S, Kheirkhah-Vakilabad A, Kargar Z. Significant correlation between high-risk HPV DNA in semen and impairment of sperm quality in infertile men. Int J Fertil Steril. 2019;12:306–9.
109. Montag M, van der Ven H, Haidl G. Recovery of ejaculated spermatozoa for intracytoplasmic sperm injection after anti-inflammatory treatment of an azoospermic patient with genital tract infection: a case report. Andrologia. 1999;31:179–81.
110. Moreira J, Peixoto TM, Siqueira AMD, Lamas CC. Sexually acquired Zika virus: a systematic review. Clin Microbiol Inf. 2017;23:296–305.
111. Moretti E, Capitani S, Figura N, Pammolli A, Federico MG, Giannerini V, Collodel G. The presence of bacteria species in semen and sperm quality. J Assist Reprod Genet. 2009;26:47–56.
112. Moskovtsev SI, Willis J, White J, Mullen JB. Leukocytospermia: relationship to sperm deoxyribonucleic acid integrity in patients evaluated for male factor infertility. Fertil Steril. 2007;88:737–40.
113. Moubasher A, Sayed H, Mosaad E, Mahmoud A, Farag F, Taha EA. Impact of leukocytospermia on sperm dynamic motility parameters, DNA and chromosomal integrity. Cent European J Urol. 2018;71:470–5.
114. Mouraviev V, McDonald M. An implementation of next generation sequencing for prevention and diagnosis of urinary tract infection in urology. Can J Urol. 2018;25:9349–56.
115. Naber KG. Management of bacterial prostatitis: what's new? BJU Int. 2008;101(Suppl 3):7–10.
116. Naber KG, Weidner W. Chronic prostatitis-an infectious disease? J Antimicrob Chemother. 2000;46:157–61.
117. Naber KG, Bergman BO, Bishop MC, Bjerklund-Johansen TE, Botto H, Lobel B, Cruz FJ, Selvaggi FP. EAU guidelines for the management of urinary and male Genital tract infections. Eur Urol. 2001;40:576–88.

118. Nickel JC, Xiang J. Clinical significance of nontraditional bacterial uropathogens in the management of chronic prostatitis. J Urol. 2008;179:1391–5.
119. Nickel JC, Downey J, Hunter D, Clark J. Prevalence of prostatitis-like symptoms in a population based study using the National Institutes of Health chronic prostatitis symptom index. J Urol. 2001;165:842–5.
120. Nickel JC, Aaron L, Barkin J, Elterman D, Nachabé M, Zorn KC. Canadian Urological Association guideline on male lower urinary tract symptoms/benign prostatic hyperplasia (MLUTS/BPH): 2018 update. Can Urol Assoc J. 2018;12:303–312. https://doi.org/10.5489/cuaj.5616.
121. Nicopoullos JD, Almeida PA, Ramsay JW, Gilling-Smith C. The effect of human immunodeficiency virus on sperm parameters and the outcome of intrauterine insemination following sperm washing. Hum Reprod. 2004;19:2289–97.
122. Nieschlag E, Behre H. Andrology. Male reproductive health and dysfunction. Berlin: Springer; 1997.
123. Nunez-Calonge R, Caballero P, Redondo C, Baquero F, Martinez-Ferrer M, Meseguer MA. Ureaplasma urealyticum reduces motility and induces membrane alterations in human spermatozoa. Hum Reprod. 1998;13:2756–61.
124. Ochsendorf FR. Sexually transmitted infections: impact on male fertility. Andrologia. 2008;40:72–5.
125. Oliva A, Multigner L. Ketotifen improves sperm motility and sperm morphology in male patients with leukocytospermia and unexplained infertility. Fertil Steril. 2006;85:240–3.
126. Onemu SO, Ibeh IN. Studies on the significance of positive bacterial semen cultures in male fertility in Nigeria. Int J Fertil Women's Med. 2001;46:210–4.
127. Parks JE, Lynch DV. Lipid composition and thermotropic phase behavior of boar, bull, stallion, and rooster sperm membranes. Cryobiology. 1992;29:255–66.
128. Pellati D, Mylonakis I, Bertoloni G, Fiore C, Andrisani A, Ambrosini G, Armanini D. Genital tract infections and infertility. Eur J Obstet Gynecol Reprod Biol. 2008;140:3–11.
129. Pentyala S, Lee J, Annam S, Alvarez J, Veerraju A, Yadlapalli N, Khan SA. Current perspectives on pyospermia: a review. Asian J Androl. 2007;9(5):593–600.
130. Pérez-Plaza M, Padrón RS, Más J, Peralta H. Semen analyses in men with asymptomatic genital gonorrhoea. Int J Androl. 1982;5:6–10.
131. Pergialiotis V, Karampetsou N, Perrea DN, Konstantopoulos P, Daskalakis G. The impact of bacteriospermia on semen parameters: A Meta-analysis. J Fam Reprod Health. 2018;12:73–83.
132. Philip J, Selvan D, Desmond AD. Mumps orchitis in the non-immune postpubertal male: a resurgent threat to male fertility? BJU Int. 2006;97:138–41.
133. Plaza Davila M, Martin Munoz P, Tapia JA, Ortega Ferrusola C, Balao da Silva CC, Pena FJ. Inhibition of mitochondrial complex I leads to decreased motility and membrane integrity related to increased hydrogen peroxide and reduced ATP production, while the inhibition of glycolysis has less impact on sperm motility. PLoS One. 2015;10:e0138777.
134. Potts JM, Sharma R, Pasqualotto F, Nelson D, Hall G, Agarwal A. Association of Ureaplasma urealyticum with abnormal reactive oxygen species levels and absence of leukocytospermia. J Urol. 2000;163:1775–8.
135. Prabha V, Sandhu R, Kaur S, Kaur K, Sarwal A, Mavuduru RS, Singh SK. Mechanism of sperm immobilization by Escherichia coli. Adv Urol. 2010;2010:240268.
136. Punab M, Loivukene K, Kermes K, Mändar R. The limit of leucocytospermia from the microbiological viewpoint. Andrologia. 2003;35:271–8.
137. Rana K, Vander H, Bhandari P, Thaper D, Prabha V. Microorganisms and male infertility: Possible pathophysiological mechanisms. Adv Clin Med Microbiol. 2016;1:002.
138. Redgrove KA, McLaughlin EA. The role of the immune response in Chlamydia trachomatis infection of the male genital tract: A double-edged sword. Front Immunol. 2014;5:534.
139. Reichart M, Kahane I, Bartoov B. In vivo and in vitro impairment of human and ram sperm nuclear chromatin integrity by sexually transmitted Ureaplasma urealyticum infection. Biol Reprod. 2000;63:1041–8.

140. Ricci S, De Giorgi S, Lazzeri E, Luddi A, Rossi S, Piomboni P, De Leo V, Pozzi G. Impact of asymptomatic genital tract infections on in vitro Fertilization (IVF) outcome. PLoS One. 2018;13:e0207684.
141. Richter C. Oxidative stress, mitochondria, and apoptosis. Restor Neurol Neurosci. 1998;12:59–62.
142. Rintala MAM, Grénman SE, Pöllänen PP, Suominen JJO, Syrjänen SM. Detection of high-risk HPV DNA in semen and its association with the quality of semen. Int J STD AIDS. 2004;15:740–3.
143. Rosser CJ, Bare RL, Meredith JW. Urinary tract infections in the critically ill patient with a urinary catheter. Am J Surg. 1999;177:287–90.
144. Rowley J, Vander Hoorn S, Korenromp E, Low N, Unemo M, Abu-Raddad LJ, Chico RM, Smolak A, Newman L, Gottlieb S, Thwin SS, Broutet N, Taylor MM. Chlamydia, gonorrhoea, trichomoniasis and syphilis: global prevalence and incidence estimates, 2016. Bull World Health Organ. 2019;97:548–562P.
145. Ruggeri M, Cannas S, Cubeddu M, Molicotti P, Piras GL, Dessole S, Zanetti S. Bacterial agents as a cause of infertility in humans. New Microbiol. 2016;39:206–9.
146. Rusz A, Pilatz A, Wagenlehner F, Linn T, Diemer T, Schuppe HC, Lohmeyer J, Hossain H, Weidner W. Influence of urogenital infections and inflammation on semen quality and male fertility. World J Urol. 2012;30:23–30.
147. Saleh RA, Agarwal A, Kandirali E, Sharma RK, Thomas AJ Jr, Nada EA, Evenson DP, Alvarez JG. Leukocytospermia is associated with increased reactive oxygen species production by human spermatozoa. Fertil Steril. 2002;78:1215–24.
148. Samplaski MK, Domes T, Jarvi KA. Chlamydial infection and its role in male infertility. Adv Androl. 2014;2014:307950.
149. Sandoval JS, Raburn D, Muasher S. Leukocytospermia: overview of diagnosis, implications, and management of a controversial finding. Middle East Fertil Soc J. 2013;18:129–34.
150. Sanocka-Maciejewska D, Ciupinska M, Kurpisz M. Bacterial infection and semen quality. J Reprod Immunol. 2005;67:51–6.
151. Sasikumar S, Dakshayani D, Franklin A, Samuel R. An in-vitro study of effectiveness of uropathogenic yeast on male infertility. Int J Curr Microbiol Appl Sci. 2013;2:233–46.
152. Saxena P, Soni R, Randhawa VS, Singh N. Can seminal IL-8 level be used as a marker of leukocytospermia and does it have any correlation with semen parameters in infertile couples? J Obstet Gynaecol India. 2019;69:451–6.
153. Schiefer HG. Microbiology of male urethroadnexitis: diagnostic procedures and criteria for aetiologic classification. Andrologia. 1998;30(Suppl 1):7–13.
154. Schulz M, Sanchez R, Soto L, Risopatron J, Villegas J. Effect of Escherichia coli and its soluble factors on mitochondrial membrane potential, phosphatidylserine translocation, viability, and motility of human spermatozoa. Fertil Steril. 2010;94:619–23.
155. Schuppe HC, Meinhardt A, Allam JP, Bergmann M, Weidner W, Haidl G. Chronic orchitis: a neglected cause of male infertility? Andrologia. 2008;40:84–91.
156. Schuppe HC, Pilatz A, Hossain H, Diemer T, Wagenlehner F, Weidner W. Urogenital infection as a risk factor for male infertility. Dtsch Arztebl Int. 2017;114:339–46.
157. Sellami H, Znazen A, Sellami A, Mnif H, Louati N, Ben Zarrouk S, Keskes L, Rebai T, Gdoura R, Hammami A. Molecular detection of Chlamydia trachomatis and other sexually transmitted bacteria in semen of male partners of infertile couples in Tunisia: the effect on semen parameters and spermatozoa apoptosis markers. PLoS One. 2014;9:e98903.
158. Sharma RK, Pasqualotto AE, Nelson DR, Thomas AJ Jr, Agarwal A. Relationship between seminal white blood cell counts and oxidative stress in men treated at an infertility clinic. J Androl. 2001;22:575–83.
159. Sharma A, Chandraker S, Patel VK, Ramteke P. Antibacterial activity of medicinal plants against pathogens causing complicated urinary tract infections. Indian J Pharm Sci. 2009;71:136–9.
160. Shi L, Wang H, Lu Z. Staphylococcal infection and infertility. In: Darwish AM, editor. Genital Infections and Infertility. IntechOpen; 2016. p. 159–75. https://doi.org/10.5772/62663.

161. Shiadeh MN, Niyyati M, Fallahi S, Rostami A. Human parasitic protozoan infection to infertility: a systematic review. Parasitol Res. 2016;115:469–77.
162. Skau PA, Folstad I. Do bacterial infections cause reduced ejaculatequality? A meta-analysis of antibiotic treatment of male infertility. Behav Ecol. 2003;14:40–7.
163. Solomon M, Henkel R. Semen culture and the assessment of genitourinary tract infections. Indian J Urol IJU: J Urological Soc India. 2017;33(3):188–93.
164. Souho T, Benlemlih M, Bennani B. Human papillomavirus infection and fertility alteration: a systematic review. PLoS One. 2015;10:e0126936.
165. Stojanov M, Baud D, Greub G, Vulliemoz N. Male infertility: the intracellular bacterial hypothesis. New Microbes New Infect. 2018;26:37–41.
166. Street E, Joyce A, Wilson J, Clinical Effectiveness Group, British Association for Sexual Health and HIV. BASHH UK guideline for the management of epididymo-orchitis, 2010. Int J STD AIDS. 2011;22:361–5.
167. Street EJ, Justice ED, Kopa Z, Portman MD, Ross JD, Skerlev M, Wilson JD, Patel R. The 2016 European guideline on the management of epididymo-orchitis. Int J STD AIDS. 2017;28:744–9.
168. Svensrup HF, Fedder J, Abraham-Peskir J, Birkelund S, Christiansen G. Mycoplasma genitalium attaches to human spermatozoa. Hum Reprod. 2003;18:2103–9.
169. Swenson CE, Toth A, Toth C, Wolfgruber L, O'Leary WM. Asymptomatic bacteriospermia in infertile men. Andrologia. 1980;12:7–11.
170. Thomas J, Fishel SB, Hall JA, Green S, Newton TA, Thornton SJ. Increased polymorphonuclear granulocytes in seminal plasma in relation to sperm morphology. Hum Reprod. 1997;12:2418–21.
171. Tomlinson MJ, White A, Barrat GLR. The removal of morphologically abnormal sperm forms by phagocytes: a positive role for seminal leukocytes. Hum Reprod. 1992;7:517–22.
172. Trojian TH, Lishnak TS, Heiman DL. Epididymitis and orchitis: an overview. Am Fam Phys. 2009;79:583–7.
173. Van Leeuwen E, Prins JM, Jurriaans S, Boer K, Reiss P, Repping S, van der Veen F. Reproduction and fertility in human immunodeficiency virus type-1 infection. Hum Reprod Update. 2007;13:197–206.
174. Villegas JV, Boguen R, Uribe P. Effect of uropathogenic Escherichia coli on human sperm function and male fertility. In: Samie A, editor. Escherichia coli – Recent advances on physiology, pathogenesis and biotechnological applications. IntechOpen Books; 2017. p. 71–80.
175. Vilvanathan S, Kandasamy B, Jayachandran AL, Sathiyanarayanan S, Tanjore Singaravelu V, Krishnamurthy V, Elangovan V. Bacteriospermia and its impact on basic semen parameters among infertile men. Interdiscip Perspect Infect Dis. 2016;2016:2614692. https://doi.org/10.1155/2016/2614692. Epub 2016 Jan 6
176. Wang Y, Liang CL, Wu JQ, Xu C, Qin SX, Gao ES. Do Ureaplasma urealyticum infections in the genital tract affect semen quality? Asian J Androl. 2006;8:562–8.
177. Weidner W, Krause W, Ludwig M. Relevance of male accessory gland infection for subsequent fertility with special focus on prostatitis. Hum Reprod Update. 1999;5:421–32.
178. Wikstrom A, Jensen JS. Mycoplasma genitalium: a common cause of persistent urethritis among men treated with doxycycline. Sex Transm Infect. 2006;82:276–9.
179. Witkin SS, Sultan KM, Neal GS, Jeremias J, Grifo JA, Rosenwaks Z. Unsuspected Chlamydia trachoma tis infection and in vitro fertilization outcome. Am J Obstet Gynecol. 1994;171:1208–14.
180. Wolff H. The biologic significance of white blood cells in semen. Fertil Steril. 1995;63:1143–57.
181. Wolff H, Panhans A, Stolz W, Meurer M. Adherence of Escherichia coli to sperm: a mannose mediated phenomenon leading to agglutination of sperm and E. coli. Fertil Steril. 1993;60:154–8.
182. World Health Organization. WHO laboratory manual for the examination and processing of human semen. 5th ed. Geneva: WHO; 2010.

183. World Health Organization (2011) Prevalence and incidence of selected sexually transmitted infections, Chlamydia trachomatis, Neisseria gonorrhoeae, syphilis and Trichomonas vaginalis: Methods and results used by WHO to generate 2005 estimates. https://apps.who.int/iris/bitstream/handle/10665/44735/9789241502450_eng.pdf;jsessionid=7565097A22E3780750B905D2C9B02E7E?sequence=1. Accessed: 29.04.2021
184. World Health Organization. WHO guidelines for the treatment of Neisseria gonorrhoeae. World Health Organization; 2016.
185. Wu H, Jiang X, Gao Y, Liu W, Wang F, Gong M, Chen R, Yu X, Zhang W, Gao B, Song C. Mumps virus infection disrupts blood-testis barrier through the induction of TNF-α in Sertoli cells. FASEB J. 2019;33:12528–40.
186. Yamamoto A, Hiei H, Katsuno S, Miyake K. Antibiotic and ejaculation treatments improve resolution rate of leukocytospermia in infertile men with prostatitis. Nagoya J Med Sci. 1995;58:41–5.
187. Yilmaz S, Koyuturk M, Kilic G, Alpak O, Aytoz A. Effects of leucocytospermia on semen parameters and outcomes of intracytoplasmic sperm injection. Int J Androl. 2005;28:337–42.
188. Zegarra Montes LR, Sanchez Mejia AA, Loza Munarriz CA, Gutierrez EC. Semen and urine culture in the diagnosis of chronic bacterial prostatitis. Int Braz J Urol. 2008;34:30–40.
189. Zegers-Hochschild F, Adamson GD, de Mouzon J, Ishihara O, Mansour R, Nygren K, Sullivan E, van der Poel S, International Committee for Monitoring Assisted Reproductive Technology; World Health Organization. The international committee for monitoring assisted reproductive technology (ICMART) and the world health organization (WHO) revised glossary on ART terminology, 2009. Hum Reprod. 2009;24:2683–7.
190. Zeighami H, Peerayeh SN, Yazdi RS, Sorouri R. Prevalence of Ureaplasma urealyticum and Ureaplasma parvum in semen of infertile and healthy men. Int J STD AIDS. 2009;20:387–90.
191. Zeyad A, Amor H, Hammadeh ME. The impact of bacterial infections on human spermatozoa. Int J Women's Health Reprod Sci. 2017;5:243–52.
192. Zeyad A, Hamad M, Amor H, Hammadeh ME. Relationships between bacteriospermia, DNA integrity, nuclear protamine alteration, sperm quality and ICSI outcome. Reprod Biol. 2018;18:115–21.
193. Zhang L, Diao RY, Duan YG, Yi TH, Cai ZM. In vitro antioxidant effect of curcumin on human sperm quality in leucocytospermia. Andrologia. 2017;49:e12760.
194. Zhou JF, Xiao WQ, Zheng YC, Dong J, Zhang SM. Increased oxidative stress and oxidative damage associated with chronic bacterial prostatitis. Asian J Androl. 2006;8:317–23.
195. Zhou YH, Ma HX, Shi XX, Liu Y. Ureaplasma spp. in male infertility and its relationship with semen quality and seminal plasma components. J Microbiol Immunol Inf. 2018;51:778–83.

Chapter 7
Bacteriospermia and Male Infertility: Role of Oxidative Stress

Sandipan Das, Shubhadeep Roychoudhury, Anwesha Dey, Niraj Kumar Jha, Dhruv Kumar, Shatabhisha Roychoudhury, Petr Slama, and Kavindra Kumar Kesari

Abstract Male infertility is one of the major challenging and prevalent diseases having diverse etiologies of which bacteriospermia play a significant role. It has been estimated that approximately 15% of all infertility cases are due to infections caused by uropathogens and in most of the cases bacteria are involved in infection and inflammation leading to the development of bacteriospermia. In response to bacterial load, excess infiltration of leukocytes in the urogenital tract occurs and concomitantly generates oxidative stress (OS). Bacteria may induce infertility either by directly interacting with sperm or by generating reactive oxygen species (ROS)

S. Das · S. Roychoudhury
Department of Life Science and Bioinformatics, Assam University, Silchar, India

A. Dey
Department of Human Physiology, Ambedkar College, Fatikroy, Tripura, India

N. K. Jha
Department of Biotechnology, School of Engineering and Technology (SET), Sharda University, Greater Noida, India

D. Kumar
Amity Institute of Molecular Medicine and Stem Cell Research, Amity University, Noida, India

S. Roychoudhury (✉)
Department of Microbiology, R. G. Kar Medical College and Hospital, Kolkata, India

Health Centre, Assam University, Silchar, India

P. Slama
Department of Animal Morphology, Physiology and Genetics, Faculty of AgriSciences, Mendel University in Brno, Brno, Czech Republic

K. K. Kesari
Department of Bioproducts and Biosystems, School of Chemical Engineering, Aalto University, Espoo, Finland

© The Author(s), under exclusive license to Springer Nature Switzerland AG 2022
K. K. Kesari, S. Roychoudhury (eds.), *Oxidative Stress and Toxicity in Reproductive Biology and Medicine*, Advances in Experimental Medicine and Biology 1358, https://doi.org/10.1007/978-3-030-89340-8_7

and impair sperm parameters such as motility, volume, capacitation, hyperactivation. They may also induce apoptosis leading to sperm death. Acute bacteriospermia is related with another clinical condition called leukocytospermia and both compromise male fertility potential by OS-mediated damage to sperm leading to male infertility. However, bacteriospermia as a clinical condition as well as the mechanism of action remains poorly understood, necessitating further research in order to understand the role of individual bacterial species and their impact in male infertility.

Keywords Uropathogens · Bacteriospermia · Infection · Inflammation · Leukocytes · ROS · Sperm parameters · Infertility

7.1 Introduction

In the third decade of twenty-first century infertility remains one of the major challenging and highly prevalent global health conditions [1]. Male infertility is a multifactorial disorder and male urogenital infection is considered one of the major contributors to male infertility accounting approximately 15% of all male infertility cases [2]. Acute or chronic male urogenital tract infection is mediated by microorganisms particularly bacteria and it affects various parts of the male reproductive system such as testis, epididymis, and male accessory sex glands leading to impairment of sperm production, maturation, and movement in the seminal tract [3]. Pathogenic bacteria gain access to the male urogenital tract by sexually transmitted infection, intracanicular spread of bacteria from urine infection or hematogenous seeding of bacteria from urogenital organs [4, 5]. Development of bacteriospermia and concomitant increment of leukocytes in the male urogenital tract due to infection and inflammation may impair the fertility potential of a man through multiple mechanisms, such as deterioration of spermatogenesis, reduction in sperm motility, genital tract obstruction and/or dysfunction, and oxidative stress (OS) [5, 6]. Excessive bacterial colonization and successive infection in the urogenital tract impair male fertility through sperm adhesion, interaction and/or forming sperm agglutination thus reducing sperm motility and lowering the chance of sperm–oocyte fusion [7]. Moreover, bacterial infection causes chronic persistent inflammation and leukocyotspermia leading to increment of pro-inflammatory cytokines and reactive oxygen species (ROS) in the urogenital tract which ultimately contribute to the development of OS-associated male infertility [8, 9]. In bacteriospermia, prevalence of bacterial species varies depending on the population types, whereas their mechanism of action in male infertility is still understood poorly. A study conducted on Canadian population found 22 species in bacteriospermic ejaculates and the most prevalent bacteria were *Enterococcus faecalis* (56%) followed by *Escherichia coli* (16%), group B *Streptococcus* (13%), and *Staphylococcus aureus* (5%). These four bacterial species contribute up to 90% of all identified bacterial species [5].

Similarly, a study from India reported a total of 7 bacterial species from bacteriospermic ejaculates, and the most dominant species was *E. faecalis* (30%) followed by Coagulase negative *Staphylococcus* (23.33%), *S. aureus* (20%), E. coli (10%), *Klebsiella pneumoniae* (6.66%), *Proteus* sp. (6.66%), and *Citrobacter* sp. (3.33%) [10]. Another study on Czech population reported predominance of three species i.e., *Staphylococcus* sp., *Streptococcus* sp., and *E. coli* from abnormal semen samples of 116 infertile men [11]. The presence of bacterial species in the urogenital tract of bacteriospermic men and leukocytes response is still poorly understood. The bacteria may utilize multiple patho-mechanisms and develop OS thus contributing to the compromise of male fertility potential. This chapter mainly focuses on the role of bacteriospermia and OS in male infertility.

7.2 Oxidative Stress (OS) and Fertility Pattern in the Male

The reproductive organs of the male including testes, epididymis, vas deferens, and accessory glands are mainly involved in the formation, storage, and ejaculation of sperm. They also produce androgens that help in the development and maintenance of male fertility potential [12]. Impairment of reproductive organs due to low hormone synthesis, Klinefelter syndrome, cryptorchidism, autoimmune disorder, exposure to radiation, altered lifestyle, infection, OS, trauma, etc., may lead to male infertility [13, 14]. Male infertility is a multifactorial disorder and OS is considered as one of the major contributors to the disease. It is defined as an imbalance between the levels of ROS and antioxidants in the semen [15]. Free radicals are the unpaired electron containing molecules which are highly reactive against lipids, amino acids, and nucleic acids [16]. Prime sources of ROS in the semen include excessive leukocytes, immature sperm, varicocele, exposure to toxins such as radiation, smoking, alcohol consumption, etc. Excess ROS in the semen can overwhelm the antioxidant defense leading to concomitant development of OS and cause sperm dysfunction and/or death, and ultimately infertility in men [13]. However, optimum levels of ROS are crucial for facilitating sperm hyperactivation, motility [17], capacitation, and acrosomal reaction [16]. Controlled production of ROS in the sperm at the time of capacitation process increases the amount of cyclic adenosine 3′,5′- monophosphate (cAMP) that facilitates the hyperactivation of the spermatozoa. Hyperactivation is very crucial because only the hyperactivated sperm have increased motility to undergo the acrosomal reaction that may lead to successful fertilization [18]. Excessive leukocytes in respone to inflammation and immature sperm are the main source of ROS generation and mature sperm are highly susceptible to ROS due to the presence of polyunsaturated fatty acids in their membranes [19]. Uncontrolled rise in the level of ROS initiates lipid peroxidation (LPO) of the sperm membrane where up to 60% of the fatty acids are reduced thus altering the membrane fluidity, disrupting the activity of enzymes and membrane receptors which may ultimately lead to abnormal fertilization [20]. OS and elevated levels of ROS have also been associated with impaired sperm parameters such as motility, concentration, and

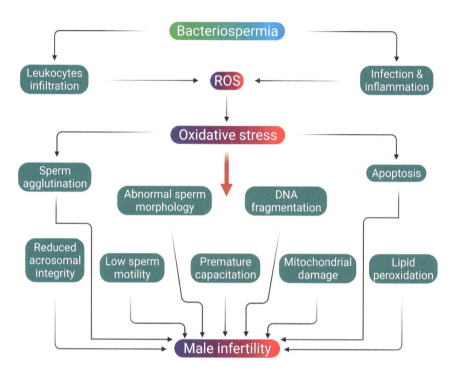

Fig. 7.1 Bacteriospermia induce infection and inflammation in the urogenital tract of men and concomitantly increases leukocytes infiltration in the infected location. Both bacteriospermia and leukocytes increase ROS in urogenital tract leading to OS. ROS can cause sperm agglutination and reduce sperm motility. On the contrary OS damages sperm morphology, sperm DNA, mitochondrial DNA and reduce sperm acrosomal integrity, cause premature capacitation, induces lipid peroxidation and apoptosis in sperm which ultimately contribute to male infertility. ROS- reactive oxygen species, OS- oxidative stress, DNA- deoxyribonucleic acid

morphology - which are the important indicators of a male's fertility potential. Diminishing sperm motility is a result of a cascade of events including LPO of the plasma membrane, which affects axonemal protein phosphorylation and sperm immobilization [21, 22]. High levels of ROS cause significant sperm morphological abnormalities like head, neck, and tail deformities and retention of cytoplasmic droplets leading to male infertility [23]. OS also facilitates sperm DNA damage by fragmenting the single- and double-stranded DNA, direct oxidation of DNA bases, and by DNA mutations [24]. Furthermore, excessive generation of ROS directly disrupts the mitochondria electron transport chain (ETC) and has the ability to damage mitochondrial DNA which ultimately activates the stress response gene by altering the mitochondrial physiology and promote apoptosis by disrupting cell division [24, 25]. Low sperm motility due to mitochondrial dysfunction can also cause asthenozoospermia [24]. OS results into functional and metabolic disorder in germ cells, too [26]. Moreover, uncontrolled ROS can modify the protein status by oxidative reaction and generate aldehydes and ketones which reflect a negative effect on spermatogenesis and overall fertility of the male [27] (Fig. 7.1).

7.3 Bacteriospermia and Male Infertility

Infertility is defined as a disease of the reproductive system characterized by the failure of a couple to establish a clinical pregnancy after one year or more of regular unprotected sexual intercourse [28]. Like females, the males are more or less equal contributors to the infertility cases with a prevalence of 30–50% of all cases reported globally [29]. However, the prevalence varies according to the geographical regions as higher prevalence has been documented from Central and Eastern Europe, Africa, the Middle East, and South and Central Asia [1, 30]. Compromised fertility potential of the male may be attributed to multiple factors including anatomical abnormalities of the reproductive system, cryptorchidism, ejaculatory duct dysfunction, genetic and hormonal imbalance, varicocele, leukocytospermia, gonadal toxicity, environmental pollutants, male urogenital tract infections, and bacteriospermia, among others [9, 31]. Male urogenital infections are caused by microorganisms that include bacteria, virus, protozoa, and fungi. Among these, infections mediated by bacteria is the most prevalent one leading to the impairment of both sperm quality and function as well as seminal tract obstruction [32]. Both infection and inflammation can reduce spermatogenesis and deteriorate sperm quality and function [29] leading to an array of clinical conditions such as oligozoospermia, asthenozoospermia, azoospermia, and dysfunction of male accessory glands [29]. Bacterial infection and inflammatory responses have been linked with poor male fertility potential; however, the exact mechanism remains inadequately understood. In case of chronic bacterial infection inflammatory responses may be asymptomatic but can still impose a long-lasting negative effect on sperm function, motility, count and spermatogenesis, and affect the permeability of the vas deferens and/or ejaculatory duct or may even induce apoptosis of spermatozoa [33–35].

According to the World Health Organization (WHO), bacteriospermia can be defined as the presence of more than 10^3 bacterial/ml of ejaculate and such condition is generally used as an indicator of active urogenital infection in men [36]. Bacteriospermia is associated with excessive generation of leukocytes in the male urogenital tract and concomitant OS. Moreover, bacteriospermia is a substantial cause of male infertility as it is associated with sperm DNA fragmentation, poor sperm motility and count, [37], genital tract dysfunction, deterioration of spermatogenesis [5], and poor assisted reproductive technology (ART) outcomes [38].

Sperm has the ability to recognize bacterial endotoxin, glycoprotein, and lipopolysaccharide by toll-like receptors (TLR-2 and TLR-4) expressed in the plasma membrane. Activated TLRs trigger local inflammatory response, create obstruction, and ultimately induce male infertility [39]. Inflammatory response also triggers excessive infiltration of leukocytes that may lead to the development of a clinical condition referred to as leukocytospermia. It is characterized by the presence of more than 1×10^6 leukocytes/ml of ejaculate [36]. Excess seminal leukocytes also trigger overproduction of ROS and introduce an imbalance between antioxidant and free radicals resulting in OS [40]. Elevated OS may cause significant biological and biochemical changes in the outer and inner mitochondrial membranes of sperm,

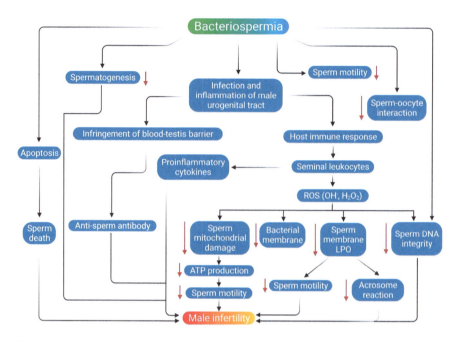

Fig. 7.2 Role of bacteriospermia in male infertility. Bacteriospermia initiates host immune response that concomitantly increases leukocytes in the male reproductive tract leading to production of reactive oxygen species (ROS) such as OH^- and H_2O_2, and proinflammatory cytokines. Some bacterial species also produce ROS leading to overproduction of ROS that cause sperm DNA damage, lipid peroxidation (LPO), and inhibits mitochondrial ATP production, which are in turn associated with reduction of sperm motility, inhibition of acrosome reaction ultimately contributing to male infertility

which may in turn negatively affect sperm morphology, acrosome integrity and promote premature capacitation [37]. Moreover, excess ROS produced by leukocytes can stimulate LPO in polyunsaturated fatty acids of sperm membrane [41]. On the contrary, ROS has been linked directly to sperm agglutination, reduced sperm motility and increased DNA fragmentation, enhanced apoptosis in mature sperm, and increased risk of compromised male fertility [5] (Fig. 7.2).

7.4 Bacteria-Associated Male Urogenital Infection

7.4.1 Urinary Tract Infection (UTI)

Urinary tract infections (UTIs) are considered as the most common bacterial infections in the urogenital tract of the male. Infection in the lower urinary tract is characterized by cystitis and in the upper urinary tract it is characterized by pyelonephritis where patients easily develop bacteremia. Cystitis is associated with dysuria and

Table 7.1 Common bacterial species implicated in bacteriospermia and their effect on semen quality

Bacteria	Effect	References
Escherichia coli	Sperm tail defect, immobilization, impaired acrosome reaction	[65, 99]
Chlamydia trachomatis	Low progressive motility and sperm DNA damage	[76]
Pseudomonas aeruginosa	Reduced sperm motility, sperm morphological deformities, necrosis, and apoptosis. Induced sperm mitochondrial damage	[69, 70, 71]
Neisseria gonorrhoeae	Low sperm motility, sperm agglutination, and apoptosis	[80, 82]
Staphylococcus aureus	Low semen volume, sperm motility, concentration, vitality, and normal morphology	[83]
Ureaplasma urealyticum	Hyperviscous semen, low pH, sperm count, and concentration	[90]
Klebsiella pneumoniae	Low progressive motility, high LPO, and apoptosis leading to sperm death	[98]
Mycoplasma genitalium	Low sperm motility	[35]
Mycoplasma hominis	Low progressive motility, total sperm motility, count, and normal morphology	[96]

pollakisuria while pyelonephritis involves flank pain and fever along with the symptoms of cystitis [42]. In case of diabetic patients, stone-associated obstructed pyelonephritis is a major risk factor for fatal septic shock [43]. Common causal agents of UTIs include uropathogens such as *E. coli*, *K. pneumoniae*, *Staphylococcus saprophyticus* [44], *Proteus mirabilis*, and *E. faecalis* [45] (Table 7.1).

7.4.2 Male Accessory Gland Infection (MAGI)

Male accessory tract infection (MAGI) occurs due to the spreading of microorganisms via epididymis, deferent duct, prostate gland, seminal vesicles, testis, and urethra [42]. Epididymitis is considered as an inflammatory condition in epididymis, which is characterized by pain and acute unilateral or bilateral swelling of the scrotum and involvement of testicular inflammation along with epididymis, which is termed as epididymo-orchitis [46]. Bacterial epididymitis is commonly associated with low sperm count, stenosis in epididymal duct, and impairment in sperm function ultimately leading to male infertility. Occurrence of azoospermia is often associated with unclear unilateral epididymitis [47]. The most common bacteria involved in such infection are *Chlamydia trachomatis*, *Escherichia coli*, and *Neisseria gonorrhoeae* [48]. Prostatitis is another common urological inflammatory disease among men of all age groups and it affects sperm motility, count and morphology [47]. Typical features of prostatitis include voiding disturbances, sexual dysfunction, and chronic pelvic pain which are most prominent signs of prostate inflammation. Prostatitis is classified into four categories - i) chronic bacterial prostatitis, ii) acute bacterial prostatitis, iii) chronic prostatitis, and iv) asymptomatic inflammatory prostatitis [49]. The most common bacteria associated with prostatitis are *Chlamydia trachomatis*, *Escherichia coli*, *Ureaplasma urealyticum*, *Nesisseria gonorrhea*, and *Klebsiella* sp. [50]. On the contrary, urethritis is the inflammation of urethra and the most prevalent bacteria associated include *Chlamydia trachomatis*, *Neisseria gonorrhoea*, and *Ureaplasma urealyticum* [51]. Urethritis can be divided into two types i) non-gonococcal urethritis, and ii) gonococcal urethritis [52]. Urethritis is commonly associated with penile itching, dysuria, urethral discharge [53], and seminiferous tubular necrosis [54].

7.4.3 Sexually Transmitted Infection (STI)

Sexually transmitted infections (STIs)-mediated male infertility depends mainly on the local prevalence of sexually transmitted diseases (STDs). Prevalence of STDs are more prominent in Africa or South-East Asian regions compared to Western countries [55]. Most common bacteria associated in STIs include *Chlamydia trachomatis*, *C. trachomatis*, *Chancroid haemophilus*, *Calymmatobacterium granulomatis*, *Neisseria gonorrhoea*, *Treponema pallidum*, and *Ureaplasma urealyticum* [55]. Men with gonorrhoeic urethritis commonly develop urethral strictures and unilateral epididymo-orchitis [54].

7.5 Prevalence of Bacteriospermia-Associated Male Infertility

Approximately 48.5 million couple face the challenge of infertility globally after having unprotected intercourse for one year or long [56]. Approximately 15% of male infertility is linked with microorganisms-mediated UTIs [2]. Among the bacterial species responsible for bateriospermic condition in male, *E. coli* is the most prevalent and frequently isolated from the semen of infertile men accounting for 65–80% cases of male infertility [2]. *E. coli* is associated with the infection of the prostate, seminal vesicle, urethra, epididymis, and testis [37]. The second most prevalent bacterium is *Chlamydia trachomatis*, an obligate pathogen involved in 30–40% of urethritis cases [57]. According to WHO, *C. trachomatis* may cause up to 92 million urogenital tract infections per year [58]. *U. urealyticum* and *Mycoplasma genitalium* are other two bacterial species frequently present in bacteriospermic semen. *U. urealyticum* can contribute to 10–40% of all cases [59]. The two most common *Mycoplasma* species are *M. hominis* and *M. genitalium* that are responsible for 10.8% and 5% of all infection associated male infertility, respectively [60]. According to WHO, sexually transmitted infections caused by *N. gonorrhoeae* represented 106.1 million of cases globally which have been associated with epididymitis, urethritis, and prostatitis along with abnormal urethral discharge [61].

7.5.1 Escherichia Coli

The most common bacteria isolated from the semen of bacteriospermic male is *E. coli*, which possibly causes asexually transmitted epididymo-orchitis and is responsible for acute or chronic prostatitis leading to male infertility, too [2]. Semen analysis of 88 infertile male patients revealed the dominance of *E. coli* (10.22%) followed by Staphylococci (9.09%), Enterococci (5.88%), *Staphylococcus aureus* (2.27%), Gonococci (2.27%), and *Klebsiella* sp. (1.13%), [62].

Binding of *E. coli* with sperm leads to agglutination of sperm and induces damage in plasma membrane resulting in swelling of midpiece and tail invagination, which promote the rate of immobilization of sperm [63]. Interaction between *E. coli* and sperm is mediated by mannose binding receptors present on *E. coli* and mannose residues present on the sperm surface that can bind with type 1 fimbriae on *E. coli* [64]. An in vitro study confirmed that incubation of sperm with *E. coli* reduces the ability of sperm acrosomal reaction [65] and mitochondrial membrane potential [66]. Thereafter, Fraczek et al. 2012 [67] confirmed that binding of sperm with *E. coli* can alter sperm membrane stability and mitochondrial activity which increase the chances of male infertility. Lipopolysaccharide and porin protein of *E. coli* can produce cellular lysis in sperm and promote infection leading to temporary sterility [68].

7.5.2 Pseudomonas Aeruginosa

Infertility in men is frequently associated with uropathogenic microbes. *Pseudomonas aeruginosa,* a Gram-negative pathogenic bacterium is a possible cause of male infertility. 3-oxododecanoyl-L-homoserine lactone, a signaling molecule secreted from *P. aeruginosa* has been reported to possess detrimental effects on spermatozoa [69]. Incubation of sperm with *P. aeruginosa* has been shown to reduce sperm motility in a dose-dependent manner, and this bacterial signaling molecule acts on the acrosome of spermatozoa and promote premature acrosomal loss through a calcium dependent mechanism. *P. aeruginosa* infection can also lead to apoptosis and necrosis of sperm without affecting immune cells [69, 70]. A cytotoxic molecule, exotoxin A, released from *P. aeruginosa* induces chromosomal aberrations and sperm abnormalities including two heads, amorphous head, head without hook, banana head, coiled tail, and divided tail, which is believed to occur due to toxic effect of the protein on the sperm tail [70]. Porin, another protein secreted from the membrane of *P. aeruginosa*, is toxic to sperm and has been shown to induce apoptosis directly in the epithelial cell line of seminal vesicles of rats. Porin causes mitochondrial damage after binding with sperm and impairs sperm motility, too [71].

7.5.3 Chlamydia Trachomatis

Chlamydia trachomatis is an obligate intracellular Gram-negative bacterium [72]. It has species-specific lipopolysaccharide (LPS) antigen, other species-specific and immune-specific antigens as investigated through immunofluorescence. Its biphasic life cycle consists of an elementary and reticulate body [73]. This pathogen causes urethritis in the male and untreated infection of *C. trachomatis* in the male leads to epididymitis and prostatitis. Most of the time *C. trachomatis* infection remains asymptomatic in the male and may contribute up to 50% of infection [72]. Co-incubation of *C. trachomatis* with sperm promotes reduction of motile sperm and increases premature sperm death. LPS from *C. trachomatis* has also been shown to generate sperm apoptosis inducing molecules and can alter all other essential sperm parameters, too [74]. IgA Chlamydial antibodies promote LPO of sperm membrane which alter membrane fluidity, membrane-associated enzyme activities, capacitation, and acrosome reaction [75]. Furthermore, Chlamydial infection increases the level of interleukin (IL-8) in semen, which acts as a biomarker of MAGI [76]. *C. trachomatis*-induced infection also increases the rate of sperm DNA fragmentation and alters sperm morphology [77]. If the infection caused by the pathogen remains untreated it may cause long-term damage to organs of the male reproductive system such as the ejaculatory ducts, seminal vesicles, and spermatogonial cells [78].

7.5.4 Neisseria Gonorrhoeae

These Gram-negative, immotile diplococci cause the common UTI and develop gonorrhea which in turn alters testicular functions and promotes male infertility. This pathogen is responsible for about 86.9 million of gonorrhea cases globally [79]. Leukocytospermia is associated with gonorrhea which enhances the cytokines and ROS resulting in the impairment of spermatogenesis and sperm function [80]. Asymptomatic infection in the male with gonorrhea does not alter sperm count, semen volume, and sperm morphology but citric acid level drops in the male with gonorrhoea [81]. Movement of this pathogen is facilitated by the presence of pili on their surface that also help them cling onto other cells [80]. Bacterial pili type IV (T4P) and LPS can bind to sperm and asialoglycoprotein receptor on the sperm surface thereby facilitating the binding. Binding of *N. gonorrhoeae* LPS on the sperm cell surface receptor can also cause sperm cell death by inducing apoptosis [82].

7.5.5 Staphylococcus Aureus

The ubiquitous Gram-positive bacterium, *Staphylococcus aureus* is mostly found in the male genital tract. *S. aureus* infection has been associated with impairment of sperm motility and semen volume and increasing semen pH [83, 84]. Dominance of this pathogen in the semen of infertile patients has been confirmed with a prevalence of 68.2–75% in some cases [85, 86]. Incubation of *S. aureus* with sperm causes sperm agglutination and reduction of motility [87]. According to a recent report, 16% of infertile men may face the challenge of infertility due to *S. aureus* infection with abnormal semen fluid density, sperm abnormal morphology, and reduced sperm motility [88].

7.5.6 Ureaplasma *sp.*

The most prevalent species of the genus Ureplasma are *U. urealyticum* and *U. parvum*, and their prevalence in the semen of infertile men is 9% and 3%, respectively [89]. *U. urealyticum* is a main causative agent of prostatitis and epididymitis in the infertile male. *U. urealyticum* infection is also associated with low semen concentration and pH, and high seminal viscosity [90]. Infection reduces the level of seminal plasma alpha-glucosidase but not the levels of acid phosphatase and fructose in the seminal plasma [91]. *U. urealyticum* may also attach massively to sperm at the midpiece leading to looped tangling of tails and multiple agglutination thereby causing sperm immobility [92]. *U. urealyticum* infection increases seminal ROS level, too, thus causing LPO and sperm DNA fragmentation. It also reduces sperm fertilization capacity [93]. Metabolic products of *U. urealyticum* are able to generate ROS such as H_2O_2 and hydroxide anion (OH-) which are highly toxic to sperm. Furthermore, *U. urealyticum* infection decreases the amounts of essential microelements such as zinc and selenium, which are crucial for antioxidant defense mechanism of semen [93].

7.5.7 Mycoplasma *sp.*

Mycoplasma genitalium is another pathogen frequently isolated from the urogenital tract and is one of the potent causative organisms of urethritis in the male [94]. This bacterium has the ability to interact directly with the sperm and render them immobile [35]. The prevalence of *M. genitalium* has been reported between 19 and 41% in patients with urethritis [95]. Another species *M. hominis* has shown strong interaction with sperm parameters such as reduced motility, deformed morphology, and low count in infected men [96]. Also, *M. genitalium* has the capability to bind with the head, neck, and tail regions of the sperm and render sperm bulky thus reducing the capacity of travelling in the reproductive tract of the female [94].

7.5.8 Klebsiella Pneumoniae

The Gram-negative bacterium *Klebsiella pneumoniae* is another important causative organism of MAGI and are responsible for 2.3% of male infertility [97]. The pathophysiology of *K. pneumoniae* infection is not clearly understood due to the lack of evidence. However, *K. pneumoniae* infection impacts sperm parameters negatively and may also be a cause of necrozoospermia [10]. *K. pneumoniae* may impair male fertility by altering progressive motility, LPO, and apoptosis leading to sperm death [98].

7.6 Bacteriospermia and ROS-Mediated Damage

Human body has mainly three defense systems for protecting against invasion by foreign particles: i) tight junctions between skin epithelium, ii) innate immune response, and iii) adaptive immune response. Bacteria potentially infect the biological system through tissue barrier [100]. In the male, bacteria are considered responsible in 50% cases of prostatitis including 10% cases of chronic prostatitis [101]. Among infertile men, 11.6–45% cases occur due to urethral discharge as a marker of infection [59]. All bacterial inflammation associated with response to influx of leukocytes result in increase in ROS formation [102]. ROS are highly reactive chemical molecules including oxygen ions, peroxides, and hydrogen peroxide (H_2O_2), which contribute to male infertility by causing damage to sperm membrane and sperm DNA [103]. The huge amount of polyunsaturated fatty acids (PUFAs) present in the plasma membrane gives membrane fluidity to spermatozoa. The ROS directly attack the unconjugated double-bond groups of the PUFAs and generate a radical chain reaction pathway [104] resulting in the formation of 4-hydroxynonenal, malondialdehyde, and acrolein.

These reactive aldehydes undergo further reaction with hydrophilic amino acids in the protein which leads to mitochondrial dysfunction and leakage of further ROS from inner membrane of mitochondria [100, 105]. The direct damage of mitochondria through ROS decreases the energy availability which deteriorates the motility of sperm and alters normal sperm morphology and induce premature capacitation [102]. ROS directly attack the protamines-coated purine, pyrimidine bases and deoxyribose bases of sperm [106], which induce apoptosis of sperm cell and cause sperm death [107]. ROS also can induce apoptosis in sperm by altering the intercellular calcium ion concentration, which ultimately leads to infertility in the male [108]. In relation to higher level of ROS in semen, cytochrome c, and caspases 9 and 3 levels also increase simultaneously indicating apoptosis in the infertile male [109]. Bacteriospermia can also induce mitochondria-dependent apoptosis in sperm which increases the percentage of fragmented DNA in sperm and decreases mitochondrial transmembrane potential. These reports indicate negative alterations in sperm density, motility and morphology which ultimately contribute to male infertility [67]. Also, anti-bacterial IgA antibody forms in response to Chlamydial infection which is associated with increased ROS [75]. When the amount of ROS exceeds the antioxidant defense mechanism of semen, sperm membrane may undergo LPO and is associated with decreased flexibility of sperm and premature capacitation. Overproduction of free radicals negatively affects spermiogenesis and promotes the release of abnormal spermatozoa with excess cytoplasmic retention from the germinal epithelium. Enzymes of additional cytoplasm activate plasma membrane redox system and promote further production of ROS resulting in the loss of sperm motility and fertilizing capacity [110].

7.7 Management of Bacteriospermia in the Infertile Male

According to European Association of Urology (EAU) guidelines, a urine culture is the first step towards the detection of bacteria or any type of microorganism present in the male urogenital tract. Besides this, history of disease, symptoms check, and physical examination are carried out as a part of routine diagnosis. Additionally, the presence of leukocytes, erythrocytes, and nitrite are investigated for better evaluation [111]. Whereas according to WHO, semen analysis is of utmost importance to detect the MAGI [36]. Four-glass and two-glass tests can also be performed along with semen analysis for the diagnosis of localization of inflammation [46, 112]. Blood count including C-reactive protein (CRP), prostate-specific antigen (PSA), and hormone status (follicle stimulating hormone - FSH, luteinizing hormone-LH, and testosterone) are also recommended for the detection of acute urogenital inflammation [46]. In the case of acute cystitis, preliminary diagnosis can be done on the basis of clinical symptoms including lower abdominal pain [113] and newly developed dysuria, polyuria, and urinary urgency [114]. For the evaluation of acute cystitis, laboratory diagnosis include urine dipstick test and antibiotic susceptibility test by using cultured pathogens and microscopic analysis [115]. Urine culture test is performed for the diagnosis of acute pyelonephritis in the laboratory; however, sometimes blood culture can be done in cases of acute pyelonephritis [116, 117]. Bacteria- or any microorganism-mediated urinary tract obstruction can also be screened by computed tomography (CT), ultrasonography, and intravenous pyelography of kidney, bladder, and ureter [118]. Acute bacterial prostatitis is diagnosed on the basis of clinical symptoms such as heat around the prostate, soft and swollen state of prostate, frequent urination, painful urination, genital pain, chill, and joint pain along with fever [119].

For the initial management of bacteriospermia in the male, clinicians usually prefer two common guidelines - the American Urological Association (AUA) guideline and the European Association of Urology (EAU) guideline [120]. However, these are inadequate to provide the complete management strategy to treat bacteriospermia in the infertile male. Most of the times the chronic or acute bacterial infections are treated using a broad spectrum of antibiotics as an only effective medication against bacterial infections, without considering the partial effects of antibiotic therapy in the human body that may include nausea, bloating, vomiting, diarrhoea, abdominal ache, low appetite, and allergic surge. Additionally, high use of antibiotics increases the risk of bacterial resistance [121]. However, for management of bacteria mediated chronic urogenital infections and inflammations oral supplementation of ciprofloxacin, norfloxacin, ofloxacin are commonly recommended by clinicians [50]. Oral administration of either ciprofloxacin or fosfomycin and cefpodoxime proxetil, cefcapene pivoxil, cefdinir, nitrofurantoin, cefditoren pivoxil, cefixime, pivmecillinam and amoxicillin/clavulanate are prescribed for domestic cases of acute uncomplicated cystitis [119]. Whereas in case of complicated cystitis caused by Gram-negative rods and Gram-positive cocci, recommended antibacterial medications include LVFX (levofloxacin), CPFX (Ciprofloxacin Hydrochloride), TFLX (Tosufloxacin), STFX (Sitafloxacin), CVA/AMPC (Clavulanate/Amoxicillin), SBTPC (Sultamicillin), and sometimes CFDN (Cefdinir), CPDX-PR(cefpodoxime proxetil), and CFPN-PI (Cefcapene pivoxil hydrochloride) can also be used as alternative to antibiotic therapy [122]. Similarly, for the treatment of acute pyelonephritis supplementation of ciprofloxacin [123], levofloxacin [124], and trimethoprim/sulfamethoxazole are recommended [125]. In order to treat specific bacterial infection, medications can be of use after diagnosis. In the case of *E. coli* infection ciprofloxacin, amoxicillin, and aminoglycosides are the commonly used antibiotics [126]. In the case of *N. gonorrhoeae* and *C. trachomatis* infection tetracyclines form the most effective and widely prescribed antibiotics [111]. An initial intravenous administration of MINO for 3 to 5 days can be used to manage severe infections caused by *C. trachomatis* [122]. *Plumbago zeylanica* [127] and *Piper lanceaefolium* can be used as alternative or supplementary herbal medicine for infection of *N. gonorrhoeae* [128]. For the treatment of *Klebsiella* sp. mediated infection, empirical therapies include the use of trimethoprim-sulfamethoxazole and fluoroquinolones [129].Moreover, extract from the plant *Aframomum melegueta* has been used against *Klebsiella* sp. infection which may also be used as a supplementary or alternative medicine after proper toxicological evaluation [130]. A wide variety of African traditional herbs including *Kigellia africana, Ballota africana, Carpobrotus edulis*, and *Pelargonium fasiculata* are used in the management of *K. pneumoniae* infection [131]. In the case of *Staphylococcus aureus* infection, nafcillin [132], and imipenem have demonstrated more effective antibacterial medicine [133]. Nigerian traditional medical practitioners also recommend the use of medicinal plants such as *Acalypha wilkesiana, Ageratum conyzoides, Bridella ferruginea, Ocimum gratissimum, Phylantus discoideus*, and *Terminalia avicennioides* against *S. aureus* infection [134]. In Persian and European traditional medicines the fruit of *Apium graveolens* is used against uncomplicated urinary infection [135]. Current findings suggested that phthalide, an active compound isolated from show strong antiadhesive activity against the uropathogen *E. coli* [135]. Essential oils from *Ocimum gratissimum, Salvia officinalis*, and *Cymbopogon citratus* have the ability to neutralize the infections mediated by *K. pneumoniae, E. coli*, and *Enterobacter* sp. Similarly, chloroform, ethanol, methanol, and petroleum ether extract of *Callistemon lanceolatus* have been found effective in cases of *S. aureus, E. faecalis, E. coli*, and *K. pneumoniae* infection [136, 137]. *Vaccinium macrocarpon* popularly known as cranberry is considered one of the potent plant products that can inhibit bacterial attachment to the uroepithelial cells thus reducing bacterial load in the urogenital tract [138]. An in vitro study from India confirmed that aqueous, ethanolic and chloroform extracts of *Hybanthus enneaspermus* possess strong antibacterial activity against common uropathogens including *E. coli, P. aeruginosa, K. pneumoniae, E. faecalis*, and *S. aureus* [139]. Another potent candidate herb is *Moringa oleifera* that is used widely in the management of various human ailments and showed strong antibacterial action against *E. coli, P. aeruginosa*, and *S. aureus* infections [140]. One of the major mechanisms of bacteriospermia associated male fertility is through the production of ROS by bacteria directly or by initiating leukocytes response at the site of infection [5]. The potent herbal candidate that may be used in the management of OS-induced damages include *Tribulus terrestris* - a highly antioxidant rich herb that can neutralize ROS action and, also prevent membrane lipid peroxidation [141]. Similarly, a bioactive molecule thymol from *Trachyspermum copticum* also exerts strong antioxidant activity against ROS-induced OS [142]. Other potent herbs that can be utilized to minimize ROS-mediated damages to the male reproductive system are *Cinnamomum verum* [143], *Terminalia chebula* [144], *Ocimum sanctum* [145], *Juniperus communis* [146], and *Taraxacum officinale* [147].

The best advantage of herbal medicine over conventional antibiotic use is that the bacteria do not develop any sort of resistance against them. Also, medicinal herbs contain a wide range of bioactive molecules which are responsible for the medicinal property and synergistic effect [138]. However, further experiments are needed to validate the effectiveness and toxicity of the herbal medicines as well as the identification of specific bioactive compounds and their exact mechanism of action for future potential use in the clinical management of bacteriospermia.

7.8 Conclusions

Male infertility is a minacious global health threat that has not been clearly understood till date and more research is needed to perceive the underlying etiologies and proper management of the disease [148]. Bacteriospermia is one of the significant etiologies of male infertility that develops as a result of chronic or mild bacterial infection of the male urogenital tract. The mechanism of infection of bacteria varies from species to species, thus in order to develop better treatment approaches, identification of virulence determinants (that help bacteria in initial attachment and disease development including adhesin molecules, siderophores, and urease) is essential. This may also help develop vaccination for preventing bacterial infection in the male reproductive system [45]. Moreover, establishment of better prevention strategy for bacteriospermia may be achieved by switching towards the use of traditional herbal medicine as alternative or complementary medicine. Traditional herbs such as *Acalypha wilkesiana*, *Ageratum conyzoides*, *Bridella ferruginea*, *Ocimum gratissimum*, *Phylantus discoideus*, *Terminalia avicennioides* [134], and *Aframomum melegueta* have been used in the management of male UTIs particularly against common uropathogens that are responsible for bacteriospermia [130]. Similarly, bacteriospermia-mediated OS can be restrained by the use of antioxidant rich extract of *Tribulus terrestris*, *Trachyspermum copticum*, *Cinnamomum verum*, *Terminalia chebula*, *Ocimum sanctum*, *Juniperus communis*, and *Taraxacum officinale*. However, detailed toxicological studies of these herbs are needed prior to use either as an alternative or complementary medicine for effective clinical management of bacteriospermia.

References

1. Inhorn MC, Patrizio P. Infertility around the globe: new thinking on gender, reproductive technologies and global movements in the 21st century. Hum Reprod Update. 2015;21(4):411–26.
2. Pellati D, Mylonakis I, Bertoloni G, Fiore C, Andrisani A, Ambrosini G, Armanini D. Genital tract infections and infertility. Eur J Obstetrics Gynecol Reproduct Biol. 2008;140(1):3–11.
3. Diemer T, Huwe P, Ludwig M, Hauck EW, Weidner W. Urogenital infection and sperm motility. Andrologia. 2003;35(5):283–7.
4. Andrada JA, Von der Walde FE, Andrada EC. Immunologic studies of male infertility. Immunol Ser. 1990;52:345–78.
5. Domes T, Lo KC, Grober ED, Mullen JB, Mazzulli T, Jarvi K. The incidence and effect of bacteriospermia and elevated seminal leukocytes on semen parameters. Fertil Steril. 2012;97(5):1050–5.
6. Sharma RK, Pasqualotto FF, Nelson DR, Agarwal A. Relationship between seminal white blood cell counts and oxidative stress in men treated at an infertility clinic. J Androl. 2001;22(4):575–83.
7. Keck C, Gerber-Schäfer C, Clad A, Wilhelm C, Breckwoldt M. Seminal tract infections: impact on male fertility and treatment options. Hum Reprod Update. 1998;4(6):891–903.
8. Askienazy-Elbhar M. Male genital tract infection: the point of view of the bacteriologist. Gynecol Obstetrique Fertilite. 2005;33(9):691–7.
9. Brunner RJ, Demeter JH, Sindhwani P. Review of guidelines for the evaluation and treatment of leukocytospermia in male infertility. World J Men's health. 2019;37(2):128–37.
10. Vilvanathan S, Kandasamy B, Jayachandran AL, Sathiyanarayanan S, Tanjore Singaravelu V, Krishnamurthy V, Elangovan V. Bacteriospermia and its impact on basic semen parameters among infertile men. Interdiscip Perspect Infect Dis. 2016;6:2016.
11. Matoušková I, Oborná I, Fingerová H, Kohnová I, Novotný J, Svobodová M, Březinová J, Vyslouzilová J, Radová L. Bacteriospermia and the production of reactive oxygen species in the semen of males from infertile couples. Klinicka mikrobiologie a infekcni lekarstvi. 2009;15(6):192–5.

12. Tiwana MS, Leslie SW.: Anatomy, Abdomen and Pelvis, Testicle. (2017)
13. Agarwal A, Roychoudhury S, Bjugstad KB, Cho CL. Oxidation-reduction potential of semen: what is its role in the treatment of male infertility? Ther Adv Urol. 2016;8(5):302–18.
14. Gurung P, Yetiskul E Jialal I.: Physiology, male reproductive. (2020)
15. Barati E, Nikzad H, Karimian M. Oxidative stress and male infertility: current knowledge of pathophysiology and role of antioxidant therapy in disease management. Cell Mol Life Sci. 2020;77(1):93–113.
16. Kefer JC, Agarwal A, Sabanegh E. Role of antioxidants in the treatment of male infertility. Int J Urol. 2009;16(5):449–57.
17. Agarwal A, Makker K, Sharma R. Clinical relevance of oxidative stress in male factor infertility: an update. Am J Reproduct Immunol. 2008;59:2–11. https://doi.org/10.1111/j.1600-0897.2007.00559.x.
18. Agarwal A, Tvrda E, Sharma R. Relationship amongst teratozoospermia, seminal oxidative stress and male infertility. Reproduct Biol Endocrinol. 2014;12(1):1–8. https://doi.org/10.1186/1477-7827-12-45.
19. Henkel RR. Leukocytes and oxidative stress: dilemma for sperm function and male fertility. Asian J Androl. 2011;13(1):43.
20. Dutta S, Majzoub A, Agarwal A. Oxidative stress and sperm function: a systematic review on evaluation and management. Arab J Urol. 2019;17(2):87–97.
21. Cocuzza M, Sikka SC, Athayde KS, Agarwal A. Clinical relevance of oxidative stress and sperm chromatin damage in male infertility: an evidence based analysis. Int Brazilian J Urol. 2007;33(5):603–21.
22. Ko EY Jr, Agarwal SS. Male infertility testing : reactive oxygen species and antioxidant capacity. Fertil Steril. 2014;102(6):1518–27. https://doi.org/10.1016/j.fertnstert.2014.10.020.
23. Rahiminia T. Etiologies of sperm oxidative stress. Iranian J Reproduct Med. 2016;14(4):231–40.
24. Ritchie C, Ko EY. Oxidative stress in the pathophysiology of male infertility. Andrologia. 2021;53(1):e13581.
25. Kumar DP, Sangeetha N. Mitochondrial DNA mutations and male infertility. Indian J Human Genet. 2009;15(3):93–7.
26. Kasperczyk MDS, Birkner SHDCE, Kasperczyk A.: Oxidative stress and motility impairment in the semen of fertile males, (2017), 1–8. https://doi.org/10.1111/and.12783.
27. Baskaran S, Finelli R, Agarwal A, Henkel R. Reactive oxygen species in male reproduction: a boon or a bane? Andrologia. 2021;53(1):e13577.
28. World Health Organization (WHO). International Classification of Diseases, 11th Revision (ICD-11) Geneva: WHO (2018)
29. Henkel R, Offor U, Fisher D. The role of infections and leukocytes in male infertility. Andrologia. 2020;53(1):e13743.
30. Mascarenhas MN, Flaxman SR, Boerma T, Vanderpoel S, Stevens GA. National, regional, and global trends in infertility prevalence since 1990: a systematic analysis of 277 health surveys. PLoS Med. 2012;9(12):e1001356.
31. Dohle GR, Weidner W, Jungwirth A, Colpi G, Papp G, Pomerol J, Hargreave TB.: Guidelines on male infertility. European Association of Urology. (2004)
32. Moretti E, Capitani S, Figura N, Pammolli A, Federico MG, Giannerini V, Collodel G. The presence of bacteria species in semen and sperm quality. J Assist Reprod Genet. 2009;26(1):47–56.
33. Eley A, Pacey AA, Galdiero M, Galdiero M, Galdiero F. Can Chlamydia trachomatis directly damage your sperm? Lancet Infectious Dis. 2005;5(1):53–7.
34. Gimenes F, Souza RP, Bento JC, Teixeira JJ, Maria-Engler SS, Bonini MG, Consolaro ME. Male infertility: a public health issue caused by sexually transmitted pathogens. Nat Rev Urol. 2014;11(12):672–87.
35. Stojanov M, Baud D, Greub G, Vulliemoz N. Male infertility: the intracellular bacterial hypothesis. New Microbes New Infect. 2018;26:37–41.

36. World Health Organization. WHO laboratory manual for the examination and processing of human semen. (2010)
37. Farsimadan M, Motamedifar M. Bacterial infection of the male reproductive system causing infertility. J Reprod Immunol. 2020;3:103183.
38. Ghasemian F, Esmaeilnezhad S, Moghaddam MJ. Staphylococcus saprophyticus and Escherichia coli: tracking from sperm fertility potential to assisted reproductive outcomes. Clin Exp Reprod Med. 2021;48(2):142.
39. Fujita Y, Mihara T, Okazaki T, Shitanaka M, Kushino R, Ikeda C, Negishi H, Liu Z, Richards JS, Shimada M. Toll-like receptors (TLR) 2 and 4 on human sperm recognize bacterial endotoxins and mediate apoptosis. Hum Reprod. 2011;26(10):2799–806.
40. Iommiello VM, Albani E, Di Rosa A, Marras A, Menduni F, Morreale G, Levi SL, Pisano B, Levi-Setti PE. Ejaculate oxidative stress is related with sperm DNA fragmentation and round cells. Int J Endocrinol. 2015;2015
41. Parks JE, Lynch DV. Lipid composition and thermotropic phase behavior of boar, bull, stallion, and rooster sperm membranes. Cryobiology. 1992;29(2):255–66.
42. La Vignera S, Condorelli RA, Vicari E, Salmeri M, Morgia G, Favilla V, Cimino S, Calogero AE. Microbiological investigation in male infertility: a practical overview. J Med Microbiol. 2014;63(1):1–4.
43. Yamamichi F, Shigemura K, Kitagawa K, Fujisawa M. Comparison between non-septic and septic cases in stone-related obstructive acute pyelonephritis and risk factors for septic shock: a multi-center retrospective study. J Infect Chemother. 2018;24(11):902–6.
44. Hayami H, Takahashi S, Ishikawa K, Yasuda M, Yamamoto S, Wada K, Kobayashi K, Hamasuna R, Minamitani S, Matsumoto T, Kiyota H. Second nationwide surveillance of bacterial pathogens in patients with acute uncomplicated cystitis conducted by Japanese surveillance committee from 2015 to 2016: antimicrobial susceptibility of Escherichia coli, Klebsiella pneumoniae, and staphylococcus saprophyticus. J Infect Chemother. 2019;25(6):413–22.
45. Flores-Mireles AL, Walker JN, Caparon M, Hultgren SJ. Urinary tract infections: epidemiology, mechanisms of infection and treatment options. Nat Rev Microbiol. 2015;13(5):269–84.
46. Schuppe HC, Pilatz A, Hossain H, Diemer T, Wagenlehner F, Weidner W. Urogenital infection as a risk factor for male infertility. Dtsch Arztebl Int. 2017;114(19):339.
47. Weidner W, Colpi GM, Hargreave TB, Papp GK, Pomerol JM. EAU working group on male infertility. EAU guidelines on male infertility. Eur Urol. 2002;42(4):313–22.
48. Henkel R, Offor U, Fisher D. The role of infections and leukocytes in male infertility. Andrologia. 2021;53(1):e13743.
49. Krieger JN, Nyberg L Jr, Nickel JC. NIH consensus definition and classification of prostatitis. JAMA. 1999;282(3):236–7.
50. Naber KG, Weidner W. Chronic prostatitis—an infectious disease? J Antimicrob Chemother. 2000;46(2):157–61.
51. Ness RB, Markovic N, Carlson CL, Coughlin MT. Do men become infertile after having sexually transmitted urethritis? An epidemiologic examination. Fertil Steril. 1997;68(2):205–13.
52. Horner PJ, Blee K, Falk L, van der Meijden W, Moi H. 2016 European guideline on the management of non-gonococcal urethritis. Int J STD AIDS. 2016;27(11):928–37.
53. Brill JR. Diagnosis and treatment of urethritis in men. Am Fam Physician. 2010;81(7):873–8.
54. Osegbe DN. Testicular function after unilateral bacterial epididymo-orchitis. Eur Urol. 1991;19:204–8.
55. Ochsendorf FR. Sexually transmitted infections: impact on male fertility. Andrologia. 2008;40(2):72–5.
56. Martinez G, Daniels K, Chandra A: Fertility of men and women aged 15–44 years in the United States: National Survey of family growth, 2006–2010. Department of Health and Human Services, Centers for Disease Control and Prevention, National Center for Health Statistics; (2012)

57. Stamm WE. Chlamydia trachomatis infections: progress and problems. J Infect Dis. 1999;179(Supplement_2):S380–3.
58. World Health Organization. Global prevalence and incidence of selected curable sexually transmitted infections: overview and estimates. (2001)
59. Solomon M, Henkel R. Semen culture and the assessment of genitourinary tract infections. Indian J Urol: IJU: J Urol Soc India. 2017;33(3):188.
60. Gdoura R, Kchaou W, Chaari C, Znazen A, Keskes L, Rebai T, Hammami A. Ureaplasma urealyticum, Ureaplasma parvum, mycoplasma hominis and mycoplasma genitalium infections and semen quality of infertile men. BMC Infect Dis. 2007;7(1):1–9.
61. World Health Organization. Global incidence and prevalence of selected curable sexually transmitted infections-2008. (2012)
62. Golshani M, Taheri S, Eslami G, Soleymani RA, Falah F, Goudarzi H.: Genital tract infection in asymptomatic infertile men and its effect on semen quality (2006): 81–84.
63. Diemer T, Weidner W, Michelmann HW, SCHIEFER HG, Rovan E, Mayer F. Influence of Escherichia coli on motility parameters of human spermatozoa in vitro. Int J Androl. 1996;19(5):271–7.
64. Wolff H, Panhans A, Stolz W, Meurer M. Adherence of Escherichia coli to sperm: a mannose mediated phenomenon leading to agglutination of sperm and E. coli. Fertil Steril. 1993;60(1):154–8.
65. El-Mulla KF, Kohn FM, Dandal M, El Beheiry AH, Schiefer HG, Weidner W, Schill WB. In vitro effect of Escherichia coli on human sperm acrosome reaction. Arch Androl. 1996;37(2):73–8.
66. Schulz M, Sánchez R, Soto L, Risopatrón J, Villegas J. Effect of Escherichia coli and its soluble factors on mitochondrial membrane potential, phosphatidylserine translocation, viability, and motility of human spermatozoa. Fertil Steril. 2010;94(2):619–23.
67. Fraczek M, Piasecka M, Gaczarzewicz D, Szumala-Kakol A, Kazienko A, Lenart S, Laszczynska M, Kurpisz M. Membrane stability and mitochondrial activity of human-ejaculated spermatozoa during in vitro experimental infection with E scherichia coli, S taphylococcus haemolyticus and B acteroides ureolyticus. Andrologia. 2012;44(5):315–29.
68. Galdiero F, Gorga F, Bentivoglio C, Mancuso R, Galdiero E, Tufano MA. The action of LPS porins and peptidoglycan fragments on human spermatozoa. Infection. 1988;16(6):349–53.
69. Rennemeier C, Frambach T, Hennicke F, Dietl J, Staib P. Microbial quorum-sensing molecules induce acrosome loss and cell death in human spermatozoa. Infect Immun. 2009;77(11):4990–7.
70. Altaee MF, Nafee SK, Hamza SJ. Evaluation for the cytotoxic effect of exotoxin a produced by Pseudomonas aeruginosa on mice by using cytogenetic parameters. Curr Microbiol. 2013;1:257–61.
71. Buommino E, Morelli F, Metafora S, Rossano F, Perfetto B, Baroni A, Tufano MA. Porin from Pseudomonas aeruginosa induces apoptosis in an epithelial cell line derived from rat seminal vesicles. Infect Immun. 1999;67(9):4794–800.
72. Malhotra M, Sood S, Mukherjee A, Muralidhar S, Bala M. Genital chlamydia trachomatis: an update. Indian J Med Res. 2013;138(3):303.
73. Saka HA, Thompson JW, Chen YS, Kumar Y, Dubois LG, Moseley MA, Valdivia RH. Quantitative proteomics reveals metabolic and pathogenic properties of chlamydia trachomatis developmental forms. Mol Microbiol. 2011;82(5):1185–203.
74. Eley A, Hosseinzadeh S, Hakimi H, Geary I, Pacey AA. Apoptosis of ejaculated human sperm is induced by co-incubation with Chlamydia trachomatis lipopolysaccharide. Human Reproduct. 2005;20(9):2601–7.
75. Segnini A, Camejo MI, Proverbio F. Chlamydia trachomatis and sperm lipid peroxidation in infertile men. Asian J Androl. 2003;5(1):47–50.
76. Kokab A, Akhondi MM, Sadeghi MR, Modarresi MH, Aarabi M, Jennings R, Pacey AA, Eley A. Raised inflammatory markers in semen from men with asymptomatic chlamydial infection. J Androl. 2010;31(2):114–20.

77. Gallegos G, Ramos B, Santiso R, Goyanes V, Gosálvez J, Fernández JL. Sperm DNA fragmentation in infertile men with genitourinary infection by chlamydia trachomatis and mycoplasma. Fertil Steril. 2008;90(2):328–34.
78. Jiminez G, Villanveva Diaz CA. Epididymal stereocilia in semen of infertile men: evidence of chronic epididymitis? Int J Androl. 2006;38:26–30.
79. Rowley J, Vander Hoorn S, Korenromp E, Low N, Unemo M, Abu-Raddad LJ, Chico RM, Smolak A, Newman L, Gottlieb S, Thwin SS. Chlamydia, gonorrhoea, trichomoniasis and syphilis: global prevalence and incidence estimates, 2016. Bull World Health Organ. 2019;97(8):548.
80. Krause W. Male accessory gland infection. Andrologia. 2008;40(2):113–6.
81. Pérez-Plaza M, Padrón RS, Más J, Peralta H. Semen analyses in men with asymptomatic genital gonorrhoea. Int J Androl. 1982;5(1):6–10.
82. Harvey HA, Jennings MP, Campbell CA, Williams R, Apicella MA. Receptor-mediated endocytosis of Neisseria gonorrhoeae into primary human urethral epithelial cells: the role of the asialoglycoprotein receptor. Mol Microbiol. 2001;42(3):659–72.
83. Onemu SO, Ibeh IN. Studies on the significance of positive bacterial semen cultures in male fertility in Nigeria. Int J Fertil Womens Med. 2001;46(4):210–4.
84. Marconi M, Pilatz A, Wagenlehner F, Diemer T, Weidner W. Impact of infection on the secretory capacity of the male accessory glands. Int Braz J Urol. 2009;35(3):299–309.
85. Emokpae MA, Uadia PO, Sadiq NM. Contribution of bacterial infection to male infertility in Nigerians. Online J Health Allied Sci. 2009;8(1)
86. Momoh AR, Idonije BO, Nwoke EO, Osifo UC, Okhai O, Omoroguiwa A, Momoh AA. Pathogenic bacteria-a probable cause of primary infertility among couples in Ekpoma. J Microbiol Biotechnol Res. 2011;1(3):66–71.
87. Kaur S, Prabha V, Shukla G, Sarwal A. Interference of human spermatozoal motility by live Staphylococcus aureus. Am J Biomed Sci. 2010;2(1):91–7.
88. Esmailkhani A, Akhi MT, Sadeghi J, Niknafs B, Bialvaei AZ, Farzadi L, Safadel N. Assessing the prevalence of Staphylococcus aureus in infertile male patients in Tabriz, Northwest Iran. Int J Reproduct BioMed. 2018 Jul;16(7):469.
89. Zeighami H, Peerayeh SN, Yazdi RS, Sorouri R. Prevalence of Ureaplasma urealyticum and Ureaplasma parvum in semen of infertile and healthy men. Int J STD AIDS. 2009;20(6):387–90.
90. Wang Y, Liang CL, Wu JQ, Xu C, Qin SX, Gao ES. Do Ureaplasma urealyticum infections in the genital tract affect semen quality? Asian J Androl. 2006;8(5):562–8.
91. Zheng J, Yu SY, Jia DS, Yao B, Ge YF, Shang XJ, Huang YF. Ureaplasma urealyticum infection in the genital tract reduces seminal quality in infertile men. Zhonghua nan ke xue=. Natl J Androl. 2008;14(6):507–12.
92. Abdel Razzak AA, Bakr SS. Role of mycoplasma in male infertility. EMHJ-Eastern Mediterranean Health J. 2000;6(1):149–55.
93. Potts JM, Sharma R, Pasqualotto F, Nelson D, Hall G, Agarwal A. Association of Ureaplasma urealyticum with abnormal reactive oxygen species levels and absence of leukocytospermia. J Urol. 2000;163(6):1775–8.
94. Svenstrup HF, Fedder J, Abraham-Peskir J, Birkelund S, Christiansen G. Mycoplasma genitalium attaches to human spermatozoa. Hum Reprod. 2003;18:2103–9.
95. Wikstrom A, Jensen JS. Mycoplasma genitalium: a common cause of persistent urethritis among men treated with doxycycline. Sex Transm Infect. 2006;82:276–9.
96. Ahmadi MH, Mirsalehian A, Sadighi Gilani MA, Bahador A, Talebi M. Asymptomatic infection with mycoplasma hominis negatively affects semen parameters and leads to male infertility as confirmed by improved semen parameters after antibiotic treatment. Urology. 2017;100:97–102.
97. Ibadin OK, Ibeh IN. Bacteriospermia and sperm quality in infertile male patient at University of Benin Teaching Hospital, Benin City, Nigeria. Malaysian J Microbiol. 2008;4(2):65–7.

98. Zuleta-González MC, Zapata-Salazar ME, Guerrero-Hurtado LS, Puerta-Suárez J, Cardona-Maya WD. Klebsiella pneumoniae and Streptococcus agalactiae: passengers in the sperm travel. Archivos espanoles de Urolgia. 2019;72(9):939–47.
99. Prabha V, Sandhu R, Kaur S, Kaur K, Sarwal A, Mavuduru RS, Singh SK. Mechanism of sperm immobilization by Escherichia coli. Adv Urol. 2010;30:2010.
100. Agarwal A, Rana M, Qiu E, AlBunni H, Bui AD, Henkel R. Role of oxidative stress, infection and inflammation in male infertility. Andrologia. 2018;50(11):e13126.
101. Schaefler AJ. Epidemiology and demographics of prostatitis. Andrologia. 2003;35(5):252–7.
102. Tremellen K. Oxidative stress and male infertility—a clinical perspective. Hum Reprod Update. 2008;14(3):243–58.
103. Cheeseman KH, Slater TF. An introduction to free radical biochemistry. Br Med Bull. 1993;49(3):481–93.
104. Aitken RJ. Reactive oxygen species as mediators of sperm capacitation and pathological damage. Mol Reprod Dev. 2017;84(10):1039–52.
105. Agarwal A, Saleh RA, Bedaiwy MA. Role of reactive oxygen species in the pathophysiology of human reproduction. Fertil Steril. 2003;79(4):829–43.
106. Oliva R. Protamines and male infertility. Hum Reprod Update. 2006;12(4):417–35.
107. Moustafa MH, Sharma RK, Thornton J, Mascha E, Abdel-Hafez MA, Thomas AJ, Agarwal A. Relationship between ROS production, apoptosis and DNA denaturation in spermatozoa from patients examined for infertility. Hum Reprod. 2004;19(1):129–38.
108. Ha HK, Park HJ, Park NC. Expression of E-cadherin and α-catenin in a varicocele-induced infertility rat model. Asian J Androl. 2011;13(3):470.
109. Sanocka D, Kurpisz M. Reactive oxygen species and sperm cells. Reprod Biol Endocrinol. 2004;2(1):1–7.
110. Aitken RJ, Sawyer D. The human spermatozoon—not waving but drowning. Adv Male Mediated Dev Toxicity. 2003:85–98.
111. Grabe M, Bjerklund-Johansen TE, Botto H, Çek M, Naber KG, Tenke P, Wagenlehner F. Guidelines on urological infections. Eur Assoc Urol. 2015;182:237–57.
112. Wagenlehner FME, Naber KG, Bschleipfer T, Brähler E, Weidner W. Prostatitis and male pelvic pain syndrome: diagnosis and treatment. Dtsch Arztebl Int. 2009;106:175–83.
113. Rowe TA, Juthani-Mehta M. Diagnosis and management of urinary tract infection in older adults. Infect Dis Clin. 2014;28(1):75–89.
114. Bent S, Nallamothu BK, Simel DL, Fihn SD, Saint S. Does this woman have an acute uncomplicated urinary tract infection? JAMA. 2002;287(20):2701–10.
115. Rubin RH, Shapiro ED, Andriole VT, Davis RJ, Stamm WE. Evaluation of new anti-infective drugs for the treatment of urinary tract infection. Clin Infect Dis. 1992;15(Supplement_1):S216–27.
116. Hooton TM. Uncomplicated urinary tract infection. N Engl J Med. 2012;366(11):1028–37.
117. Lee HN, Yoon H. Management of antibiotic-resistant acute pyelonephritis. Urogenital Tract Infect. 2017;12(3):95–102.
118. Roy C, Pfleger DD, Tuchmann CM, Lang HH, Saussine CC, Jacqmin D. Emphysematous pyelitis: findings in five patients. Radiology. 2001;218(3):647–50.
119. Kang CI, Kim J, Park DW, Kim BN, Ha US, Lee SJ, Yeo JK, Min SK, Lee H, Wie SH. Clinical practice guidelines for the antibiotic treatment of community-acquired urinary tract infections. Infect Chemotherapy. 2018;50(1):67–100.
120. Trost LW, Nehra A. Guideline-based management of male infertility: why do we need it? Indian J Urol: IJU: J Urol Soc India. 2011;27(1):49.
121. Mohsen S, Dickinson JA, Somayaji R. Update on the adverse effects of antimicrobial therapies in community practice. Can Fam Physician. 2020;66(9):651–9.
122. Yamamoto S, Ishikawa K, Hayami H, Nakamura T, Miyairi I, Hoshino T, Hasui M, Tanaka K, Kiyota H, Arakawa S. JAID/JSC guidelines for clinical management of infectious disease 2015– urinary tract infection/male genital infection. J Infect Chemother. 2017;23(11):733–51.

123. Talan DA, Klimberg IW, Nicolle LE, Song J, Kowalsky SF, Church DA. Once daily, extended release ciprofloxacin for complicated urinary tract infections and acute uncomplicated pyelonephritis. J Urol. 2004;171(2):734–9.
124. Klausner HA, Brown P, Peterson J, Kaul S, Khashab M, Fisher AC, Kahn JB. A trial of levofloxacin 750 mg once daily for 5 days versus ciprofloxacin 400 mg and/or 500 mg twice daily for 10 days in the treatment of acute pyelonephritis. Curr Med Res Opin. 2007;23(11):2637–45.
125. Talan DA, Stamm WE, Hooton TM, Moran GJ, Burke T, Iravani A, Reuning-Scherer J, Church DA. Comparison of ciprofloxacin (7 days) and trimethoprim-sulfamethoxazole (14 days) for acute uncomplicated pyelonephritis in women: a randomized trial. JAMA. 2000;283(12):1583–90.
126. Adamus-Białek W, Wawszczak M, Arabski M, Majchrzak M, Gulba M, Jarych D, Parniewski P, Głuszek S. Ciprofloxacin, amoxicillin, and aminoglycosides stimulate genetic and phenotypic changes in uropathogenic Escherichia coli strains. Virulence. 2019;10(1):260–76.
127. Gundidza M, Manwa G. Activity of chloroform extract from Plumbago zeylanica against Neisseria gonorrhoeae. Fitoterapia. 1990;61(1):47–9.
128. Ruddock PS, Charland M, Ramirez S, López A, Towers GN, Arnason JT, Liao M, Dillon JA. Antimicrobial activity of flavonoids from Piper lanceaefolium and other Colombian medicinal plants against antibiotic susceptible and resistant strains of Neisseria gonorrhoeae. Sex Transm Dis. 2011;38(2):82–8.
129. Bouza E, Cercenado E: Klebsiella and enterobacter: antibiotic resistance and treatment implications. InSeminars in respiratory infections 2002; (Vol. 17, No. 3, pp. 215–230).
130. Doherty VF, Olaniran OO, Kanife UC. Antimicrobial activities of Aframomum melegueta (Alligator pepper). Int J Biol. 2010;2(2):126–31.
131. Cock IE, Van Vuuren SF. The potential of selected South African plants with anti-Klebsiella activity for the treatment and prevention of ankylosing spondylitis. Inflammopharmacology. 2015;23(1):21–35.
132. Chang FY, Peacock JE Jr, Musher DM, Triplett P, MacDonald BB, Mylotte JM, O'Donnell A, Wagener MM, Victor LY. Staphylococcus aureus bacteremia: recurrence and the impact of antibiotic treatment in a prospective multicenter study. Medicine. 2003;82(5):333–9.
133. Isaiah IN, Nche BT, Nwagu IG, Nnanna II. Current studies on bacterospermia the leading cause of male infertility: a protégé and potential threat towards mans extinction. N Am J Med Sci. 2011;3(12):562.
134. Akinyemi KO, Oladapo O, Okwara CE, Ibe CC, Fasure KA. Screening of crude extracts of six medicinal plants used in South-West Nigerian unorthodox medicine for anti-methicillin resistant Staphylococcus aureus activity. BMC Complement Altern Med. 2005;5(1):1–7.
135. Grube K, Spiegler V, Hensel A. Antiadhesive phthalides from Apium graveolens fruits against uropathogenic E. coli. J Ethnopharmacol. 2019;237:300–6. system. StatPearls
136. Noormandi A, Dabaghzadeh F. Effects of green tea on Escherichia coli as a uropathogen. J Tradit Complement Med. 2015;5(1):15–20.
137. Das S. Natural therapeutics for urinary tract infections—a review. Future J Pharm Sci. 2020;6(1):1–13.
138. Shaheen G, Akram M, Jabeen F, Ali Shah SM, Munir N, Daniyal M, Riaz M, Tahir IM, Ghauri AO, Sultana S, Zainab R. Therapeutic potential of medicinal plants for the management of urinary tract infection: a systematic review. Clin Exp Pharmacol Physiol. 2019;46(7):613–24.
139. Sahoo S, Kar DM, Mohapatra S, Rout SP, Dash SK. Antibacterial activity of Hybanthus enneaspermus against selected urinary tract pathogens. Indian J Pharm Sci. 2006;68(5)
140. Amabye TG, Tadesse FM. Phytochemical and antibacterial activity of moringa oleifera available in the market of Mekelle. J Analyt Pharm Res. 2016;2(1):1–4.
141. Zheleva-Dimitrova D, Obreshkova D, Nedialkov P. Antioxidant activity of tribulus terrestris—a natural product in infertility therapy. Int J Pharm Pharm Sci. 2012;4(4):508–11.
142. Nickavar B, Adeli A, Nickavar A. TLC-bioautography and GC-MS analyses for detection and identification of antioxidant constituents of Trachyspermum copticum essential oil. Iranian J Pharm Res: IJPR. 2014;13(1):127.

143. Mathew S, Abraham TE. In vitro antioxidant activity and scavenging effects of Cinnamomum verum leaf extract assayed by different methodologies. Food Chem Toxicol. 2006;44(2):198–206.
144. Saha S, Verma RJ. Antioxidant activity of polyphenolic extract of Terminalia chebula Retzius fruits. J Taibah Univ Sci. 2016;10(6):805–12.
145. Hakkim FL, Shankar CG, Girija S. Chemical composition and antioxidant property of holy basil (Ocimum sanctum L.) leaves, stems, and inflorescence and their in vitro callus cultures. J Agric Food Chem. 2007;55(22):9109–17.
146. Elmastaş M, Gülçin İ, Beydemir Ş, İrfan Küfrevioğlu Ö, Aboul-Enein HY. A study on the in vitro antioxidant activity of juniper (Juniperus communis L.) fruit extracts. Anal Lett. 2006;39(1):47–65.
147. Ghaima KK, Hashim NM, Ali SA. Antibacterial and antioxidant activities of ethyl acetate extract of nettle (Urtica dioica) and dandelion (Taraxacum officinale). J Appl Pharm Sci. 2013;3(5):96.
148. Kumar N, Singh AK. Trends of male factor infertility, an important cause of infertility: a review of literature. J Human Reprod Sci. 2015 Oct;8(4):191.

Chapter 8
Oxidant-Sensitive Inflammatory Pathways and Male Reproductive Functions

Sulagna Dutta, Pallav Sengupta, and Srikumar Chakravarthi

Abstract Male component is the major contributing factor in over half of all cases of infertility, with over 25% of infertile males having no recognised underlying cause of infertility. In around 40–50% of male infertility cases, oxidative stress (OS)-related processes have been found to be responsible for fertility impairment. Inflammation is a major stress signal leading to OS. Redox imbalance occurs when endogenous antioxidant network fails to curb the excess generation of reactive oxygen species (ROS), leading to activation of stress-sensitive intracellular signalling pathways directed to cellular damage. Oxidant-sensitive-inflammatory pathways are intricate vicious intracellular networking loops that initiate and exaggerate cellular damage, including chronic impact on male reproductive tissues. These mechanisms, however, are poorly known in connection to male reproductive abnormalities. Thus, the goal of this chapter is to explain the oxidant-sensitive-inflammatory pathways in male reproductive organs in a succinct manner, as well as their potential influence on male fertility.

Keywords Infection · Inflammation · Male infertility · Oxidative stress

8.1 Introduction

Infertility affects 10–15% couples of reproductive age across the world [1, 2]. It is described as a couple's inability to conceive naturally after a year of regular, unprotected sexual contact [3]. Infertility has a psychological impact on the affected couples, and it can lead to depression, and other physical and psychological problems [4]. Although fertility declines with age, it is most commonly caused by anatomical abnormalities, endocrinopathies, immunological issues, genetic and epigenetic causes, radiation, chemotherapy and exposure to environmental and occupational

---

S. Dutta (✉)
Faculty of Dentistry, MAHSA University, Jenjarom, Selangor, Malaysia

P. Sengupta · S. Chakravarthi
Faculty of Medicine, Biosciences and Nursing, MAHSA University, Jenjarom, Selangor, Malaysia

© The Author(s), under exclusive license to Springer Nature Switzerland AG 2022
K. K. Kesari, S. Roychoudhury (eds.), *Oxidative Stress and Toxicity in Reproductive Biology and Medicine*, Advances in Experimental Medicine and Biology 1358, https://doi.org/10.1007/978-3-030-89340-8_8

endocrine disrupting chemicals (EDCs) [5–8]. Male factor is the main or significant contributing factor in nearly half of all cases of infertility, with no identified reason discovered in over 25% of infertile men [9, 10]. Oxidative stress (OS)-associated processes are shown to be accountable for the damage of sperm function and fertility in about 40–50% of male infertility cases [11, 12]. The capacity of the organism to withstand the detrimental impacts of reactive oxygen species (ROS) through neutralisation by antioxidant defence mechanisms is disrupted in OS [12]. ROS, which includes superoxide anion ($O_2\bullet-$), hydroxyl radical (HO•) and hydrogen peroxide (H_2O_2), are highly reactive oxidizing molecules generated unceasingly with various metabolic processes [11]. Oxidative mechanisms in spermatozoa are especially fascinating owing to their dual role in these cells. Fertilisation is impossible without physiological impacts of ROS to control crucial redox-sensitive processes including capacitation and hyperactivation [13]. Although its supraphysiological level of ROS impairs regular sperm functions such as motility, capacity, acrosome reaction, oocyte penetration and sperm head decondensation, which are important for conception [11, 12]. In the seminiferous tubules, haploid spermatozoa are produced during spermatogenesis, which is a metabolically active biological process. As a natural by-product of cellular respiration, O_2 is produced during this process. Sertoli cells, which contain endogenous antioxidant enzymes (superoxide dismutase (SOD), catalase, transferase, peroxidase and glutathione reductase) in high levels, protect germ cells (differentiating to spermatids) from OS in testes [14, 15]. The spermatozoa become sensitive to oxidative damage once they are propelled out from the germinal epithelium because they are no longer protected by the Sertoli cell defence system [11, 16]. ROS in excess can damage DNA, lipids and proteins in cells, causing cellular injury [17–19]. Therefore, for the optimum sperm functions, ROS must be maintained at physiological levels to preserve cellular homeostasis and to regulate the redox-sensitive signal-transduction processes influencing fertility [13].

OS is closely associated with inflammation [20]. The oxidant-sensitive inflammatory pathways are complex and follow a vicious loop of intracellular networking [21]. The result is exaggeration of cellular damage and chronic diseases, which include disruption of male reproductive tissues affecting male fertility [8, 20]. However, these pathways in relation to male reproductive disruptions are poorly understood. Thus, the present chapter aims to concisely portray the oxidant-sensitive inflammatory pathways in the male reproductive tissues and their possible impact upon male fertility.

8.2 ROS and Oxidative Stress

In biology and medicine, there are three main types of reactive species: (a) reactive species of oxygen (ROS), (b) reactive nitrogen species (RNS), and (c) reactive chlorine species (RCS). In terms of structure, a reactive species can be either a free radical or a non-radical [22]. ROS consists of both oxygen radicals and some non-radicals, which are either oxidising agents or are effortlessly transformed to radicals. Radical and non-radical ROS include superoxide (O_2) and hydrogen peroxide (H_2O_2), respectively. Likewise, another collective term, reactive nitrogen species is used for nitric oxide radicals and non-radicals (peroxynitrite, ONOO−); while the reactive chlorine species refers to atomic chlorine radicals and non-radicals (hypochlorous acid, HOCl) [23, 24].

Superoxide anion is a crucial primary reactive species generated in the cells. Other reactive species with pathophysiological relevance, such as H_2O_2, hydroxyl radical (OH·), and ONOO−, are also generated via the downstream oxidative reaction cascades [24, 25]. In mammals, the endogenous sources of ROS include the mitochondrial electron transport chain, NADPH oxidases, cyclooxygenases, xanthine oxidases, nitric oxide synthases, lipoxygenases and cytochrome P450s [26, 27].

8.3 Oxidative Tissue Injury

ROS-facilitated oxidative damage to cellular macromolecules can be induced in a variety of ways. One of these pathways is initiated by interaction between two usually found free radicals: $O_2\bullet- + NO\bullet = ONNO^-$ (peroxynitrite). Peroxynitrite is the product, which protonates quickly to peroxynitrous acid (ONOOH) at physiological pH. Proteins, lipids, and DNA can all be damaged by this potent oxidising and nitrating agent [23].

Nitrotyrosine is produced by nitration of protein tyrosine residues and is frequently utilised as a biomarker for nitrosative and oxidative stress. However, because there are numerous additional nitrating agents in vivo, nitrotyrosine cannot be used as specific marker of peroxynitrite production [28]. Protein nitration, on the other hand, is extremely harmful to the cell or organism. Filament assemblies with important pathogenic implications can be disrupted by nitration of structural proteins, comprising actin and neurofilaments [29]. Nitration of signalling molecules or transcription factors, on the other hand, might substantially change the physiological functions of those proteins [30]. In addition, peroxynitrite promotes calcium-based mitochondrial malfunction and cell death through calpain activation [31].

Lipid peroxidation and DNA hydroxylation caused by hydroxyl free radicals are two more significant oxidative damage mechanisms. In chemistry, these radical species are the extremely reactive, since they can impact almost every cellular molecule [32]. They are able to react with the DNA guanine ring structure to form an adduct, 8-hydroxy-2′-deoxyguanosine (8-OHdG) radical that can generate and propel a chain reaction inducing chemical alteration in DNA bases as well as can cause DNA strand breaks. Improper DNA repair may cause mutations, cell growth arrest or apoptosis [33]. By interacting with membrane lipids, the radical can potentially start a chain reaction, resulting in lipid peroxidation. Lipid peroxidation has the overall impact of decreasing membrane fluidity, increasing membrane leakiness, and damaging membrane proteins, enzymes, deactivating receptors and ion channels [23].

8.4 Oxidative Stress and Male Infertility

Aetiology of male infertility is yet unknown in roughly half of the cases, but it is evident that about 30–80% of infertile men bear high ROS in their ejaculate. Because of the significant connection between OS and male infertility, Agarwal et al. have coined the term 'Male Oxidative Stress Infertility (MOSI)' to define OS-related male infertility [34].

Leukocytes in seminal fluid and immature sperm with a morphologically defective head and cytoplasmic retention are the two primary sources of endogenous ROS in human semen [35–37]. Extrinsic ROS are generated during male genital tract infection and the chemotaxis and activation of leukocytes drive further inflammatory responses. Leukocytes activate the myeloperoxidase system to break down pathogens and in turn ROS is generated [38]. Excessive ROS generation by leukocytes can result in OS in the seminal fluid. On the other hand, abnormal and immature spermatozoa are the sources of intrinsic ROS. During the normal spermiogenesis process, cytoplasm accumulates in the mid-piece fall off, causing cell elongation and condensation. Immature spermatozoa with morphological abnormalities retain the excess residual body, which contains high levels of cytosolic glucose-6-phosphate dehydrogenase (G6PD) enzyme and generate intracellular nicotinamide adenine dinucleotide phosphate (NADPH). NADPH is then converted to ROS by the intramembrane NADPH oxidase NOX5 [39].

When highly reactive ROS outstrip antioxidant defence mechanisms, the homeostatic equilibrium between ROS and antioxidants is disrupted, which can contribute to the development of OS. It can have damaging effects on sperm, such as lipid peroxidation (LPO), sperm DNA fragmentation (SDF) and germ cell apoptosis.

8.4.1 Lipid Peroxidation

Plasma membrane of spermatozoa have abundant polyunsaturated fatty acids (PUFAs), especially docosahexaenoic acid, with its non-conjugated methylene groups having six double bonds. Increased generation of ROS promotes sperm membrane PUFA peroxidation which results in cellular dysfunctions owing to impaired membrane integrity and fluidity that are needed for effective fusion of sperm–oocytes following capacitation and biochemical acrosome reaction cascade [11, 12]. Besides damaging the cell membrane of sperm, LPO by-products damage the mitochondrial proteins of the electron transport chain, resulting in electron leakage and a decline in mitochondrial membrane potential, ATP generation and sperm motility [40, 41]. The first phase of LPO is 'initiation', referring to hydrogen atoms extraction in carbon-carbon double bonds from an unsaturated fatty acid to produce free radicals. The second step is 'propagation', which involves the production of lipid radicals and their rapid interaction with oxygen to produce peroxyl radicals [42]. When metals like copper and iron are present, peroxyl radicals can snip a hydrogen atom from an unsaturated fatty acid when are present, resulting in a lipid radical and lipid hydrogen peroxide [43]. The final phase is called 'termination', in which the generated radicals react with subsequent lipids to produce damaging aldehydes and other end products. Malondialdehyde (MDA), 4-hydroxynonenal (4-HNE) and acrolein are the primary products of LPO. MDA is a vital biomarker in PUFA peroxidation analysis and monitoring [44, 45].

8.4.2 Sperm DNA Fragmentation

Excess ROS generation and low antioxidant capacities in sperm might potentially result in SDF [17]. Via sperm caspase activation and induction of endonuclease, OS can directly or indirectly damage sperm DNA. SDF is triggered by DNA susceptibility due to a chromatin condensation error in the course of spermiogenesis, which leads to chromatin structure substitution failure from histone to protamine. It is reported that following spermiation, during the migration of spermatozoa through rete testis, from seminiferous tubules to the cauda epididymis, excess of ROS exposure causes DNA damage [13, 46]. This results in the formation of 8-OH-guanine and 8-OH-2′-deoxyguanosine (8-OHdG),75, oxidised guanine adducts. Increased 8-OHdG levels are linked to DNA fragmentation and strand breaks.

In both single- and double-stranded (ds-) forms, DNA fragmentation may occur [17, 41]. The DNA repair can take place only at specific spermiogenesis stages, and during the nuclear condensation of the epididymis it is no longer activated. The human oocyte, provides the next opportunity for ss-DNA break repair; however, the efficacy to SDF repair diminishes with advanced maternal age [47]. In the absence of repair, a ds-DNA break causes genomic instability and cell death [48]. The existence of unrepaired SDF over the crucial threshold, also known as the 'late paternal effect', is said to have a negative impact on embryo development and pregnancy outcome [49]. At the second day of human embryo development (4-cell stage), significant activation of embryonic genome expression occurs in a cleavage-stage embryo, and embryogenesis shifts from maternal factor dependency to embryo's own genome dependence [50]. As a result, after fertilisation, a spermatozoon containing SDF has a detrimental impact on blastulation, implantation and pregnancy outcomes. In addition, Kuroda and colleagues found that OS had a negative impact on cleavage embryo development, a phenomenon known as the 'early paternal effect' [51]. A number of studies have looked into the link between ART outcomes and SDF [52, 53]. A meta-analysis showed inverse correlation between SDF with pregnancy outcome and positive correlation with miscarriage [54]. Appropriate measurement and managing can lessen the problem on couples since SDF is one of the causatives of recurrent pregnancy loss.

8.4.3 Apoptosis

Apoptosis is recognised as physiologically planned cell death owing to fragmentation of DNA via various cell death signalling and regulatory mechanisms [55]. Breaks of ds-DNA caused by ROS may lead to apoptosis. ROS also interrupts mitochondrial membranes such that it releases cytochrome-C signalling molecule, capable of activating apoptotic caspases and phosphatidylserine-binding of annexin-V [56]. Significant damage to mitochondria in infertile people may be induced by excessive amounts of cytochrome-C in seminal plasma [57].

8.5 Stress Response Pathways in the Male Germ Cells

Eukaryotic cells have a highly conserved signalling mechanism to adapt and respond to stress conditions, called the unfolded protein response (UPR), which promotes cell survival. Apoptosis signalling is triggered in the case of chronic stress or UPR dysfunction. HSR, UPR^{mt} and UPR^{ER} are three transcription factors and molecular chaperones that, reliant on the site of UPR-mediated protein aggregates and the kind of stress, activate various other molecular chaperones and transcription factors that rectify protein-folding. If UPR mechanisms fail to restore protein conformation and proteins continue to aggregate in the cell, degradation pathways such as ubiquitin-proteasome-mediated proteolysis (ERAD) and autophagy may be triggered. There is some debate over the presence of an active proteasome and alternative mechanisms for the destruction of low-quality and/or misfolded proteins in sperm. However, recent research suggests that autophagy is present and activated in ejaculated human spermatozoa and germ cells, and that it is involved in cell survival and motility control [58], as well as reaction to chemicals like Cadmium [59]. Furthermore, constituents of the ubiquitin-proteasome system, notably ubiquitin enzymes, have been found in mature spermatozoa [60]. But on the other hand, proteasomes in spermatozoa are poorly understood. These components are important in sperm DNA repair, sperm capacitation and acrosome response, and accumulating data indicated that an aberrant ubiquitin-proteasome system causes sperm deformity and male infertility [60–62].

Sperm cells are susceptible to environmental changes once they leave the testes. The female reproductive system can reach high temperatures [63], and tobacco, alcohol, and a variety of environmental toxins can all have an effect on sperm functions and fertility [64, 65]. These cellular stresses frequently cause a disruption in protein turnover and folding, resulting in an accumulation of faulty proteins or proteins that would otherwise be removed during the maturation process. The mechanism by which the UPR is activated in response to this event in ejaculated human sperm is unknown. This is owing to the fact that sperm are dormant cells in terms of its central dogma, and that several of these signalling pathways lead to the transcription of numerous genes important in maintaining proteostasis. Nevertheless, latest evidence pertaining to translation of nuclear-encoded protein in mature spermatozoa is coming to the fore [66, 67]. Gur and Breitbart published a pioneering revelation that during sperm capacitation labelled amino acids get incorporated into polypeptides, and the use of particular protein translation inhibitors has a substantial influence on sperm motility, capacitation and in vitro fertilisation outcomes [66].

8.6 Sperm Mitochondrial Response to Stress

With mitochondrial matrix stress, c-Jun N-terminal kinase (JNK2) activates the transcription factor c-Jun, which links the elements of AP-1 and stimulates transcription of the C/EBP Homologous Protein (CHOP) and CCAAT/enhancer-binding protein beta (CEBPB) transcription factors. CHOP and CEBPB dimers are linked to the CHOP element in the UPR^{mt} gene promoter which encodes mitochondrial quality regulator proteins, such as HSPD1 (HSP60), caseinolytic peptidase P protection protease (P) and mitochondrial components of import. There is presently no information on the signal inputs which initiate stress in the mitochondrial matrix and cause c-Jun N-terminal kinases (JNK2) activation, although the detection of unfolded proteins seems to be mediated by Heat Shock Protein Family D (HSPD) 1 by triggering eukaryotic Initiation Factor 2 (eIF2) phosphorylation, the double-strand RNA–activated protein kinase (PKR) suppresses global cytosolic protein translation. ROS activate protein kinase-B (AKT1) in the intermembrane space (IMS), which phosphorylates oestrogen receptor-α (ESR1). By boosting the activity of the proteasome 26S and promoting the transcription of the IMS protease high-temperature requirement (HtrA)-2 and the mitochondrial regulator Nuclear Respiratory Factor 1 (NRF1), activated ESR1 promotes IMS protein quality control.

8.7 Oxidative Stress and Inflammation: A Vicious Loop

Inflammation is usually thought of as a multimodal response in vascularised connective tissue to external and endogenous stimuli. This protective response's ultimate objective is to rid the organism of both the initial source of cell damage and the effects of that injury. Excessive or uncontrolled protracted inflammation, on the other hand, can cause tissue damage and is the source of many chronic illnesses [68]. Infiltration of the inflammatory cells such as neutrophils, monocytes and lymphocytes to the location of the stimulation is a crucial component of inflammation. Leucocytes infiltrate the inflammatory region and are strongly coordinated by the margination, rolling and adhesion of leucocytes along vascular endothelium, and migration via chemotactic stimuli. Functions of adhesion machineries, such as intercellular adhesion molecule-1 (ICAM-1), selectins and vascular cell adhesion molecule-1 (VCAM-1), along with their respective leukocyte receptors, and chemokines like interleukin-8 (IL-8) or monocyte chemoattractant protein 1 (MCP-1) are required for inflammatory cellular infiltration [68]. Numerous enzymes (elastase, collagenase, proteases, acid hydrolases, lipases, phosphatases etc.), ROS and other inflammatory mediators (complement components, eicosanoids, nitric oxide, cytokines, chemokines etc.) are released by activated inflammatory cells at the site of inflammation [68].

ROS are products of normal cellular metabolism that play an important role in the activation of signalling pathways in plant and animal cells in response to variations in intracellular and extracellular environmental circumstances [69]. The mitochondrial respiratory chain generates the majority of ROS in cells, as previously stated [70]. Moreover, it is also being discussed earlier that when ROS production exceeds the threshold of physiological buffering capacity, as in case of inflammation, they trigger OS and uncontrolled oxidative reactions with non-target intracellular components, damaging cellular membranes, nucleic acids, lipids and various proteins. The modifications in these macromolecules, especially in the proteins, lipids and DNA, may also lead to mutagenesis [21, 71]. OS can activate a vast spectra of transcription factors that overlap with inflammatory signalling pathways, namely, the nuclear factor kappa light chain enhancer of activated B cells (NF-κB), p53, HIF-hypoxia-inducible factor 1α, activator protein 1 (AP-1), peroxisome proliferator-activated receptor γ (PPAR-γ), Nrf2 and β-catenin/Wnt. These activated transcription factors may be responsible for more than 500 different gene expressions, which contribute to the progression of inflammatory pathways [72, 73].

Thus, chronic OS can induce chronic inflammation, which can lead to a variety of chronic illnesses. The consensus of whether OS triggers inflammation or it is the other-way-round is difficult to reach, since OS and inflammation are interdependent pathophysiological processes. Thus, inflammation can also be the primary trigger for generation of the highly toxic ROS and pro-inflammatory mediators, cytokines, bioactive lipids and enzymes, modulating cellular and vascular functions, activating the immune cells and leading to adverse consequences. Pervasive apoptosis, vascular disruption, tissue remodelling and fibrosis are the main consequences of a local inflammatory response, all of which are detrimental to the tissues involved. Nevertheless, it is becoming patently apparent that systemic inflammation and illness, even including numerous inflammatory diseases that do not entail infection or overt injury to the reproductive organs, can affect male reproductive function [74, 75]. The association of OS and inflammation is one of the most fundamental conundrums in male reproductive biology, with the underlying processes still elusive.

8.8 OS and Inflammation in Testes

Until the advent in male infertility research in the last few decades, the mechanisms of inflammation-mediated male reproductive disruptions remained as ignored notions. O'Bryan et al. in 2000 had demonstrated that systemic inflammation in male rat models, induced via bacterial virulent factor (lipopolysaccharide, LPS) can be detrimental to male reproductive functions [76]. Followed by this observation, there had been several studies that explored the possibilities of inflammation-induced male infertility [8, 20, 77, 78]. Inflammatory mediators such as the ROS, pro-inflammatory cytokines, interleukin-1β (IL1β) and tumour necrosis factorα (TNFα), nitric oxide and prostaglandins inhibit hypothalamic luteinizing hormone (LH)-releasing hormone secretion, pituitary LH production and Leydig cell cholesterol mobilisation, resulting in loss of steroidogenic functions. The steroidogenic activity of Leydig cells is also inhibited by inflammation-activated neural pathways and corticosteroids, which are produced in response to inflammation [79, 80]. However, these studies provide only inadequate, and sometimes conflicting, fragments of the puzzle.

Recognition of particular motifs, or pathogen-associated molecular patterns (PAMPs), found on bacterial, viral, fungal and protozoan pathogens, activates immune response and inflammation during infection. This recognition is mediated by unique pattern recognition receptors (PRRs). Toll-like receptors (TLR) are the most well-studied of these receptors, and they detect bacterial and viral nucleic acids, as well as other pathogen-specific molecules including LPS, peptidoglycans and bacterial lipopeptides [81, 82]. TLRs are primarily expressed by myeloid cells (macrophages, monocytes, and dendritic cells), although they are also present in epithelial and connective tissue cells. Among the testicular cells, TLRs expression is the highest in the Sertoli cells. Stimulation of Sertoli cells by TLR ligands may trigger inflammatory signalling pathways, such as the Myeloid differentiation primary response (MyD88), that lead to mitogen-activated protein (MAP) kinases and inflammatory transcription factors such as the NFkB and interferon regulatory factor 3 (IRF3) [83, 84]. This activates a gene expression profile typically triggered by inflammation, including the IL1 alpha (the cell-associated form of IL1B), IL6 and nitric oxide synthase 2 (NOS2 or inducible NOS), as well as the immunoregulatory cytokine activin A [83, 85, 86]. Many of these endogenous inflammatory mediators, including TNF, type 1 and type 2 interferons, IL1A and IL1B, nitric oxide and transforming growth factor beta 3 (TGFβ3), can also be responded to by Sertoli cells [87, 88]. There is substantial evidence that these molecules, which are commonly linked with inflammation, and the pathways via which they exert their effects, play critical roles in facilitating intercellular communication within the seminiferous epithelium [73, 89]. Throughout the cycle of the seminiferous epithelium, these inflammatory mediators control spermatogenic cell mitosis and meiosis, the formation of the Sertoli cell cytoskeleton and intercellular connections and different Sertoli cell functions. As a result, pathogenic molecules, cytokines and other inflammatory mediators generated inside the testis or entering the blood during systemic inflammation interfere with the normal activities of Sertoli cells and spermatogenic cells, causing spermatogenesis to be disrupted. Leydig cells also express a variety of PRRs which recognise pathogenic constituents and trigger inflammatory pathways. Among these PRRs, TLRs were the first discovered receptors and were first found in Leydig cells of mouse. The Leydig cells possess high expression levels of TLR3 and TLR4 [90]. The TLR3-mediated innate immune pathway in these cells leads to activation of NF-κB and IRF3. This is followed by induction of proinflammatory cytokines, including IL-6 and TN-α, as well as IFN-α and β. Both TLR3 and TLR4 pathways activation reportedly suppress testosterone synthesis in the Leydig cells, which is caused by the action of the TLR-induced high levels of cytokines, TNF-α and IL-6 [91, 92]. TLRs are present throughout the reproductive tract epithelium, including the epididymis, in addition to the testis. However, evidence from many investigations in rodents show locational variations in PRRs expression, with TLRs 1–6 being expressed highly in the testis and TLRs 7, 9 and 11 being expressed more strongly in the epididymis, vas deferens or both [83–86].

As per the evidences presented above, TLRs primarily activate the NF-κB pathway, cause inflammatory responses, and can initiate OS. During inflammation, the testicular macrophages play a critical role in stimulating inflammatory factors and ROS, that consequently disrupt gonadal steroidogenesis, reduction of testosterone levels and thereby also affect normal spermatogenesis [93]. In physiological condition, testicular macrophages contribute to testicular immune privilege and display diminished inflammatory functions [94]. In men with infertility-related orchitis or other inflammatory conditions, the number of macrophages with high phagocytic activity in testicular interstitial tissue is reduced, while the number of macrophages secreting copious inflammatory cytokines is raised. These macrophages respond to TLRs-induced activation of p38 MAPK and NF-κB followed by the release of inflammatory agents and establishment of OS causing testicular oxidative damage [95]. Leydig cell malfunction disrupted mitochondrial physiology, the inability of steroidogenic acute regulatory protein (StAR) to stimulate cholesterol transport into mitochondria, and suppression of steroid hormone synthesis are all caused by the inflammation-induced oxidative damage [96]. Nitric oxide production is also influenced. GCHI (guanosine triphosphate cyclohydrolase I) is a rate-limiting enzyme in the production of tetrahydrobiopterin, which is required for inducible nitric oxide synthetase activation (iNOS). The NF-κB pathways also control the transcription of iNOS [97]. ROS are primarily produced by nicotinamide adenine dinucleotide phosphate (NADPH) oxidase, and the activity of NF-κB is controlled by the direct interaction of TLR4 with NADPH oxidase 2 (Nox2), one of the NAPDH oxidase subunits [98].

8.9 Summary

Inflammation and OS follow a vicious loop of intracellular pathway that engages in cellular damage and chronic diseases, including disruption of male reproductive tissues and impacts on male fertility (Fig. 8.1). Pathogenic virulent factors are recognised by PRRs, which are differentially expressed in male reproductive tissues, and may trigger MyD88 and MAPK pathways as discussed above, thereby activating the inflammatory transcription factor, NF-kβ and IRF3 [8, 99].

This in turn upregulates transcription of inflammatory gene subsets [99]. These inflammatory mediators induce infiltration of activated leukocytes causing further increase of immune biomolecules as well as an exaggerated generation of ROS inducing testicular OS. Sertoli cells are sensitive to endogenous inflammatory

8 Oxidant-Sensitive Inflammatory Pathways and Male Reproductive Functions   175

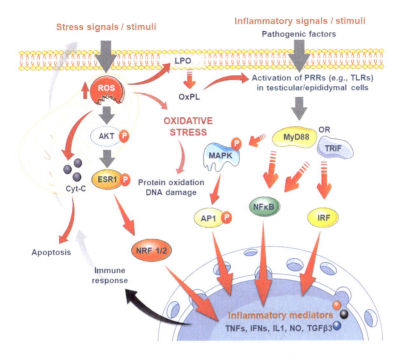

Fig. 8.1 Mechanisms of oxidant-sensitive pathways in relation to inflammation and male infertility. Inflammatory stimuli activate the pattern recognition receptors (PRRs) in the testicular and epididymal cells. The PRRs trigger downstream signalling via Myeloid differentiation primary response (MyD88) and mitogen-activated protein (MAP) kinases pathways to activate the transcription factors, such as the nuclear factor kappa light chain enhancer of activated B cells (NF-κB), activator protein 1 (AP1) and interferon regulatory factor 3 (IRF3). On the other hand, oxidative stress (OS) can be induced by excess ROS generation, causing oxidation of membrane phospholipids and intracellular proteins which again can trigger the PRRs-inflammatory pathway. Moreover, ROS can act through AKT (Protein kinase B) and oestrogen receptor (ESR) 1 to activate the transcription factor, Nuclear Respiratory Factor (NRF) 1 and 2. These pathways-induced activated transcription factors enhance the expressions of various inflammatory mediators, such as the tumour necrosis factors (TNFs), interferons (IFNs), interleukin (IL) 1, nitric oxide (NO) and tumour growth factor (TGF) β3, which lead to exaggerated inflammation as well as can act as OS stimuli, thereby operating in a loop. In addition, OS can trigger apoptotic cascades by aiding the release of cytochrome-c (Cyt-c) from mitochondria, thereby inducing cellular apoptosis

mediators, most notably IL1A, IL1B, TNF, nitric oxide, transforming growth factor β3 (TGFβ3) and type 1 and type 2 interferons [99]. These molecules may extend their effects on Sertoli cells as well as mediate intercellular communication within the seminiferous epithelium [100]. Consequently, increased levels of inflammatory cytokines produced within the testis, or getting access in the testis from blood circulation in case of severe systemic inflammation, affect the Sertoli cells, damage spermatogenic cells, and impair spermatogenesis [8]. The excessive ROS production acting through PRRs activation continues a positive feedback circuit of OS and inflammation [101]. The chronic inflammatory responses can potentially cause orchitis which also can lead to induction of OS. Finally, OS-induced male infertility and its mechanisms are extensively documented. OS alters semen parameters and affect sperm functions and morphology by the processes of lipid peroxidation of sperm membrane, intracellular oxidative damage to spermatozoa, sperm DNA damage and induction of apoptotic pathways in the germ cells [16]. However, further in-depth research is needed to reveal several gaps in the oxidant-sensitive inflammatory pathways to underpin the pathogenesis of inflammation or OS-induced male infertility.

Conflict of Interest The authors declare that they have no conflict of interest.

Ethical Approval Not Applicable.

Funding None.

References

1. Louis JF, Thoma ME, Sørensen DN, McLain AC, King RB, Sundaram R, et al. The prevalence of couple infertility in the United States from a male perspective: evidence from a nationally representative sample. Andrology. 2013;1(5):741–8.
2. Chandra A, Copen CE, Stephen EH. Infertility service use in the United States: data from the National Survey of Family Growth, 1982–2010. Hyattsville: US Department of Health and Human Services, Centers for Disease Control and Prevention; 2014.
3. Catanzariti F, Cantoro U, Lacetera V, Muzzonigro G, Polito M. Comparison between WHO (World Health Organization) 2010 and WHO 1999 parameters for semen analysis–interpretation of 529 consecutive samples. Arch Ital Urol Androl. 2013;85(3):125–9.
4. Chachamovich JR, Chachamovich E, Ezer H, Fleck MP, Knauth D, Passos EP. Investigating quality of life and health-related quality of life in infertility: a systematic review. J Psychosom Obstet Gynecol. 2010;31(2):101–10.
5. Sengupta P. Environmental and occupational exposure of metals and their role in male reproductive functions. Drug Chem Toxicol. 2013;36(3):353–68.
6. Sengupta P, Banerjee R. Environmental toxins: alarming impacts of pesticides on male fertility. Hum Exp Toxicol. 2014;33(10):1017–39.
7. Ahmad G, Agarwal A. Ionizing radiation and male fertility. In: Male infertility. Cham: Springer; 2017. p. 185–96.
8. Sengupta P, Dutta S, Alahmar AT, D'souza UJA. Reproductive tract infection, inflammation and male infertility. Chem Biol Lett. 2020;7(2):75–84.
9. Kumar N, Singh AK. Trends of male factor infertility, an important cause of infertility: a review of literature. J Hum Reprod Sci. 2015;8(4):191.
10. Sengupta P, Dutta S, Krajewska-Kulak E. The disappearing sperms: analysis of reports published between 1980 and 2015. Am J Mens Health. 2017;11(4):1279–304.
11. Agarwal A, Leisegang K, Sengupta P. Oxidative stress in pathologies of male reproductive disorders. In: Pathology. Amsterdam: Elsevier; 2020. p. 15–27.
12. Agarwal A, Sengupta P. Oxidative stress and its association with male infertility. In: Male infertility. Cham: Springer; 2020. p. 57–68.
13. Dutta S, Henkel R, Sengupta P, Agarwal A. Physiological role of ROS in sperm function. In: Male infertility. Cham: Springer; 2020. p. 337–45.
14. Izuka E, Menuba I, Sengupta P, Dutta S, Nwagha U. Antioxidants, anti-inflammatory drugs and antibiotics in the treatment of reproductive tract infections and their association with male infertility. Chem Biol Lett. 2020;7(2):156–65.

15. Barati E, Nikzad H, Karimian M. Oxidative stress and male infertility: current knowledge of pathophysiology and role of antioxidant therapy in disease management. Cell Mol Life Sci. 2020;77(1):93–113.
16. Dutta S, Majzoub A, Agarwal A. Oxidative stress and sperm function: a systematic review on evaluation and management. Arab J Urol. 2019;17(2):87–97.
17. Selvam MKP, Sengupta P, Agarwal A. Sperm DNA fragmentation and male infertility. In: Genetics of male infertility. Cham: Springer; 2020. p. 155–72.
18. Alahmar AT, Calogero AE, Sengupta P, Dutta S. Coenzyme Q10 improves sperm parameters, oxidative stress markers and sperm DNA fragmentation in infertile patients with idiopathic oligoasthenozoospermia. World J Men's Health. 2021;39(2):346.
19. Alahmar AT, Sengupta P, Dutta S, Calogero AE. Coenzyme Q10, oxidative stress markers, and sperm DNA damage in men with idiopathic oligoasthenoteratospermia. Clin Exp Reprod Med. 2021;48(2):150.
20. Azenabor A, Ekun AO, Akinloye O. Impact of inflammation on male reproductive tract. J Reprod Infertil. 2015;16(3):123.
21. Reuter S, Gupta SC, Chaturvedi MM, Aggarwal BB. Oxidative stress, inflammation, and cancer: how are they linked? Free Rad Biol Med. 2010;49(11):1603–16.
22. Halliwell B, Whiteman M. Measuring reactive species and oxidative damage in vivo and in cell culture: how should you do it and what do the results mean? Brit J Pharmacol. 2004;142(2):231–55.
23. Halliwell B. Reactive species and antioxidants. Redox biology is a fundamental theme of aerobic life. Plant Physiol. 2006;141(2):312–22.
24. Dutta S, Sengupta P. Role of nitric oxide in male and female reproduction, Malaysia: Malaysian Journal of Medical Sciences; 2021.
25. Münzel T, Afanas'ev IB, Kleschyov AL, Harrison DG. Detection of superoxide in vascular tissue. Arterioscler Thromb Vasc. Biol. 2002;22(11):1761–8.
26. Schnackenberg CG. Oxygen radicals in cardiovascular–renal disease. Curr Opin Pharmacol. 2002;2(2):121–5.
27. Fleming I, Michaelis UR, Bredenkötter D, Fisslthaler B, Dehghani F, Brandes RP, et al. Endothelium-derived hyperpolarizing factor synthase (Cytochrome P450 2C9) is a functionally significant source of reactive oxygen species in coronary arteries. Circ Res. 2001;88(1):44–51.
28. Halliwell B. What nitrates tyrosine? Is nitrotyrosine specific as a biomarker of peroxynitrite formation in vivo? FEBS Lett. 1997;411(2–3):157–60.
29. Beckman JS, Koppenol WH. Nitric oxide, superoxide, and peroxynitrite: the good, the bad, and ugly. Am J Physiol Cell Physiol. 1996;271(5):C1424–C37.
30. Biswas SK. Lopes de Faria JB. Does peroxynitrite sustain nuclear factor-κB? Cardiovas Res. 2005;67(4):745–6.
31. Whiteman M, Armstrong JS, Cheung NS, Siau JL, Rose P, Schantz JT, et al. Peroxynitrite mediates calcium-dependent mitochondrial dysfunction and cell death via activation of calpains. FASEB J. 2004;18(12):1395–7.
32. Halliwell B. Tell me about free radicals, doctor: a review. J R Soc Med. 1989;82(12):747–52.
33. Evans MD, Dizdaroglu M, Cooke MS. Oxidative DNA damage and disease: induction, repair and significance. Mutat Res. 2004;567(1):1–61.
34. Agarwal A, Parekh N, Selvam MKP, Henkel R, Shah R, Homa ST, et al. Male oxidative stress infertility (MOSI): proposed terminology and clinical practice guidelines for management of idiopathic male infertility. World J. Men's Health. 2019;37(3):296.
35. Aitken RJ, Clarkson JS. Cellular basis of defective sperm function and its association with the genesis of reactive oxygen species by human spermatozoa. Reproduction. 1987;81(2):459–69.
36. Sharma RK, Pasqualotto FF, Nelson DR, Agarwal A. Relationship between seminal white blood cell counts and oxidative stress in men treated at an infertility clinic. J Androl. 2001;22(4):575–83.
37. Koppers AJ, Garg ML, Aitken RJ. Stimulation of mitochondrial reactive oxygen species production by unesterified, unsaturated fatty acids in defective human spermatozoa. Free Rad Biol Med. 2010;48(1):112–9.

38. Iwasaki A, Gagnon C. Formation of reactive oxygen species in spermatozoa of infertile patients. Fertil Steril. 1992;57(2):409–16.
39. Williams A, Ford W. Functional significance of the pentose phosphate pathway and glutathione reductase in the antioxidant defenses of human sperm. Biol Reprod. 2004;71(4):1309–16.
40. Bui A, Sharma R, Henkel R, Agarwal A. Reactive oxygen species impact on sperm DNA and its role in male infertility. Andrologia. 2018;50(8):e13012.
41. Sengupta P, Durairajanayagam D, Agarwal A. Fuel/energy sources of spermatozoa. In: Male infertility. Cham: Springer; 2020. p. 323–35.
42. Aitken J, Fisher H. Reactive oxygen species generation and human spermatozoa: the balance of benefit and risk. BioEssays. 1994;16(4):259–67.
43. Tavilani H, Goodarzi MT, Vaisi-Raygani A, Salimi S, Hassanzadeh T. Activity of antioxidant enzymes in seminal plasma and their relationship with lipid peroxidation of spermatozoa. Int Braz J Urol. 2008;34(4):485–91.
44. Jones R, Mann T, Sherins R. Peroxidative breakdown of phospholipids in human spermatozoa, spermicidal properties of fatty acid peroxides, and protective action of seminal plasma. Fertil Steril. 1979;31(5):531–7.
45. Aitken R, Harkiss D, Buckingham D. Analysis of lipid peroxidation mechanisms in human spermatozoa. Mol Reprod Dev. 1993;35(3):302–15.
46. Sengupta P, Arafa M, Elbardisi H. Hormonal regulation of spermatogenesis. In: Molecular signaling in spermatogenesis and male infertility. Boca Raton: CRC Press; 2019. p. 41–9.
47. Wells D, Bermudez M, Steuerwald N, Thornhill A, Walker D, Malter H, et al. Expression of genes regulating chromosome segregation, the cell cycle and apoptosis during human preimplantation development. Hum Rreprod. 2005;20(5):1339–48.
48. Sakkas D, Alvarez JG. Sperm DNA fragmentation: mechanisms of origin, impact on reproductive outcome, and analysis. Fertil Steril. 2010;93(4):1027–36.
49. Tesarik J, Greco E, Mendoza C. Late, but not early, paternal effect on human embryo development is related to sperm DNA fragmentation. Hum Reprod. 2004;19(3):611–5.
50. Tesařík J, Kopečný V, Plachot M, Mandelbaum J. Activation of nucleolar and extranucleolar RNA synthesis and changes in the ribosomal content of human embryos developing in vitro. Reproduction. 1986;78(2):463–70.
51. Kuroda S, Takeshima T, Takeshima K, Usui K, Yasuda K, Sanjo H, et al. Early and late paternal effects of reactive oxygen species in semen on embryo development after intracytoplasmic sperm injection. Sys Biol Reprod Med. 2020;66(2):122–8.
52. Guerin P, Matillon C, Bleau G, Levy R, Menezo Y. Impact of sperm DNA fragmentation on ART outcome. Gynecol Obstet Fertil. 2005;33(9):665–8.
53. Simon L, Zini A, Dyachenko A, Ciampi A, Carrell DT. A systematic review and meta-analysis to determine the effect of sperm DNA damage on in vitro fertilization and intracytoplasmic sperm injection outcome. Asian J Androl. 2017;19(1):80.
54. Zhao J, Zhang Q, Wang Y, Li Y. Whether sperm deoxyribonucleic acid fragmentation has an effect on pregnancy and miscarriage after in vitro fertilization/intracytoplasmic sperm injection: a systematic review and meta-analysis. Fertil Steril. 2014;102(4):998–1005.
55. Trevelyan SJ, Brewster JL, Burgess AE, Crowther JM, Cadell AL, Parker BL, et al. Structure-based mechanism of preferential complex formation by apoptosis signal–regulating kinases. Sci Signal. 2020;13(622):eaay6318.
56. Shukla KK, Mahdi AA, Rajender S. Apoptosis, spermatogenesis and male infertility. Front Biosci. 2012;4(1):746–54.
57. Latchoumycandane C, Vaithinathan S, D'Cruz S, Mathur PP. Apoptosis and male infertility. In: Male infertility. Cham: Springer; 2020. p. 479–86.
58. Aparicio I, Espino J, Bejarano I, Gallardo-Soler A, Campo M, Salido G, et al. Autophagy-related proteins are functionally active in human spermatozoa and may be involved in the regulation of cell survival and motility. Sci Rep. 2016;6(1):1–19.
59. Li R, Luo X, Zhu Y, Zhao L, Li L, Peng Q, et al. ATM signals to AMPK to promote autophagy and positively regulate DNA damage in response to cadmium-induced ROS in mouse spermatocytes. Environ Pollut. 2017;231:1560–8.

60. Hou C-C, Yang W-X. New insights to the ubiquitin–proteasome pathway (UPP) mechanism during spermatogenesis. Mol Biol Rep. 2013;40(4):3213–30.
61. Zigo M, Kerns K, Sutovsky M, Sutovsky P. Modifications of the 26S proteasome during boar sperm capacitation. Cell Tissue Res. 2018;372(3):591–601.
62. Kerns K, Morales P, Sutovsky P. Regulation of sperm capacitation by the 26S proteasome: an emerging new paradigm in spermatology. Biol Reprod. 2016;94(5):117. 1–9
63. Ng KYB, Mingels R, Morgan H, Macklon N, Cheong Y. In vivo oxygen, temperature and pH dynamics in the female reproductive tract and their importance in human conception: a systematic review. Hum Reprod Update. 2018;24(1):15–34.
64. Mínguez-Alarcón L, Hauser R, Gaskins AJ. Effects of bisphenol a on male and couple reproductive health: a review. Fertil Steril. 2016;106(4):864–70.
65. Henriques MC, Loureiro S, Fardilha M, Herdeiro MT. Exposure to mercury and human reproductive health: a systematic review. Reprod Toxicol. 2019;85:93–103.
66. Gur Y, Breitbart H. Mammalian sperm translate nuclear-encoded proteins by mitochondrial-type ribosomes. Genes Dev. 2006;20(4):411–6.
67. Zhu Z, Umehara T, Okazaki T, Goto M, Fujita Y, Hoque S, et al. Gene expression and protein synthesis in mitochondria enhance the duration of high-speed linear motility in boar sperm. Front Physiol. 2019;10:252.
68. Cotran RS. Acute and chronic inflammation. In: Robbins pathological basis of disease. Philadelphia: Saunders; 1999. p. 50–88.
69. Jabs T. Reactive oxygen intermediates as mediators of programmed cell death in plants and animals. Biochem Pharmacol. 1999;57(3):231–45.
70. Poyton RO, Ball KA, Castello PR. Mitochondrial generation of free radicals and hypoxic signaling. Trends Endocrinol Metab. 2009;20(7):332–40.
71. Fridovich I. The biology of oxygen radicals. Science. 1978;201(4359):875–80.
72. Dreger H, Westphal K, Wilck N, Baumann G, Stangl V, Stangl K, et al. Protection of vascular cells from oxidative stress by proteasome inhibition depends on Nrf2. Cardiovasc Res. 2010;85(2):395–403.
73. Hedger MP. Toll-like receptors and signalling in spermatogenesis and testicular responses to inflammation—a perspective. J Reprod Immunol. 2011;88(2):130–41.
74. Baker HG. Reproductive effects of nontesticular illness. Endocrinol Metab Clin N Am. 1998;27(4):831–50.
75. Dong Q, Hawker F, McWilliam D, Bangah M, Burger H, Handelsman DJ. Circulating immunoreactive inhibin and testosterone levels in men with critical illness. Clin Endocrinol. 1992;36(4):399–404.
76. O'Bryan M, Schlatt S, Gerdprasert O, Phillips DJ, de Kretser DM, Hedger MP. Inducible nitric oxide synthase in the rat testis: evidence for potential roles in both normal function and inflammation-mediated infertility. Biol Reprod. 2000;63:1285–93.
77. Dutta S, Sengupta P, Chhikara BS. Reproductive inflammatory mediators and male infertility. Chem Biol Lett. 2020;7(2):73–4.
78. Bhattacharya K, Sengupta P, Dutta S, Karkada IR. Obesity, systemic inflammation and male infertility. Chem Biol Lett. 2020;7(2):92–8.
79. Gow RM, O'Bryan M, Canny B, Ooi GT, Hedger M. Differential effects of dexamethasone treatment on lipopolysaccharide-induced testicular inflammation and reproductive hormone inhibition in adult rats. J Endocrinol. 2001;168(1):193–202.
80. Ogilvie KM, Held Hales K, Roberts ME, Buchanan Hales D, Rivier C. The inhibitory effect of intracerebroventricularly injected interleukin 1β on testosterone secretion in the rat: role of steroidogenic acute regulatory protein. Biol Reprod. 1999;60(2):527–33.
81. Dutta S, Sengupta P, Hassan MF, Biswas A. Role of toll-like receptors in the reproductive tract inflammation and male infertility. Chem Biol Lett. 2020;7(2):113–23.
82. Kawai T, Akira S. Toll-like receptors and their crosstalk with other innate receptors in infection and immunity. Immunity. 2011;34(5):637–50.
83. Riccioli A, Starace D, Galli R, Fuso A, Scarpa S, Palombi F, et al. Sertoli cells initiate testicular innate immune responses through TLR activation. J Immunol. 2006;177(10):7122–30.

84. Bhushan S, Tchatalbachev S, Klug J, Fijak M, Pineau C, Chakraborty T, et al. Uropathogenic Escherichia coli block MyD88-dependent and activate MyD88-independent signaling pathways in rat testicular cells. J Immunol. 2008;180(8):5537–47.
85. Wu H, Wang H, Xiong W, Chen S, Tang H, Han D. Expression patterns and functions of toll-like receptors in mouse sertoli cells. Endocrinology. 2008;149(9):4402–12.
86. Winnall WR, Muir JA, Hedger MP. Differential responses of epithelial Sertoli cells of the rat testis to Toll-like receptor 2 and 4 ligands: implications for studies of testicular inflammation using bacterial lipopolysaccharides. Innate Immun. 2011;17(2):123–36.
87. Lui W-Y, Wong C-H, Mruk DD, Cheng CY. TGF-β3 regulates the blood-testis barrier dynamics via the p38 mitogen activated protein (MAP) kinase pathway: an in vivo study. Endocrinology. 2003;144(4):1139–42.
88. Okuma Y, Saito K, O'Connor A, Phillips DJ, de Kretser DM, Hedger MP. Reciprocal regulation of activin A and inhibin B by interleukin-1 (IL-1) and follicle-stimulating hormone (FSH) in rat Sertoli cells in vitro. J Endocrinol. 2005;185(1):99–110.
89. O'Bryan MK, Hedger MP. Inflammatory networks in the control of spermatogenesis. In: Molecular mechanisms in spermatogenesis. New York: Springer; 2009. p. 92–114.
90. Shang T, Zhang X, Wang T, Sun B, Deng T, Han D. Toll-like receptor-initiated testicular innate immune responses in mouse Leydig cells. Endocrinology. 2011;152(7):2827–36.
91. Samir MS, Glister C, Mattar D, Laird M, Knight PG. Follicular expression of pro-inflammatory cytokines tumour necrosis factor-α (TNFα), interleukin 6 (IL6) and their receptors in cattle: TNFα, IL6 and macrophages suppress thecal androgen production in vitro. Reproduction. 2017;154(1):35–49.
92. Ding D-C, Liu H-W, Chu T-Y. Interleukin-6 from ovarian mesenchymal stem cells promotes proliferation, sphere and colony formation and tumorigenesis of an ovarian cancer cell line SKOV3. J Cancer. 2016;7(13):1815.
93. Frungieri MB, Calandra RS, Lustig L, Meineke V, Köhn FM, Vogt H-J, et al. Number, distribution pattern, and identification of macrophages in the testes of infertile men. Fertil Steril. 2002;78(2):298–306.
94. Fehervari Z. Testicular macrophage origin. Nat Immunol. 2017;18(10):1067.
95. Aslani F, Schuppe HC, Guazzone VA, Bhushan S, Wahle E, Lochnit G, et al. Targeting high mobility group box protein 1 ameliorates testicular inflammation in experimental autoimmune orchitis. Hum Reprod. 2015;30(2):417–31.
96. Allen JA, Diemer T, Janus P, Hales KH, Hales DB. Bacterial endotoxin lipopolysaccharide and reactive oxygen species inhibit Leydig cell steroidogenesis via perturbation of mitochondria. Endocrine. 2004;25(3):265–75.
97. Choi YY, Kim MH, Han JM, Hong J, Lee T-H, Kim S-H, et al. The anti-inflammatory potential of Cortex Phellodendron in vivo and in vitro: down-regulation of NO and iNOS through suppression of NF-κB and MAPK activation. Int Immunopharmacol. 2014;19(2):214–20.
98. Kim SY, Jeong J-M, Kim SJ, Seo W, Kim M-H, Choi W-M, et al. Pro-inflammatory hepatic macrophages generate ROS through NADPH oxidase 2 via endocytosis of monomeric TLR4–MD2 complex. Nat Commun. 2017;8(1):1–15.
99. Schuppe H-C, Meinhardt A. Immune privilege and inflammation of the testis. In: Immunology of gametes and embryo implantation, vol. 88. Basel: Karger Publishers; 2005. p. 1–14.
100. Hedger MP. Immunophysiology and pathology of inflammation in the testis and epididymis. J Androl. 2011;32(6):625–40.
101. Delgado-Roche L, Mesta F. Oxidative stress as key player in severe acute respiratory syndrome coronavirus (SARS-CoV) infection. Arch Med Res. 2020;51(5):384–7.

Chapter 9
Oxidative Stress and Idiopathic Male Infertility

Pallav Sengupta, Shubhadeep Roychoudhury ⓘ, Monika Nath, and Sulagna Dutta

Abstract Idiopathic male infertility (IMI) refers to the condition where semen quality declines, but exact causatives are not identified. This occurs in almost 30–40% of infertile men. Traditional semen analyses are extensively used for determining semen quality, but these bear critical shortcomings such as poor reproducibility, subjectivity, and reduced prediction of fertility. Oxidative stress (OS) has been identified as the core common mechanism by which various endogenous and exogenous factors may induce IMI. Male oxidative stress infertility (MOSI) is a term used to describe infertile males with abnormal semen parameters and OS. For the treatment of MOSI, antioxidants are mostly used which counteract OS and improve sperm parameters with appropriate combinations, dosage, and duration. Diagnosis and management of male infertility have witnessed a substantial improvement with the advent in the omics technologies that address at genetic, molecular, and cellular levels. Incorporation of oxidation-reduction potential (ORP) can be a useful clinical biomarker for MOSI. Moreover, various modulations of male fertility status can be achieved via stem cell and next-generation sequencing (NGS) technologies. However, several challenges must be overcome before the advanced techniques can be utilized to address IMI, including ethical and religious considerations, as well as the possibility of genetic abnormalities. Considering the importance of robust understanding of IMI, its diagnosis, and possible advents in management, the present article reviews and updates the available information in this realm, emphasizes various facets of IMI, role of OS in its pathophysiology, and discusses the novel concept of MOSI with a focus on its diagnostic and therapeutic aspects.

P. Sengupta
Faculty of Medicine, Bioscience and Nursing, MAHSA University, Jenjarom, Malaysia

S. Roychoudhury (✉) · M. Nath
Department of Life Science and Bioinformatics, Assam University, Silchar, India

S. Dutta
Faculty of Dentistry, MAHSA University, Jenjarom, Malaysia

© The Author(s), under exclusive license to Springer Nature Switzerland AG 2022
K. K. Kesari, S. Roychoudhury (eds.), *Oxidative Stress and Toxicity in Reproductive Biology and Medicine*, Advances in Experimental Medicine and Biology 1358, https://doi.org/10.1007/978-3-030-89340-8_9

Keywords Antioxidants · Male infertility · Oxidation-reduction potential · Oxidative stress

9.1 Introduction

Male infertility is defined as a male partner's inability to make a healthy fertile female conceive even after one year of unprotected coitus [2]. As much as about 186 million people in the world are infertile [3], among which male factor infertility is involved in 50% of the cases [4]. This further comprises of 50% of unexplained male infertility and those with known causes, while 50% is idiopathic male infertility (IMI) in which semen quality declines, but exact causatives are not identified [5]. Oxidative stress (OS) is reportedly responsible for 30–40% of unexplained male infertility [6] and about 80% of IMI cases [5, 7–9]. An overview of the global incidence of male infertility is presented in Fig. 9.1.

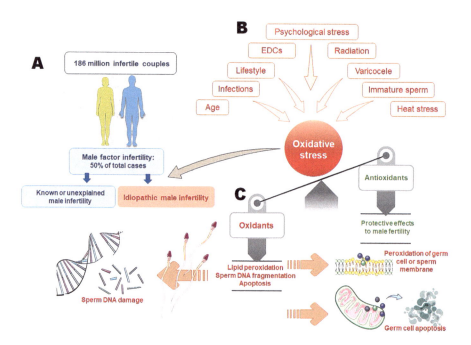

Fig. 9.1 Global incidence of male factor infertility and oxidative stress as its key mechanism. About half of the global infertility cases are due to male factor. Oxidative stress is involved in both unexplained and idiopathic male infertility (**a**). There are numerous endogenous and exogenous factors that can cause the production of reactive oxygen species resulting in oxidative stress (**b**). Oxidative stress can disrupt male reproductive functions by inducing lipid peroxidation and apoptosis of germ cells and mature spermatozoa and also by causing sperm DNA damage (**c**). EDC: endocrine disrupting chemical

Male infertility can be caused by a variety of factors, including reversible and irreversible conditions such as endocrine disruptions (usually due to hypogonadism) at a rate of 2–5%, sperm transport disorders (such as vasectomy) at 5%, and primary testicular abnormalities (includes abnormal sperm parameters without a known cause) at 65–80% [10]. Other factors such as age, medical and surgical history, occupational and environmental exposure to endocrine disruptors, genetic and epigenetic factors, and systemic and urogenital infections could also influence each of the partners [11–14]. Unexplained male infertility and IMI are examples of infertility that arises suddenly or as a result of obscure or inexplicable reasons [15]. Unexplained male infertility is defined as infertility of unknown cause in which normal sperm and female infertility factors have been eliminated [16]. Idiopathic infertility affects males who have an unexplained decrease in sperm quality with no previous history of reproductive issues and normal physical examination and endocrine laboratory testing findings [17]. Males with idiopathic infertility have fewer spermatozoa (oligozoospermia), lower motility (asthenozoospermia), or a higher percentage of morphologically abnormal sperm (teratozoospermia) [17].

The focal point of this chapter is to emphasize the various aspects of IMI and its pathophysiological relationship with OS. We have also tried to present a brief discussion on the major sources of ROS in body and its influence on sperm parameters. Finally, the chapter explores the novel concept of male oxidative stress infertility (MOSI), with particular attention to its diagnostic and therapeutic approach.

9.2 Idiopathic Male Infertility

Idiopathic male infertility, also known as idiopathic oligoasthenoteratozoospermia (IOAT), is a condition in which men's sperm quality declines for no apparent reason [18]. No male infertility factor has been identified in approximately 30–40% of infertile men [1]. These men do not have any prior history of infertility, and normal physical examination, and endocrine laboratory evaluation. Semen analysis of such men shows reduced number of spermatozoa and sperm motility with much abnormality in spermatozoa. These conditions commonly appear together and are termed as oligoasthenoteratozoospermia or OAT syndrome. IMI can be accredited to endocrine disruptors, ROS, or genetic aberrations [1, 19–21]. In idiopathic infertile men, the basis of altered semen parameters cannot be determined even after complete diagnostic investigations [1]. IMI is a complex condition influenced by environmental, hormonal, and genetic factors [22] together with ethnic characteristics [23]. OS seems to be one of the underlying mechanisms of IMI, although the molecular basis has not clearly been established [22]. ROS are required for proper sperm functions including capacitation and acrosome reaction, but excessive ROS can disrupt sperm functions. ROS can damage membranes and DNA through oxidation and peroxidation reactions as plasma membranes contain a large quantity of polyunsaturated fatty acids (PUFAs) [24]. Depending on their concentration, ROS can have both positive and negative impacts on sperm activity [24, 25]. To sustain proper physiological functioning, ROS and RNS should be maintained at optimal levels. ROS are by-products of normal metabolic activities in stem cells, sperm, and ovum; and these reactive species are required for key reproductive processes including capacitation, acrosomal reaction, and hyperactivation. Capacitation of sperm is a critical step in achieving fertilization capability. To complete the process, reactive species such as superoxide, H_2O_2, nitric oxide (NO), and superoxide anion radicals are required [25, 26].

9.3 Pathophysiology of Oxidative Stress and Male Infertility

9.3.1 *Inflammation and OS*

Inflammation is a normal response of the organism against the microbial infection or injury of the tissue [27, 28]. Neutrophils are the initial inflammatory immune cells, but macrophages play an essential role in inflammatory response [29, 30]. A considerable amount of prostaglandin E_2 (PGE_2), cytokines, and NO are released by macrophages and other activated inflammatory cells [29, 31]. However, inflammation is a major factor in steroidogenesis and spermatogenesis. In addition, testosterone and luteinizing hormone (LH) levels in the blood in response to inflammation are significantly reduced [32]. Inflammation also inhibits sperm production and development and overall sperm maturation. Spermatocytes and spermatids act as the major sites for inflammation. Epididymitis also makes prone to inflammatory attacks on the testes. The inflammatory reaction can occur through leukocytes that penetrate semen and produce anti-sperm antibodies. The efficacy of inflammation is increased by reducing the lipid components of the membrane of sperm flagellum, therefore reducing sperm motility and resulting in asthenozoospermia [29, 33]. These events complicate maturation processes such as the reactivity of acrosome, capacitation, and sperm penetration of oocytes [34]. Inflammation is highly associated with OS. Sperm of infertile men comprise elevated levels of ROS besides high proinflammatory and cytokine levels. Although certain exogenous pathogens, including bacteria, can produce ROS, leukocytes are still the most significant source of seminal ROS [11, 35]. Leukocytes indirectly enhance ROS by generating proinflammatory cytokines that function as signals to control cell reactions, for example, inflammation [29, 31]. These proteins can trigger the xanthine oxidase system; they increase ROS levels, and thus initiate OS [36]. On the other hand, active phagocytes generate high levels of ROS via phagocytosis that react with sperm membrane and raise the oxidant-to-antioxidant ratio. Even after the eradication of infections, the OS generated by these responses persists [32]. Sperm count and motility, as well as the morphological defects, are negatively related to seminal concentrations of TNF-α. Proliferation and differentiation of beta (β) cells, T cells, and natural lethal cells cause apoptosis as they increase seminal cytoplasm levels. Interleukin-1α (IL-1α) and IL-1β also induce apoptosis through proliferation and differentiation of β cells, chemical absorption of leukocytes into inflammation site, induction of neutrophils, and production of monocytes. These cytokines also have harmful effects on the quality of semen. Increased levels of IL-1β are associated with decreased sperm motility [37, 38] which is accompanied by a simultaneous increase in ROS and malonaldehyde (MDA) levels [36]. With infection, cytokines, which function as immune-regulating components for the gonads, elevate, causing destruction to seminal tissue. Leukocytospermia is linked to the cytokines generated during inflammation. White blood cells (WBCs) in the semen penetrate the inflammatory site during inflammation; however, leukocytes do not enter the seminal vesicle in normal circumstances. The arteries that supply blood to the testes dilate in inflammatory areas, allowing leukocytes to go to the vascular areas and depart the vascular endothelium. In addition to the aforementioned alterations, greater permeability causes fluid to collect in the reproductive tracts, causing swelling and discomfort as well as the accumulation of immunoglobulins and other seminal proteins. The crosslinking reactions that are initiated by the release of inflammatory mediators are required for the removal of leukocytes from seminal vesicles. As a result of the reproductive tract inflammation, phagocytic cells are absorbed into the damage site. TNF-α stimulates neutrophils, causing them to generate free radicals as well as discharge their granular content, resulting in host defense and tissue damage. As male infertility is frequently linked to infection or damage to the testes, and the inflammatory response has a negative impact on sperm parameters and fertility by producing ROS, all possible measures to eliminate or reduce inflammation should be taken to control OS and improve pregnancy outcome [39].

9.3.2 Varicocele and OS

Varicocele is one of the most common causes of infertility in men, with a prevalence of 15% in the general male population and 40% in infertile men [40]. OS plays a significant role in the pathophysiology of this disease. The possibility of developing varicocele increases with simultaneous increase in ROS level and decrease in concentrations of antioxidants [41]. Under heat stress and hypoxia, three kinds of reproductive system cells are responsible for the production of ROS in males with varicocele. Epididymal cells, endothelial cells in the dilated pampiniform plexus, and testicular cells such as developing germ cells, Leydig cells, macrophages, and peritubular cells are the most common [36]. Heat and hypoxic stress activate Complex III of the electron transport chain in mitochondria, which can lead to the production of ROS. Furthermore, NO produced from testicular and endothelial cells in varicocele-affected testes can nitrosylate complexes I and IV, simulating complex III and releasing ROS [42]. OS has a significant impact on testicular tissues. Infertile males with varicoceles had greater amounts of ROS and MDA, as well as free radicals such as NO, H_2O_2, and hexanoyllysine (another product of lipid peroxidation), as compared to idiopathic infertile men [43]. The spermatic veins are the key elements in the pathophysiology of varicocele. Certain stimuli can also cause endothelial cells to generate ROS. Infertile males with varicocele had greater NO and iNOS levels in their spermatic veins than in other veins [44]. Infertile males with varicocele had lower seminal TAC than healthy men, according to enzyme and non-enzymatic antioxidant measurements [43]. However, it has been discovered that an early rise in antioxidant concentrations such as SOD and CAT is related to a disease-prevention defensive mechanism [45]. ROS level has been high in both infertile men with and without varicocele, but it is significantly greater in the first group, suggesting that varicocele enhances ROS generation and hence oxidative damage. Varicocele is also linked to the severity of OS; therefore, the most severe varicocele indicates the most severe OS [43].

9.3.3 Obesity and OS

Obesity is a condition that is brought on by a combination of environmental, genetic, lifestyle, and physical activity variables [46]. Around 1.9 billion people on the planet are overweight or obese. Obesity has a detrimental impact on male reproductive function [47]. In obese or overweight persons, semen quality determinants such as sperm concentration and motility are lower, and DNA damage is higher than in those with normal BMIs [48]. As a result, the risk of infertility in such men increases by more than 20% [49]. Obesity along with chronic inflammation increases the metabolic rate as well as increases ROS production in the testes tissue and sperm [50]. OS in sperm and testis is positively correlated with increasing BMI and sperm DNA damage but shows a negative relation with reduced sperm motility and acrosome reaction [51]. In addition, gonadal temperature may also impair sperm parameters in obese men. Higher temperatures affect spermatogenesis as optimum temperature for this process ranges between 34 and 35 °C. In obese men, internal temperature of gonads increases due to continuous deposition of fats in the scrotum [52]. Naturally, increasing testicular temperature will significantly intensify OS, thus lowering the sperm motility and concentration and increasing DNA damage [53]. Obesity raises serum-free fatty acid levels, and unsaturated fatty acids are more susceptible to ROS assaults, lowering SOD and raising MDA levels through peroxidation. As a result, the long-term impact of obesity on men's fertility can be detrimental because of higher OS [47]. OS is a very usual complication in several ailments such as male infertility, as spermatogenesis and reproduction are very much associated with ROS. Therefore, normal sperm function requires the equilibrium of ROS and antioxidants, thus natural antioxidant usage, healthy lifestyles, regular exercise, and appropriate food habit may decrease excess weight and gradually lower BMI, reducing OS and promoting normal healthy sperm [54, 55].

9.4 Sources of ROS

Several factors are responsible for the generation of ROS in the body which can ultimately lead to oxidative stress. These factors include infection in genitalia, varicocele, metabolic syndrome, smoking, alcohol and drugs, ionizing radiation, use of mobile phones, psychological stress, severe exercise, and environmental pollution. These factors can be considered as sources of ROS and can be classified as endogenous or exogenous sources [20, 25].

9.4.1 Endogenous Sources

Immature Sperm

During spermatogenesis, sperm are released which may be accompanied by additional amount of cytoplasm from germinal epithelium, where sperm usually remain immature and non-functional in nature. Immature sperm are characterized by cytoplasmic residues in the mid-piece. The sperm-mid-piece cytoplasm includes a glucose-6-phosphate dehydrogenase enzyme (G6PD), which regulates the synthesis, via a shunt called hexose monophosphate, of intracellular β-nicotinamide adenine dinucleotide phosphate (NADPH). NADPH oxidase, which is found in the sperm membrane, provides energy for the generation of ROS [56, 57]. Furthermore, the sperm mid-piece has a high number of mitochondria, which serve as an energy store system for sperm movement. The diaphorase enzyme, which is found in the mitochondrial respiratory chain, maintains the ratio between reduced and oxidized forms of NADH, and therefore contributes to the sperm energy balance. This oxidoreductase enzyme is found in the mitochondrial respiratory chain and generates superoxide anions, which influences the levels of ROS [58]. This may lead to mitochondrial dysfunction causing the generation of ROS production which can further disrupt sperm mitochondrial integrity. Mitochondrial membrane gets damaged by ROS which leads to increased production of ROS. In immature sperm, mitochondrial and cell membrane indeed are two main sites for ROS production [59].

Leukocyte

Natural ejaculation contains a small number of leukocytes. Leukocytospermia is a condition in which WBC count is more than one million per ml of semen [60]. In patients with leukocytospermia, semen shows an increased amount of ROS resulting in damage to sperm DNA [29, 61]. The reproductive system having inflammation or irritation contains a high number of leukocytes in their seminal plasma. By enhancing the production of NADPH through the transfer of hexose monophosphate, peroxidase-positive leukocytes can generate 1000-fold ROS than immature sperm [62]. Infertile men produce more ROS than fertile men, as there is a link between OS and increased leukocyte count [63]. Inflammation or infection can stimulate the body to produce leukocytes or WBCs which can produce up to 100-times more ROS than inactive cells [11, 29]. When the myeloperoxidase (MPO) system in PMN cells and macrophages is activated, it causes a respiratory burst and an increase in ROS generation. Thus, in infertile men, inflammation or infection should be considered clinically, as it might develop into significant health issues caused by OS [64].

9.4.2 Exogenous Sources

Smoking

Cigarette smoking is one of the leading causes of male infertility [65]. Harmful chemicals present in cigarettes such as nicotine can disturb hormonal levels which in turn affect semen parameters [66]. Smoking results in accumulation of ROS in the testis which can have negative impacts on sperm DNA. Cigarette smoke contains a variety of hazardous chemicals, carcinogens, and mutagenic substances, as well as stable and unstable free radicals and ROS. These chemicals promote the formation of superoxide anions and H_2O_2, which induce oxidative damage to cell lipid membranes, proteins, enzymes, and DNA, resulting in male infertility [67]. Creatine kinase (CK) is a protein found in sperm that serves as an energy reserve for fast ATP buffering and rebuilding. It also aids sperm motility. Cigarette smoking also reduces the CK activity [68]. In addition, damage caused to mitochondrial DNA due to increased level of ROS also reduces ATP production and available energy, affecting sperm motility [69]. Moreover, cigarette smoking alters semen characteristics such as semen quality, acrosin activity, and protein phosphorylation, disruption in the expression of micro-ribonucleic acids (miRNAs) and the histone-to-protamine transition resulting in male infertility [70]. Indirect exposure to cigarette smoke can cause DNA damage and affect methylation patterns caused by high ROS levels in tissues and may be considered as an additional side effect of cigarettes [49].

Alcohol Consumption

Both quantity and quality of sperm are deleteriously affected by excessive alcohol consumption. Alcohol intake has been shown to have an inverse relationship with sperm parameters, resulting in decreased sperm motility and concentration, as well as a lower proportion of sperm with normal morphology [71]. Alcohol affects sperm motility, nuclear maturation, and DNA integrity by causing spermatic chromatin abnormalities through apoptosis [59]. After ethanol consumption, its metabolic activities in the liver increase ROS production which leads to structural and functional changes in mitochondria and lowers ATP production [72]. Alcohol consumption enhances the activity of cytochrome P450 enzymes (CYP2E), which boosts NADPH oxidase even further, resulting in altered physiological concentration of certain metals (specifically Cu^{2+} and Fe^{3+}). As a result, the production of superoxide anions increases [72] in men who regularly consume alcohol; generation of nitric oxide (NO) is also increased by iNOS [73]. The mediators of mitochondrial dysfunction include NO and its metabolite, peroxynitrite (ONOO–) [26, 73].

Radiation

Another external source of ROS that affects male fertility is radiation, which has biological effects that vary depending on the kind of radiation, the quantity of energy generated, and the length of exposure to radiation [74]. Ionizing and non-ionizing radiation severely affect the process of sperm formation. Thermal, radioactive, radio frequency (RF), and other harmful radiations have a significant impact on male fertility [75]. In recent years, the usage of mobile phones, computers, wireless systems, and microwave ovens has increased, releasing electromagnetic radiation that has sparked worries regarding male infertility. By inducing OS, electromagnetic radiation causes a variety of alterations in reproductive parameters. Mobile phone radiation has been shown to reduce spermatogenic cells, produce sperm membrane alterations, increase ROS and lipid peroxidation (LPO), as well as decrease sperm count and morphology [76]. Cell phone radiation has the potential to damage plasma membranes and activates plasma membrane NADH oxidase, which is involved in a variety of harmful cellular effects, including OS [77]. Electromagnetic radiation damages the mitochondrial DNA, causing disruptions in the electron transport chain (ETC) and leading to OS [78]. Even little variations in ROS levels can have an impact on sperm capacitation, acrosome reaction, and fertilization, while OS produced by radiofrequency can cause serious damage to spermatozoa [75]. Furthermore, exposure to microwave radiation for two hours per day for 35 days has been shown to cause oxidative changes [79]. Radiofrequency radiation also induces OS which shows a deleterious effect on male fertility by reducing glutathione level and disturbing sperm membrane integrity [80].

Environmental Sources

Genital heat stress is one of the most prominent sources of ROS. Long-term exposure to heat radiation causes scrotal hyperthermia, which raises ROS production substantially [81]. Heat stress has an impact on spermatogenesis, spermatozoa motility, sperm concentration, and sperm viability [82]. Subsequent increase in ROS caused by heat stress may cause over-expression of caspase 3 which leads to apoptosis in several cell types, including Sertoli and Leydig cells [83]. Pollution, which includes phthalate-like compounds, air pollution, and heavy metals, is another important source of ROS in the environment. Phthalates are synthetic compounds found in personal care products, plastics, and food packaging materials, among other places. Increased ROS generation, a lack of testicular antioxidants, and a reduction in hormone levels are all likely effects of phthalate exposure [84]. Phthalates can promote LPO, which can raise the level of OS in the testis, resulting in mitochondrial malfunction and sperm function reduction [85]. Air pollution can elevate OS by generating free radicals through damage of sperm lipid membrane, thereby affecting sperm parameters [86]. In addition, heavy metals such as cadmium and lead are considered another source of ROS which may cause testicular OS and subsequently damage sperm DNA and reduction of sperm parameters [12, 21].

9.5 Impact of Oxidative Stress on Spermatozoa

Homeostatic balance between oxidants and reductants is affected when ROS exceed the antioxidant protection, which can further lead to the occurrence of OS. OS may cause LPO, DNA breakage, and germ cell apoptosis.

9.5.1 Lipid Peroxidation

Sperm are particularly vulnerable to ROS damage because their plasma membranes contain PUFAs with numerous double bonds on the one hand, while cytoplasmic antioxidant enzyme concentrations in sperm are extremely low on the other. Excessive levels of ROS cause LPO cascades to be triggered in OS-induced sperm, resulting in sperm function being impaired [78]. LPO changes the structure and dynamics of sperm lipid membranes, resulting in the generation of a substantial quantity of ROS [87]. A strong source of LPO is hydroxylic radical (·OH). The majority of sperm membrane fatty acids are unsaturated with unconjugated double bonds separated by groups of methylene, which weakens methylene H-C bonds to increase opportunities for hydrogen separation. The formation of free radicals is caused by the separation of hydrogen bonds, and these radicals are stabilized by altering the position of double bonds. Two double bonds are separated by a single bond in this radical. As a result of the enormous number of double bonds responsible for radical production in lipids, they are extremely susceptible to peroxidation. Lipid peroxyl radicals (ROO) react quickly with conjugated radicals, separating hydrogen atoms from other lipid molecules and forming lipid hydroperoxide (ROOH) [88]. A sequence of events, including oxidation of sulfhydryl groups (SH–), decreases axonal protein phosphorylation and raises ROS, lowering sperm motility. Hydrogen peroxide (a kind of ROS) enters the cytoplasm through the sperm membrane and inhibits the activity of specific enzymes, notably G6PD. The glucose entrance into the hexose monophosphate shunt (pentose phosphate pathway) is regulated by this enzyme. This process generates reduced NADPH for cellular reduction reactions under normal physiological circumstances, and blockage of this system reduces the synthesis of NADPH as a reduction equivalent in sperm. Reduced glutathione (SH-G) is used by glutathione peroxidase (GPx), a key antioxidant enzyme in sperm, to decrease ROS. Reduced glutathione is therefore transformed to oxidized glutathione (G-S-S-G). Because NADPH is required for the reduction of oxidized glutathione, a drop in NADPH levels caused by inhibition of the G6PD enzyme reduces the activity of glutathione peroxidase as an antioxidant defense. As a result, the quantity of phospholipid peroxidation rises, reducing membrane fluidity and affecting sperm motility. In different biochemical assays, malondialdehyde (MDA), a by-product of LPO, is utilized to evaluate the amount of peroxidative damage produced to spermatozoa [78]. In addition, loss of electrons from the plasma membrane lipid by ROS may lead to LPO, a series of reduction–oxidation processes, leading to highly mutagenic and genotoxic aldehydes such as MDA, 4-hydroxynonenal (4-HNE), and acrolein [89]. High ROS levels, which lead to the activation of caspases and eventually apoptosis, disrupt mitochondrial membranes as well. During apoptosis, cytochrome-c is constantly generated, increasing ROS levels and perhaps accelerating the apoptotic cycle by increasing DNA damage and fragmentation [7]. As a result, the sperm plasma membrane might be regarded a primary target for ROS, which can damage the genetic makeup of these membranes by activating cascade signaling [90].

9.5.2 Sperm DNA Fragmentation

Increased ROS generation and lower antioxidant levels in the sperm induce SDF. Through the activities of sperm caspase and endonuclease, OS can damage sperm DNA both directly and indirectly [91]. SDF is mostly caused by DNA damage during spermiogenesis as a result of chromatin compaction errors, which results in chromatin structural substitution failure from histone to protamine. This damage is mostly caused by ROS exposure during spermiation or during spermatozoa comigration from seminiferous tubules to cauda epididymis via the rete testes. 8-OH-guanine and 8-OH-2′-deoxyguanosine (8-OHdG) are formed as a result of this reaction [92]. DNA fragmentation and strand breakage are strongly linked with increased concentrations of 8-OHdG. As DNA has a double-helix structure, it can be fragmented in both single-stranded (ss-) and double-stranded (ds-) forms. DNA repair may occur only during particular phases of spermiogenesis, and repair mechanisms do not operate during nuclear condensation in the epididymis. Although the capacity to repair SDF reduces with increasing maternal age, human oocytes can repair ss-DNA breakage, which is a critical stage in embryo development [93]. ds-DNA breakage causes genomic instability and apoptosis in the absence of repair [94]. The unstable SDF affects the development of embryos and pregnancy outcome – also known as the "late paternal effects" [95]. Major activation of embryonic genome expression begins on the second day of embryo development (the 4-cell stage), and embryogenesis then shifts from maternal factor dependency to embryogenesis reliant on the genome of the embryo [95]. As a result, spermatozoa containing SDF have a negative impact on post-fertilization processes such as blastulation, implantation, and pregnancy outcomes. OS also has an influence on the cleavage of embryos, which is known as the "early paternal effect" [96]. SDF negatively associates with pregnancy outcome while having positive correlation with miscarriage. Because SDF can result in recurrent pregnancy loss, appropriate monitoring and management are necessary to reduce the risk [91].

9.5.3 Apoptosis

Apoptosis can be described as non-inflammatory reaction to a sequence of morphological and biochemical changes marked by tissue damages [57]. It plays a vital function in the removal of a defective sperm and hence preserves Sertoli cells' nursing capacity [97]. Increased ROS levels disrupt inner and outer mitochondrial membranes, causing cytochrome-c to be released and caspases to be activated, resulting in apoptosis. ROS-independent mechanisms involving cell surface proteins, Fas, may potentially cause apoptosis in sperm [98]. Fas is a membrane protein that causes apoptosis and belongs to the tumor necrosis factor-nerve growth factor receptor family [99]. When Fas ligand or agonistic anti-Fas binds to Fas, apoptosis is triggered [100]. Bcl-2, on the other hand, which serves as an apoptosis inhibitor gene, protects the cell by producing ROS [101]. Although Fas protein causes apoptosis in most of the cells, abortive apoptosis may allow certain Fas-labeled cells to avoid the process. Because all spermatozoa are not ejaculated, the number of defective spermatozoa in the semen increases. Failure to clear Fas-positive spermatozoa might be caused by a malfunction at one or more stages. To begin with, spermatozoa production may not be adequate to induce apoptosis in men with hypospermatogenesis because they may evade the apoptotic signal. Second, difficulties triggering Fas-mediated apoptosis may result in Fas-positive spermatozoa. Apoptosis is aborted in this scenario, and spermatozoa that are designated for apoptosis are not cleared [102]. When mitochondria are exposed to ROS, apoptosis inducing factor (AIF) is produced, which interacts directly with DNA, causing DNA fragmentation [103].

9.6 Male Oxidative Stress Infertility (MOSI) and Its Diagnosis

Male oxidative stress infertility (MOSI) is a term used to characterize infertile males who have abnormal semen parameters and OS [5]. Traditional semen analysis, which was first developed a century ago, is widely used for determining the quantity and quality of produced sperm. However, determining male infertility, in this approach is not appropriate [104], as it has several flaws, including low reproducibility, subjectivity, and poor fertility prediction [105, 106]. Incorporation of oxidation-reduction potential (ORP) has been recommended as a helpful clinical biomarker for MOSI in males with aberrant semen analysis, given the limited clinical efficiency of standard semen analysis, as well as the universality of OS among the subfertile male population [107, 108]. ORP may be used to assess the balance of antioxidants and oxidants in a wide range of biological fluids, and it is a key component of sperm analysis because of its significant link to sperm dysfunction [109]. OS is measured using a variety of methods, including chemiluminescence for ROS, total antioxidant capacity for antioxidants, and the MDA test for post-hoc LPO damage [110]. Despite their efficacy, these tests are difficult to implement because they are too expensive, complicated, and time-sensitive, and they may also need sophisticated apparatus, and significant technical expertise [110]. Another limitation of these analyses is that they implicate only a single OS marker that either includes the levels of oxidant or antioxidant, or post-hoc damage [104, 106]. ROS measurement in semen is not performed very often these days since it can be subject to intra- and inter-laboratory variability, as well as a long turnaround time and greater expenses [110]. The development of new technologies that can quickly detect seminal OS utilizing a bench-top analyzer to measure ORP in a reproducible manner allows for an accurate and cost-effective diagnosis of MOSI [107]. The male infertility oxidative system (MiOXSYS), a newly designed test for the detection of ORP in semen, is a simple and rapid approach [107]. The ORP test, which is based on a galvanostatic measure of electrons, is a unique approach in the field of infertility. ORP levels have been demonstrated to negatively impact sperm concentration, sperm motility, normal morphology, total motile count as well as sperm DNA fragmentation (SDF), but normal SDF levels can still suggest OS existence [107, 111].

9.7 Antioxidants as Possible Treatment to MOSI

ROS mediate both physiological as well as pathological roles in male reproduction. In case of excess ROS generation, spermatozoa are incapable of repairing oxidative damage because of a lack of cytoplasmic enzymes. Antioxidants are widely applied in andrology as they provide protection to spermatozoa from excess ROS produced by abnormal spermatozoa, infiltrated and resident leukocytes thereby preventing oxidative damage to sperm. Antioxidants have been shown to enhance semen quality in smokers, reduce sperm cryodamage, prevent release of immature sperm, and have positive impact on assisted reproductive technique (ART) outcomes. Three distinct protective systems of antioxidants contribute interdependently to OS reduction in males: endogenous, nutritional, and metal-binding antioxidants [95, 112]. Endogenous antioxidants comprise of those present in spermatozoa and seminal plasma. The main antioxidant enzymes present in semen are superoxide dismutase (SOD), catalase, and glutathione peroxidase/glutathione reductase (GPX/GRD).

However, there are several non-enzymatic antioxidants present in the seminal plasma, for example, urate, ascorbate, pyruvate, vitamin E, vitamin A, GSH, taurine, albumin, ubiquitol, and hypotaurine. Spermatozoa mostly have enzyme antioxidant, the most common of which is SOD. Dietary or nutritional antioxidants are consumed as β-carotenes, carotenoids, vitamins C and E, and flavonoids. Antioxidants that are in the form of metal-binding proteins include ceruloplasmin, ferritin, albumin, transferrin, metallothionein, and myoglobin and they reduce OS via inactivation of free radical generating transition metal ions [95]. The metal chelators found in human semen also regulate the sperm plasma membrane LPO to maintain its integrity, such as lactoferrin, transferrin, and ceruloplasmin [113].

SOD metalloenzymes are key first-order antioxidant defense in the body to protect against ROS and mediate conversion of superoxide anion to H_2O_2. Glutathione peroxidase or catalase activity removes the generated H_2O_2. SOD protects sperm from OS-induced damage, inhibits sperm damage by O_2 toxicity and LPO [114]. Peroxisomal catalase has an essential role in sperm capacitation [115], while the glutathione peroxidase aids conversion of H_2O_2 to either water or alcohol, rendering protection to the sperm encountering low oxidative damage [116]. Reduced glutathione in semen causes motility defect and instability in sperm [117]. GSH supplement finds importance in the treatment of male infertility, mainly in cases of unilateral varicoceles or reproductive tissue inflammation, and it improves sperm parameters in terms of sperm count, mobility, and morphology [118].

Carotenoids are natural antioxidants that preserve cell membrane integrity, control epithelial cells differentiation, and help regulate and sustain spermatogenesis. Deficiency in dietary carotenoids has been shown to reduce sperm motility [115], and low serum levels of retinol also correspond to sperm quality [119], reversed by vitamin A supplementation [120]. Vitamin C (ascorbic acid) content of seminal plasma is ten times higher than serum [121]. Increase in vitamin C levels reportedly correlates positively to percentage of normal sperm [122] and negatively to sperm DNA fragmentation [123], which suggests its beneficial effects in male infertility [124]. Vitamin E also safeguards sperm membrane from oxidative damage and mitigates hydroxyl and superoxide free radicals [115]. Coenzyme Q10 (CoQ10), a fat-soluble antioxidant, also efficiently prevents OS-induced sperm disruptions and modulates mitochondrial electron transfer chain to reduce generation of free radicals [125, 126]. CoQ10 has been shown to have profound ameliorative impacts upon the overall sperm parameters [127, 128]. CoQ10 treatment to men with varicocele could significantly enhance fertility parameters [126]. Interestingly, it also aids production of other antioxidants, vitamins E and C [129]. Selenium is a vital nutrient for proper testicular development, spermatogenesis, sperm motility, and sperm functions and its inadequacy leads to impaired sperm maturation, seminiferous tubule epithelial atrophy [130]. For proper sperm capacitation and acrosome reactions, low ROS levels are needed. Therefore, excessive consumption of antioxidants may lead to extreme diminished ROS levels which also may have adverse effects on fertility [131]. Nevertheless, antioxidants counteract OS and sperm parameters do improve with appropriate dosage and combinations of antioxidants [36].

9.7.1 Role of Antioxidants in Sperm Motility

Antioxidants mainly vitamins E and C, glutathione, hypotaurine, albumin, taurine, as well as SOD, and catalase act in preventing sperm motility reduction, while CoQ10 and N-acetyl cysteine increase sperm motility [63]. Effects of individual antioxidants and few of their combinations on sperm motility are discussed here. Vitamin E oral administration has been shown to reduce seminal concentration of MDA, which is the major LPO marker and also to improve sperm motility [132]. Oral carnitine supplement also enhances sperm motility and concentration [117]. Sperm motility is greatly improved when sperm are incubated with D-penicillamine [133]. When sperm preparation medium is supplemented with vitamins C and E combinations, sperm produce less ROS [134]. Superoxide supplementation influences acrosome reaction rate and maintenance of sperm motility [135].

9.7.2 Role of Antioxidants in Preventing Cryodamage

Issue with cryopreservation of sperm emerges if there is a need of frequent freezing and thawing as these procedures adversely impact and irreversibly affect sperm motility and metabolism as well as disrupt the sperm membrane [136]. Vitamin E and rebamipide have been shown to reduce cryodamage caused by freeze-thaw procedures and enhance sperm motility thereafter [137].

9.7.3 Role of Antioxidants in Preventing DNA Damage

Antioxidants effectively combat OS and protect sperm from DNA fragmentation [63]. It is evident that vitamins C and E oral supplements, used on a daily basis, could decrease the percentage of "terminal deoxynucleotidyl transferase biotin-dUTP nick end labeling" (TUNEL)-positive sperm, while the percentage of normal sperm remained unaltered [95].

9.8 Conclusions

Idiopathic male infertility should be tested for MOSI employing reliable, cost-effective, extremely sensitive, and specific ORP test, for example, by using MiOXSYS, which is more convenient than other elaborate methods [138]. Infertile men diagnosed with MOSI should have a thorough evaluation to screen for any curable conditions and be instructed to take the necessary actions to reduce the known causes of OS that include lifestyle factors, exposure to radiations or toxins, smoking, and alcohol consumption [139]. In case the causes of OS are not known, ORP test should be done for those infertile men in every three months and suitable management plans should be followed. After all other causes of OS have been eliminated, appropriate antioxidants treatment with proper dose should be given to these patients for at least three months. During the antioxidant treatment, ORP testing should be performed regularly to confirm treatment compliance and to observe the treatment effectiveness [138]. Measurement of ORP and stratification of male fertility/infertility on the basis of ORP can be an important tool in the management of infertile couples [138]. As knowledge and awareness of MOSI as a unique male infertility diagnosis expands, it is essential to develop different approaches targeting the underlying causes of IMI, while balancing the risk factors and advantages of various treatments.

9.9 Future Perspectives

Male infertility diagnosis and management have witnessed a substantial improvement with the advent in the omics technologies that address at genetic, molecular, and cellular levels. Disease-targeted sequencing, entire exome and genome sequencing, and sperm epigenetic analysis are examples of NGS technologies that have been demonstrated to be effective in genetic testing [140]. Novel candidate genes linked with male infertility disorders such as oligozoospermia, azoospermia, and IMI have been identified using NGS [98, 141, 142]. Another interesting field of investigation in the case of IMI is a metabolic fingerprinting analysis of seminal plasma [143]. Changes in sperm epigenetics in infertile males (whose sperm appear normal in semen analysis) have a correlation with seminal metabolic profile caused by ROS [144, 145]. Recent developments in proteomic research have identified numerous proteins as biomarkers for a variety of male infertility causes, including OS-mediated sperm abnormalities, varicocele, asthenozoospermia, globozoospermia, and testicular cancer [33, 40, 129, 146–149]. If andrology and artificial intelligence are integrated, utilizing extensive machine learning, future diagnosis, and management of male infertility will become more advanced. Algorithms are constructed to detect males with azoospermia and men who may need genetic testing, sperm selection for ART, as well as embryo selection for IVF [150]. In recent years, next-generation therapies based on stem cells have made significant progress in research. For the effective generation of spermatozoa, several *in vitro* techniques and organ models employing embryonic stem cells induced pluripotent stem cells, and glioblastoma stem cells have been established [151]. Human pluripotent stem cells might be utilized to treat male infertility, reconstruct spermatogenesis, and correct genetic defects using the CRISPR-Cas9 gene-editing method [152]. Exosomes generated from pluripotent stem cells may also have therapeutic relevance in restoring spermatogenic activity in individuals who have undergone chemotherapy or radiation [152]. Similarly, spermatogonial stem cells' regenerative and self-renewal capabilities have inspired new approaches in the treatment of male infertility [153].

Autografting cryobanked spermatogonial tissue as a potential fertility preservation approach for pediatric patients who have undergone gonadotoxic treatment has been proposed [154]. However, numerous obstacles must be addressed before these technologies can be used to modulate male infertility, including ethical and religious concerns as well as the risk of genetic problems being passed down to offspring.

References

1. Jungwirth A, Giwercman A, Tournaye H, Diemer T, Kopa Z, Dohle G, et al. European Association of urology guidelines on male infertility: the 2012 update. Eur Urol. 2012;62(2):324–32.
2. Leslie S, Siref L, Soon-Sutton T, Khan MA. Male infertility. StatPearls; 2021.
3. Rutstein SO, Shah IH. Infecundity, infertility, and childlessness in developing countries. DHS comparative reports no. 9. Calverton: ORC Macro and the World Health Organization. 2004.
4. Agarwal A, Mulgund A, Hamada A, Chyatte MR. A unique view on male infertility around the globe. Reprod Biol Endocrinol. 2015;13:37.
5. Agarwal A, Parekh N, Selvam MKP, Henkel R, Shah R, Homa ST, et al. Male oxidative stress infertility (MOSI): proposed terminology and clinical practice guidelines for management of idiopathic male infertility. World J Men's Health. 2019;37(3):296–312.
6. Agarwal A, Virk G, Ong C, Du Plessis SS. Effect of oxidative stress on male reproduction. World J Men's Health. 2014;32(1):1–17.
7. Wagner H, Cheng JW, Ko EY. Role of reactive oxygen species in male infertility: an updated review of literature. Arab J Urol. 2018;16(1):35–43.
8. Ko EY, Sabanegh ES Jr, Agarwal A. Male infertility testing: reactive oxygen species and antioxidant capacity. Fertil Steril. 2014;102(6):1518–27.
9. Natioanl Institute of Health (NIH). How common is male infertility, and what are its causes? Available from: https://www.nichd.nih.gov/health/topics/menshealth/conditioninfo/infertility. 2016.
10. Winters BR, Walsh TJ. The epidemiology of male infertility. Urol Clin North Am. 2014;41(1):195–204.
11. Sengupta P, Dutta S, Alahmar AT, D'souza UJA. Reproductive tract infection, inflammation and male infertility. Chem Biol Lett. 2020;7(2):75–84.
12. Sengupta P. Environmental and occupational exposure of metals and their role in male reproductive functions. Drug Chem Toxicol. 2013;36(3):353–68.
13. Mazur DJ, Lipshultz LI. Infertility in the aging male. Curr Urol Rep. 2018;19(7):1–9.
14. Cheung S, Parrella A, Rosenwaks Z, Palermo GD. Genetic and epigenetic profiling of the infertile male. PLoS One. 2019;14(3):e0214275.
15. Wallach EE, Moghissi KS. Unexplained infertility. Fertil Steril. 1983;39(1):5–21.
16. Hamada A, Esteves SC, Agarwal A. Unexplained male infertility: potential causes and management. Hum Androl. 2011;1(1):2–16.
17. Dohle G. Male factors in couple's infertility. In: Clinical uro-andrology. Berlin: Springer; 2015. p. 197–201.
18. Ko EY, Siddiqi K, Brannigan RE, Sabanegh ES Jr. Empirical medical therapy for idiopathic male infertility: a survey of the American Urological Association. J Urol. 2012;187(3):973–8.
19. Sengupta P, Dutta S. Metals. In: Encyclopedia of reproduction. Elsevier; 2018.
20. Darbandi M, Darbandi S, Agarwal A, Sengupta P, Durairajanayagam D, Henkel R, et al. Reactive oxygen species and male reproductive hormones. Reprod Biol Endocrinol. 2018;16(1):1–14.
21. Sengupta P, Banerjee R. Environmental toxins: alarming impacts of pesticides on male fertility. Hum Exp Toxicol. 2014;33(10):1017–39.

22. Aktan G, Doğru-Abbasoğlu S, Küçükgergin C, Kadıoğlu A, Özdemirler-Erata G, Koçak-Toker N. Mystery of idiopathic male infertility: is oxidative stress an actual risk? Fertil Steril. 2013;99(5):1211–5.
23. Ambasudhan R, Singh K, Agarwal J, Singh S, Khanna A, Sah R, et al. Idiopathic cases of male infertility from a region in India show low incidence of Y-chromosome microdeletion. J Biosci. 2003;28(5):605–12.
24. Dutta S, Henkel R, Sengupta P, Agarwal A. Physiological role of ROS in sperm function. Male infertility. Springer; 2020. p. 337–45.
25. Agarwal A, Sengupta P. Oxidative stress and its association with male infertility. Male infertility: Springer; 2020. p. 57–68.
26. Dutta S, Sengupta P. Role of nitric oxide on male and female reproduction. Malays. J Med Sci. 2021;
27. Adewoyin M, Mohsin SM, Arulselvan P, Hussein MZ, Fakurazi S. Enhanced anti-inflammatory potential of cinnamate-zinc layered hydroxide in lipopolysaccharide-stimulated RAW 264.7 macrophages. Drug Des Dev Ther. 2015;9:2475–84.
28. Dutta S, Sengupta P, Chhikara BS. Reproductive inflammatory mediators and male infertility. Chem Biol Lett. 2020;7(2):73–4.
29. Theam OC, Dutta S, Sengupta P. Role of leucocytes in reproductive tract infections and male infertility. Chem Biol Lett. 2020;7(2):124–30.
30. Dutta S, Sengupta P, Hassan MF, Biswas A. Role of toll-like receptors in the reproductive tract inflammation and male infertility. Chem Biol Lett. 2020;7(2):113–23.
31. Irez T, Bicer S, Sahin E, Dutta S, Sengupta P. Cytokines and adipokines in the regulation of spermatogenesis and semen quality. Chem Biol Lett. 2020;7(2):131–9.
32. Sarkar O, Bahrainwala J, Chandrasekaran S, Kothari S, Mathur PP, Agarwal A. Impact of inflammation on male fertility. Front Biosci. 2011;3:89–95.
33. Zhang X, Diao R, Zhu X, Li Z, Cai Z. Metabolic characterization of asthenozoospermia using nontargeted seminal plasma metabolomics. Clin Chim Acta. 2015;450:254–61.
34. Liew SH, Meachem SJ, Hedger MP. A stereological analysis of the response of spermatogenesis to an acute inflammatory episode in adult rats. J Androl. 2007;28(1):176–85.
35. Pasqualotto FF, Sharma RK, Potts JM, Nelson DR, Thomas AJ, Agarwal A. Seminal oxidative stress in patients with chronic prostatitis. Urology. 2000;55(6):881–5.
36. Barati E, Nikzad H, Karimian M. Oxidative stress and male infertility: current knowledge of pathophysiology and role of antioxidant therapy in disease management. Cell Mol Life Sci. 2020;77(1):93–113.
37. Zamani-Badi T, Karimian M, Azami-Tameh A, Nikzad H. Association of C3953T transition in interleukin 1β gene with idiopathic male infertility in an Iranian population. Hum Fertil. 2019;22(2):111–7.
38. Zamani-Badi T, Nikzad H, Karimian M. IL-1RA VNTR and IL-1α 4845G>T polymorphisms and risk of idiopathic male infertility in Iranian men: a case-control study and an in silico analysis. Andrologia. 2018;50(9):e13081.
39. Azenabor A, Ekun AO, Akinloye O. Impact of inflammation on male reproductive tract. J Reprod Infertil. 2015;16(3):123–9.
40. Agarwal A, Esteves SC. Varicocele and male infertility: current concepts and future perspectives. Asian J Androl. 2016;18(2):161–2.
41. Mostafa T, Anis T, El Nashar A, Imam H, Osman I. Seminal plasma reactive oxygen species-antioxidants relationship with varicocele grade. Andrologia. 2012;44(1):66–9.
42. Agarwal A, Hamada A, Esteves SC. Insight into oxidative stress in varicocele-associated male infertility: part 1. Nat Rev Urol. 2012;9(12):678–90.
43. Hamada A, Esteves SC, Agarwal A. Insight into oxidative stress in varicocele-associated male infertility: part 2. Nat Rev Urol. 2013;10(1):26–37.
44. Türkyilmaz Z, Gülen S, Sönmez K, Karabulut R, Dinçer S, Can Başaklar A, et al. Increased nitric oxide is accompanied by lipid oxidation in adolescent varicocele. Int J Adrol. 2004;27(3):183–7.

45. Altunoluk B, Efe E, Kurutas EB, Gul AB, Atalay F, Eren M. Elevation of both reactive oxygen species and antioxidant enzymes in vein tissue of infertile men with varicocele. Urol Int. 2012;88(1):102–6.
46. Bhattacharya K, Sengupta P, Dutta S, Bhattacharya S. Pathophysiology of obesity: endocrine, inflammatory and neural regulators. Res J Pharm Technol. 2020;13(9):4469–78.
47. Dutta S, Biswas A, Sengupta P. Obesity, endocrine disruption and male infertility. Asian Pac J Reprod. 2019;8(5):195.
48. Soubry A, Guo L, Huang Z, Hoyo C, Romanus S, Price T, et al. Obesity-related DNA methylation at imprinted genes in human sperm: results from the TIEGER study. Clin Epigenetics. 2016;8:51.
49. Cui X, Jing X, Wu X, Wang Z, Li Q. Potential effect of smoking on semen quality through DNA damage and the downregulation of Chk1 in sperm. Mol Med Rep. 2016;14(1):753–61.
50. Bhattacharya K, Sengupta P, Dutta S, Karkada IR. Obesity, systemic inflammation and male infertility. Chem Biol Lett. 2020;7(2):92–8.
51. Bakos HW, Mitchell M, Setchell BP, Lane M. The effect of paternal diet-induced obesity on sperm function and fertilization in a mouse model. Int J Androl. 2011;34(5 Pt 1):402–10.
52. Garolla A, Torino M, Miola P, Caretta N, Pizzol D, Menegazzo M, et al. Twenty-four-hour monitoring of scrotal temperature in obese men and men with a varicocele as a mirror of spermatogenic function. Hum Reprod. 2015;30(5):1006–13.
53. Shiraishi K, Takihara H, Matsuyama H. Elevated scrotal temperature, but not varicocele grade, reflects testicular oxidative stress-mediated apoptosis. World J Urol. 2010;28(3):359–64.
54. Adewoyin M, Ibrahim M, Roszaman R, Isa MLM, Alewi NAM, Rafa AAA, et al. Male infertility: the effect of natural antioxidants and phytocompounds on seminal oxidative stress. Diseases. 2017;5(1)
55. Izuka E, Menuba I, Sengupta P, Dutta S, Nwagha U. Antioxidants, anti-inflammatory drugs and antibiotics in the treatment of reproductive tract infections and their association with male infertility. Chem Biol Lett. 2020;7(2):156–65.
56. Gomez E, Buckingham DW, Brindle J, Lanzafame F, Irvine DS, Aitken RJ. Development of an image analysis system to monitor the retention of residual cytoplasm by human spermatozoa: correlation with biochemical markers of the cytoplasmic space, oxidative stress, and sperm function. J Androl. 1996;17(3):276–87.
57. Said TM, Agarwal A, Sharma RK, Mascha E, Sikka SC, Thomas AJ Jr. Human sperm superoxide anion generation and correlation with semen quality in patients with male infertility. Fertil Steril. 2004;82(4):871–7.
58. Golas A, Malek P, Piasecka M, Styrna J. Sperm mitochondria diaphorase activity – a gene mapping study of recombinant inbred strains of mice. Int J Dev Biol. 2010;54(4):667–73.
59. Sabeti P, Pourmasumi S, Rahiminia T, Akyash F, Talebi AR. Etiologies of sperm oxidative stress. Int J Reprod Biomed. 2016;14(4):231–40.
60. Cooper TG, Noonan E, von Eckardstein S, Auger J, Baker HW, Behre HM, et al. World Health Organization reference values for human semen characteristics. Hum Reprod Update. 2010;16(3):231–45.
61. Fariello RM, Del Giudice PT, Spaine DM, Fraietta R, Bertolla RP, Cedenho AP. Effect of leukocytospermia and processing by discontinuous density gradient on sperm nuclear DNA fragmentation and mitochondrial activity. J Assist Reprod Genet. 2009;26(2–3):151–7.
62. Sharma R, Gupta S, Henkel R. Relevance of leukocytospermia and semen culture and its true place in diagnosing and treating male infertility. 2021.
63. Makker K, Agarwal A, Sharma R. Oxidative stress & male infertility. Indian J Med Res. 2009;129(4):357–67.
64. Hamada A, Agarwal A, Sharma R, French DB, Ragheb A, Sabanegh ES Jr. Empirical treatment of low-level leukocytospermia with doxycycline in male infertility patients. Urology. 2011;78(6):1320–5.
65. Aboulmaouahib S, Madkour A, Kaarouch I, Sefrioui O, Saadani B, Copin H, et al. Impact of alcohol and cigarette smoking consumption in male fertility potential: Looks at lipid peroxidation, enzymatic antioxidant activities and sperm DNA damage. Andrologia. 2018;50(3)

66. Brand JS, Chan MF, Dowsett M, Folkerd E, Wareham NJ, Luben RN, et al. Cigarette smoking and endogenous sex hormones in postmenopausal women. J Clin Endocrinol Metab. 2011;96(10):3184–92.
67. Valavanidis A, Vlachogianni T, Fiotakis K. Tobacco smoke: involvement of reactive oxygen species and stable free radicals in mechanisms of oxidative damage, carcinogenesis and synergistic effects with other respirable particles. Int J Env Res Pub Health. 2009;6(2):445–62.
68. Ghaffari MA, Rostami M. The effect of cigarette smoking on human sperm creatine kinase activity: as an ATP buffering system in sperm. Int J Fertil Steril. 2013;6(4):258–65.
69. Gogol P, Szcześniak-Fabiańczyk B, Wierzchoś-Hilczer A. The photon emission, ATP level and motility of boar spermatozoa during liquid storage. Reprod Biol. 2009;9(1):39–49.
70. Hamad MF, Shelko N, Kartarius S, Montenarh M, Hammadeh ME. Impact of cigarette smoking on histone (H2B) to protamine ratio in human spermatozoa and its relation to sperm parameters. Andrology. 2014;2(5):666–77.
71. Guthauser B, Boitrelle F, Plat A, Thiercelin N, Vialard F. Chronic excessive alcohol consumption and male fertility: a case report on reversible azoospermia and a literature review. Alcohol Alcohol. 2014;49(1):42–4.
72. Manzo-Avalos S, Saavedra-Molina A. Cellular and mitochondrial effects of alcohol consumption. Int J Env Res Pub Health. 2010;7(12):4281–304.
73. Bailey SM, Robinson G, Pinner A, Chamlee L, Ulasova E, Pompilius M, et al. S-adenosylmethionine prevents chronic alcohol-induced mitochondrial dysfunction in the rat liver. Am J Physiol Gastrointest Liver Physiol. 2006;291(5):G857–67.
74. Angelopoulou R, Lavranos G, Manolakou P. ROS in the aging male: model diseases with ROS-related pathophysiology. Reprod Toxicol. 2009;28(2):167–71.
75. Kesari KK, Agarwal A, Henkel R. Radiations and male fertility. Reprod Biol Endocrinol. 2018;16(1):118.
76. Gautam R, Singh KV, Nirala J, Murmu NN, Meena R, Rajamani P. Oxidative stress-mediated alterations on sperm parameters in male Wistar rats exposed to 3G mobile phone radiation. Andrologia. 2019;51(3):e13201.
77. Desai NR, Kesari KK, Agarwal A. Pathophysiology of cell phone radiation: oxidative stress and carcinogenesis with focus on male reproductive system. Reprod Biol Endocrinol. 2009;7:114.
78. Aitken RJ, Gibb Z, Baker MA, Drevet J, Gharagozloo P. Causes and consequences of oxidative stress in spermatozoa. Reprod Fertil Dev. 2016;28(1–2):1–10.
79. Chauhan P, Verma HN, Sisodia R, Kesari KK. Microwave radiation (2.45 GHz)-induced oxidative stress: whole-body exposure effect on histopathology of Wistar rats. Electromag Biol Med. 2017;36(1):20–30.
80. Kesari KK, Kumar S, Behari J. 900-MHz microwave radiation promotes oxidation in rat brain. Electromag Biol Med. 2011;30(4):219–34.
81. Gracia CR, Sammel MD, Coutifaris C, Guzick DS, Barnhart KT. Occupational exposures and male infertility. Am J Epidemiol. 2005;162(8):729–33.
82. Sabés-Alsina M, Tallo-Parra O, Mogas MT, Morrell JM, Lopez-Bejar M. Heat stress has an effect on motility and metabolic activity of rabbit spermatozoa. Anim Reprod Sci. 2016;173:18–23.
83. Zhang M, Jiang M, Bi Y, Zhu H, Zhou Z, Sha J. Autophagy and apoptosis act as partners to induce germ cell death after heat stress in mice. PLoS One. 2012;7(7):e41412.
84. Pereira C, Mapuskar K, Rao CV. Chronic toxicity of diethyl phthalate in male Wistar rats – a dose-response study. Regul Toxicol Pharmacol. 2006;45(2):169–77.
85. Pant N, Shukla M, Kumar Patel D, Shukla Y, Mathur N, Kumar Gupta Y, et al. Correlation of phthalate exposures with semen quality. Toxicol Appl Pharmacol. 2008;231(1):112–6.
86. Radwan M, Jurewicz J, Polańska K, Sobala W, Radwan P, Bochenek M, et al. Exposure to ambient air pollution – does it affect semen quality and the level of reproductive hormones? Ann Hum Biol. 2016;43(1):50–6.

87. Aitken RJ. Free radicals, lipid peroxidation and sperm function. Reprod Fertil Dev. 1995;7(4):659–68.
88. Saleh RA, Agarwal A. Oxidative stress and male infertility: from research bench to clinical practice. J Androl. 2002;23(6):737–52.
89. Bui AD, Sharma R, Henkel R, Agarwal A. Reactive oxygen species impact on sperm DNA and its role in male infertility. Andrologia. 2018;50(8):e13012.
90. Moretti E, Collodel G, Fiaschi AI, Micheli L, Iacoponi F, Cerretani D. Nitric oxide, malondialdheyde and non-enzymatic antioxidants assessed in viable spermatozoa from selected infertile men. Reprod Biol. 2017;17(4):370–5.
91. Takeshima T, Usui K, Mori K, Asai T, Yasuda K, Kuroda S, et al. Oxidative stress and male infertility. Reprod Med Biol. 2021;20(1):41–52.
92. Fernández JL, Muriel L, Rivero MT, Goyanes V, Vazquez R, Alvarez JG. The sperm chromatin dispersion test: a simple method for the determination of sperm DNA fragmentation. J Androl. 2003;24(1):59–66.
93. Cariati F, Jaroudi S, Alfarawati S, Raberi A, Alviggi C, Pivonello R, et al. Investigation of sperm telomere length as a potential marker of paternal genome integrity and semen quality. Reprod Biomed Online. 2016;33(3):404–11.
94. Sakkas D, Alvarez JG. Sperm DNA fragmentation: mechanisms of origin, impact on reproductive outcome, and analysis. Fertil Steril. 2010;93(4):1027–36.
95. Greco E, Iacobelli M, Rienzi L, Ubaldi F, Ferrero S, Tesarik J. Reduction of the incidence of sperm DNA fragmentation by oral antioxidant treatment. J Androl. 2005;26(3):349–53.
96. Kuroda S, Takeshima T, Takeshima K, Usui K, Yasuda K, Sanjo H, et al. Early and late paternal effects of reactive oxygen species in semen on embryo development after intracytoplasmic sperm injection. Syst Biol Reprod Med. 2020;66(2):122–8.
97. Lone S, Shah N, Yadav HP, Wagay MA, Singh A, Sinha R. Sperm DNA damage causes, assessment and relationship with fertility: a review. Theriogenol Insight. 2017;7(1):13–20.
98. Chen CH, Lee SS, Chen DC, Chien HH, Chen IC, Chu YN, et al. Apoptosis and kinematics of ejaculated spermatozoa in patients with varicocele. J Androl. 2004;25(3):348–53.
99. Krammer PH, Behrmann I, Daniel P, Dhein J, Debatin K-M. Regulation of apoptosis in the immune system. Curr Opinion Immunol. 1994;6(2):279–89.
100. Suda T, Takahashi T, Golstein P, Nagata S. Molecular cloning and expression of the Fas ligand, a novel member of the tumor necrosis factor family. Cell. 1993;75(6):1169–78.
101. Kane DJ, Sarafian TA, Anton R, Hahn H, Gralla EB, Valentine JS, et al. Bcl-2 inhibition of neural death: decreased generation of reactive oxygen species. Science. 1993;262(5137):1274–7.
102. Sakkas D, Mariethoz E, John JCS. Abnormal sperm parameters in humans are indicative of an abortive apoptotic mechanism linked to the Fas-mediated pathway. Exp Cell Res. 1999;251(2):350–5.
103. Paasch U, Sharma RK, Gupta AK, Grunewald S, Mascha EJ, Thomas AJ Jr, et al. Cryopreservation and thawing is associated with varying extent of activation of apoptotic machinery in subsets of ejaculated human spermatozoa. Biol Reprod. 2004;71(6):1828–37.
104. Esteves SC, Miyaoka R, Agarwal A. An update on the clinical assessment of the infertile male. Clinics. 2011;66(4):691–700.
105. Medina S, Domínguez-Perles R, Cejuela-Anta R, Villaño D, Martínez-Sanz JM, Gil P, et al. Assessment of oxidative stress markers and prostaglandins after chronic training of triathletes. Prost Lipid Mediat. 2012;99(3–4):79–86.
106. Dutta S, Majzoub A, Agarwal A. Oxidative stress and sperm function: a systematic review on evaluation and management. Arab J Urol. 2019;17(2):87–97.
107. Agarwal A, Wang SM. Clinical relevance of oxidation-reduction potential in the evaluation of male infertility. Urology. 2017;104:84–9.
108. Tanaka T, Kobori Y, Terai K, Inoue Y, Osaka A, Yoshikawa N, et al. Seminal oxidation–reduction potential and sperm DNA fragmentation index increase among infertile men with varicocele. Hum Fertil. 2020:1–5.

109. Okouchi S, Ohnami H, Shoji M, Ohno Y, Ikeda S, Agishi Y, et al. Effects of electrolyzed-reduced water as artificial hot spring water on human skin and hair. 2005.
110. Robert KA, Sharma R, Henkel R, Agarwal A. An update on the techniques used to measure oxidative stress in seminal plasma. Andrologia. 2021;53(2):e13726.
111. Homa ST, Vassiliou AM, Stone J, Killeen AP, Dawkins A, Xie J, et al. A comparison between two assays for measuring seminal oxidative stress and their relationship with sperm DNA fragmentation and semen parameters. Genes. 2019;10(3):236.
112. Agarwal A, Nallella KP, Allamaneni SS, Said TM. Role of antioxidants in treatment of male infertility: an overview of the literature. Reprod Biomed Online. 2004;8(6):616–27.
113. Sanocka D, Kurpisz M. Reactive oxygen species and sperm cells. Reprod Biol Endocrinol. 2004;2:12.
114. Fujii J, Iuchi Y, Matsuki S, Ishii T. Cooperative function of antioxidant and redox systems against oxidative stress in male reproductive tissues. Asian J Androl. 2003;5(3):231–42.
115. Walczak-Jedrzejowska R, Wolski JK, Slowikowska-Hilczer J. The role of oxidative stress and antioxidants in male fertility. Centr Eur J Urol. 2013;66(1):60.
116. Valko M, Rhodes CJ, Moncol J, Izakovic M, Mazur M. Free radicals, metals and antioxidants in oxidative stress-induced cancer. Chem Biol Int. 2006;160(1):1–40.
117. Lenzi A, Lombardo F, Sgrò P, Salacone P, Caponecchia L, Dondero F, et al. Use of carnitine therapy in selected cases of male factor infertility: a double-blind crossover trial. Fertil Steril. 2003;79(2):292–300.
118. Opuwari CS, Henkel RR. An update on oxidative damage to spermatozoa and oocytes. Biomed Res Int. 2016;2016:9540142.
119. Maya-Soriano MJ, Taberner E, Sabés-Alsina M, López-Béjar M. Retinol might stabilize sperm acrosomal membrane in situations of oxidative stress because of high temperatures. Theriogenology. 2013;79(2):367–73.
120. Comhaire F, Mahmoud A. The andrologist's contribution to a better life for ageing men: part 2. Andrologia. 2016;48(1):99–110.
121. Jacob RA, Pianalto FS, Agee RE. Cellular ascorbate depletion in healthy men. J Nutr. 1992;122(5):1111–8.
122. Thiele JJ, Friesleben HJ, Fuchs J, Ochsendorf FR. Ascorbic acid and urate in human seminal plasma: determination and interrelationships with chemiluminescence in washed semen. Hum Reprod. 1995;10(1):110–5.
123. Song GJ, Norkus EP, Lewis V. Relationship between seminal ascorbic acid and sperm DNA integrity in infertile men. Int J Androl. 2006;29(6):569–75.
124. Eskenazi B, Kidd SA, Marks AR, Sloter E, Block G, Wyrobek AJ. Antioxidant intake is associated with semen quality in healthy men. Hum Reprod. 2005;20(4):1006–12.
125. Alahmar AT, Sengupta P. Impact of coenzyme Q10 and selenium on seminal fluid parameters and antioxidant status in men with idiopathic infertility. Biol Trace Elem Res. 2021;199(4):1246–52.
126. Alahmar AT, Calogero AE, Singh R, Cannarella R, Sengupta P, Dutta S. Coenzyme Q10, oxidative stress, and male infertility: a review. Clin Exp Reprod Med. 2021;48(2):97.
127. Alahmar AT, Sengupta P, Dutta S, Calogero AE. Coenzyme Q10, oxidative stress markers, and sperm DNA damage in men with idiopathic oligoasthenoteratospermia. Clin Exp Reprod Med. 2021;48(2):150.
128. Alahmar AT, Calogero AE, Sengupta P, Dutta S. Coenzyme Q10 improves sperm parameters, oxidative stress markers and sperm DNA fragmentation in infertile patients with idiopathic oligoasthenozoospermia. World J Men's Health. 2021;39(2):346.
129. Mancini A, Conte G, Milardi D, De Marinis L, Littarru GP. Relationship between sperm cell ubiquinone and seminal parameters in subjects with and without varicocele. Andrologia. 1998;30(1):1–4.
130. Atig F, Raffa M, Ali HB, Abdelhamid K, Saad A, Ajina M. Altered antioxidant status and increased lipid per-oxidation in seminal plasma of tunisian infertile men. Int J Biol Sci. 2012;8(1):139–49.

131. Agarwal A, Sekhon LH. Oxidative stress and antioxidants for idiopathic oligoasthenoteratospermia: Is it justified? Indian J Urol. 2011;27(1):74–85.
132. Suleiman SA, Ali ME, Zaki ZM, el-Malik EM, Nasr MA. Lipid peroxidation and human sperm motility: protective role of vitamin E. J Androl. 1996;17(5):530–7.
133. Wroblewski N, Schill WB, Henkel R. Metal chelators change the human sperm motility pattern. Fertil Steril. 2003;79(Suppl 3):1584–9.
134. Donnelly ET, McClure N, Lewis SE. Antioxidant supplementation in vitro does not improve human sperm motility. Fertil Steril. 1999;72(3):484–95.
135. Griveau JF, Le Lannou D. Reactive oxygen species and human spermatozoa: physiology and pathology. Int J Androl. 1997;20(2):61–9.
136. Meseguer M, Garrido N, Martínez-Conejero JA, Simón C, Pellicer A, Remohí J. Role of cholesterol, calcium, and mitochondrial activity in the susceptibility for cryodamage after a cycle of freezing and thawing. Fertil Steril. 2004;81(3):588–94.
137. Park NC, Park HJ, Lee KM, Shin DG. Free radical scavenger effect of rebamipide in sperm processing and cryopreservation. Asian J Androl. 2003;5(3):195–201.
138. Agarwal A, Sharma R, Roychoudhury S, Du Plessis S, Sabanegh E. MiOXSYS: a novel method of measuring oxidation reduction potential in semen and seminal plasma. Fertil Steril. 2016;106(3):566–73.e10.
139. Hendin BN, Kolettis PN, Sharma RK, Thomas AJ Jr, Agarwal A. Varicocele is associated with elevated spermatozoal reactive oxygen species production and diminished seminal plasma antioxidant capacity. J Urol. 1999;161(6):1831–4.
140. Thirumavalavan N, Gabrielsen JS, Lamb DJ. Where are we going with gene screening for male infertility? Fertil Steril. 2019;111(5):842–50.
141. Araujo TF, Friedrich C, Grangeiro CHP, Martelli LR, Grzesiuk JD, Emich J, et al. Sequence analysis of 37 candidate genes for male infertility: challenges in variant assessment and validating genes. Andrology. 2020;8(2):434–41.
142. Fakhro KA, Elbardisi H, Arafa M, Robay A, Rodriguez-Flores JL, Al-Shakaki A, et al. Point-of-care whole-exome sequencing of idiopathic male infertility. Genet Med. 2018;20(11):1365–73.
143. Jafarzadeh N, Mani-Varnosfaderani A, Minai-Tehrani A, Savadi-Shiraz E, Sadeghi MR, Gilany K. Metabolomics fingerprinting of seminal plasma from unexplained infertile men: a need for novel diagnostic biomarkers. Mol Reprod Dev. 2015;82(3):150.
144. Darbandi M, Darbandi S, Agarwal A, Baskaran S, Dutta S, Sengupta P, et al. Reactive oxygen species-induced alterations in H19-Igf2 methylation patterns, seminal plasma metabolites, and semen quality. J Assist Reprod Genet. 2019;36(2):241–53.
145. Darbandi M, Darbandi S, Agarwal A, Baskaran S, Sengupta P, Dutta S, et al. Oxidative stress-induced alterations in seminal plasma antioxidants: is there any association with keap1 gene methylation in human spermatozoa? Andrologia. 2019;51(1):e13159.
146. Agarwal A, Durairajanayagam D, Halabi J, Peng J, Vazquez-Levin M. Proteomics, oxidative stress and male infertility. Reprod Biomed Online. 2014;29(1):32–58.
147. Swain N, Samanta L, Agarwal A, Kumar S, Dixit A, Gopalan B, et al. Aberrant upregulation of compensatory redox molecular machines may contribute to sperm dysfunction in infertile men with unilateral varicocele: a proteomic insight. Antioxid Redox Signal. 2020;32(8):504–21.
148. Alvarez Sedó C, Rawe VY, Chemes HE. Acrosomal biogenesis in human globozoospermia: immunocytochemical, ultrastructural and proteomic studies. Hum Reprod. 2012;27(7):1912–21.
149. Panner Selvam MK, Agarwal A, Pushparaj PN. A quantitative global proteomics approach to understanding the functional pathways dysregulated in the spermatozoa of asthenozoospermic testicular cancer patients. Andrology. 2019;7(4):454–62.
150. Benson M. Clinical implications of omics and systems medicine: focus on predictive and individualized treatment. J Int Med. 2016;279(3):229–40.

151. Chu KY, Nassau DE, Arora H, Lokeshwar SD, Madhusoodanan V, Ramasamy R. Artificial intelligence in reproductive urology. Curr Urol Rep. 2019;20(9):52.
152. Fang F, Li Z, Zhao Q, Li H, Xiong C. Human induced pluripotent stem cells and male infertility: an overview of current progress and perspectives. Hum Reprod. 2018;33(2):188–95.
153. Pourmoghadam Z, Aghebati-Maleki L, Motalebnezhad M, Yousefi B, Yousefi M. Current approaches for the treatment of male infertility with stem cell therapy. J Cell Physiol. 2018;233(10):6455–69.
154. Neuhaus N, Schlatt S. Stem cell-based options to preserve male fertility. Science. 2019;363(6433):1283–4.

Chapter 10
Oxidative Stress and Varicocele-Associated Male Infertility

Terence Chun-Ting Lai, Shubhadeep Roychoudhury, and Chak-Lam Cho

Abstract Despite being regarded as one of the most common causes of male subfertility, the pathophysiology of varicocele remains largely unknown. Recently, oxidative stress (OS) is proposed to be the mediator in how varicocele may negatively impact fertility. The imbalance of reactive oxygen species (ROS) and seminal antioxidants results in damage to sperm DNA and lipid membrane. There is evidence demonstrating higher OS level in men with varicocele which is also positively correlated with clinical grading of varicocele. Moreover, a number of studies have revealed the negative correlation between OS and conventional semen parameters. Furthermore, various interventions have shown their potential in alleviating OS in men with varicocele-associated infertility. Although direct evidence on improving pregnancy rate is not available at the moment, varicocelectomy has demonstrated promising results in relieving OS. Oral antioxidants represent another option with a favourable safety profile. The supplement can be used alone or as adjunct to varicocelectomy. However, most of the studies are hampered by heterogenous dose regime and high-level evidence is lacking.

Keywords Varicocele · Pathophysiology · ROS · Semen parameters · Infertility · Antioxidant supplementation · Varicocelectomy

T. C.-T. Lai
Department of Surgery, Li Ka Shing Faculty of Medicine, The University of Hong Kong, Hong Kong, Hong Kong

S. Roychoudhury
Department of Life Science and Bioinformatics, Assam University, Silchar, India

C.-L. Cho (✉)
S. H. Ho Urology Centre, Department of Surgery, The Chinese University of Hong Kong, Hong Kong, Hong Kong

10.1 Introduction

Varicocele is an abnormality of the venous drainage system of the testicle(s). Clinical varicocele is presented as dilated and tortuous veins of the pampiniform plexus, which is visible or palpable; whilst subclinical varicocele can only be detected on duplex ultrasonography. It is a common disease, affecting 15–20% of healthy adult male population. Varicocele is accountable for 69% of the secondary infertility, while it is the cause of primary infertility in 50% of cases [1]. It is the major correctable disease in male factor subfertility. In the 1950s, Tullock first demonstrated that high ligation of varicocele could improve semen parameters in infertile men with varicocele [2]. Substantial evidence shows that varicocele intervention significantly improved semen parameters [3]. The odds of spontaneous pregnancy are 2.4 after surgical intervention on men with clinical varicocele and abnormal semen parameters. The result is encouraging but inconclusive due to the low quality of evidence [4].

Despite being regarded as the major cause of male-factor infertility, the pathophysiology of varicocele is not completely clear. The proposed mechanisms include scrotal hyperthermia, testicular hypoxia, backflow of toxic metabolites, hormonal disturbance, and oxidative stress (OS) [5]. OS is believed to be a major mechanism in how varicocele may impact fertility. It is well proven that OS, due to an imbalance of reactive oxygen species (ROS) and seminal antioxidants, leads to spermatozoa damage. There is evidence showing negative correlation of semen parameters and level of OS in men with varicocele. Sperm concentration, total sperm motility, and normal sperm morphology were shown to correlate negatively with seminal ROS, malondialdehyde (MDA), and hydrogen peroxide, and correlate positively with seminal antioxidants [6].

However, OS alone cannot fully explain the variable effect of varicocele on spermatogenesis and fertility, as not every man with varicocele, even with increased OS level, suffers from infertility.

In this chapter, we discuss the possible sources of ROS in varicocele and the correlation of OS level and varicocele-associated infertility. We will also investigate how surgical and medical therapy may alleviate the OS in varicocele.

10.2 Pathophysiology of Varicocele-Induced Oxidative Stress

The aetiology and pathophysiology of varicocele is believed to be multifactorial. The aetiology may be explained by three possible theories that are not mutually exclusive [5]. Firstly, insufficient, or absent venous valves that result in retrograde flux [7]. Secondly, insertion of the left testicular vein into the left renal vein at right-angle consequently elevating the hydrostatic pressure which is transmitted to the pampiniform plexus [7]. And, finally, partial obstruction of the blood flow because of the compression of the left renal vein between the superior mesenteric artery and abdominal aorta [8]. Although up to 90% of men with varicocele are diagnosed with a unilateral, left-sided varicocele [8], abnormal venous reflux has been detected on the right side, too [9]. In addition, tallness [10], an upright posture [8], and hereditary factors [11] have been associated with greater prevalence of varicoceles. Varicocele is believed to involve one or more of pathophysiological mechanisms such as a state of energy deprivation, hypoxia, transient scrotal hyperthermia which may exert negative effect on testicular function [12]. However, the exact pathophysiological mechanism by which varicocele can lead to male subfertility remains largely unknown [13].

In patients with varicocele, the antioxidant buffering capacity decreases that perturbs the fine-tuned oxidant–antioxidant balance [14]. ROS may be released due to an increased pressure on venous walls [15]. Recently, it was reported that the basal ROS production by spermatozoa is elevated in patients with varicocele [16]. Free radicals such as nitric oxide (NO) and non-radical groups such as hydrogen peroxide (H_2O_2) are elevated in the semen of infertile men with varicocele [17–19]. In the cascade of ROS-related damage, overproduction of the superoxide anion (●OH) by spermatozoa is another important step that results in altered semen parameters in varicocele patients [20].

The three major sites of ROS generation include the principal cells in the epididymis, the endothelial cells in the dilated pampiniform plexus, and the testicular cells such as the developing germ cells, Leydig cells, macrophages, and peritubular cells. In the mitochondria, the electron transport chain (ETC) can be directly activated by heat and hypoxic stress to release ROS [8]. Exposure to heat, hypoxia and toxic adrenal and renal metabolites are the major extrinsic stimulators of ROS in patients with varicocele [5]. Moreover, cadmium (Cd) deposition in the testes of patients with varicocele also induces oxidative stress (OS) [18]. The pathophysiology of varicocele-associated OS in male infertility is presented in Fig. 10.1.

Multiple molecular pathways have been proposed correlating varicocele and elevated OS. Heat stress is associated with production of ROS from mitochondrial membranes, cytoplasm, and peroxisomes [8, 9, 21]. Venous reflux in patients with varicocele increases scrotal temperature and impairs normal spermatogenesis [22]. Testicular heat stress is believed to impair spermatogenesis through diminishing the overall protein synthesis and particularly the crucial enzymes including topoisomerase I, DNA polymerase, and heat shock proteins (HSPs) [23]. The release of NOS and xanthine oxidase within the dilated spermatic veins also lead to an increased OS level in subfertile varicocele patients as detected from their peripheral blood samples. In the spermatic vein, NOS has been suggested to induce the generation of local testicular ROS [24, 25]. Also, seminal concentrations of heat-labile enzymes – catalase (CAT) and glutathione peroxidase (GPx) – activity are considerably reduced in infertile men with varicocele [6, 26, 27]. Testicular heat stress can induce oxidative damage leading to impairment of sperm quality parameters in infertile varicocele patients. This has been associated with the activation of hypoxia-inducible factor subunit-1α (HIF-1α) and p53 [28]. Hypoxia is believed to be another important stimulator of ROS in varicocele patients, and it might increase OS through inflammatory reactions [8, 9, 21]. Interleukin-1 (IL-1) is a potent

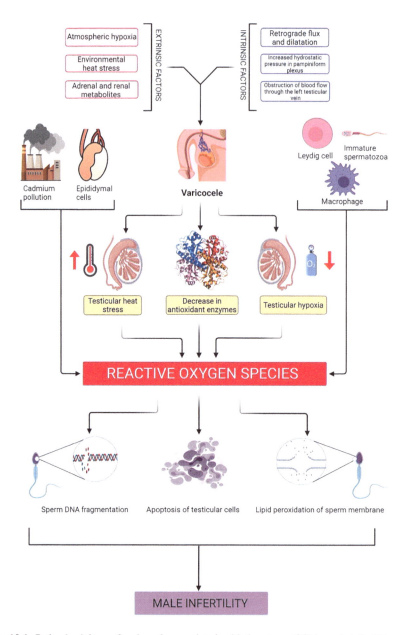

Fig. 10.1 Pathophysiology of varicocele-associated oxidative stress (OS) in male infertility

pro-inflammatory activator that has been found to be upregulated during experimental varicocele leading to increased production of ROS [29]. In dilated veins of varicocele patients, markers of hypoxia particularly the expression of HIF-1α and subsequently vascular endothelial growth factor (VEGF) were elevated [30, 31].

Recently, higher involvement of hypoxia pathway (as revealed by the markers HIF-1α and p53) has been reported in the pathogenesis of varicocele-associated male infertility than the inflammation pathway (TLR-2, TLR-4 and TNF-α) [28]. In fact, ROS production by complex I and III in the mitochondrial ETC is stimulated by hypoxia. During hypoxia, superoxide generation by the ETC also increases [32]. Hence, defective sperm function in varicocele patients may be attributed to hypoxia-associated mitochondrial dysfunction and elevated generation of mitochondrial $O_2^{\bullet-}$/ROS by the spermatozoa [20]. Furthermore, OS may be enhanced by the reflux of renal and adrenal metabolites. Elevated hydrostatic pressure can lead to the reflux of adrenal and renal metabolites into the internal spermatic vein and, consequently, into the testes. This in turn causes vasoconstriction of testicular arterioles resulting in hypoxia and altered spermatogenesis [8, 21] consequently resulting in impaired fertility in patients with varicocele. In some patients with varicocele, a bilateral elevation of the testicular Cd level has been associated with increased apoptosis which was related inversely to the postoperative improvement in their semen parameters [33].

Overproduction of ROS and deficiency in the pro-oxidant defence system are believed to be the major molecular causes contributing to the varicocele-associated male infertility [34]. Recently, a decreased expression of antioxidant enzymes superoxide dismutase (SOD 1 and 2) and glutathione S-transferase omega-2 (GSTO2) was observed in varicocele sperm [16]. Oxidative stress can result in germ cells damage either directly or indirectly by affecting the non-spermatogenic cells and the basal lamina of the seminiferous tubules leading to apoptosis [9]. Also, ROS directly attacks spermatozoa DNA and lipid membranes that may also result in lipid peroxidation, altered expression of proteins, and oxidative DNA damage including base modification, DNA strand breaks, and chromatin cross-linking [9, 35, 36]. All these findings support the hypothesis that increased generation of ROS plays a key role in the alteration of testicular function in varicocele patients leading to infertility [21].

Accurate laboratory assessment of seminal OS is critical in the identification and monitoring of patients who may benefit from male infertility treatment [37, 38]. Direct assays measure the ROS and mainly include chemiluminescence, nitroblue tetrazolium test (NBT), cytochrome c reduction, flow cytometry, and oxidation-reduction potential (ORP) assay [38, 39]. Indirect methods measure the oxidized products and include myeloperoxidase or Endtz test as well as lipid peroxidation levels [38, 40]. Total antioxidant capacity (TAC) measures the ability of seminal antioxidants to scavenge free radicals, whereas individual antioxidant enzymes (superoxide dismutase – SOD, CAT, and GPx) also provide a good measure of seminal ROS-scavenging ability [38, 40].

Currently, there is ample evidence to suggest that varicocele represents a major contributor of OS. OS has been shown to be a key player in the pathophysiology between varicocele and male infertility. The development of various assays in the measurement of OS allows its clinical application in patients with varicocele and male subfertility.

10.3 Correlations Among OS, Varicocele, and Infertility

Evidence has shown that infertile men with varicocele have higher OS levels than fertile men without varicocele [6, 19]. However, an increase of OS can be observed in infertile men without varicocele as well. Studies have therefore been conducted to investigate the causal relationship between varicocele and OS. Hereby, we review the evidence on the correlations among OS, varicocele, and male infertility.

10.3.1 Varicocele Grade and OS

Besides a few studies which showed no difference in OS across varicocele grade [41–43], majority of studies showed the higher the varicocele grade, the higher seminal and testicular ROS (Table 10.1) [26, 44–48]. Cocuzza et al. measured seminal ROS by luminol-dependent chemiluminescence (CL) method [41], and men with Grade III varicocele were shown to have higher seminal ROS than those with lower grade varicocele [47]. Alkan et al. made use of lucigen-dependent CL method, a more specific way in assaying extracellular ROS, also showed a positive correlation between seminal ROS and varicocele grade, regardless of the fertility status [48]. The level of antioxidant has been shown to be inversely correlated with the varicocele grade [26, 44].

10.3.2 Varicocele-Associated OS and Fertility

Fertile Men with or Without Varicocele

It is likely that men with clinical varicocele, even with preserved fertility, have higher OS than their normal counterparts (Table 10.2). Several studies have shown that seminal ROS was higher and antioxidant level was lower in fertile men with varicocele than in healthy fertile men [6, 49, 50]. Ni et al. showed that normozoospermic men with clinical, but not subclinical varicocele, has significantly higher seminal MDA and DNA fragmentation than healthy controls, regardless of grade [50]. However, Cocuzza et al. compared 33 fertile men with varicocele to 81 fertile men without varicocele and found that their seminal ROS level has no significant difference [41]. This finding was echoed by Shiraishi et al., showing no statistically significant correlation between testicular volume and the generation of 4-hydroxy-2-nonenal (4-HNE)-modified protein in testicular tissue in fertile men with or without varicocele [51]. A subgroup analysis of the study by Sakamoto et al., which compared normozoospermic men with or without varicocele, showed that besides a higher seminal plasma NO and HEL concentration in the varicocele group, their antioxidant level, in terms of SOD activity, was also higher [19]. Whether there is any other protective mechanism that prevents fertile men with varicocele from developing infertility is yet to be studied.

Table 10.1 Correlation of OS and grading of varicocele

Study	Study group (n)	Marker(s) of OS	Results
Koksal et al. [46] (2000)	Infertile men with varicocele [15] Control group: infertile men without varicocele [10]	Testicular MDA	Significantly higher in Grade III compared to subclinical, Grade I or II ($p < 0.05$)
Allamaneni et al. [45] (2004)	Infertile men with clinical left varicocele [45]	Seminal ROS level	Significantly higher in Grade II and III compared to Grade I ($p = 0.02$)
Cocuzza et al. [41] (2008)	Fertile men without varicocele [79] and fertile men with varicocele [32] Control group: infertile men [29]	Seminal ROS level	No significant difference between Grade I and Grade II/III No correlation between ROS level and varicocele grade
Shiraishi et al. [43] (2010)	Men with left varicoceles [31] Control group: fertile men [8]	Testicular 4-HNE-modified proteins	No significant difference between varicocele grades
Abd-Elmoaty et al. [44] (2010)	Infertile men with varicocele [35] Control group: fertile men without varicocele [18]	Seminal oxidants: MDA, NO Antioxidants: SOD, GPX, CAT, vitamin C	Grade I vs II: significantly lower antioxidant levels (SOD and vitamin C ($p < 0.05$)) in grade II Grade I vs III: significantly higher oxidants levels (MDA and NO ($p < 0.001$)) and lower antioxidant levels (CAT and SOD ($p < 0.01$) and GPX ($p = 0.01$)) in Grade III Grade II vs III: significantly higher oxidant levels (MDA and NO ($p < 0.01$)) and lower antioxidant levels (CAT and Vitamin C ($p < 0.05$), SOD and GPX ($p = 0.01$)) in Grade III
Blumer et al. [42] (2012)	Men with Grade II or III varicocele [29] Control group: men without varicocele [31]	Seminal MDA	No significant difference between Grade II and III
Mostafa et al. [26] (2012)	Infertile men with clinical varicocele [87] Control group: fertile men without varicocele [20]	Seminal oxidants: MDA, H_2O_2 Antioxidants: SOD, GPX, CAT, vitamin C	Significant higher seminal oxidant levels (MDA and H_2O_2) and lower antioxidant levels (SOD, CAT, GPX, and vitamin C) in Grade II/III compared to Grade I

(continued)

Table 10.1 (continued)

Study	Study group (n)	Marker(s) of OS	Results
Cocuzza et al. [47] (2012)	Fertile men with clinical varicocele (156) Control group: fertile men without varicocele (113)	Seminal ROS level	Significantly higher in Grade III compared to Grade I or II ($p = 0.015$) No significant difference between Grade I and II
Alkan et al. [48] (2018)	Men with Grade II-III varicocele Control group: men without varicocele [13]	Seminal ROS level Superoxide anion (lucigenin-dependent CL)	Significantly higher total ROS levels ($p = 0.006$) and superoxide anion levels ($p = 0.002$) in Grade III compared to Grade II

Abbreviations: *MDA* malondialdehyde, *ROS* reactive oxygen species, *4-HNE* 4-hydroxy-2-nonenal, *NO* nitric oxide, *SOD* superoxide dismutase, *CAT* catalase, *GPX* glutathione peroxidase, H_2O_2 hydrogen peroxide

Infertile Men with or Without Varicocele

Men with varicocele-associated infertility were shown to have higher seminal ROS than men with infertility of other causes (Table 10.3) [19, 52]. Mehraban et al. compared seminal NO level in 40 infertile men with varicocele to that in 40 infertile men without varicocele and found that the NO level is significantly higher in the varicocele group [52]. Sakamoto et al. examined the NO and hexanoyl-lysine levels in infertile men with or without varicocele and showed that the levels were higher in the former group. They also reported a higher SOD level in infertile men with varicocele [19]. Several studies support the findings of higher TAC in infertile men with varicocele than those without [53, 54]. Coenzyme Q_{10} (CoQ_{10}), a lipid-type molecule and a potent antioxidant, is shown to be more abundant in infertile men with varicocele than those without [55]. The increase in both OS and antioxidant may be explained by an ineffective utilization and neutralization system of antioxidant. However, one study showed significant increase in seminal ROS and a significant decrease in seminal antioxidants in infertile men with varicocele compared with infertile men without [6]. Nonetheless, the TAC of infertile men with varicocele is still significantly lower than fertile men without varicocele [56]. In summary, while level of OS is high in infertile men with and without varicocele, the level is higher in the former group. This suggests that although infertility is associated with increased OS, varicocele is associated with a further increase of it.

Fertility in Men with Varicocele

Evidence of OS level in fertile and infertile men with varicocele is somehow mixed (Table 10.4). Earlier studies showed seminal ROS and TAC were not significantly different between fertile and infertile men with varicocele [49, 57], while some showed significantly higher seminal ROS in men with varicocele and infertility

Table 10.2 OS in fertile men with or without varicoceles

Study	Study group (n)	Control group (n)	Marker(s) of OS	Results
Hendin et al. [49] (1999)	Fertile men with varicocele [15]	Fertile men without varicocele [17]	Seminal ROS level Seminal TAC	Significantly higher seminal ROS level ($p = 0.02$) and lower TAC ($p = 0.05$) in men with varicocele
Sakamoto et al. [19] (2008)	Fertile men with varicocele [15]	Fertile men without varicocele [15]	Seminal oxidants: NO, 8-OHdG, HEL Antioxidant: SOD	Significantly higher NO and HEL ($p < 0.05$) and lower SOD activity ($p < 0.005$) in men with varicocele. No significant difference in 8-OHdG
Cocuzza et al. [41] (2008)	Fertile men with clinical varicocele [32]	Fertile men without varicocele [79>]	Seminal ROS levels	No significant difference
Mostafa et al. [6] (2009)	Fertile men with varicocele [44]	Fertile men without varicocele [44]	Seminal oxidants: MDA, H_2O_2 Antioxidants: SOD, CAT, GPX, vitamin and E	Significantly higher seminal oxidant levels (MDA and H_2O_2) and antioxidant levels (SOD, CAT, GPX, vitamin C and E) ($p < 0.05$)
Shiraishi et al. [51] (2009)	Fertile men with Grade I–III left varicocele (100)	Fertile men without varicocele (100)	Testicular 4-HNE-modified proteins	No significant difference
Ni et al. [50] (2016)	Normozoospermic men with clinical varicocele [22]	Fertile men without varicocele [25]	Seminal MDA	Significantly higher in men with clinical varicocele regardless of grading ($p < 0.05$)

Abbreviations: *ROS* reactive oxygen species, *TAC* total antioxidant capacity, *NO* nitric oxide, *8-OHdG* 8-hydroxy-2'-deoxyguanosine, *HEL* hexanoyl-lysine, *SOD* superoxide dismutase, *MDA* malondialdehyde, H_2O_2 hydrogen peroxide, *CAT* catalase, *GPX* glutathione peroxidase, *4-HNE* 4-hydroxy-2-nonenal

[41, 47]. Mostafa et al. studied the OS of fertile men with/without varicocele and infertile men with oligoasthenozoospermia with/without varicocele, by two ROS parameters (MDA, hydrogen peroxide) and five antioxidants (SOD, CAT, GPx, vitamins E and C). They showed that infertile men with varicocele have higher levels of all two ROS parameters and lower levels of all five antioxidant parameters, when compared with the fertile control with varicocele [6]. Shirashi et al. also showed a significant increase in 4-HNE-modified proteins in infertile men with varicocele when compared with fertile men with varicocele [51]. These contradictory results show that OS may not be the sole cause of varicocele-related infertility.

Table 10.3 OS in infertile men with or without varicoceles

Study	Study group (n)	Control group (n)	Marker(s) of OS	Results
Balercia et al. [55] (2002)	Men with varicocele-associated asthenozoospermia [12]	Men with idiopathic asthenozoospermia [12]	Sperm and seminal CoQ_{10}	Men with varicocele had significantly higher sperm ($p < 0.05$) and seminal ($p < 0.001$) CoQ_{10}
Meucci et al. [54] (2003)	Infertile men with oligozoospermia and varicocele [9]	Infertile men with idiopathic oligozoospermia [7]	Seminal TAC	Significantly higher in men with varicocele ($p < 0.05$)
Mehraban et al. [52] (2005)	Infertile men with varicocele [39]	Infertile men without varicocele [39]	Seminal NO	Significantly higher in men with varicocele ($p = 0.001$)
Mancini et al. [53] (2007)	Men with oligozoospermia and varicocele [12]	Infertile men with idiopathic oligozoospermia [10]	Seminal TAC	Significantly higher in men with varicocele ($p < 0.05$)
Sakamoto et al. [19] (2008)	Oligozoospermic [15] and normozoospermic [15] men with varicocele	Oligozoospermic [15] and normozoospermic [15] men without varicocele [15]	Seminal oxidants: NO, 8-OHdG, HEL Antioxidant: SOD	Men with varicocele had significantly higher NO, HEL and SOD activity ($p < 0.05$)
Mostafa et al. [6] (2009)	Infertile men with oligoasthenozoospermia and varicocele [41]	Infertile men with oligoasthenozoospermia without varicocele [43]	Seminal oxidants: MDA, H_2O_2 Antioxidants: SPD, CAT, GPX, vitamin C and E	Significantly higher oxidant levels (MDA and H_2O_2) and lower antioxidant levels (SOD, and GPX) ($p < 0.05$) No significant difference in CAT, vitamin C and E levels

Abbreviations: *CoQ10* coenzyme Q_{10}, *TAC* total antioxidant capacity, *NO* nitric oxide, *8-OHdG* 8-hydroxy-2′-deoxyguanosine, *HEL* hexanoyl-lysine, *MDA* malondialdehyde, H_2O_2 hydrogen peroxide, *SOD* superoxide dismutase, *CAT* catalase, *GPX* glutathione peroxidase

Table 10.4 OS men with varicoceles

Study	Study group (n)	Control group (n)	Marker(s) of OS	Results
Hendin et al. [49] (1999)	Infertile men with varicocele [21]	Fertile men with varicocele [15]	Seminal ROS levels	No significant difference
Pasqualotto et al. [57] (2008)	Infertile men with varicocele [21]	Fertile men with varicocele [15]	Seminal ROS levels and TAC	No significant difference in seminal ROS level, TAC and ROS-TAC score
Cocuzza et al. [41] (2008)	Fertile men with clinical varicocele [32]	Infertile men with clinical varicocele [29]	Seminal ROS levels	Significantly higher in infertile men with varicocele ($p < 0.0001$)
Mostafa et al. [6] (2009)	Fertile men with varicocele [44]	Infertile men with oligoasthenozoospermia and varicocele [41]	Seminal oxidants: MDA, H_2O_2 Antioxidants: SPD, CAT, GPX, vitamin C and E	Infertile men had significantly higher MDA ($p < 0.05$) and lower SOD, CAT, GPX, and vitamin C ($p < 0.05$) No significant difference in H_2O_2 and vitamin E
Shiraishi et al. [51] (2009)	Infertile men with varicocele [29]	Fertile men with varicocele [12]	Testicular 4-HNE-modified proteins	Significantly higher in infertile men ($p < 0.01$)
Cocuzza et al. [47] (2012)	Fertile men with clinical varicocele (156)	Infertile men with clinical varicocele [37]	Seminal ROS levels	Significantly higher in infertile men with varicocele ($p < 0.0001$)

Abbreviations: *ROS* reactive oxygen species, *TAC* total antioxidant capacity, *MDA* malondialdehyde, H_2O_2 hydrogen peroxide, *SOD* superoxide dismutase, *CAT* catalase, *GPX* glutathione peroxidase, *4-HNE* 4-hydroxy-2-nonenal

10.3.3 Varicocele in Adolescent

The prevalence of varicocele is around 15% in the adolescent population [58]. OS by varicocele is evident as early as in the adolescent period. Romeo et al. first investigated the production of NO by estimating the serum concentration of L-hydroxyarginine (L-NHA), a by-product of NO synthase. In a group of ten adolescents (9–18 years of age) with grade II or III left varicoceles, L-NHA concentration was found to be enhanced by 50-folds in their spermatic vein as compared to their peripheral veins [59]. This is further supported by the findings of a significant increase in spermatic vein nitrite/nitrate (NOx) and MDA levels when compared with peripheral levels in another study of 13 adolescents [60]. Bertolla et al. studied the rate of DNA fragmentation in sperm, by Comet assay, in adolescents with and without bilateral Grade II or III varicoceles. They demonstrated that even with similar semen parameters on semen analysis, adolescents with varicocele had significantly higher percentage of sperm with class III (meaningful DNA fragmentation) or class IV (high DNA fragmentation) fragmentation. This may be attributed to apoptosis or OS [61].

10.3.4 Varicocele and Secondary Male Infertility

Varicocele is a more common cause of secondary infertility than primary infertility [1]. The prevalence of varicocele increases with age [58]. This suggests that varicocele in some men is a progressive condition. The findings that ROS is only elevated in spermatic veins in adolescents with varicocele but in systemic circulation in adults also support that varicocele is a progressive disease. Ageing is also associated with increased mitochondrial production of ROS and a reduction of antioxidant activity [62]. This explains why some men with varicocele are fertile at earlier stage of life but develop secondary infertility at older age. This progressive increase of OS due to varicocele raises the question if early intervention, either surgical repair or antioxidant, can prevent future damage by OS.

In conclusion, an increased generation of OS is evident in men with varicocele. The level of OS is directly correlated with the grading of varicocele. An increase in OS can be observed in men with preserved fertility, and as early as in the adolescent period. Some men with varicocele and increased OS have their fertility preserved. This suggests that OS may be one of the major mechanisms, but not the sole pathophysiology, in varicocele-related infertility. It may be due to a homeostasis between OS and antioxidants or other unknown protective mechanisms as well. As men age, such protective mechanisms may be impaired, and the homeostasis will be disrupted. This explains the progressive nature of varicocele and the development of secondary infertility [59–61, 63].

10.4 Role of Interventions and to Reduce OS in Men with Varicocele-Associated Infertility

10.4.1 Varicocelectomy

Varicocelectomy in Adult Men

Evidence suggests that varicocelectomy can lower the level of OS and improve semen parameters and pregnancy rate in infertile men with clinical varicocele [2–4, 63–67]. Level of seminal MDA [63], NO [19, 63], and 8-OHdG [19, 64] has been shown to decrease after varicocele repair in infertile men. This effect may also be observed in fertile men with clinical varicocele, as Dada et al. performed varicocelectomy in 11 fertile men, and their seminal ROS significantly reduced in 1 month after operation, and further reduced at a slower rate in 3–6 months [65]. However, contradictory evidence is also present. Yesilli et al. failed to show a significant improvement in post-operative seminal MDA in 26 infertile men who underwent subinguinal microsurgical varicocelectomy, despite having a significantly higher seminal MDA than the control non-varicocele group before the operation. Nonetheless, there was a significant increase in sperm HspA2 activities after the operation, indicating an improvement in sperm maturation [66].

Several studies have demonstrated that the TAC and level of antioxidants increase after varicocelectomy. Mostafa et al. examined six antioxidants (SOD, CAT, GPx, vitamin C, vitamin E) in infertile men with varicocelectomy. Four out of the six antioxidants, namely SOD, CAT, GPx, and vitamin C, significantly increased 3 and 6 months after surgery. Albumin did not significantly increase in post-operative 3 months, but a significant increase was observed in 6 months. Interestingly, level of vitamin E was observed to significantly reduce after 3 and 6 months of operation [63]. Sakamoto et al., however, showed that the SOD and HEL activities, which was high before surgery, normalized after varicocelectomy [19]. Mancini et al. also failed to show a significant difference in TAC after varicocelectomy in both oligozoospermic and normozoospermic men [67]. They also showed that total seminal plasma and cellular CoQ_{10} significantly decrease after varicocelectomy in oligozoospermic men, but not in asthenozoospermic or normozoospermic men [68]. The evidence of the effect of varicocelectomy on OS is summarized in Table 10.5.

The improvement in OS after varicocelectomy is also time dependent. Mostafa et al. observed that both the markers of seminal OS (MDA and H_2O_2) were significantly reduced and those of antioxidants (SOD, CAT, GPx, and vitamin C) significantly elevated in both 3 and 6 months after varicocelectomy [63]. Similar phenomenon was observed in the study by Ni et al., where there was a significant reduction of seminal MDA 3 months after varicocelectomy, and a further reduction after 6 months, almost reaching the level of the normozoospermic non-varicocele control group [50]. Therefore, many clinicians allow a post-operative period of 6 months to evaluate any meaningful improvement of OS by varicocelectomy.

Collectively, there is some evidence that varicocelectomy may alleviate OS in infertile men with varicocele. However, most studies were uncontrolled ones, with a small sample size and lacked long-term follow-up. In addition, it is still uncertain if this alleviation of OS leads to an improvement in semen parameters or pregnancy rate, as evidence on the direct association of change of OS, semen parameters and pregnancy rate is lacking.

Varicocelectomy in Adolescent

Semen parameters, including sperm count, motility, and morphology, significantly improve after varicocelectomy in adolescents with clinical varicocele. Testicular catch-up growth rates are also observed after varicocele treatment, with the rate

Table 10.5 Effects of varicocele repair on OS

Study	Patients (n)	Post-operative markers of OS	Post-operative antioxidants	Conclusion
Mostafa et al. [63] (2001)	Infertile men undergone varicocelectomy [66]	At 3 months and 6 months, significant reduction in: MDA (both $p < 0.0001$) H_2O_2 (both $p < 0.0001$) NO ($p = 0.0002$ and $p = 0.0014$)	At 3 months and 6 months, significant increase in: SOD (both $p < 0.0001$) CAT (both $p < 0.0001$) GPX (both $p < 0.0001$) Vitamin C (both $p < 0.0001$) Vitamin E (both $p < 0.0001$) No significant increase in albumin at 3 months but significant in 6 months ($p < 0.0001$)	Varicocelectomy reduces OS at 3 and 6 months
Mancini et al. [67] (2004)	Infertile oligozoospermic [6] and normozoospermic [8] men undergone varicocelectomy	N/A	At 10–24 months, no significant difference in seminal TAC in either oligozoospermic or normozoospermic men	Varicocelectomy does not improve antioxidant capacity at 24 months
Yesilli et al. [66] (2005)	Infertile men undergone left [19] and bilateral [7] MSV	At 6 months, no significant difference in MDA	N/A	Varicocelectomy does not reduce OS at 6 months
Mancini et al. [68] (2005)	Infertile oligoasthenozoospermic [11], asthenoszoospermic [14] and normozoospermic [8] men undergone varicocelectomy	N/A	At 6–8 months, significant increase in total and cellular CoQ10 only in oligozoospermic men	Varicocelectomy improves antioxidant capacity in oligozoospermic men at 6–8 months

(continued)

Table 10.5 (continued)

Study	Patients (n)	Post-operative markers of OS	Post-operative antioxidants	Conclusion
Sakamoto et al. [19] (2008)	Oligozoospermic men with Grade II–III varicocele undergone MSV [15]	At 6 months, significant reduction in: NO ($p < 0.001$) 8-OHdG ($p < 0.001$) HEL ($p = 0.005$) Seminal NO and HEL not significantly higher than control group (normozoospermic men with no varicocele)	At 6 months, significant reduction in SOD activity ($p = 0.01$) No significant difference in SOD activity compared to control group	Varicocelectomy reduces OS at 6 months
Chen et al. [64] (2008)	Infertile men undergone varicocelectomy [29]	At 6 months, significant reduction in 8-OHdG ($p < 0.001$)	At 6 months, significant increase in protein thiols and vitamin C ($p < 0.001$)	Varicocelectomy reduces OS at 6 months
Dada et al. [65] (2010)	Men with clinical varicocele [11]	At 1 month and 3 months, significant reduction in seminal ROS levels ($p < 0.001$)	N/A	Varicocelectomy reduces OS at 1 and 3 months
Ni et al. [50] (2016)	Infertile astheno/oligozoospermic men with clinical varicocele undergone MSV [53]	At 3 months, significant reduction in seminal MDA in men with grade II ($p < 0.01$) and III ($p < 0.05$) varicocele At 6 months, significant reduction in seminal MDA in men with grade I ($p < 0.05$), II and III (both $p < 0.01$)	N/A	Varicocelectomy reduces OS at 3 and 6 months

Abbreviations: *MDA* malondialdehyde, H_2O_2 hydrogen peroxide, *NO* nitric oxide, *SOD* superoxide dismutase, *CAT* catalase, *GPX* glutathione peroxidase, *MSV* microsurgical subinguinal varicocelectomy, *8-OHdG* 8-hydroxy-2′-deoxyguanosine, *HEL* hexanoyl-lysine, *ROS* reactive oxygen species, *TAC* total antioxidant capacity, CoQ_{10} coenzyme Q_{10}

from 62.8% to 100% for interventional varicocele treatment and varicocelectomy [69]. Çayan et al. studied adolescents with testicular hypotrophy and > 50% of cases underwent bilateral varicocelectomy and showed that varicocelectomy group had a paternity rate of 77.3%, significantly higher than 48.4% in the control group [70]. However, this improvement in paternity rate is not supported by another study with sclerotherapy as the treatment modality [71].

The effect of varicocelectomy on OS may be less pronounced in the adolescent group. Lacerda et al. showed that varicocelectomy in boys aged 14–19 improved the sperm DNA integrity and mitochondrial activity, but there was no significant difference in pre- and post-operative seminal MDA [72]. On the contrary, Cervellione et al. demonstrated that varicocelectomy significantly reduced the peripheral venous peroxidative level (TBARS) and increased the plasma peroxidation susceptibility lag time in 1 year [73].

10.4.2 Exogenous Antioxidants

Administration of exogenous antioxidants is an attractive option due to its non-invasiveness. Adverse events related to the use of exogenous antioxidants are reported to be low, with the majority being gastrointestinal side-effects [74]. High-quality evidence is lacking due to the diversity of antioxidants, non-standardized dosage use, small number of participants in each study, and different outcome measures [75]. Multiple classes of antioxidants have been studied in varicocele-associated infertility. These include single agent with known antioxidizing mechanism (e.g. L-carnitine, CoQ_{10}), or combination therapies of such agents, commercially available products with a mixture of multiple ingredients, and novel agents such as herbal medicine. Hereby, we investigate the use of antioxidants as a sole therapy or as an adjunct to varicocelectomy.

Sole Therapy

Different exogenous antioxidants have been used in an attempt to alleviate varicocele-related OS (Table 10.6). Cinnoxicam, an anti-inflammatory agent which inhibits ROS and prostaglandin synthesis, was studied by Cavallini et al. They used cinnoxicam 30 mg suppository every 4 days for 12 months, and a significant improvement of sperm quality was observed for moderate-grade varicocele in 2 months, best in 4 months and stable till 12 months, whereas subinguinal microsurgical varicocelectomy improved semen parameters in moderate- and high-grade varicocele in 4 months. Moreover, the semen parameters returned to baseline after cessation of cinnoxicam [76]. The same group has studied the use of L-carnitine/acetyl-L-carnitine, which play an important role in spermatozoa energy metabolism, in the treatment of idiopathic and varicocele-associated oligoasthenozoospermia. The participants were randomized into three groups. Group 1 used a placebo,

Table 10.6 Use of exogenous antioxidants as sole therapy in men with varicocele

Study	Patients	Group labels (n)	Dose and duration	Outcome parameters	Outcomes
Cavallini et al. [76] (2003)	Men with oligoasthenozoospermia and Grade 3–5 (5-grade classification) left varicocele	MSV (Group 1) [40]	–	Semen parameters	Significant improvement in sperm concentration 2 and 4 months, motility 4 and 8 months, and morphology 2 and 4 months for Grade 3 (5-grade classification) after drug delivery. Effects stable till 12 months. Parameters back to baseline 2 months after cessation of drug No significant effects for Grade 4 or 5
		Cinnoxicam (Group 2) [59]	30 mg suppository every 4 days × 12 months		
		Control placebo (Group 3) [52]	Glycerine suppository every 4 days × 12 months		

(continued)

Table 10.6 (continued)

Study	Patients	Group labels (n)	Dose and duration	Outcome parameters	Outcomes
Cavallini et al. [77] (2004)	Infertile men with OAT	Control placebo (Group 1) [69]	1 × 500-mg starch tablet twice daily and 1 glycerine suppository every 4 days × 6 months	Semen parameters	For grade 1–3 varicoceles (5-grade classification): significant increased sperm concentration, motility and morphology at 3–6 months for Group 2 and 3 For Grade 4 varicocele: significant increased sperm concentration, motility and morphology at 3–6 months for Group 3 For Grade 5 varicocele: no significant effects All effects back to baseline 3 months after drug cessation
		L-carnitine + acetyl L-carnitine (Group 2) [60]	L-carnitine (2 g per day) + acetyl L-carnitine (1 g per day) × 6 months		
		L-carnitine + acetyl L-carnitine + cinnoxicam (Group 3) [60]	L-carnitine (2 g per day) + acetyl L-carnitine (1 g per day) + cinnoxicam 30 mg per day suppository every 4 days × 6 months	Pregnancy rate	Group 2 vs 1: significantly higher in Group 2 ($p < 0.01$) Group 3 vs 2: significantly higher in Group 3 ($p < 0.05$)

Kiliç et al. [82] (2005)	Normozoospermic men with varicocele	MPFF [16]	1000 mg per day × 6 months	Relief of varicocele-related pain	Significant improvement ($p < 0.001$)
				Semen parameters	Significant improvement in motility ($p = 0.009$) No significant effects on semen volume, sperm concentration and morphology
				Colour Doppler parameters	Significant decrease in reflux time for left varicocele ($p = 0.003$)
Oliva et al. [79] (2009)	Infertile men with Grade 3 or above varicocele (5-grade classification)	Pentoxifylline + zinc + folic acid [35]	Pentoxifylline (1200 mg) + zinc (66 mg) + folic acid (5 mg) × 3 months	Semen parameters	Significant improvement in sperm morphology 4, 8 and 12 weeks Significant improvement in semen volume (if baseline volume <2 mL) 8 and 12 weeks Effects persist at least 4 weeks after cessation of drug No significant effects on sperm concentration and motility

(continued)

Table 10.6 (continued)

Study	Patients	Group labels (n)	Dose and duration	Outcome parameters	Outcomes
Zampieri et al. [81] (2010)	Adolescents with left subclinical varicocele	O-β-hydroxyethylrutoside × 3 cycles (Group 1) [35] O-β-hydroxyethylrutoside × 1–2 cycles (Group 1) [36] Control (95)	1 cycle: 1000 mg per day × 3 months then 3 treatment-free months	Resolution and progression of varicocele	41% of Group 1 had resolution of reflux (ns) Significantly higher rate of stable reflux and lower rate of progression to clinical varicocele in Group 1 ($p < 0.05$)
Fang et al. [85] (2010)	Infertile men with varicocele	AescuvenForte (Escin group) (106)	150 mg (equivalent to 30 mg of escin) every 12 h × 2 months	Semen parameters	Escin group: significant improvement in sperm density ($p < 0.05$), no effect on sperm motility Surgery group: significant improvement in both sperm density and motility ($p < 0.05$)
		High ligation (surgery group) [46]	–		
		Vitamin E + pentoxifylline + clomiphene (control group) [63]	Vitamin E 20 mg + pentoxifylline 400 mg + clomiphene 50 mg per day × 2 months	Colour Doppler parameters	Significant decrease in spermatic vein diameter for moderate varicocele ($p < 0.05$)

Soylemez et al. [83] (2012)	Normozoospermic men with varicocele	MPFF (treatment group) [20]	1000 mg per day × 6 months	Relief of varicocele-related pain	Significant improvement ($p < 0.001$)
		Pancreatin + Metylpolysiloksan (Pankreoflat®) (placebo) [20]	2 g granule per day × 6 months	Semen parameters	Significant improvement in motility at 6 months ($p = 0.038$) No significant effects on semen volume, sperm concentration and morphology
				Colour Doppler parameters	Significant decrease in reflux time for left varicocele ($p < 0.01$)
Festa et al. [80] (2014)	Infertile men with varicocele	Coenzyme Q_{10} [37]	100 mg per day × 3 months	Semen parameters	Significant increase in sperm density ($p = 0.03$) and forward motility ($p = 0.03$) No significant effect on semen volume and morphology
				Seminal TAC	Significant increase in TAC ($p < 0.01$)

(continued)

Table 10.6 (continued)

Study	Patients	Group labels (n)	Dose and duration	Outcome parameters	Outcomes
Busetto et al. [78] (2018)	Infertile men with oligo- and/or astheno- and/or teratozoospermia	Proxeed plus (supplement group) [6]	1000 mg L-carnitine, 725 mg fumarate, 500 mg acetyl-L-carnitine, 1000 mg fructose, 20 mg coenzyme Q_{10}, 90 mg vitamin C, 10 mg zinc, 200 μg folic acid and 1.5 μg vitamin B12 × 6 months	Semen parameters	Significant improvement in sperm concentration ($p = 0.0403$), total sperm count ($p = 0.0009$), progressive motility ($p = 0.0149$) and total motility ($p = 0.0065$). No significant effect on semen volume
		Placebo Group [6]	Sucrose, silica (anti-caking), lemon flavour, acesulfame K (E950 sweetener) × 6 months		

Abbreviations: *MSV* microsurgical subinguinal varicocelectomy, *MPFF* micronized purified flavonoid fraction, *OAT* oligoasthenoteratozoospermia, *TAC* total antioxidant capacity group 2 used oral L-carnitine (2 g/d) and acetyl-L-carnitine (1 g/d), group 3 used L-carnitine/acetyl-L-carnitine and 1 x 30 mg cinnoxicam suppository every 4 days, for a duration of 6 months. Both treatment groups showed significant improvements in sperm concentration, morphology, and total count 3 and 6 months after initiation of the therapy for moderate-grade varicocele, with the result more pronounced in the combination group. The semen parameters returned to baseline after 3 months of drug suspension. Pregnancy rate was significantly improved in both treatment groups at the end of 9 months [77]. In a randomized controlled trial, 104 men were randomized into placebo control group and L-carnitine/acetyl-L-carnitine and micronutrient group. The regimen they used was daily 1000 mg l-carnitine, 725 mg fumarate, 500 mg acetyl-L-carnitine, 1000 mg fructose, 20 mg CoQ_{10}, 90 mg vitamin C, 10 mg zinc, 200 µg folic acid, and 1.5 µg vitamin B12 for 6 months. The supplemented group had significantly higher sperm concentration, progressive and total motility, and total count [78].

Oliva et al. used a combination of daily 1200 mg (600 mg every 12 h) of pentoxifylline, 5 mg of folic acid and 66 mg of zinc sulphate for 12 weeks. Thirty-six men with infertility, clinical varicocele and abnormal semen parameters were recruited. Significant improvement in sperm with normal morphology was observed, and the effect persisted for at least 4 weeks after treatment cessation [79].

Festa et al. evaluated the use of CoQ_{10} 100 mg per day for 12 weeks in patients with male infertility associated with low-grade varicocele. They found significant increase in TAC and improvements in sperm density and forward sperm motility; however, whether the effect was persistent after drug suspension was not reported [80].

Bioflavonoid, a plant pigment, is another class of antioxidants. Zámpeiri et al. conducted a longitudinal observational study with adolescents aged 10–14. They were prescribed the semisynthetic derivative of bioflavonoid, O-β-hydroxyethylrutoside, at a daily dose of 1000 mg/day for 3 months followed by three treatment-free months for 1 year. There was a significant decrease in clinical progression of subclinical to clinical varicocele (11% vs. 31%). There was higher rate of stable vein reflux (47 vs. 38%) and higher resolution rate of varicocele (41 vs. 31%), but statistical significance was not reached [81]. Another derivative of bioflavonoid, micronized purified flavonoid fraction (MPFF), commonly available as Daflon, with the dosage of 1 g per day for 6 months, was shown to achieve significant improvement in sperm motility and colour Doppler parameters, and relief of varicocele-related pain [82, 83].

Chinese herbal medicine may as well be a source of exogenous antioxidants. The *escin*, which has an anti-inflammatory effect by enhancing the release of glucocorticoids (GCs) and prostaglandin-F2α (PGF2α) [84], was administered to men with varicocele-associated infertility at the daily dosage of 60 mg (30 mg every 12 h) for 2 months. There was a significant improvement of sperm density, but not motility, in the *escin* group when compared with the control group. Nonetheless, varicocelectomy is superior in improving semen parameters than the *ecsin* and control groups [85].

Antioxidants as an Adjunct Therapy to Varicocelectomy

Given the inferior result with antioxidant as a sole therapy when compared with varicocelectomy, there is a trend of studying the role of exogenous antioxidants as an adjunct therapy (Table 10.7). A randomized controlled trial (RCT) with L-carnitine-based combination supplement for 6 months after varicocelectomy showed similar significant improvement of semen parameters, and significantly higher pregnancy rate compared with the non-supplemented group (29% vs. 17.9%) [86]. On the contrary, Pourmand et al. failed to show a significant difference in semen parameters or DNA damage with the regime of L-carnitine 750 mg per day (250 mg 3 times a day), for 6 months after varicocelectomy [87]. The dosage of L-carnitine they used was, however, lower than the common dosage of 1 g per day. Barekat et al. found that the use of N-acetyl-L-cysteine 200 mg daily for 3 months after varicocelectomy did not result in a significant improvement in semen analysis, as well as percentage of ROS-negative sperm and intensity of sperm ROS. However, they found significantly higher percentage of improvement of normal protamine content and DNA integrity in the N-acetyl-L-cysteine group than in the placebo group [88].

Vitamin C has also been studied as an adjunct therapy to varicocelectomy. In an RCT of 115 men with infertility, abnormal semen analysis, and clinical varicocele who underwent varicocelectomy, vitamin C of 250 mg twice per day was given to the treatment group for 3 months after surgery. The treatment group had significantly better normal motility and morphology, but not sperm count, than the placebo group [89].

A Chinese herbal medicine, *Jingling* oral liquid, was used to compare with intramuscular injection of human chorionic gonadotropin (hCG), as a control group, after varicocelectomy. Semen quality significantly improved in both groups, and the improvement was better in the treatment group. In the treatment group, there was a significant increase in seminal SOD activity and zinc and decrease in Cd [90].

In a systemic review by Agarwal et al. which pooled 11 studies on antioxidant supplementation in varicocele, improvements of semen parameters and sperm function were reported in 75% and 83% of studies, respectively. However, the result is not statistically significant given the poor quality of studies [75].

There lacks high-level evidence on the use of exogenous antioxidants in varicocele intervention. Most of the studies have a small number of participants and inconsistent outcome measures. The combination and dosage of antioxidants are heterogenous. In conclusion, there is some evidence that the use of exogenous antioxidants may alleviate varicocele-associated OS and improve semen quality in Grade I-II varicocele, but the effect is not long-lasting and inferior to varicocelectomy. The use of antioxidants after varicocelectomy may further augment the improvement in semen parameters and sperm quality. Evidence on the effect of pregnancy outcome with antioxidants in varicocele-associated infertility is lacking. Nonetheless, exogenous antioxidants in general have a good safety profile. However, the effective bioavailability of many of the antioxidants is unknown, and there is potential harmful effect with overdosage due to the induced reductive stress.

Table 10.7 Use of exogenous antioxidants as adjunct therapy to varicocele repair in men with varicocele

Study	Patients	Group labels (n)	Dose and duration	Outcome parameters	Outcomes
Yan et al. [90] (2004)	Infertile men undergone varicocelectomy	Jingling oral liquid [29]	Not reported	Semen parameters	Significant improvement in semen quality ($p < 0.01$) compared to control group
		Intramuscular hCG (control) [29]			
Pourmand et al. [87] (2014)	Infertile men undergone left inguinal varicocelectomy	L-carnitine [48]	750 mg per day (250 mg 3 times per day) × 6 months after varicocelectomy	Semen parameters	No significant effects on sperm counts, motility and morphology
		Control group [48]			
Cyrus et al. [89] (2015)	Infertile men with Grade II–III varicocele undergone open inguinal (Ivanissevich) varicocelectomy	Vitamin C [45]	250 mg twice per day × 3 months after varicocelectomy	Semen parameters	Supplement group has significant improvement in mean normal motility ($p = 0.041$), mean normal morphology ($p < 0.001$), normal count ($p = 0.025$) and normal motility ($p = 0.033$) No significant effect on sperm count
		Control placebo group [67]	Starch-filled capsule × 3 months after varicocelectomy		
Barekat et al. [88] (2016)	Infertile men with left Grade II–III varicocele undergone MSV	N-acetyl-L-cysteine [15]	200 mg per day × 3 months after MSV	Semen parameters	Between supplement and control group, no significant difference in improvement of sperm concentration, percentage of sperm motility and abnormal morphology
		Control group [20]	No drug	Percentage of ROS positive sperm, protamine content and DNA integrity	No significant difference in percentage of ROS-positive sperm and intensity of sperm ROS Significantly higher percentages of improvement of normal protamine content ($p < 0.05$) and DNA integrity ($p < 0.01$) in supplement group

(continued)

Table 10.7 (continued)

Study	Patients	Group labels (n)	Dose and duration	Outcome parameters	Outcomes
Kizilay et al. [86] (2019)	Infertile men with clinical varicocele undergone MSV	L-carnitine-based therapy [62]	2000 mg of L-carnitine fumarate, 1000 mg of acetyl-L-carnitine HCl, 2000 mg of fructose, 100 mg of citric acid, 180 mg of vitamin C, 20 mg of zinc, 400 mcg of folic acid, 100 mcg of selenium, 4 mg of coenzyme Q_{10}, and 3 mcg of vitamin B12 per day × 6 months after MSV	Semen parameters	At 6 months, significant improvement with supplement group, compared with control group, in total sperm count ($p = 0.001$), sperm concentration ($p = 0.008$), morphology ($p < 0.001$), total motility ($p = 0.024$) and progressive motility ($p < 0.001$)
		Control group [27]	No drug	Pregnancy rate	Significantly higher pregnancy rate (29% vs 17.9%, $p = 0.029$)

Abbreviations: *hCG* human chorionic gonadotropin, *ROS* reactive oxygen species, *MSV* microsurgical subinguinal varicocelectomy

10.5 Conclusions

Not a single proposed pathophysiology can fully explain how varicocele leads to male infertility, but OS plays a central role in it. Regardless of fertility status, varicocele is associated with increased seminal ROS level. For instance, fertile or infertile men with varicocele demonstrate higher OS level than their counterparts without varicocele. The higher the grading of varicocele, the higher the OS level is. Surgical repair of varicocele can alleviate the OS, and potentially prevent its further harmful effect on spermatogenesis given the progressive nature of varicocele. Exogenous antioxidants may have a role especially as an adjunct to surgical repair, yet which and how they should be given has to be further studied.

References

1. Witt MA, Lipshultz LI. Varicocele: a progressive or static lesion? Urology. 1993;42(5):541–3.
2. Tulloch WS. Varicocele in subfertility; results of treatment. Br Med J. 1955;2(4935):356–8.
3. Schauer I, Madersbacher S, Jost R, Hubner WA, Imhof M. The impact of varicocelectomy on sperm parameters: a meta-analysis. J Urol. 2012;187(5):1540–7.
4. Abdel-Meguid TA, Al-Sayyad A, Tayib A, Farsi HM. Does varicocele repair improve male infertility? An evidence-based perspective from a randomized, controlled trial. Eur Urol. 2011;59(3):455–61.
5. Agarwal A, Hamada A, Esteves SC. Insight into oxidative stress in varicocele-associated male infertility: part 1. Nat Rev Urol. 2012;9(12):678–90.
6. Mostafa T, Anis T, Imam H, El-Nashar AR, Osman IA. Seminal reactive oxygen species-antioxidant relationship in fertile males with and without varicocele. Andrologia. 2009;41(2):125–9.
7. Miyaoka R, Esteves SC. A critical appraisal on the role of varicocele in male infertility. Adv Urol. 2012;2012:597495.
8. Hamada A, Esteves SC, Agarwal A. Varicocele and male infertility: origin and pathophysiology. Cham: Springer; 2016. p. 5–17. ISBN 978-3-319-24936-0. https://doi.org/10.1007/978-3-319-24936-0_2.
9. Jensen CFS, Ostergren P, Dupree JM, Ohl DA, Sonksen J, Fode M. Varicocele and male infertility. Nat Rev Urol. 2017;14(9):523–33.
10. Bae K, Shin HS, Jung H-J, Kang SH, Jin BS, Park JS. Adolescent varicocele: are somatometric parameters a cause? Korean J Urol. 2014;55(8):533–5.
11. Benoff S, Gilbert BR. Varicocele and male infertility: part I. Preface. Hum Reprod Update. 2001;7(1):47–54.
12. Agarwal A, Sharma R, Samanta L, Durairajanayagam D, Sabanegh E. Proteomic signatures of infertile men with clinical varicocele and their validation studies reveal mitochondrial dysfunction leading to infertility. Asian J Androl. 2016;18(2):282–91.
13. Cho C-L, Esteves SC, Agarwal A. Indications and outcomes of varicocele repair. Panminerva Med. 2019;61(2):152–63.
14. Tavilani H, Doosti M, Saeidi H. Malondialdehyde levels in sperm and seminal plasma of asthenozoospermic and its relationship with semen parameters. Clin Chim Acta. 2005;356(1–2):199–203.
15. Krzysciak W, Kozka M. Generation of reactive oxygen species by a sufficient, insufficient and varicose vein wall. Acta Biochim Pol. 2011;58(1):89–94.
16. Malivindi R, De Rose D, Gervasi MC, Cione E, Russo G, Santoro M, Aquila S. Influence of all-trans retinoic acid on sperm metabolism and oxidative stress: its involvement in the physiopathology of varicocele-associated male infertility. J Cell Physiol. 2018;233(12):9526–37.
17. Aksoy Y, Ozbey I, Aksoy H, Polat O, Akcay F. Seminal plasma nitric oxide concentration in oligo- and/or asthenozoospermic subjects with/without varicocele. Arch Androl. 2002;48(3):181–5.
18. Mehraban D, Ansari M, Keyhan H, Gilani MS, Naderi G, Esfehani F. Comparison of nitric oxide concentration in seminal fluid between infertile patients with and without varicocele and normal fertile men. Urol J. 2005;2(2):106–10.
19. Sakamoto Y, Ishikawa T, Kondo Y, Yamaguchi K, Fujisawa M. The assessment of oxidative stress in infertile patients with varicocele. BJU Int. 2008;101(12):1547–52.
20. Alkan I, Yuksel M, Canat HL, Atalay HA, Can O, Ozveri H, et al. Superoxide anion production by the spermatozoa of men with varicocele: relationship with varicocele grade and semen parameters. World J Mens Health. 2018;36(3):255–62.
21. Cho C-L, Esteves SC, Agarwal A. Novel insights into the pathophysiology of varicocele and its association with reactive oxygen species and sperm DNA fragmentation. Asian J Androl. 2016;18(2):186–93.

22. Naughton CK, Nangia AK, Agarwal A. Pathophysiology of varicoceles in male infertility. Hum Reprod Update. 2001;7(5):473–81.
23. Hosseinifar H, Gourabi H, Salekdeh GH, Alikhani M, Mirshahvaladi S, Sabbaghian M, Modarresi T, Gilani MAS. Study of sperm protein profile in men with and without varicocele using two-dimensional gel electrophoresis. Urology. 2013;81(2):293–300.
24. Mostafa T, Anis TH, Ghazi S, El-Nashar AR, Imam H, Osman IA. Reactive oxygen species and antioxidants relationship in the internal spermatic vein blood of infertile men with varicocele. Asian J Androl. 2006;8(4):451–4.
25. Ozbek E, Ilbey YYO, Simsek A, Cekmen M, Balbay MD. Preoperative and postoperative seminal nitric oxide levels in patients with infertile varicocele. Arch Ital Urol Androl. 2009;81(4):248–50.
26. Mostafa T, Anis T, El Nashar A, Imam H, Osman I. Seminal plasma reactive oxygen species-antioxidants relationship with varicocele grade. Andrologia. 2012;44(1):66–9.
27. Abd-Elmoaty MA, Saleh R, Sharma R, Agarwal A. Increased levels of oxidants and reduced antioxidants in semen of infertile men with varicocele. Fertil Steril. 2010 Sep;94(4):1531–4.
28. Ghandehari-Alavijeh R, Tavalaee M, Zohrabi D, Foroozan-Broojeni S, Abbasi H, Nasr-Esfahani MH. Hypoxia pathway has more impact than inflammation pathway on etiology of infertile men with varicocele. Andrologia. 2019;51(2):e13189.
29. Pasqualotto FF, Sharma RK, Kobayashi H, Nelson DR, Thomas AJ Jr, Agarwal A. Oxidative stress in normospermic men undergoing infertility evaluation. J Androl. 2001;22(2):316–22.
30. Kilinc F, Kayaselcuk F, Aygun F, Aygun C, Guvel S, Egilmez T, Ozkardes H. Experimental varicocele induces hypoxia inducible factor-1α, vascular endothelial growth factor expression and angiogenesis in the rat testis. J Urol. 2004;172(3):1188–91.
31. Paick JS, Park K, Kim SW, Park JW, Kim JJ, Kim MS, Park JY. Increased expression of hypoxia-inducible factor-1α and connective tissue growth factor accompanied by fibrosis in the rat testis of varicocele. Actas Urol Esp. 2012;36(5):282–8.
32. Fernandez-Aguera MC, Gao L, Gonzalez-Rodriguez P, Pintado CO, Arias-Mayenco I, Garcia-Flores P, Garcia-Perganeda A, Pascual A, Ortega-Saenz P, Lopez-Barneo J. Cell Metab. 2015;22(5):825–37.
33. Benoff SH, Millan C, Hurley IR, Napolitano B, Marmar JL. Bilateral increased apoptosis and bilateral accumulation of cadmium in infertile men with left varicocele. Hum Reprod. 2004;19(3):616–27.
34. Ata-abadi NS, Mowla SJ, Aboutalebi F, Dormiani K, Kiani-Esfahani A, Tavalaee M, Nasr-Esfahani MH. Hypoxia-related long noncoding RNAs are associated with varicocele-related male infertility. PLoS One. 2020;15(4):e0232357.
35. Choi WS, Kim SW. Current issues in varicocele management: a review. World J Men's Health. 2013;31(1):12–20.
36. Hassanin AM, Ahmed HH, Kaddah AN. A global view of the pathophysiology of varicocele. Andrology. 2018;6(5):654–61.
37. Agarwal A, Roychoudhury S, Bjugstad KB, Cho C-L. Oxidation-reduction potential of semen: what is its role in the treatment of male infertility? Ther Adv Urol. 2016;8(5):302–18.
38. Sharma R, Roychoudhury S, Singh N, Sarda Y. Methods to measure reactive oxygen species (ROS) and total antioxidant capacity (TAC) in the reproductive system. In: Agarwal A, Sharma R, Gupta S, Harlev A, Ahmad G, du Plessis SS, Esteves SC, Wang SM, Durairajanayagam D, editors. Oxidative stress in human reproduction. Cham: Springer; 2017. p. 17–46. ISBN 978-3-319-48427-3.
39. Finelli R, Panner Selvam MK, Agarwal A. Oxidative stress testing: direct tests. In: Agarwal A, Henkel R, Majzoub A, editors. Manual of sperm function testing in human assisted reproduction. Cambridge: Cambridge University Press; 2021. p. 111–22. ISBN 9781108878715.
40. Sharma R, Robert K, Agarwal A. Oxidative stress testing: indirect tests. In: Agarwal A, Henkel R, Majzoub A, editors. Manual of sperm function testing in human assisted reproduction. Cambridge: Cambridge University Press; 2017. p. 123–41. ISBN 9781108878715.

41. Cocuzza M, Athayde KS, Agarwal A, Pagani R, Sikka SC, Lucon AM, et al. Impact of clinical varicocele and testis size on seminal reactive oxygen species levels in a fertile population: a prospective controlled study. Fertil Steril. 2008;90(4):1103–8.
42. Blumer CG, Restelli AE, Giudice PT, Soler TB, Fraietta R, Nichi M, et al. Effect of varicocele on sperm function and semen oxidative stress. BJU Int. 2012;109(2):259–65.
43. Shiraishi K, Takihara H, Matsuyama H. Elevated scrotal temperature, but not varicocele grade, reflects testicular oxidative stress-mediated apoptosis. World J Urol. 2010;28(3):359–64.
44. Abd-Elmoaty MA, Saleh R, Sharma R, Agarwal A. Increased levels of oxidants and reduced antioxidants in semen of infertile men with varicocele. Fertil Steril. 2010;94(4):1531–4.
45. Allamaneni SS, Naughton CK, Sharma RK, Thomas AJ Jr, Agarwal A. Increased seminal reactive oxygen species levels in patients with varicoceles correlate with varicocele grade but not with testis size. Fertil Steril. 2004;82(6):1684–6.
46. Koksal IT, Tefekli A, Usta M, Erol H, Abbasoglu S, Kadioglu A. The role of reactive oxygen species in testicular dysfunction associated with varicocele. BJU Int. 2000;86(4):549–52.
47. Cocuzza M, Athayde KS, Alvarenga C, Srougi M, Hallak J. Grade 3 varicocele in fertile men: a different entity. J Urol. 2012;187(4):1363–8.
48. Alkan I, Yuksel M, Canat HL, Atalay HA, Can O, Ozveri H, et al. Superoxide anion production by the spermatozoa of men with varicocele: relationship with varicocele grade and semen parameters. World J Men's Health. 2018;36(3):255–62.
49. Hendin BN, Kolettis PN, Sharma RK, Thomas AJ Jr, Agarwal A. Varicocele is associated with elevated spermatozoal reactive oxygen species production and diminished seminal plasma antioxidant capacity. J Urol. 1999;161(6):1831–4.
50. Ni K, Steger K, Yang H, Wang H, Hu K, Zhang T, et al. A comprehensive investigation of sperm DNA damage and oxidative stress injury in infertile patients with subclinical, normozoospermic, and astheno/oligozoospermic clinical varicocoele. Andrology. 2016;4(5):816–24.
51. Shiraishi K, Takihara H, Naito K. Testicular volume, scrotal temperature, and oxidative stress in fertile men with left varicocele. Fertil Steril. 2009;91(4):1388–91.
52. Mehraban D, Ansari M, Keyhan H, Sedighi Gilani M, Naderi G, Esfehani F. Comparison of nitric oxide concentration in seminal fluid between infertile patients with and without varicocele and normal fertile men. Urol J. 2005;2(2):106–10.
53. Mancini A, Milardi D, Bianchi A, Festa R, Silvestrini A, De Marinis L, et al. Increased total antioxidant capacity in seminal plasma of varicocele patients: a multivariate analysis. Arch Androl. 2007;53(1):37–42.
54. Meucci E, Milardi D, Mordente A, Martorana GE, Giacchi E, De Marinis L, et al. Total antioxidant capacity in patients with varicoceles. Fertil Steril. 2003;79:1577–83.
55. Balercia G, Arnaldi G, Fazioli F, Serresi M, Alleva R, Mancini A, et al. Coenzyme Q10 levels in idiopathic and varicocele-associated asthenozoospermia. Andrologia. 2002;34(2):107–11.
56. Agarwal A, Prabakaran S, Allamaneni SS. Relationship between oxidative stress, varicocele and infertility: a meta-analysis. Reprod Biomed Online. 2006;12(5):630–3.
57. Pasqualotto FF, Sundaram A, Sharma RK, Borges E Jr, Pasqualotto EB, Agarwal A. Semen quality and oxidative stress scores in fertile and infertile patients with varicocele. Fertil Steril. 2008;89(3):602–7.
58. Canales BK, Zapzalka DM, Ercole CJ, Carey P, Haus E, Aeppli D, et al. Prevalence and effect of varicoceles in an elderly population. Urology. 2005;66(3):627–31.
59. Romeo C, Ientile R, Santoro G, Impellizzeri P, Turiaco N, Impala P, et al. Nitric oxide production is increased in the spermatic veins of adolescents with left idiophatic varicocele. J Pediatr Surg. 2001;36(2):389–93.
60. Türkyilmaz Z, Gülen Ş, Sönmez K, Karabulut R, Dinçer S, Can Başaklar A, et al. Increased nitric oxide is accompanied by lipid oxidation in adolescent varicocele. Int J Androl. 2004;27(3):183–7.
61. Bertolla RP, Cedenho AP, Hassun Filho PA, Lima SB, Ortiz V, Srougi M. Sperm nuclear DNA fragmentation in adolescents with varicocele. Fertil Steril. 2006;85(3):625–8.

62. Beckman KB, Ames BN. The free radical theory of aging matures. Physiol Rev. 1998;78(2):547–81.
63. Mostafa T, Anis TH, El-Nashar A, Imam H, Othman IA. Varicocelectomy reduces reactive oxygen species levels and increases antioxidant activity of seminal plasma from infertile men with varicocele. Int J Androl. 2001;24(5):261–5.
64. Chen SS, Huang WJ, Chang LS, Wei YH. Attenuation of oxidative stress after varicocelectomy in subfertile patients with varicocele. J Urol. 2008;179(2):639–42.
65. Dada R, Shamsi MB, Venkatesh S, Gupta NP, Kumar R. Attenuation of oxidative stress & DNA damage in varicocelectomy: implications in infertility management. Indian J Med Res. 2010;132:728–30.
66. Yesilli C, Mungan G, Seckiner I, Akduman B, Acikgoz S, Altan K, et al. Effect of varicocelectomy on sperm creatine kinase, HspA2 chaperone protein (creatine kinase-M type), LDH, LDH-X, and lipid peroxidation product levels in infertile men with varicocele. Urology. 2005;66(3):610–5.
67. Mancini A, Meucci E, Milardi D, Giacchi E, Bianchi A, Pantano AL, et al. Seminal antioxidant capacity in pre- and postoperative varicocele. J Androl. 2004;25(1):44–9.
68. Mancini A, Milardi D, Conte G, Festa R, De Marinis L, Littarru GP. Seminal antioxidants in humans: preoperative and postoperative evaluation of coenzyme Q10 in varicocele patients. Horm Metab Res. 2005;37(7):428–32.
69. Silay MS, Hoen L, Quadackaers J, Undre S, Bogaert G, Dogan HS, et al. Treatment of varicocele in children and adolescents: a systematic review and meta-analysis from the European Association of Urology/European Society for Paediatric Urology Guidelines Panel. Eur Urol. 2019;75(3):448–61.
70. Cayan S, Sahin S, Akbay E. Paternity rates and time to conception in adolescents with varicocele undergoing microsurgical varicocele repair vs observation only: a single institution experience with 408 patients. J Urol. 2017;198(1):195–201.
71. Bogaert G, Orye C, De Win G. Pubertal screening and treatment for varicocele do not improve chance of paternity as adult. J Urol. 2013;189(6):2298–303.
72. Lacerda JI, Del Giudice PT, da Silva BF, Nichi M, Fariello RM, Fraietta R, et al. Adolescent varicocele: improved sperm function after varicocelectomy. Fertil Steril. 2011;95(3):994–9.
73. Cervellione RM, Cervato G, Zampieri N, Corroppolo M, Camoglio F, Cestaro B, et al. Effect of varicocelectomy on the plasma oxidative stress parameters. J Pediatr Surg. 2006;41(2):403–6.
74. Smits RM, Mackenzie-Proctor R, Yazdani A, Stankiewicz MT, Jordan V, Showell MG. Antioxidants for male subfertility. Cochrane Database Syst Rev. 2019;3:CD007411.
75. Agarwal A, Leisegang K, Majzoub A, Henkel R, Finelli R, Panner Selvam MK, et al. Utility of antioxidants in the treatment of male infertility: clinical guidelines based on a systematic review and analysis of evidence. World J Men's Health. 2021;39(2):233–90.
76. Cavallini G, Biagiotti G, Ferraretti AP, Gianaroli L, Vitali G. Medical therapy of oligoasthenospermia associated with left varicocele. BJU Int. 2003;91(6):513–8.
77. Cavallini G, Ferraretti AP, Gianaroli L, Biagiotti G, Vitali G. Cinnoxicam and L-carnitine/acetyl-L-carnitine treatment for idiopathic and varicocele-associated oligoasthenospermia. J Androl. 2004;25(5):761–70. discussion 71-2
78. Busetto GM, Agarwal A, Virmani A, Antonini G, Ragonesi G, Del Giudice F, et al. Effect of metabolic and antioxidant supplementation on sperm parameters in oligo-astheno-teratozoospermia, with and without varicocele: a double-blind placebo-controlled study. Andrologia. 2018;50(3) https://doi.org/10.1111/and.12927.
79. Oliva A, Dotta A, Multigner L. Pentoxifylline and antioxidants improve sperm quality in male patients with varicocele. Fertil Steril. 2009;91(4 Suppl):1536–9.
80. Festa R, Giacchi E, Raimondo S, Tiano L, Zuccarelli P, Silvestrini A, et al. Coenzyme Q10 supplementation in infertile men with low-grade varicocele: an open, uncontrolled pilot study. Andrologia. 2014;46(7):805–7.
81. Zampieri N, Pellegrino M, Ottolenghi A, Camoglio FS. Effects of bioflavonoids in the management of subclinical varicocele. Pediatr Surg Int. 2010;26(5):505–8.

82. Kiliç S, Günes A, Ipek D, Dusak A, Günes G, Balbay MD, et al. Effects of micronised purified flavonoid fraction on pain, spermiogram and scrotal color Doppler parameters in patients with painful varicocele. Urol Int. 2005;74(2):173–9.
83. Soylemez H, Kilic S, Atar M, Penbegul N, Sancaktutar AA, Bozkurt Y. Effects of micronised purified flavonoid fraction on pain, semen analysis and scrotal color Doppler parameters in patients with painful varicocele: results of a randomized placebo-controlled study. Int Urol Nephrol. 2012;44(2):401–8.
84. Wang H, Zhang L, Jiang N, Wang Z, Chong Y, Fu F. Anti-inflammatory effects of escin are correlated with the glucocorticoid receptor/NF-κB signaling pathway, but not the COX/PGF2α signaling pathway. Exp Ther Med. 2013;6(2):419–22.
85. Fang Y, Zhao L, Yan F, Xia X, Xu D, Cui X. Escin improves sperm quality in male patients with varicocele-associated infertility. Phytomedicine. 2010;17(3–4):192–6.
86. Kizilay F, Altay B. Evaluation of the effects of antioxidant treatment on sperm parameters and pregnancy rates in infertile patients after varicocelectomy: a randomized controlled trial. Int J Impot Res. 2019;31(6):424–31.
87. Pourmand G, Movahedin M, Dehghani S, Mehrsai A, Ahmadi A, Pourhosein M, et al. Does L-carnitine therapy add any extra benefit to standard inguinal varicocelectomy in terms of deoxyribonucleic acid damage or sperm quality factor indices: a randomized study. Urology. 2014;84(4):821–5.
88. Barekat F, Tavalaee M, Deemeh MR, Bahreinian M, Azadi L, Abbasi H, et al. A preliminary study: N-acetyl-L-cysteine improves semen quality following varicocelectomy. Int J Fertil Steril. 2016;10(1):120–6.
89. Cyrus A, Kabir A, Goodarzi D, Moghimi M. The effect of adjuvant vitamin C after varicocele surgery on sperm quality and quantity in infertile men: a double-blind placebo controlled clinical trial. Int Braz J Urol. 2015;41(2):230–8.
90. Yan LF, Jiang MF, Shao RY. Clinical observation on effect of jingling oral liquid in treating infertile patients with varicocele after varicocelectomy. Zhongguo Zhong Xi Yi Jie He Za Zhi. 2004;24(3):220–2.

Chapter 11
Oxidative Stress in Men with Obesity, Metabolic Syndrome and Type 2 Diabetes Mellitus: Mechanisms and Management of Reproductive Dysfunction

Kristian Leisegang

Abstract Reactive oxygen species (ROS) are critical physiological mediators of cellular function, including male fertility. When ROS exceed antioxidant regulation, oxidative stress occurs which is detrimental to cellular function. Oxidative stress has been found to be a central mediator of obesity, metabolic syndrome (MetS) and type 2 diabetes mellitus (T2DM), as well as with male infertility. Human studies have correlated testicular oxidative stress in obese males, and animal studies have further provided insight into potential mechanisms of action. Management of oxidative stress is not well defined. Appropriate nutrition and exercise can be recommended for all diabetic patients, and weight loss for obese patients with MetS and T2DM. Consideration of dietary supplements including micronutrients, antioxidants or medicinal herbs are recommended. Metformin may also offer benefits on testicular oxidative stress and fertility parameters. Significantly more research on causation, mechanisms, clinical assessments and appropriate management of infertility on obesity, MetS and T2DM is still required.

Keywords Obesity · Metabolic syndrome · Type 2 diabetes mellitus · Oxidative stress · Inflammation · Male infertility · Semen parameters · Spermatogenesis

---

K. Leisegang (✉)
School of Natural Medicine, University of the Western Cape,
Bellville, Cape Town, South Africa
e-mail: kleisegang@uwc.ac.za

© The Author(s), under exclusive license to Springer Nature Switzerland AG 2022
K. K. Kesari, S. Roychoudhury (eds.), *Oxidative Stress and Toxicity in Reproductive Biology and Medicine*, Advances in Experimental Medicine and Biology 1358, https://doi.org/10.1007/978-3-030-89340-8_11

11.1 Introduction

Reactive oxygen species (ROS) have been found to be important regulators of physiological function in all biological systems, particularly as signalling molecules that are regulated by various hormones and cytokines [1]. This activity is mediated through ROS activation of phosphorylation/dephosphorylation (redox) switches [2]. As ROS are potentially toxic to cells, endogenous antioxidant defences neutralise and regulate redox biology. Endogenous antioxidants include catalase (CAT), glutathione peroxidase (GPx), superoxide dismutase (SOD) family, peroxiredoxins and thioredoxins. Additional antioxidant activity is from exogenous molecules obtained through nutritional sources, predominantly including Vitamin A and other β-carotenes, Vitamin C and Vitamin E, alongside numerous phytonutrients [3].

In excessive concentrations that overcome antioxidant regulation defences, ROS damage lipids and proteins, including cell membranes, mitochondrial membranes, enzymes and DNA [1]. An increase in ROS and/or a decrease in antioxidant defences defines a state of oxidative stress, resulting in cellular injury and damage to organelles and molecules [2]. Oxidative stress has been implicated in numerous degenerative chronic pathologies, including complications of obesity, metabolic syndrome (MetS) and type 2 diabetes mellitus (T2DM) [4, 5]. Oxidative stress is also well defined to negatively affect male fertility across numerous different causes, which also include obesity, MetS and T2DM [6].

11.2 Obesity, Metabolic Syndrome and Type 2 Diabetes Mellitus

An increase in adiposity, particularly abdominal white adipose tissue, is well established to cause debilitating health effects, increasing morbidity and mortality [7]. Obesity is a clinical condition in which there is an excessive accumulation of fat that adversely affects health outcomes [8]. Closely related but considered a distinct clinical entity, the metabolic syndrome (MetS) is clinically defined as a clustering or constellation of increased abdominal adiposity, increased blood pressure, reduced HDL cholesterol, increased serum triglycerides and increased glucose concentrations [9, 10]. This syndrome describes a subgroup of patients who are at increased risk for the development of cardiovascular disease (CVD) and T2DM [11]. Obesity is a common feature of MetS [9], and closely associated with insulin resistance and the development of T2DM [7]. T2DM is considered a complication of obesity and MetS, with similar aetiologies and underlying pathophysiology, characterised clinically by hyperglycaemia due to insulin resistance [12].

Common risk factors for the development of obesity, MetS and T2DM include an increased energy-dense and nutrient-poor nutritional consumption and reduced physical activity. This includes exposure to environmental toxins, endocrine-disrupting chemicals and pharmacological interventions, alongside genetic and epigenetic changes [12–14]. Metabolic and co-morbidity risk factors include dyslipidaemia, hypertension, hyperinsulinaemia (insulin resistance), chronic inflammation, oxidative stress, obstructive sleep apnoea and micronutrient deficiencies [15].

An exponential increase in obesity prevalence globally has been observed in adults, adolescents and children in recent decades [9]. This is reported as up to 27% of adults and 47% of children [16]. Although there is a clearly defined increased risk in disadvantaged communities [8], this increase in the obesity pandemic is reported in both sexes and across all geographical regions, ethnic groups and socioeconomic groups [17]. Similarly to obesity, MetS has emerged as a global health problem, with the incidence parallel to that of obesity [13]. The International Diabetes Federation (IDF) reports a global incidence of 8,8% for MetS, representing more than 415 million people [13]. However, MetS is reported to be over 40% of the population in some regions, including Iran and Tunisia [9, 18]. This is set to increase into the future globally, with the most significant growths expected in Africa and the Middle East [13]. MetS is further estimated to be 3 times more common than diabetes [18]. Based on data available from the Global Burden of Disease, T2DM affected 8,8% of the global population in 2017, which is approximately 462 million people. This is across all age groups, including 4,4% in ages 15–19 years, 15% in ages 50–69 years, and 22% in those aged over 70 years, and incidence peaking at 55 years of age [18, 19]. T2DM is considered the ninth leading cause of death globally, with over one million deaths annually [18].

11.3 Clinical Assessment of Obesity, Metabolic Syndrome and Type 2 Diabetes Mellitus

BMI is the standard tool used to clinically define bodyweight categories and estimate risk of complications in underweight, overweight and obese patients. Based on this classification system, obesity is defined as >30 Kg/m^2 (Table 11.1) by the World Health Organisation (WHO) [8, 20, 21]. This is considered a useful predictor of mortality over and under an apparent optimal BMI range of 22,5–25 Kg/m^2 [21]. Mortality associated with increased BMI is due predominantly due to vascular disease complications, with a 30% increased risk for vascular mortality and 60–120% increased risk for diabetic, renal or hepatic complications for every additional 5 Kg/m^2 above normal [21].

BMI has been shown to have limitations, as it does not distinguish variations in body fat percentage and fat distribution across all BMI categories (Table 11.1), or distinguish between protective subcutaneous adipose tissue and detrimental visceral (abdominal) adipose tissue accumulation [7]. Cardiometabolic complications are associated predominantly with increased visceral adipose tissue [17]. BMI use alone has been estimated to miss up to half of all patients with cardiometabolic risk due to its high specificity with relatively low sensitivity [22]. Although body fat percentage may offer additional information, the determination of the waist

Table 11.1 Body mass index (BMI) categories and risk of co-morbidities

BMI	Classification	Risk of co-morbidities	Consequences
< 18.5	Underweight (undernourished)	Increased	Immunodeficiency-related infectious disease and malignancies
18.5–24.9	Normal weight	Low	Uncommon related to BMI
25.0–29.9	Overweight	Mild	NCCDs: CVD; T2DM; malignancies; degenerative disease
30.0–34.9	Class I obesity	Moderate	
35.0–39.9	Class II obesity	Severe	
> 40	Class III obesity	Very severe	

NCCDs non-communicable chronic disease, *CVD* cardiovascular disease, *T2DM* type 2 diabetes mellitus [20, 21, 131]

Table 11.2 The metabolic syndrome diagnostic criteria in males. A diagnosis is made with any three of the five criteria [10]

Criteria	Categorical cut off points
Waist circumference	Sub-Saharan African ≥94 cm; Caucasian ≥94 cm; Asian ≥90 cm
Blood pressure (or relevant medication)	Systolic ≥130 mmHg and/or diastolic ≥85 mmHg
Fasting triglycerides (or relevant medication)	> 150 mg/dL or 1.70 mmol/L
HDL cholesterol (or relevant medication)	< 40 mg/dL or 1.00 mmol/L
Fasting glucose (or relevant medication)	> 100 mg/dL or 5.5 mmol/L

HDL = high-density lipoprotein

circumference (WC) is seen as a simple alternative with better prediction of complications and mortality than BMI [22, 23]. In overweight and obese men, a WC > 102 cm significantly increases mortality risk compared to those with WC < 102 cm, and these risk thresholds are reduced in different ethnic backgrounds Asian and Europoid ethnic backgrounds [23]. However, lower thresholds with ethnic variation are provided in the MetS diagnostic criteria thresholds [10].

MetS is diagnosed based on the presence of any three of the five defining features, specifically increased waist circumference (abdominal adiposity), blood pressure, serum triglycerides, serum glucose, and/or reduced HDL cholesterol. Cut-off values for waist circumference and blood pressure vary with gender and/or ethnic background (Table 11.2) [10].

T2DM screening and diagnostic evaluation can be done using $HbA1_c$ criteria (> 6.5%), fasting plasma glucose (> 126 mg/dL or 7.0 mmol/L) or the 2 h 75 g oral glucose tolerance test (> 200 mg/dL or 11.1 mmol/L). In patients diagnosed with T2DM, additional CVD risk factors should be clinically evaluated [24].

11.4 Inflammation and Oxidative Stress in the Pathophysiology of Obesity, Metabolic Syndrome and Type 2 Diabetes Mellitus

Obesity, MetS and T2DM arise from complex interactions of polygenetic predisposition, socioeconomic factors and cultural influences [16]. This complex pathophysiology includes phenomena such as insulin resistance, systemic inflammation with altered cytokine and adipokine expressions and oxidative stress. Endocrine changes in males also include reduced testosterone and progesterone, with increased gonadotropins, oestrogen and prolactin [4, 25–27]. These systemic biochemical changes are summarised in Table 11.3.

Although adipose tissue is considered an important source of inflammation and ROS in obesity, oxidative stress and inflammation have been identified in numerous different tissues and organs in obesity [25, 28]. Oxidative stress and inflammation mediate obesity and related metabolic disorders through cellular dysfunction, loss of cellular energy metabolism, altered cellular signalling and control of growth cycles, and DNA damage and mutations [29]. This leads to increased lipid peroxidation, protein carbonylation, and reduced endogenous antioxidant expression, which are considered important mechanisms in the development of T2DM, CVD and malignancies [29]. As systemic inflammation is associated with reproductive tract inflammation in obesity and metabolic syndrome; oxidative stress in this context is increasingly considered detrimental to male reproductive tissues [25]. Oxidative stress and inflammation can be triggered by excessive macronutrients intake [30], including high fat and/or high carbohydrate diets [29]. This increased energy intake increases NADPH oxidase and reduces endogenous antioxidant defence [31, 32].

Inflammation is widespread in multiple tissues in obesity, MetS and T2DM. This includes adipose tissue, skeletal and cardiac muscle, liver, pancreas, brain and male reproductive tissues, associated with increased TNFα, IL-1β, IL-6 and IL-8 and reduced IL-2, IL-10 [25, 33]. This is mediated by a shift from an immune regulating phenotype to an inflammatory phenotype, with the infiltration of macrophages and lymphocytes predominantly into affected tissues [25, 26, 33]. Furthermore, inflammation is known to be involved in the pathogenesis of insulin resistance in obesity and MetS, increasing the risk of T2DM [25, 26, 34].

Inflammation is also closely associated with increased ROS production and oxidative stress, mediated in part through mitochondrial dysfunction [35]. Overnutrition (excess nutrient supply) results in mitochondrial dysfunction and increased ROS generation, which can further exacerbate the inflammatory response and increase cellular apoptosis [35]. Oxidative stress in adipose tissue also results in adipokine dysregulation, with increased leptin, orexin obestatin and plasminogen activating factor-1, and decreased adipokinine [36, 37]. Increased leptin, ghrelin and orexin correlates with oxidative stress, whereas adiponectin negatively correlates with oxidative stress [37].

There is a significant correlation between systemic oxidative stress markers and increasing BMI, blood glucose, smoking, T2DM and CVD [38]. Oxidative stress in

Table 11.3 Correlated molecular changes of obesity and metabolic syndrome in males [4]

Category	Marker
Endocrine	↓ Total and free testosterone
	↓ Progesterone
	↑ Oestrogen
	↑ Gonadotropins
	↓ SHBG
	↑ Prolactin
	↑ Insulin
Cytokines and adipokines and inflammation	↓ Adiponectin
	↑ Resistin
	↑ Platelet activating inhibitor-1
	↑ Angiotensin II
	↑ Leptin
	↑ TNFα, IL-1β, IL-6, IL-8
	↓ IL-2, IL-10
	↑ Hs-CRP
Oxidative stress	↓ TAC
	↑ Malondialdehyde (MDA)
	↑ LDLox
	↑ F2 isoprostanes (F2-IsoP)
	↑ 8-iso Prostaglandin F2a (8-isoPGF2a)
	↑ Protein carbonylation
	↓ Cu-Zn-SOD
	↓ Mn-SOD
	↓ Catalase (CAT)
	↓ Glutathione peroxidase (GPx)
Micronutrients	↓ Vitamins A, B1, B9, B12, C, D, E
	↓ Selenium
	↓ Zinc
	↓ Iron

obesity mediates the development of MetS, where accumulation of adipose tissues correlates with oxidative stress *in vitro* and *in vivo* (animals and humans). In adult humans, these markers include increased F2α-isoprostane (F2-IsoP), 8-isoprostaglandin F2a (8-isoPGF2a), malondialdehyde (MDA), oxidised LDL cholesterol (Ox-LDL), and reduced PON1 activity and total antioxidant capacity (TAC) [39]. There is also a reduction in endogenous antioxidant defence enzymes, including GPx, CAT and SOD [39].

Genetic predisposition to obesity through single nucleotide polymorphisms (SNPs) includes genes involved with antioxidant defence systems. SNPs associated with increased oxidative stress in obesity includes GPX1 (594C/T and Ala^5/Ala^6+), GPX7 (G/A), CAT (−89A/T), CAT (−844A/G), CAT (−20C/T), PON1 (Q192R and

L55M), SOD2 (Ala16Val), p22phox (−930A/G) and PPARγ (Pro12Ala), amongst others [39]. These SNPs are further associated with an increased risk inflammation and insulin resistance [39].

Paradoxically, obesity is also associated with micronutrient deficiency [40, 41]. Common deficiencies include vitamin D, selenium, vitamin C, zinc, B-vitamins, carotenes, manganese, chromium and copper. These deficiencies contribute significantly to oxidative stress, as well as defective insulin signalling, reduced tyrosine kinase activity, β-cell dysfunction, loss of muscle mass, increased protein kinase C and intracellular calcium [42]. Ultimately, through dysfunction insulin signalling and metabolic activity, these deficiencies contribute significantly to the pathophysiology of MetS and T2DM [41].

Insulin resistance is closely associated with obesity and metabolic syndrome and has a central role in the pathogenesis of T2DM. This change in insulin signalling function is complex, but mediated through oxidative stress and inflammation [43]. Up to 30% of obese patients however do not have insulin resistance, and these patients are associated with reduced visceral fat and intima-media thickness of the carotid artery compared to obese patients with metabolic derangements [9]. Pathways for insulin resistance are also mediated through mitochondrial dysfunction, and well as lipotoxicity and endothelial stress induced through oxidative stress [44].

Hyperglycaemia from both types of diabetes mellitus increases oxidative stress through increased oxidation of glucose, glycation of proteins and oxidative metabolism of these glycated proteins. This oxidative stress has been closely associated with the development of complications [45]. Oxidative stress in obesity has also been shown to emerge from endothelial dysfunction induced by hyperleptinaemia [46]. Furthermore, oxidative stress has mediated the development of T2DM as well as vascular damage [28]. Complications of obesity are partly due to lipotoxicity, where there are increased lipids in peripheral tissues that outweigh the storage and antioxidant protection capacities. Increased intracellular lipids cause inhibition of GLUT4, leading to insulin resistance, reduced glucose uptake, oxidative stress, ER stress and apoptosis [47].

11.5 Impact of Obesity, Metabolic Syndrome and Type 2 Diabetes Mellitus on Male Fertility

The association between male obesity, MetS and T2DM with reproductive dysfunction has been well established [4, 25, 48–50]. This includes reduced sperm parameters and spermatogenesis, increased DNA fragmentation and apoptosis, and changes in reproductive hormones that include hypogonadism and hyperoestrogenaemia. Obesity and MetS are further associated with low-grade chronic prostatitis, benign prostatic hyperplasia (BPH) and prostatic carcinoma. Importantly, sperm DNA damage and epigenetic changes can increase the risk of pregnancy complications and poor health outcomes of the offspring, including metabolic and neurodevelopmental disorders [4, 25, 48–50].

Exposure to an obesogenic environment has been described to explain the relationship between environmental factors and infertility. Many of these obesogens may disrupt reproductive hormone activity, acting as endocrine-disrupting compounds (ECDs). These lipophilic compounds accumulate in adipose tissue, modulating reproductive function and reducing spermatogenesis. Furthermore, these obesogens can also transfer to the offspring through epigenetic modifications [51].

Diabetes, through chronic hyperglycaemia, is known to negatively affect sperm parameters, spermatogenesis, Leydig cell function and reductive hormone balance, as well as induce sexual dysfunction. It is postulated that the mitochondrial dysfunction and oxidative stress are key mediators in this negative impact of hyperglycaemia on reproductive dysfunction [52].

11.6 Pathophysiology of Inflammation and Oxidative Stress in Metabolic Dysfunction on Male Fertility

As with numerous other tissues, ROS are important regulators and mediators of male fertility in normal redox physiology. ROS are critical in spermatogenesis, chromatin condensation, maturation in the epididymis, and involved in post-ejaculatory functions such as hyperactivation, acrosomal reaction and oocyte fusion [53, 54]. ROS, particularly H_2O_2, is generated from spermatozoa and leucocytes [55, 56]. Additional sources of ROS and antioxidants in seminal fluid are through prostate and seminal vesicle secretions [53, 54]. However, oxidative stress results in abnormal spermatogenesis, lipid peroxidation and DNA fragmentation, alongside mitochondrial dysfunction [57].

Obesity and MetS have been associated with increased oxidative stress and inflammation in the male reproductive tract [25, 58]. Alongside T2DM, obesity and metabolic syndrome also negatively affect MMP, alongside increased DNA fragmentation and markers of apoptosis [25, 59–61]. Excess adipose tissue produces excessive adipokines (cytokines derived from adipocytes), which in turn leads to inflammation and oxidative stress in the male reproductive tract [50]. Obesity is also associated with increased seminal leptin and insulin concentrations [59], and MetS also has increased seminal inflammatory cytokines and insulin [62].

There is also a positive association between adiposity, leptin, sperm parameter abnormalities and oxidative stress in obese men. Furthermore, leptin may increase oxidative stress directly in obese males [63]. Excess adipose tissue produces excessive adipokines (cytokines derived from adipocytes), which in turn leads to inflammation and oxidative stress in the male reproductive tract [50]. This impairs testicular and epididymal tissues [50]. Diabetes, through chronic hyperglycaemia, is known to negatively affect sperm parameters, spermatogenesis, Leydig cell function and reductive hormone balance, as well as induce sexual dysfunction. It is postulated that the mitochondrial dysfunction and oxidative stress are key mediators in this negative impact of hyperglycaemia on reproductive dysfunction [52].

This has been shown in a number of observational studies in humans. Obese, non-obese T2DM and obese T2DM males all showed increased oxidative stress, reduced TAC, increased DNA fragmentation and apoptosis in spermatozoa, as well as reduced testosterone [64]. In male patients with excellent and mediocre semen quality, an increased fertility potential was associated with reduced iron and increased antioxidants, specifically CuZn-SOD, CAT and GPx [65]. In infertile males, a positive correlation was reported for BMI, increased intestinal permeability, metabolic endotoxaemia, sperm DNA oxidative damage and DNA fragmentation [66]. In obese, non-obese infertile males and non-obese fertile males, the obese infertile males showed increased MDA alongside reduced sperm concentration, viability and acrosome intactness test compared to both groups, and reduced nuclear chromatin decondensation test (NCD) compared to non-obese fertile control [67]. A positive correlation between BMI and ROS was also reported in male patients, with a negative correlation between BMI and testosterone and sperm concentration [68]. Obese males that had increased ROS and DNA fragmentation alongside reduced sperm parameters compared with fertile normal weight men had a BMI positively correlated with seminal ROS concentrations [69]. Furthermore, proteomic analysis of seminal plasma of obese males has identified elevated GPx, mitochondrial glutathione reductase, ceruloplasmin, clusterin, haptoglobin, S100A9 and ADP ribosyl cyclase, which are all involved in antioxidant activity where the overexpression may be an attempt to counterbalance excessive ROS. This included identification of proteins that activate the inflammatory pathways and intrinsic apoptosis [70]. Although these studies in humans remain limited, they strongly suggest a relationship between oxidative stress and reduced semen parameters in obese males predominantly, with less direct evidence of reproductive oxidative stress available for MetS and T2DM.

Potential mechanisms of action of high energy diets and obesity-related metabolic derangements are provided in animal studies. High energy diets lead to disruption of reproductive function and testicular histology, disrupting reproductive pathways and negatively affecting spermiogenesis and sperm maturation. This results in abnormal sperm parameters and sperm defections, changes to sperm membrane lipids and sperm morphology and decreased acrosome reaction and increased apoptosis [50, 71–73]. This further induces mitochondrial dysfunction which exacerbates increased ROS production [74, 73]. There is reduced testicular mRNA expressions of endogenous antioxidant enzyme activity, including CAT, SOD, GSH-Px and Nrf2 [75–77], and increased mRNA expression of NF-κB, TNFα and IL-1b, and decreased IL-10 [76, 77]. Inflammatory cytokines can further damage the seminiferous tubules and the epididymal epithelium through the induction of ROS [68, 78]. Increased testicular and serum MDA, SOD and PC (protein carbonyl) has also been reported [72, 79], mediated partly by a reduced Nrf2/HO-1 pathway [72]. With increased apoptosis, there is also increased mRNA levels of p53, Bax/BCl-2, caspase-9, caspase-8 and caspase-3 in the testes [76]. Furthermore, this is associated with decreased serum-free testosterone with increased SHBG [71–73, 75], mediated by inhibition of steroidogenic enzymes in testosterone synthesis [77]. Histology of testicular tissue further shows seminiferous tubules vacuolar changes and increased spermatogenic cell apoptosis in testicular tissue [71, 73]. This was also associated with pathological damage and increased apoptosis in Leydig cells [75]. High energy diets resulting in hyperglycaemia and hypoinsulinaemia also increase testicular oxidative stress through an increased intratesticular lactate/alanine ratio [80]. Sperm cell epigenetic changes can also transfer metabolic and reproductive co-morbidities across the generations [50, 81]. Maternal obesity in rats has also been shown to increase oxidative stress in the reproductive system of male offspring, as well as causing impaired fertility. This included a reduction in serum LH and testosterone, reduced sperm concentration, viability, motility, and normal morphological forms [82]. This is supported by reduced sperm quality and antioxidant enzyme activity in offspring from maternal obese rats [83].

Proteomic investigation of obesity in mice on a high-fat diet (inducing hyperinsulinaemia and hyperglycaemia) alongside human testicular biopsies identified 102 proteins affecting the testis. This included blood-testes barrier structural proteins (filamin A), oxidative stress response proteins (spermatogenesis associated 20), proteins involved in lipid homeostasis (sterol regulatory element-binding protein 2 and apolipoprotein A1), and proteins interacting with androgen receptors (paraspeckle component 1). Additional altered protein pathways identified include GABA receptor signalling, Rho GTPase pathway, pentose phosphate pathway, L-carnitine synthesis, AMPK signalling and BRCA1 in DNA damage [84]. Hyperglycaemia in type 1 diabetes mellitus has been shown to result in reproductive dysfunction. In most animal studies, the mechanisms of hyperglycaemia impact on male reproduction may include increased oxidative stress, ER stress, mitochondrial dysfunction, changes in advanced glycaemic end-products increased DNA damage and alterations in the HPG axis [85].

11.7 Assessment and Management of Oxidative Stress in Male Infertility

Testing of oxidative stress in male infertility assessments consists of direct and indirect assessment in seminal fluid. Direct tests aim to measure a difference in ROS production and antioxidants. These include specific ROS chemiluminescence or fluorescence, nitroblue tetrazolium (NBT) test, or flow cytometry for intracellular SOD or H_2O_2 [86–91].. Unfortunately these assays have been limited to research currently, and are not widely used in clinical practice [90, 92]. Indirect tests determine molecular consequences of oxidative stress in seminal fluid. These include total antioxidant capacity, redox potential, malondialdehyde (MDA), 8-hydroxy-2-deoxyguanosine and sperm DNA damage [6, 90, 91, 93].

11.7.1 Management

Management of obesity, MetS and T2DM remains fairly elusive, mostly due to the complex pathophysiology associated with these conditions [94, 95]. The impact of these pathologies on fertility is also complex and multifactorial, which further challenges appropriate management in these patients [94]. Therefore, management strategies need to be multifactorial, and target underlying mechanisms such as inflammation and oxidative stress in multiple organ systems [95, 96].

Long-term reduction in body weight depends on permanent changes in nutritional intake and physical activity [8]. An appropriate healthy lifestyle and weight management is a critical consideration in the management of obesity, MetS and T2DM. This includes appropriate quality and quality to caloric intake and increased physical activity [97]. In humans, weight management can improve male fertility parameters, as well as circulating reproductive hormones and improved sexual dysfunction [98–100]. In high-fat diet-induced obese rats, weight loss improves testicular oxidative stress through improved MDA, PC and SOD levels [79].

11.7.2 Lifestyle: Nutrition and Exercise

The study of human nutrition is complex, relevant to the regulation of energy, cellular stress, genetic predispositions (SNPs) and epigenetics [101]. Appropriate nutrition can reduce oxidative stress and inflammation, and improve male fertility and sperm quality. A diet rich in dietary fibre, fruits and vegetables, nuts and seeds, oils, unsaturated fatty acids and omega-3 fatty acids, micronutrients and phytonutrients improve inflammatory and oxidative stress markers, alongside male sperm parameters [102–105]. This should further include an active lifestyle, stress avoidance and where possible avoidance or mitigation of environmental or occupational toxin exposures can improve male infertility [106]. In this context, the Mediterranean diet has been found to improve oxidative stress and male fertility parameters [104]. These diets are also rich in antioxidants, particularly obtained through micronutrients and phytochemicals. The Mediterranean dietary principles with antioxidant-rich foods have been shown to improve fertility outcomes in non-obese and obese males. Important nutritional micronutrients for fertility include vitamin A and β-carotenes, vitamin D and E, vitamin C, folate, zinc and lycopene [102].

Exercise may acutely increase oxidative stress, but adaption to regular exercise improves adaption to oxidative stress, although these mechanisms remain poorly understood [107]. In a mouse model of testicular oxidative stress and inflammation in high-fat diet-induced obesity, moderate exercise load was reported to improve testicular oxidative stress (through increased endogenous antioxidants and reduced ROS) and inflammation (through reduced NF-κB mRNA expression and transcription), increase testosterone synthesis (through increased mRNA expression and transcription of steroidogenic enzymes), and improved sperm parameters and sperm apoptosis compared to no exercise and intense exercise groups, although both moderate and intense exercise resulted in improved adiposity [77]. In reduced fertility found in offspring of maternal obesity rat models, exercise improved the reduced sperm quality and antioxidant enzymes, alongside adiposity and gonadal fat found in the offspring [83].

Caloric restriction and intermittent fasting have been shown to reduce inflammation and oxidative stress associated with overnutrition and obesity, as well as improve obesity and metabolic complications [30, 33, 107]. This is done through limiting total calorie intake below optimal, without inducing micronutrient deficiencies [108]. Caloric restriction may improve male fertility parameters, although this remains unclear. Although there is a proposed trade-off between longevity associated with caloric restriction and fertility, long-term restriction in Rhesus monkeys shows minimal negative effect on fertility parameters. Further studies are required on the impact of caloric restriction on male fertility [109].

11.7.3 Nutritional Supplements and Antioxidants

Non-enzymatic exogenous antioxidants can be obtained through nutritional sources, particularly micronutrients and phytochemicals [37]. Supplementation with antioxidants may offer benefits on semen parameters, but the exact indications, correct form, dosage and duration of therapy are currently unclear [37, 92]. The use of antioxidants has also been reported to be useful in diabetes complications, mitigating the important role of oxidative stress in this pathogenesis [110, 111].

Rich nutritional consumption of antioxidants is recommended. Vitamin E and C, carotenoids and carnitine have been found to be beneficial in restoring redox balance in males with infertility [106]. Vitamin D supplementation is also reported to improve oxidative stress markers in obesity [107]. Additional supplementation for male infertility includes Alpha-lipoic acid, lycopene, co-enzyme Q10 and melatonin [112–116]. Melatonin, as an antioxidant, also improved testicular oxidative stress in high-fat diet-induced obese rats. This included improved MDA, SOD and PC levels [79]. However, further research is warranted to determine any benefit on fertility parameters in obesity, MetS and T2DM.

11.7.4 Herbal Medicines

Secondary metabolites found in medicinal plants, such as polyphenols, flavonoids and catechins, are rich in antioxidant activity [117]. Studies have suggested that increased consumption of phytonutrients reduce obesity and associated metabolic derangements, reducing morbidity and mortality [118]. Herbal products have also been shown to improve oxidative stress in the male reproductive system including *Vitis vinifera* (grape seed), *Origanum majorana* essential oil, *Nigella sativa,*

Eurycoma longifolia Jack, *Withania somnifera*, *Androgrraphis paniculata* and *Tribulus terrestris* [106]. This has been demonstrated in animal models specifically on obesity-induced oxidative stress. Soy isoflavones at 150 and 450 mg/Kg improved obesity-induced testicular oxidative stress, germ cell apoptosis and reduced serum testosterone. This was mediated through increased Nrf2/HO-1 pathway that regulates Bcl-2, BAX and caspase-3 expressions [72]. Proanthocyanidin extract from *Vitis vinifera* (grape seed) in high-fat diet-induced obese rats significantly improved serum testosterone, sperm parameters, histological changes in testes, as well as improved markers of testicular oxidative stress (increased SOD, GSH, and GSH-Px and decreased MDA) and reduction in spermatogenic cell apoptosis [73].

11.7.5 Pharmaceuticals

Non-steroidal anti-inflammatory drugs (salicylates) are associated with reduced inflammation and serum glucose in obesity, insulin resistance and T2DM [119–123]. Low-dose methotrexate has failed to show a reduction in CRP, IL-1β or IL-6, does not reduce CVD incidence and may increase adverse events in patients with MetS or T2DM [124]. Colchicine may lower CVD risk in patients following myocardial infarction, but this effect is yet to be investigated in obesity, MetS and T2DM [125]. Further immune-regulating drug targets include TNFα, IL-1β and IL-18 antagonists, but do not show consistent effect on reducing T2DM. Chemokine antagonists for CCR2 and CCR5 are reported to reduce inflammation and improve insulin resistance in obese animal models [33]. However, any impact on oxidative stress of these immune regulating approaches is not clear, nor any positive impact on male reproduction in obesity, MetS and T2DM patients. The treatment of obese mice with NADPH oxidase inhibitor is reported to reduce ROS and regulate adipokine production in adipose tissue, improving metabolic derangements and hepatic steatosis [36]. Treating mice with antioxidants may also prevent the progression of obesity into T2DM [28].

Metformin is a commonly used hypoglycaemic agent in the management of T2DM, primarily to improve insulin signalling and cellular uptake of glucose. Further mechanisms of action include a reduction in inflammation and oxidative stress, mediated in part through reduced AMPK activation [126, 127]. Metformin has been suggested to also improve testicular oxidative stress and inflammation, with improved sperm parameters, in T2DM [128]. In an obese mouse model, metformin has been shown to improve fertility, reduce oxidative stress and recover NF-κB activity in Sertoli cells (SCs) in the reproductive system, mediated through improving blood-testis barrier integrity. Metformin further reduced lipid deposition in the testis and increased serum FSH [129].

Testosterone administration can also downregulate systemic inflammation and oxidative stress in obesity, MetS and T2DM. This may also improve insulin signalling, adipokine secretion, dyslipidaemia and blood pressure [130]. However, excessive testosterone administration is detrimental for male fertility, and should not be considered as part of fertility treatment in these patients [94, 130]. As an alternative approach to hypogonadism in these males, aromatase inhibitors may improve sperm parameters in males. However, there is currently no significant evidence base for these medicines for the management of male infertility [94].

11.8 Conclusions

Although ROS are important physiological mediators of cellular function, including male fertility, excessive ROS and subsequent oxidative stress can be detrimental to cell function. Obesity, MetS and T2DM are associated with oxidative stress and inflammation as important mechanisms of disease, which is also found to affect testicular tissue and contribute to male infertility. Human and animal studies increasingly establish a direct association between this systemic oxidative stress and male infertility. Management of this oxidative stress is not well defined; however, improved nutrition and exercise can be recommended for improvement in oxidative stress and male fertility parameters. Consideration of dietary supplements with micronutrients, antioxidants or medicinal herbs is recommended, however, further research on efficacy and dosage is required. Metformin may also offer benefits on testicular oxidative stress and fertility parameters in obesity, MetS and T2DM. However, significantly more research on causation, mechanisms, clinical assessments and appropriate management of infertility on obesity, MetS and T2DM is still required.

References

1. Tafuri S, Ciani F, Iorio EL, Esposito L, Cocchia N. Reactive Oxygen Species (ROS) and male fertility. In: New discoveries in embryology. InTech; 2015. p. 19–40.
2. Sies H. Hydrogen peroxide as a central redox signaling molecule in physiological oxidative stress: Oxidative eustress. Redox Biol. 2017;11:613–9.
3. Höhn A, Weber D, Jung T, et al. Happily (n)ever after: aging in the context of oxidative stress, proteostasis loss and cellular senescence. Redox Biol. 2017;11:482–501.
4. Leisegang K. Malnutrition and obesity. In: Oxidants, antioxidants and impact of the oxidative status in male reproduction. Elsevier; 2019. p. 117–34.
5. Leisegang K, Henkel R, Agarwal A. Redox regulation of fertility in aging male and the role of antioxidants: a savior or stressor. Curr Pharm Des. 2016;23(30):2017–8.
6. Leisegang K, Henkel R. Oxidative stress: relevance, evaluation, and management. In: Male infertility in reproductive medicine. CRC Press; 2019. p. 119–28.
7. Dulloo AG, Jacquet J, Solinas G, Montani J-P, Schutz Y. Body composition phenotypes in pathways to obesity and the metabolic syndrome. Int J Obes. 2010;34(Suppl 2):S4–17.
8. Haslam DW, James WPT. Obesity. Lancet. 2005;366(9492):1197–209.
9. Engin A. The definition and prevalence of obesity and metabolic syndrome. Adv Exp Med Biol. 2017;960:1–17.
10. Alberti KGMM, Eckel RH, Grundy SM, et al. Harmonizing the metabolic syndrome a joint interim statement of the international diabetes federation task force on epidemiology

and prevention; national heart, lung, and Blood Institute; American Heart Association; World Heart Federation; International Association for the Study of Obesity. Circulation. 2009;120(16):1640–50.
11. Huang PL. A comprehensive definition for metabolic syndrome. Dis Model Mech. 2009;2(5–6):231–7.
12. Chen L, Magliano DJ, Zimmet PZ. The worldwide epidemiology of type 2 diabetes mellitus – present and future perspectives. Nat Rev Endocrinol. 2012;8(4):228–36.
13. Saklayen MG. The global epidemic of the metabolic syndrome. Curr Hypertens Rep. 2018;20(2)
14. Han TS, Lean ME. A clinical perspective of obesity, metabolic syndrome and cardiovascular disease. JRSM Cardiovasc Dis. 2016;5:2048004016633337.
15. Kumar Khemka V, Banerjee A. Metabolic risk factors in obesity and diabetes mellitus: implications in the pathogenesis and therapy. Integr Obes Diabetes. 2017;3(3)
16. Apovian CM. Obesity: definition, comorbidities, causes, and burden. Am J Manag Care. 2016;22(7):s176–85.
17. Chooi YC, Ding C, Magkos F. The epidemiology of obesity. Metabolism. 2019;92:6–10.
18. Khan MAB, Hashim MJ, King JK, Govender RD, Mustafa H, Al KJ. Epidemiology of type 2 diabetes – global burden of disease and forecasted trends. J Epidemiol Glob Health. 2020;10(1):107–11.
19. Ogurtsova K, da Rocha Fernandes JD, Huang Y, et al. IDF diabetes atlas: global estimates for the prevalence of diabetes for 2015 and 2040. Diabetes Res Clin Pract. 2017;128:40–50.
20. WHO. Obesity: preventing and managing the global epidemic. Report of a WHO consultation. World Health Organ Tech Rep Ser 2000;894: i–xii, 1–253.
21. MacMahon S, Baigent C, Duffy S, et al. Body-mass index and cause-specific mortality in 900 000 adults: Collaborative analyses of 57 prospective studies. Lancet. 2009;373(9669):1083–96.
22. Okorodudu DO, Jumean MF, Montori VM, et al. Diagnostic performance of body mass index to identify obesity as defined by body adiposity: a systematic review and meta-analysis. Int J Obes. 2010;34(5):791–9.
23. Han TS, Sattar N, Lean M. Assessment of obesity and its clinical implications. BMJ. 2006;333(7570):695–8.
24. Association AD. 2. Classification and diagnosis of diabetes. Diabetes Care. 2015;38(Supplement 1):S8–S16.
25. Leisegang K, Henkel R, Agarwal A. Obesity and metabolic syndrome associated with systemic inflammation and the impact on the male reproductive system. Am J Reprod Immunol. 2019;82(5):e13178.
26. Esser N, Legrand-Poels S, Piette J, Scheen AJ, Paquot N. Inflammation as a link between obesity, metabolic syndrome and type 2 diabetes. Diabetes Res Clin Pract. 2014;105(2):141–50.
27. Emanuela F, Grazia M, Marco DR, Maria Paola L, Giorgio F, Marco B. Inflammation as a link between obesity and metabolic syndrome. J Nutr Metab. 2012;2012
28. Matsuda M, Shimomura I. Increased oxidative stress in obesity: Implications for metabolic syndrome, diabetes, hypertension, dyslipidemia, atherosclerosis, and cancer. Obes Res Clin Pract. 2013;7(5)
29. Rani V, Deep G, Singh RK, Palle K, Yadav UCS. Oxidative stress and metabolic disorders: pathogenesis and therapeutic strategies. Life Sci. 2016;148:183–93.
30. Biobaku F, Ghanim H, Batra M, Dandona P. Macronutrient-mediated inflammation and oxidative stress: relevance to insulin resistance, obesity, and atherogenesis. J Clin Endocrinol Metab. 2019;104(12):6118–28.
31. Anderson EJ, Lustig ME, Boyle KE, et al. Mitochondrial H2O2 emission and cellular redox state link excess fat intake to insulin resistance in both rodents and humans. J Clin Invest. 2009;119(3):573–81.
32. Emami SR, Jafari M, Haghshenas R, Ravasi A. Impact of eight weeks endurance training on biochemical parameters and obesity-induced oxidative stress in high fat diet-fed rats. J Exerc Nutr Biochem. 2016;20(1):30–6.

33. Wu H, Ballantyne CM. Metabolic inflammation and insulin resistance in obesity. Circ Res. Published online. 2020;1549–1564.
34. Luft VC, Schmidt MI, Pankow JS, et al. Chronic inflammation role in the obesity-diabetes association: a case-cohort study. Diabetol Metab Syndr. 2013;5(1):1–8.
35. de Mello AH, Costa AB, Engel JDG, Rezin GT. Mitochondrial dysfunction in obesity. Life Sci. 2018;192:26–32.
36. Furukawa S, Fujita T, Shimabukuro M, et al. Increased oxidative stress in obesity and its impact on metabolic syndroame. J Clin Invest. 2004;114(12):1752–61.
37. Darbandi M, Darbandi S, Agarwal A, et al. Reactive oxygen species and male reproductive hormones. Reprod Biol Endocrinol. 2018;16(1):1–14.
38. Keaney JF, Larson MG, Vasan RS, et al. Obesity and systemic oxidative stress: clinical correlates of oxidative stress in the Framingham study. Arterioscler Thromb Vasc Biol. 2003;23(3):434–9.
39. Rupérez AI, Gil A, Aguilera CM. Genetics of oxidative stress in obesity. Int J Mol Sci. 2014;15(2):3118–44.
40. Lapik IA, Galchenko AV, Gapparova KM. Micronutrient status in obese patients: a narrative review. Obes Med. 2020;18:100224.
41. Via M. The malnutrition of obesity: micronutrient deficiencies that promote diabetes. ISRN Endocrinol. 2012;2012:1–8.
42. Christopher EE. Micronutrient deficiency, a novel nutritional risk factor for insulin resistance and Syndrom X. Arch Food Nutr Sci. 2018;2(1):016–30. https://doi.org/10.29328/journal.afns.1001013.
43. Barazzoni R, Gortan Cappellari G, Ragni M, Nisoli E. Insulin resistance in obesity: an overview of fundamental alterations. Eat Weight Disord. 2018;23(2):149–57. https://doi.org/10.1007/s40519-018-0481-6.
44. Yazıcı D, Sezer H. Insulin resistance, obesity and lipotoxicity. Adv Exp Med Biol. 2017;960:277–304.
45. Maritim AC, Sanders RA, Watkins JB. Diabetes, oxidative stress, and antioxidants: a review. J Biochem Mol Toxicol. 2003;17(1):24–38.
46. Korda M, Kubant R, Patton S, Malinski T. Leptin-induced endothelial dysfunction in obesity. Am J Physiol Heart Circ Physiol. 2008;295(4):H1514–21.
47. Chavez JA, Summers SA. Lipid oversupply, selective insulin resistance, and lipotoxicity: molecular mechanisms. Biochim Biophys Acta Mol Cell Biol Lipids. 2010;1801(3):252–65.
48. Du Plessis SS, Cabler S, McAlister DA, Sabanegh E, Agarwal A. The effect of obesity on sperm disorders and male infertility. Nat Rev Urol. 2010;7(3):153–61.
49. Lotti F, Marchiani S, Corona G, Maggi M. Molecular Sciences metabolic syndrome and reproduction. Published online. 2021.
50. Liu Y, Ding Z. Obesity, a serious etiologic factor for male subfertility in modern society. Reproduction. 2017;154(4):R123–31. https://doi.org/10.1530/REP-17-0161.
51. Cardoso AM, Alves MG, Mathur PP, Oliveira PF, Cavaco JE, Rato L. Obesogens and male fertility. Obes Rev. 2017;18(1):109–25.
52. Ramalho-Santos J, Amaral S, Oliveira P. Diabetes and the impairment of reproductive function: possible role of mitochondria and reactive oxygen species. Curr Diabetes Rev. 2008;4(1):46–54.
53. Agarwal A, Sharma R, Roychoudhury S, Du Plessis S, Sabanegh E. MiOXSYS: a novel method of measuring oxidation reduction potential in semen and seminal plasma. Fertil Steril. 2016;106(3):566–573.e10.
54. Agarwal A, Parekh N, Panner Selvam MK, et al. Male oxidative stress infertility (MOSI): proposed terminology and clinical practice guidelines for management of idiopathic male infertility. World J Men?s Health. 2019;37(3):296–312.
55. Baker MA, Aitken RJ. Reactive oxygen species in spermatozoa: methods for monitoring and significance for the origins of genetic disease and infertility. Reprod Biol Endocrinol. 2005;3:67.

56. Tremellen K. Oxidative stress and male infertility – a clinical perspective. Hum Reprod Update. 2008;14(3):243–58.
57. Sabeti P, Pourmasumi S, Rahiminia T, Akyash F, Talebi AR. Etiologies of sperm oxidative stress. Int J Reprod Biomed. 2016;14(4):231–40.
58. Agarwal A, Rana M, Qiu E, AlBunni H, Bui AD, Henkel R. Role of oxidative stress, infection and inflammation in male infertility. Andrologia. 2018;50(11)
59. Leisegang K, Bouic PJD, Menkveld R, Henkel RR. Obesity is associated with increased seminal insulin and leptin alongside reduced fertility parameters in a controlled male cohort. Reprod Biol Endocrinol. 2014;12(1) https://doi.org/10.1186/1477-7827-12-34.
60. La VS, Condorelli RA, Vicari E, Calogero AE. Negative effect of increased body weight on sperm conventional and nonconventional flow cytometric sperm parameters. J Androl. 2012;33(1):53–8.
61. Leisegang K, Udodong A, Bouic PJD, Henkel RR. Effect of the metabolic syndrome on male reproductive function: a case-controlled pilot study. Andrologia. 2014;46(2):167–76. https://doi.org/10.1111/and.12060.
62. Leisegang K, Bouic PJD, Henkel RR. Metabolic syndrome is associated with increased seminal inflammatory cytokines and reproductive dysfunction in a case-controlled male cohort. Am J Reprod Immunol. 2016;76(2):155–63.
63. Malik I, Durairajanayagam D, Singh H. Leptin and its actions on reproduction in males. Asian J Androl. 2019;21(3):296–9.
64. Abbasihormozi S, Babapour V, Kouhkan A, et al. Stress hormone and oxidative stress biomarkers link obesity and diabetes with reduced fertility potential. Cell J. 2019;21(3):307–13.
65. Dobrakowski M, Kaletka Z, Machoń-Grecka A, et al. The role of oxidative stress, selected metals, and parameters of the immune system in male fertility. Oxidative Med Cell Longev. 2018;2018
66. Pearce KL, Hill A, Tremellen KP. Obesity related metabolic endotoxemia is associated with oxidative stress and impaired sperm DNA integrity. Basic Clin Androl. 2019;29(1)
67. Najafi M, Sreenivasa G, Aarabi M, Dhar M, Babu S, Malini S. Seminal malondialdehyde levels and oxidative stress in obese male infertility. J Pharm Res. 2012;5(7 Cop):3597–601.
68. Tunc O, Bakos HW, Tremellen K. Impact of body mass index on seminal oxidative stress. Andrologia. 2011;43(2):121–8. Accessed 10 May 2021. https://onlinelibrary.wiley.com/doi/full/10.1111/j.1439-0272.2009.01032.x
69. Taha EA, Sayed SK, Gaber HD, et al. Does being overweight affect seminal variables in fertile men? Reprod Biomed Online. 2016;33(6):703–8.
70. Ferigolo PC, Ribeiro de Andrade MB, Camargo M, et al. Sperm functional aspects and enriched proteomic pathways of seminal plasma of adult men with obesity. Andrology. 2019;7(3):341–9.
71. Jia YF, Feng Q, Ge ZY, et al. Obesity impairs male fertility through long-term effects on spermatogenesis. BMC Urol. 2018;18(1)
72. Luo Q, Li Y, Huang C, et al. Soy isoflavones improve the spermatogenic defects in diet-induced obesity rats through Nrf2/HO-1 pathway. Molecules. 2019;24(16):2966.
73. Wang EH, Yu ZL, Bu YJ, Xu PW, Xi JY, Liang HY. Grape seed proanthocyanidin extract alleviates high-fat diet induced testicular toxicity in rats. RSC Adv. 2019;9(21):11842–50.
74. Rato L, Alves MG, Cavaco JE, Oliveira PF. High-energy diets: a threat for male fertility? Obes Rev. 2014;15(12):996–1007. https://doi.org/10.1111/obr.12226.
75. Zhao J, Zhai L, Liu Z, Wu S, Xu L. Leptin level and oxidative stress contribute to obesity-induced low testosterone in murine testicular tissue. Oxidative Med Cell Longev. 2014;2014
76. Suleiman JB, Nna VU, Zakaria Z, Othman ZA, Bakar ABA, Mohamed M. Obesity-induced testicular oxidative stress, inflammation and apoptosis: protective and therapeutic effects of orlistat. Reprod Toxicol. 2020;95:113–22.
77. Yi X, Tang D, Cao S, et al. Effect of different exercise loads on testicular oxidative stress and reproductive function in obese male mice. Oxidative Med Cell Longev. 2020;2020

78. Cannarella R, Crafa A, Barbagallo F, et al. Seminal plasma proteomic biomarkers of oxidative stress. Int J Mol Sci. 2020;21(23):1–13.
79. Atilgan D, Parlaktas BS, Uluocak N, et al. Weight loss and melatonin reduce obesity-induced oxidative damage in rat testis. Adv Urol. 2013;2013
80. Rato L, Alves MG, Dias TR, et al. High-energy diets may induce a pre-diabetic state altering testicular glycolytic metabolic profile and male reproductive parameters. Andrology. 2013;1(3):495–504.
81. Zhou Y, Wu H, Huang H. Epigenetic effects of male obesity on sperm and offspring. J Bio-X Res. 2018;1(3):105–10.
82. Rodríguez-González GL, Vega CC, Boeck L, et al. Maternal obesity and overnutrition increase oxidative stress in male rat offspring reproductive system and decrease fertility. Int J Obes. 2015;39(4):549–56.
83. Santos M, Rodríguez-González GL, Ibáñez C, Vega CC, Nathanielsz PW, Zambrano E. Adult exercise effects on oxidative stress and reproductive programming in male offspring of obese rats. Am J Physiol Regul Integr Comp Physiol. 2015;308(3):R219–25.
84. Jarvis S, Gethings LA, Samanta L, et al. High fat diet causes distinct aberrations in the testicular proteome. Int J Obes. 2020;44(9):1958–69.
85. Maresch CC, Stute DC, Alves MG, Oliveira PF, de Kretser DM, Linn T. Diabetes-induced hyperglycemia impairs male reproductive function: a systematic review. Hum Reprod Update. 2018;24(1):86–105.
86. Mahfouz R, Sharma R, Lackner J, Aziz N, Agarwal A. Evaluation of chemiluminescence and flow cytometry as tools in assessing production of hydrogen peroxide and superoxide anion in human spermatozoa. Fertil Steril. 2009;92(2):819–27.
87. Esfandiari N, Sharma RK, Saleh RA, Thomas AJ, Agarwal A. Utility of the nitroblue tetrazolium reduction test for assessment of reactive oxygen species production by seminal leukocytes and spermatozoa. J Androl. 2003;24(6):862–70. https://doi.org/10.1002/j.1939-4640.2003.tb03137.x.
88. Gosálvez J, Coppola L, Fernández JL, et al. Multi-centre assessment of nitroblue tetrazolium reactivity in human semen as a potential marker of oxidative stress. Reprod Biomed Online. 2017;34(5):513–21.
89. Agarwal A, Allamaneni SSR, Said TM. Chemiluminescence technique for measuring reactive oxygen species. Reprod Biomed Online. 2004;9(4):466–8.
90. Ko EY, Sabanegh ES, Agarwal A. Male infertility testing: reactive oxygen species and antioxidant capacity. Published online. 2014.
91. Agarwal A, Majzoub A. Laboratory tests for oxidative stress. Indian J Urol. 2017;33(3):199–206.
92. Agarwal A, Finelli R, Panner Selvam MK, et al. A global survey of reproductive specialists to determine the clinical utility of oxidative stress testing and antioxidant use in male infertility. World J Mens Health. 2021;39
93. Agarwal A, Leisegang K, Sengupta P. Oxidative stress in pathologies of male reproductive disorders. In: Pathology. Elsevier; 2020. p. 15–27.
94. Stokes VJ, Anderson RA, George JT. How does obesity affect fertility in men – and what are the treatment options? Clin Endocrinol. 2015;82(5):633–8.
95. Singh H, Pragasam SJ, Venkatesan V. Emerging therapeutic targets for metabolic syndrome: lessons from animal models. Endocr Metab Immune Disord Drug Targets. 2018;19(4):481–9.
96. Leisegang K, Dutta S. Lifestyle management approaches to male infertility. Male Infertil Reprod Med. Published online October 28, 2019:141–51. https://doi.org/10.1201/9780429485763-15.
97. Martins AD, Majzoub A, Agawal A. Metabolic syndrome and male fertility. World J Men?s Health. 2019;37(2):113–27.
98. Håkonsen L, Thulstrup A, Aggerholm A, et al. Does weight loss improve semen quality and reproductive hormones? Results from a cohort of severely obese men. Reprod Health. 2011;8(1) https://doi.org/10.1186/1742-4755-8-24.

99. Maiorino MI, Bellastella G, Esposito K. Lifestyle modifications and erectile dysfunction: what can be expected? Asian J Androl. 2015;17(1):5–10.
100. Jaffar M, Ashraf M. Does weight loss improve fertility with respect to semen parameters – results from a large cohort study. Int J Infertil Fetal Med. 2017;8(1):12–7.
101. Bjørklund G, Chirumbolo S. Role of oxidative stress and antioxidants in daily nutrition and human health. Nutrition. 2017;33:311–21.
102. Giahi L, Mohammadmoradi S, Javidan A, Sadeghi MR. Nutritional modifications in male infertility: a systematic review covering 2 decades. Nutr Rev. 2016;74(2):118–30.
103. Salas-Huetos A, Bulló M, Salas-Salvadó J. Dietary patterns, foods and nutrients in male fertility parameters and fecundability: a systematic review of observational studies. Hum Reprod Update. 2017;23(4):371–89.
104. Karayiannis D, Kontogianni MD, Mendorou C, Douka L, Mastrominas M, Yiannakouris N. Association between adherence to the Mediterranean diet and semen quality parameters in male partners of couples attempting fertility. Hum Reprod. 2017;32(1):215–22.
105. Belobrajdic DP, Lam YY, Mano M, Wittert GA, Bird AR. Cereal based diets modulate some markers of oxidative stress and inflammation in lean and obese Zucker rats. Nutr Metab. 2011;8
106. Adewoyin M, Ibrahim M, Roszaman R, et al. Male infertility: the effect of natural antioxidants and phytocompounds on seminal oxidative stress. Diseases. 2017;5(1):9.
107. Huang CJ, McAllister MJ, Slusher AL, Webb HE, Mock JT, Acevedo EO. Obesity-related oxidative stress: the impact of physical activity and diet manipulation. Sport Med Open. 2015;1(1):1–12.
108. Heilbronn LK, De Jonge L, Frisard MI, et al. Effect of 6-month calorie restriction on biomarkers of longevity, metabolic adaptation, and oxidative stress in overweight individuals: a randomized controlled trial. J Am Med Assoc. 2006;295(13):1539–48.
109. Sitzmann BD, Brown DI, Garyfallou VT, et al. Impact of moderate calorie restriction on testicular morphology and endocrine function in adult rhesus macaques (Macaca mulatta). Age (Omaha). 2014;36:183–97. https://doi.org/10.1007/s11357-013-9563-6.
110. Rahimi R, Nikfar S, Larijani B, Abdollahi M. A review on the role of antioxidants in the management of diabetes and its complications. Biomed Pharmacother. 2005;59(7):365–73.
111. Osawa T, Kato Y. Protective role of antioxidative food factors in oxidative stress caused by hyperglycemia. Ann NY Acad Sci. 2005;1043:440–51.
112. Akbari M, Ostadmohammadi V, Tabrizi R, et al. The effects of alpha-lipoic acid supplementation on inflammatory markers among patients with metabolic syndrome and related disorders: a systematic review and meta-analysis of randomized controlled trials. Nutr Metab. 2018;15(1)
113. Cheng HM, Koutsidis G, Lodge JK, Ashor A, Siervo M, Lara J. Tomato and lycopene supplementation and cardiovascular risk factors: a systematic review and meta-analysis. Atherosclerosis. 2017;257:100–8.
114. Fan L, Feng Y, Chen GC, Qin LQ, Fu C, ling, Chen LH. Effects of coenzyme Q10 supplementation on inflammatory markers: a systematic review and meta-analysis of randomized controlled trials. Pharmacol Res. 2017;119:128–36.
115. Akbari M, Ostadmohammadi V, Tabrizi R, et al. The effects of melatonin supplementation on inflammatory markers among patients with metabolic syndrome or related disorders: a systematic review and meta-analysis of randomized controlled trials. Inflammopharmacology. 2018;26(4):899–907.
116. Agarwal A, Leisegang K, Majzoub A, et al. Utility of antioxidants in the treatment of male infertility: clinical guidelines based on a systematic review and analysis of evidence. World J Mens Health. 2021;39(2):1–58.
117. Leisegang K. Herbal pharmacognosy: an introduction. In: Herbal medicine in andrology. Elsevier; 2021. p. 17–26.

118. Mitjavila MT, Moreno JJ. The effects of polyphenols on oxidative stress and the arachidonic acid cascade. implications for the prevention/treatment of high prevalence diseases. Biochem Pharmacol. 2012;84(9):1113–22.
119. Donath MY, Dinarello CA, Mandrup-Poulsen T. Targeting innate immune mediators in type 1 and type 2 diabetes. Nat Rev Immunol. 2019;19(12):734–46.
120. Goldfine AB, Silver R, Aldhahi W, et al. Use of salsalate to target inflammation in the treatment of insulin resistance and type 2 diabetes. Clin Transl Sci. 2008;1(1):36–43.
121. Fleischman A, Shoelson SE, Bernier R, Goldfine AB. Salsalate improves glycemia and inflammatory parameters in obese young adults. Diabetes Care. 2008;31(2):289–94.
122. Goldfine AB, Shoelson SE. Therapeutic approaches targeting inflammation for diabetes and associated cardiovascular risk. J Clin Invest. 2017;127(1):83–93.
123. Faghihimani E, Aminorroaya A, Rezvanian H, Adibi P, Ismail-Beigi F, Amini M. Reduction of insulin resistance and plasma glucose level by salsalate treatment in persons with prediabetes. Endocr Pract. 2012;18(6):826–33.
124. Ridker PM, Everett BM, Pradhan A, et al. Low-dose methotrexate for the prevention of atherosclerotic events. N Engl J Med. 2019;380(8):752–62.
125. Tardif J-C, Kouz S, Waters DD, et al. Efficacy and safety of low-dose colchicine after myocardial infarction. N Engl J Med. 2019;381(26):2497–505.
126. Saisho Y. Metformin and inflammation: its potential beyond glucose-lowering effect. Endocr Metab Immune Disord Targets. 2015;15(3):196–205.
127. De Araújo AA, Pereira ADSBF, De Medeiros CACX, et al. Effects of metformin on inflammation, oxidative stress, and bone loss in a rat model of periodontitis. PLoS One. 2017;12(8):e0183506.
128. Alves MG, Martins AD, Vaz CV, et al. Metformin and male reproduction: effects on Sertoli cell metabolism. Br J Pharmacol. 2014;171(4):1033–42.
129. Ye J, Luo D, Xu X, et al. Metformin improves fertility in obese males by alleviating oxidative stress-induced blood-testis barrier damage. Oxidative Med Cell Longev. 2019;2019
130. Winter AG, Zhao F, Lee RK. Androgen deficiency and metabolic syndrome in men. Transl Androl Urol. 2014;3(1):50–8. https://doi.org/10.3978/j.issn.2223-4683.2014.01.04.
131. James PT, Leach R, Kalamara E, Shayeghi M. The worldwide obesity epidemic. Obes Res. November 2001;9(Suppl 4):228S–33S.

Chapter 12
Metabolic Dysregulation and Sperm Motility in Male Infertility

Sujata Maurya, Kavindra Kumar Kesari, Shubhadeep Roychoudhury, Jayaramulu Kolleboyina, Niraj Kumar Jha, Saurabh Kumar Jha, Ankur Sharma, Arun Kumar, Brijesh Rathi, and Dhruv Kumar

Abstract Nowadays, about 14% of couples have difficulty in conceiving, and half of the cases are attributed to men. Asthenozoospermia or poor sperm motility is considered as the cause of infertility in males which is most common. Even though energy metabolism is considered the main reason for the etiology of asthenospermia, few attempts are made to determine the pathway of its metabolic potential. Recognition of cellular as well as molecular pathways that lead to reduced sperm

---

S. Maurya · D. Kumar (✉)
Amity Institute of Molecular Medicine and Stem Cell Research (AIMMSCR), Amity University Uttar Pradesh, Noida, India
e-mail: dkumar13@amity.edu

K. K. Kesari
Department of Bioproducts and Biosystems, School of Chemical Engineering, Aalto University, Espoo, Finland

S. Roychoudhury
Department of Life Science and Bioinformatics, Assam University, Silchar, India

J. Kolleboyina
Department of Chemistry, Indian Institute of Technology Jammu, Jammu, Jammu & Kashmir, India

N. K. Jha · S. K. Jha
Department of Biotechnology, School of Engineering & Technology (SET), Sharda University, Greater Noida, India

A. Sharma
Department of Life Science, School of Basic Science & Research (SBSR), Sharda University, Greater Noida, India

A. Kumar
Mahavir Cancer Sansthan and Research Centre, Patna, Bihar, India

B. Rathi
Laboratory for Translational Chemistry and Drug Discovery, Department of Chemistry, Hansraj College, University of Delhi, New Delhi, India

© The Author(s), under exclusive license to Springer Nature Switzerland AG 2022
K. K. Kesari, S. Roychoudhury (eds.), *Oxidative Stress and Toxicity in Reproductive Biology and Medicine*, Advances in Experimental Medicine and Biology 1358, https://doi.org/10.1007/978-3-030-89340-8_12

motility may lead to the implementation of new therapeutic strategies to eliminate low sperm motility in people with asthenozoospermia. This review article discusses the key causes of decreased sperm motility and some of the muted genes and metabolic causes of the same.

Keywords Male infertility · Asthenozoospermia · Glycolysis · OXPHOS · ATP production · cAMP/PRKA pathway · ROS · Autophagy · Curcumin

12.1 Introduction

India has a population of over 1.3 billion people and stands worldwide second after China [8]. Still, overpopulation is not the only topic to focus but for the past few years Indian people are also facing the problem of infertility which is not easily talked about. Although, the Indian Society of Assisted Reproduction reported 10–14% of the Indian population suffers from infertility [28]. According to some estimates, this amounts to roughly one in every six couples in urban India [26]. According to the report of Ernst and Young in 2015 they estimated that 27.5 million couples in India wanted to become parents but were unable to do so naturally.

Infertility is defined as having unprotected sex for more than a year still the couple is unable to conceive. Although, infertility is often considered a female problem marital infertility can also be caused by problems in the male reproductive system [28]. The total fertility rate (TFR) is calculated based on the number of children born to each woman and does not take into account any female mortality, taking into account the country's overall fertility rate. In 2016, TFR fell from 2 out of 12 US states. The total TFR of India is 2.3. In rural areas, the ratio is slightly higher—2.5, while in urban areas the overall ratio is 1.8 [51]. Although male and female infertility are common in India, the share of male infertility has been increasing in recent years. Sperm motility, concentration, and morphology are key parameters that affect semen quality; so routine evaluation should be performed in a clinical settings [59]. It has been reported that the abnormal morphology (teratozoospermia), low sperm count (oligozoospermia), and poor sperm motility (asthenozoospermia) are the most common causes of male factors, out of which 20% are asthenozoospermic.

12.2 Asthenozoospermia

Reduced or absent sperm motility is called asthenozoospermia. There are live sperm in the semen, but they cannot move or are not able to travel the distance from the vagina to the fallopian tube. Therefore, no further fertilization process may take place. According to WHO guidelines, the total motility of asthenozoospermic sperm samples is <40% and progressive motility (forward motility) is <32%. Acute asthenozoospermia is also characterized by total sperm immotility or extremely low motile sperm in semen sample. It appears that complete immotility of spermatozoa is associated with genetic factors. The cellular as well as molecular processes that lead to sperm motility should be minimized to assist researchers in solving reduced motility problems and accurate diagnostic results. Also, the muted genes are needed to investigate for better clinical results. Therefore, extensive research at the cellular and molecular level as well as its signaling pathways is needed [36].

12.3 Cellular and Molecular Factors for Sperm Motility

12.3.1 Role of the Cilia in Sperm Motility as Well as Its Defect

The sperm tail is an essential structure of sperm movement. The flagella form a very large part of a sperm tail structure of a mammalian sperm, and the sperm tail structure is responsible for sperm movement. The tail and its structure of the human sperm facilitate the movement [31]. Sperm motility could be defined as the hydrodynamic impulse that pushes sperm into oocytes through the female reproductive tract. This type of pulse propagates when transverse waves propagate along the flagella in the proximal-distal direction [31, 62].

The cilia can sense environmental signals, control sperm movement, and mediate the transmission of cilia signals. The ciliary structure consists of a matrix, a membrane structure, and axons. In most active cilia, there are two central microtubules in the axons, consisting of 9 pairs of microtubules +2 central microtubules, after meiosis [33]. In each of these pairs without microtubules are placed axonal proteins, called dyneins, and used as "arms". The movement of the flagella is based on the displacement of these dynein's pair of axon microtubules, which mechanically strengthen the axons. To move the sperm in a harsh environment, this automatic reinforcement is necessary. Sperm flagella are activated, forming asymmetrical shapes of lower amplitude, and the sperms move in a very straight line [37].

As tail structure plays a major role in sperm motility, so any defect directly affects flagella structure, sperm motility, and fertilization result. Currently, there are 24 mutated genes (Fig. 12.1) that are related to structural defects of sperm flagella, which may be associated with other ciliary diseases and fertilization failure too [56].

When further studies done, the mutations in the ciliary gene cause most of these defects, name "Immotile Cilia Syndrome" or "Primary Ciliary Dyskinesia" (PCD)" was proposed [24, 53]. Currently, over 40 genes were identified whose mutations cause PCD in human. But infertility in males due to PCD is not well recorded in the literature. However, only 50% of PCD pathogenic genes were studied that cause asthenozoospermia in infertile males [64]. Some of such genes of PCD are listed in Table 12.1.

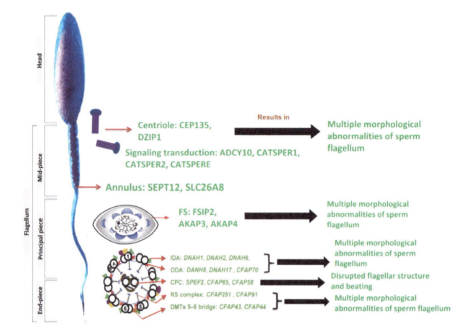

Fig. 12.1 Genes linked to asthenozoospermia in male. Outer dynein arms (ODAs) and inner dynein arms (IDAs) are responsible for maintaining flagellum/ciliary beat frequency and normal waveform. RS and central pair complex (CPC) are essential for the beating of flagella/cilia by regulating the activity of the dyneins. Mutations in FS-related protein and centrosome-related proteins cause MMAF.

Table 12.1 PCD pathogenic genes known to cause asthenozoospermia

Genes	Functional category	Cause	References
SPAG6	CPC component	PCD and asthenozoospermia	Wu et al. [60]
RSPH3	RS component	PCD and asthenozoospermia	Jeanson et al. [25]
DNAI1	ODA intermediate chain	PCD and asthenozoospermia	Guichard et al. [22]
DNAH9	ODA heavy chain	PCD and asthenozoospermia	Loges et al. [34], Fassad et al. [18]
DNAAF2	Assembly of dynein arm complexes	PCD and asthenozoospermia	Omran et al. [44]
LRRC6	Assembly of dynein arm complexes	PCD and asthenozoospermia	Kott et al. [27]
HYDIN	C2b projection	PCD and asthenozoospermia	Schou et al. [54]
DNAJB13	C2b projection or a structural RS component	PCD and severe oligoastheno-teratozoospermia	El Khouri et al. [17]

12.3.2 Sperm Motility, Energy Driving as Well as Its Defect

ATP is used as a source of energy through the sperm in many cellular processes, including capacitation, movement, hyperactivity, acrosome reaction, and maintenance of the cell environment. Glycolysis as well as oxidative phosphorylation (OXPHOS) are two metabolic pathways responsible for the production of ATP in sperm cells. It happens in a central part of the flagella, mitochondria, and head, respectively [19]. It is essential to be familiar with that the ATP production occurs just through the glycolysis in the central part of the flagella and that there are no respiratory enzymes in the head. In this regard, many glycolytic enzymes have been found in a fibrous sheath (FS) of sperm. These are lactate dehydrogenase, phosphofructokinase, hexokinase, glyceraldehyde-3-phosphate dehydrogenase (GAPD), and phosphoglucokinase isomerases [21].

When a chemical energy formed by the hydrolysis of ATP is converted to mechanical energy by dynein protein, and force is simply generated. But it is through mitochondrial respiration that the ATP can be efficiently synthesized. Therefore, the question arises as to whether mitochondrial ATP can provide energy for rapid proliferation through rapid and sufficient proliferation in all flagella? The main characteristics of immobile sperm are the disruption of the integrity of the mitochondrial membrane and also the disruption of its sheath function, as a part of the energy for movement is given by the mitochondria [13, 14]. Sperm motility may also be affected by changes in the activity of the mitochondrial chain enzymes. Experimental results on the specific activity of the mitochondrial enzymes as well as sperm motility suggest that impaired mitochondrial function might lead to idiopathic asthenozoospermia [7].

At a molecular level, many studies have shown that changes in the mtDNA (including deletions that affect cell homeostasis) are a way to decrease sperm function and male infertility. Moreover, including motility as sperm parameters where motility and sperm concentration are related with the activity of the sperm mitochondrial enzymes, with electron transfer chain complexes (ETCs). Also, movement is associated with mitochondrial respiration efficiency as well as oxygen consumption in the sperm mitochondria and several diverse ETC inhibitors may adversely affect sperm motility [55]. Mammalian sperm can use a variety of carbohydrates as substrates for ATP production. Evidence recommends that glycolysis plays an essential role in sperm motility, even in the presence of mitochondrial substrates; inhibition of glycolysis may lead to a decrease sperm motility. Therefore, glycolysis is measured to be the source of the energy, especially for the sperm motility function. An interesting research or study showed that in the presence of isolated oxidative phosphorylation agents, sperm can maintain normal motility. This suggests that mitochondrial oxidative phosphorylation may not be essential for flagellar function because of ATP, which contributes a little to oxidative phosphorylation to flagellar energy [37]. However, glycolytic enzyme complexes that are distributed throughout the tail can provide all the necessary ATPs to facilitate movement through the adequate fluid intake. Moreover, circulating AMP is important to

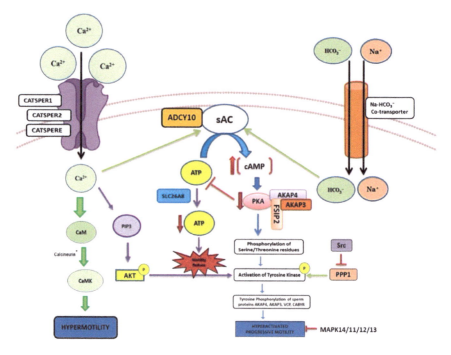

Fig. 12.2 Schematic presentation of genetic relationship in signaling pathways of human sperm motility. Genes reported in human and related to the cAMP/PRKA and calcium pathways are indicated in the figure and also hyperactivated motility is induced when the latter pathway is activated to a much higher level, in combination with the activation of the calmodulin kinase (CaMK) pathway. Also, Ca^{2+} activates PIK3C, which forms PIP3 that in turn activates AKT by phosphorylation. AKT activates tyrosine kinases by phosphorylation. Src inhibits PPP1, which allows an increase of tyrosine kinases phosphorylation. Tyrosine kinases phosphorylate key proteins inducing hyperactivated progressive motility. *P* phosphorylation, *Prg* progesterone

regulate sperm motility, because a decrease in CAMP levels is associated with a decrease in sperm motility. The mechanism that explains this inhibition is the reduction in protein phosphorylation caused by Ca^{2+} (as a result of conformational alteration or depletion of the substrate), which prevents the substrates from interacting with kinases (Fig. 12.2) [46]. Indeed, the resolution of sperm metabolic motility requires much debate, as well as strong proof about the major provider of ATP.

12.3.3 Signaling Pathway and Genes Involve in Sperm Motility

Sperm cells are transcriptionally inactive that's why their motility depend on the activation and inhibition of key signaling pathways. The two common signaling pathways that regulate its motility are calcium and cAMP/PRKA pathway. The influx of Ca^{2+} ions is permitted by Catsper channels that get activated either by progesterone or alkalization of sperm cytosol. Mutations in CatSper cause male infertility due to an inability of sperm to hyperactivate [45]. Simultaneously Na⁺ and HCO₃⁻ co-transporter increase the concentration of Na⁺ and HCO₃⁻ ions in sperm cell, which maintain the pH of spermatozoa [10, 49]. Both Ca^{2+} (Ca2þ) and HCO₃⁻ activate intracellular soluble adenylyl cyclase (sAC) to increase the cyclic AMP (cAMP) levels. Activation of sAC and constant supply of cAMP is necessary for sperm motility but mutation in ADCY10 gene, which encodes for sAC, causes severe asthenozoospermia, due to the absence of progressive motility and segregating with absorptive hypercalciuria [48]. cAMP, in turn, activates sperm protein kinase A (PKA) (PRKA sperm PKA) in the principal piece of sperm flagella. The onset of PKA substrates phosphorylation is followed by activation of unidentified tyrosine kinases and the promotion of tyrosine phosphorylation of sperm proteins (e.g., AKAP4, AKAP3, VCP, and CABYR), which results in hyperactivated progressive motile sperm (Fig. 12.2) ([30, 40]).

The anchoring of PKA with fibrous sheath (FS) is done by structural proteins (i.e., AKAP3 and APAK4) via regulatory subunit of the kinase. AKAP3 may organize the basic of FS structure, initiate its formation, and incorporate AKAP4 into the FS during later spermiogenesis while AKAP4 may act in completing the FS assembly. Lack of AKAP3 causes immotility of sperm due to accumulation of RNA metabolism and translation factors, displacement of PKA subunits, and misregulated PKA activity. In humans, mutations in both AKAP3 and AKAP4 were reported in extremely short tails [20, 61]. Mutations in another FS-related gene FSIP2 were identified in complete disorganization of the FS and absence of components of CPC, IDA, and ODA. Surprisingly, FSIP2 could directly interact with AKAP4 and the patients whose FSIP2 is mutated, they lacked AKAP4 protein – indicating that FSIP2 may be essential for the maintenance of sperm function by anchoring cAMP-dependent PKA to AKAP4 [43]. However, this genetic relationship and mechanisms of these genes in asthenozoospermia require further investigation.

12.3.4 Role of Reactive Oxygen Species (ROS) in the Acquisition and Control of Sperm Motility

Scientist has found that the free movement of electrons is caused by the ROS molecules from actively respiring spermatozoa followed by the intracellular redox reaction. ROS is produced by two processes in spermatozoa: [i] the nicotinamide adenine dinucleotide phosphate oxidase system at the level of the sperm plasma membrane and/or [ii] the nicotinamide adenine dinucleotide-dependent oxidoreductase reaction at the mitochondrial level [1]. However, it is seen that ROS is mostly produced by the second process. Spermatozoa are highly motile cells due to their richness in mitochondria which provide constant energy to its tail but sometimes the energy is less in dysfunctional spermatozoa in semen due to the high level of ROS present. This elevated level of ROS in spermatozoa affects the function of mitochondria and results in low or absence of motility in sperm cells [42].

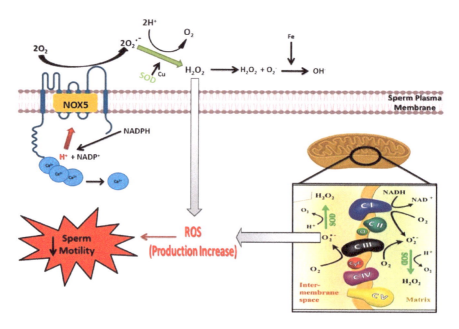

Fig. 12.3 Possible mechanisms by which sperm cells may generate reactive oxygen species (ROS): (1) Ros production by an NOX system, such as NADPH-oxidase isoform 5 (NOX5), which is embedded in the sperm plasma membrane and is activated through an EF-hand Ca^{2+} binding domains; (2) ROS production by the mitochondrial ETC (electron-transport chain), with the electron leakage within the ubiquinone binding sites in complex I (CI) and in complex III (CIII) being the most important mechanisms

O_2^- is a major free radical which is produced by ROS in spermatozoa and this O_2^- (electron – reduce product) reacts itself via dismutation to produce another ROS compound H_2O_2. After this both O_2^- and H_2O_2 undergo the Haber-Weiss reaction in the presence of iron and copper as transitional metals to produce OH^- which is highly reactive and destructive (Fig. 12.3). Not only this O_2^- initiates LPO cascade, which is highly potent but it also leads to disruption of sperm membrane fluidity and its function [47].

Recent studies that describe the production of O_2^- in sperm cells also revealed the presence of NOX5 (a calcium-dependent NADPH oxidase) mainly in the midpiece and acrosomal area of human spermatozoa. Initially, NOX5 is found in the human testis and gets activated by the binding of Ca_2 to its cytosolic N-terminal EF-hand domain [39]. This activated NOX5 causes conformational changes in the spermatozoa by inducing oxidative stress (OS). This investigation strongly prove that NOX5 is a major source of ROS generation in the sperm cells of human. But still, overexpression of NOX5 associated with OS in infertile male is yet to be studied more.

12.3.5 Role of Metabolic Pathways and the Gene Responsible for Sperm Motility

After performing several studies on asthenozoospermia at the molecular level, now studying its metabolic pathway is equally important and needed. Asgari and his team reported the reconstructed SpermNet using all proteome data and the mCADRE algorithm, which is the first sperm cell proteome scale model [5]. Then they used the COBRA toolbox to analyze the updated model, and to study its effect on ATP production in the model. In total, 78 genes significantly increase the rate of ATP production, most of which symbolize oxidative phosphorylation, fatty acid oxidation, the Krebs cycle, and the carrier components of the products of 25 family members. Table 12.2 lists 11 new genes that they have identified in 29 genes. These genes were not previously associated with sperm cell energy exchange and may therefore be associated with asthenozoospermia. They further tested the azoosperm-related genes that are now known using computer knockout methods, but these genes were not predicted, so they further examined the reconstructed model. The pathways affected by the knockout of these genes are also related to energy metabolism and therefore confirms previous results. Therefore, their model not only predicts known pathways, but also identifies several non-metabolic genes for insufficient energy exchange in asthenozoospermia. Finally, their model supports the idea that in addition to glycolysis, metabolic pathways (such as phosphorylation and oxidation of fatty acids) are important for the exchange of sperm energy, which can be the basis for fertility restoration [5].

Table 12.2 Novel predicted genes as potential biomarkers

S.no.	Entrez gene ID	Gene name
1.	2592	GALT
2.	55,577	NAGK
3.	5238	PGM3
4.	2650	GNT1
5.	51,727	CMPK
6.	9583	ENTPD4
7.	2720	GLB1
8.	5476	CTSA
9.	11,046	SLC35D2
10.	7355	SLC35A2
11.	4668	NAGA

12.3.6 Sperm Motility, Molecular Pathways, and Lifestyle Factors

In addition to the defects of genetic diseases that can lead to absent or decreased sperm motility, certain physiological processes during sperm maturation can also affect sperm motility. Although, these physiological phenomena become uncontrolled and physiologically increased, though the more negative effect on sperm motility becomes more pronounced. Lifestyle factors, air pollution, and exposures to chemical pesticides can simply disrupt sperm motility [46].

These physiological phenomena are thought to be the reason for ROS production during sperm maturation, pregnancy, and fertilization. Excessive ROS production can lead to structural damage and initiate many pathways that lead to sperm loss during normal fertilization. Similarly, some studies have shown that ROS has a detrimental effect on sperm function and maturation. ROS are produced by sperm mitochondria, and sperm mitochondria has been considered as an important source of ROS, in a form of sperm cytoplasmic droplets (excessive cytoplasm) [13, 14]. Pathological conditions, for example, leukocytospermia, varicocele, genital infections, moreover chronic inflammation can cause ROS and sperm can't move. Higher O_2 is detrimental to sperm function, while higher H_2O_2 can disrupt sperm motility parameters. Diffusion of H_2O_2 from the membrane into the cell interferes with the activity of few important enzymes, for example, glucose-6-phosphate dehydrogenase (G6PD), which can reduce sperm motility. G6PD regulates intracellular dose and NADPH utilization through hexose monophosphate shunts [6].

Lifestyle risk factors likewise play a major role in male infertility as well as sperm quality. These include poor diet, smoking, excessive drinking, and mental stress. Other factors, such as exposure to environmental pollutants and age, have a positive relationship with oxidative stress, while they have a positive relationship with sperm correction. Although several clinical studies have not been conducted in such areas, indeed results of this study suggests that the dietary status is associated with sperm motility in semen samples [12, 57].

Smoking is another important factor that damages semen parameters, particularly sperm motility. From review article in a meta-analysis, the results showed that smoking had a negative and significant impact on both general and advanced exercise [23].

Apoptosis is another physiological phenomenon that can lead to decreased sperm motility. Human sperm may have indicators of apoptosis and affect sperm functions or remove sperm that have been damaged by DNA in a woman's reproductive system. Or, it may result from residual apoptosis in the testis, which may be linked to male infertility. Sperm with low motility cannot participate in fertilization of the egg and cell apoptosis to achieve its physiological purpose. When different inducers increase apoptosis uncontrolled way, it affects sperm motility and leads to poor sperm motility and asthenozoospermia [2].

12.4 Therapeutic Approach for Athenozoospermia

Due to genetic reasons, there is currently no cure for sperm motility. However, one can protect sperm quality including sperm motility. As mentioned above, unhealthy lifestyle habits (lack of physical activity, excessive usage of electronic gadgets and recreation toxics), specific environmental phenomena as well as several pathologies linked to endocrine as well as cardiovascular diseases are associated with their oxidative stress [15]. Use of antibiotics such as corticosteroids (antibodies against sperm after chronic inflammation of the genitals) on time to treat genital infections and surgery (varicocele) can also be used to treat other diseases that increase sperm oxidative stress [32].

For example, vitamin E is the radical peroxide cleanser and can be used as a serial antioxidant. This secures and prevents the spread of free radicals in the membrane and plasma lipoproteins and reduces the level of malondialdehyde (an organic compound that is used as an indicator of oxidative stress) and thus improves sperm viability [38]. Furthermore, by processing vitamin E, it can prevent DNA damage caused by H_2O_2 free radicals. This molecule is broadly used in prophylactic treatment [29]. However, antioxidants have received a lot of attention in recent days. Vitamin C, co-enzymes Q10, glutathione, HGH (the growth hormone of humans), and carnitine are used in preventing the overproduction of ROS in males (Fig. 12.4). Asthenozoospermia, which majorly causes male infertility, is treated mostly through antioxidants but it is helpful only to some limits and the reason lies in the inefficiency of the natural substances [50]. Therefore, it is important to take therapies of some natural products to cure asthenozoospermia.

Preventing oxidative stress is the main objective of therapy for the correction of asthenozoospermia. So, the plants having these series of constituents are important toward effective therapies. Curcumin, a natural polyphenolic compound, is categorized as diferuloylmethane which is again produced from the rhizome of the turmeric plant (*Curcuma longa* Lin). It is mostly known for its effectiveness as an anti-inflammatory, antioxidant, and is also known for being the scavenger of free radicals in some cases [4]. But it was found out that curcumin was capable of improving the spermatozoa of Bull (Fig. 12.4) but however it has been seen to induce oxidative stress of rats' testis. Besides this, it improves the cryopreservation of the boar spermatozoa. Infertile men and patients with leukocytospermia are highly benefitted with curcumin. This is because curcumin records improve sperm motility. Curcumin is also recorded as one of the most capable non-steroidal contraceptive solutions. It works by blocking the motility of sperm. These two functions of curcumin act opposite to one another [63].

Myo-inositol or MYO is generally a therapeutic option available for infertility of females and in normal cases, it is connected to insulin resistance. Several data have recently accumulated regarding the possible application of MYO for treating male infertility. In research carried out by R.A. CONDORELLI, 2017 shows the possible antioxidant and prokinetic effect of MYO and its significance in hormonal regulation and modulation. A proposal for a clinical algorithm was reserved and proposed

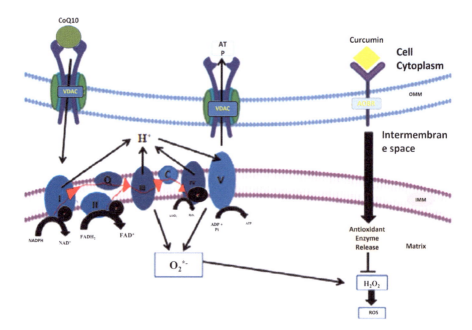

Fig. 12.4 Mechanism of (1) coenzyme Q10 and (2) curcumin effect in sperm mitochondria. *CoQ10* Coenzyme Q 10, *VADC* voltage-dependent anion channel, *ATP* adenosine triphosphate, *ADP* adenosine diphosphate, *Pi* inorganic phosphate; I, II, III, IV, and V: respiratory chain complexes, H^+ proton, e^- electron, Q coenzyme Q cycle, C cytochrome c, *NADH* reduced nicotinamide adenine dinucleotide, FAD^+ nicotinamide adenine dinucleotide, $FADH_2$ reduced flavin adenine dinucleotide, *FAD* flavin adenine dinucleotide, O_2^{*-} superoxide radical, H_2O_2 hydrogen peroxide, *ROS* reactive oxygen species, *AOBR* antioxidant binding receptor, *OMM* outer mitochondrial membrane, *INN* inner mitochondrial membrane

for patients suffering from low sperm motility or asthenozoospermia at the last part of the script. MYO was expected to impart some relevant pharmacological effect [11].

Proposed Mechanism coQ10 Mitochondrial CoQ functions include regulation of electron transport in the respiratory chain, receiving electrons from complex I and complex II and passing them to complex III, and transfer of protons from fatty acids to the matrix. An alternative function of CoQ may also be the regulation of permeability transition pore opening and nutrition uptake through the voltage-dependent anion channel (VDAC) of the outer mitochondrial membrane (OMM).

Proposed Mechanism Curcumin The antioxidant, curcumin, is reported to bind promoters of antioxidant genes to promote antioxidant enzymes expression. Thus, curcumin plays an important role as an antioxidant and is very important for maintaining spermatozoa motility through suppressing oxidative stress.

12.5 Conclusion

Decreased sperm motility or asthenozoospermia has been extensively studied, but now several researchers have focused on possible causes of asthenozoospermia and infertility. Many important cellular along with molecular factors related to the pathogenesis of asthenozoospermia have been reported, although genes related to metabolic pathways still need to be investigated. Structures such as SpermNet can predict several related genes in asthenozoospermia and confirm the idea that asthenozoosperm involves multiple non-glycolytic pathways be taken. The best therapy for asthenozoospermia has not yet been established. Therefore, immediate preventive measures and some empirical therapies are needed to develop. Safe and targeted therapies should be based on the genetic causes of sperm motility and ways to control sperm immotility.

Recent advancements in sequencing technologies have greatly accelerated the discovery of the genes that cause AZS, which has improved our understanding of the physiopathology of AZS to provide candidate biomarkers in treatment in AZS patients. However, considering the range of phenotypic and genetic heterogeneity of AZS, only a limited number of mutations have been identified, indicating that additional studies are still needed.

Authors' Contribution All the authors contributed to the writing of the manuscript, and all authors approved the final version.

Consent for Publication Not applicable.

Funding None.

Conflict of Interest The authors declare no conflict of interest, financial or otherwise.

Acknowledgments We thank our lab members and collaborators for carefully reading the manuscript and contributing valuable inputs for improving the manuscript.

References

1. Agarwal A, Virk G, Ong C, du Plessis SS. Effect of oxidative stress on male reproduction. World J Men's Health. 2014;32(1):1–17. https://doi.org/10.5534/wjmh.2014.32.1.1.
2. Aitken RJ, Koppers AJ. Apoptosis and DNA damage in human spermatozoa. Asian J Androl. 2011;13(1):36–42. https://doi.org/10.1038/aja.2010.68.
3. Alahmar AT. Role of oxidative stress in male infertility: an updated review. J Hum Reprod Sci. 2019;12(1):4–18. https://doi.org/10.4103/jhrs.JHRS_150_18.
4. Amalraj A, Pius A, Gopi S, Gopi S. Biological activities of curcuminoids, other biomolecules from turmeric and their derivatives – a review. J Tradit Complement Med. 2016;7(2):205–33. https://doi.org/10.1016/j.jtcme.2016.05.005.
5. Asghari A, Marashi SA, Ansari-Pour N. A sperm-specific proteome-scale metabolic network model identifies non-glycolytic genes for energy deficiency in asthenozoospermia. Syst Biol Reprod Med. 2017;63(2):100–12. https://doi.org/10.1080/19396368.2016.1263367.

6. Barati E, Nikzad H, Karimian M. Oxidative stress and male infertility: current knowledge of pathophysiology and role of antioxidant therapy in disease management. Cell Mol Life Sci. 2020;77(1):93–113. https://doi.org/10.1007/s00018-019-03253-8.
7. Barbagallo F, La Vignera S, Cannarella R, Aversa A, Calogero AE, Condorelli RA. Evaluation of sperm mitochondrial function: a key organelle for sperm motility. J Clin Med. 2020;9(2):363. https://doi.org/10.3390/jcm9020363.
8. Bavel JV. The world population explosion: causes, backgrounds and -projections for the future. Facts Views Vision ObGyn. 2013;5(4):281–91.
9. Cannarella R, Maniscalchi ET, Condorelli RA, Scalia M, Guerri G, La Vignera S, Bertelli M, Calogero AE. Ultrastructural sperm flagellum defects in a patient with CCDC39 compound heterozygous mutations and primary ciliary dyskinesia/situs viscerum inversus. Front Genet. 2020;11:974. https://doi.org/10.3389/fgene.2020.00974.
10. Chang JC, Oude-Elferink RPJ. Role of the bicarbonate-responsive soluble adenylyl cyclase in pH sensing and metabolic regulation. Front Physiol. 2014; https://doi.org/10.3389/fphys.2014.00042.
11. Condorelli RA, La Vignera S, Mongioì LM, Vitale SG, Laganà AS, Cimino L, Calogero AE. Myo-inositol as a male fertility molecule: speed them up! Eur Rev Med Pharmacol Sci. 2017;21(2 Suppl):30–5.
12. Danielewicz A, Przybyłowicz KE, Przybyłowicz M. Dietary patterns and poor semen quality risk in men: a cross-sectional study. Nutrients. 2018;10(9):1162. https://doi.org/10.3390/nu10091162.
13. Du Plessis SS, Agarwal A, Halabi J, Tvrda E. Contemporary evidence on the physiological role of reactive oxygen species in human sperm function. J Assist Reprod Genet. 2015a;32(4):509–20. https://doi.org/10.1007/s10815-014-0425-7.
14. Du Plessis SS, Agarwal A, Mohanty G, van der Linde M. Oxidative phosphorylation versus glycolysis: what fuel do spermatozoa use? Asian J Androl. 2015b;17(2):230–5. https://doi.org/10.4103/1008-682X.135123.
15. Durairajanayagam D. Lifestyle causes of male infertility. Arab J Urol. 2018;16(1):10–20. https://doi.org/10.1016/j.aju.2017.12.004.
16. Dutta S, Majzoub A, Agarwal A. Oxidative stress and sperm function: a systematic review on evaluation and management. Arab J Urol. 2019;17(2):87–97. https://doi.org/10.1080/2090598X.2019.1599624.
17. El Khouri E, Thomas L, Jeanson L, Bequignon E, Vallette B, Duquesnoy P, Montantin G, Copin B, Dastot-Le Moal F, Blanchon S, Papon JF, Lorès P, Yuan L, Collot N, Tissier S, Faucon C, Gacon G, Patrat C, Wolf JP, Dulioust E, et al. Mutations in DNAJB13, encoding an HSP40 family member, cause primary ciliary dyskinesia and male infertility. Am J Hum Genet. 2016;99(2):489–500. https://doi.org/10.1016/j.ajhg.2016.06.022.
18. Fassad MR, Shoemark A, Legendre M, Hirst RA, Koll F, le Borgne P, Louis B, Daudvohra F, Patel MP, Thomas L, Dixon M, Burgoyne T, Hayes J, Nicholson AG, Cullup T, Jenkins L, Carr SB, Aurora P, Lemullois M, Aubusson-Fleury A, et al. Mutations in outer dynein arm heavy chain DNAH9 cause motile cilia defects and situs inversus. Am J Hum Genet. 2018;103(6):984–94. https://doi.org/10.1016/j.ajhg.2018.10.016.
19. Ferramosca A, Zara V. Bioenergetics of mammalian sperm capacitation. Biomed Res Int. 2014;2014:902953. https://doi.org/10.1155/2014/902953.
20. Fiedler SE, Dudiki T, Vijayaraghavan S, Carr DW. Loss of R2D2 proteins ROPN1 and ROPN1L causes defects in murine sperm motility, phosphorylation, and fibrous sheath integrity. Biol Reprod. 2013;88(2):41. https://doi.org/10.1095/biolreprod.112.105262.
21. Goldberg E, Eddy EM, Duan C, Odet F. LDHC: the ultimate testis-specific gene. J Androl. 2010;31(1):86–94. https://doi.org/10.2164/jandrol.109.008367.
22. Guichard C, Harricane MC, Lafitte JJ, Godard P, Zaegel M, Tack V, Lalau G, Bouvagnet P. Axonemal dynein intermediate-chain gene (DNAI1) mutations result in situs inversus and primary ciliary dyskinesia (Kartagener syndrome). Am J Hum Genet. 2001;68(4):1030–5. https://doi.org/10.1086/319511.

23. Harlev A, Agarwal A, Gunes SO, Shetty A, du Plessis SS. Smoking and male infertility: an evidence-based review. World J Men's Health. 2015;33(3):143–60. https://doi.org/10.5534/wjmh.2015.33.3.143.
24. Inaba K, Mizuno K. Sperm dysfunction and ciliopathy. Reprod Med Biol. 2015;15(2):77–94. https://doi.org/10.1007/s12522-015-0225-5.
25. Jeanson L, Copin B, Papon JF, Dastot-Le Moal F, Duquesnoy P, Montantin G, Cadranel J, Corvol H, Coste A, Désir J, Souayah A, Kott E, Collot N, Tissier S, Louis B, Tamalet A, de Blic J, Clement A, Escudier E, Amselem S, et al. RSPH3 mutations cause primary ciliary dyskinesia with central-complex defects and a near absence of radial spokes. Am J Hum Genet. 2015;97(1):153–62. https://doi.org/10.1016/j.ajhg.2015.05.004.
26. Katole A, Saoji AV. Prevalence of primary infertility and its associated risk factors in urban population of Central India: a community-based cross-sectional study. Indian J Community Med. 2019;44(4):337–41.
27. Kott E, Duquesnoy P, Copin B, Legendre M, Dastot-Le Moal F, Montantin G, Jeanson L, Tamalet A, Papon JF, Siffroi JP, Rives N, Mitchell V, de Blic J, Coste A, Clement A, Escalier D, Touré A, Escudier E, Amselem S. Loss-of-function mutations in LRRC6, a gene essential for proper axonemal assembly of inner and outer dynein arms, cause primary ciliary dyskinesia. Am J Hum Genet. 2012;91(5):958–64. https://doi.org/10.1016/j.ajhg.2012.10.003.
28. Kumar N, Singh AK. Trends of male factor infertility, an important cause of infertility: a review of literature. J Hum Reprod Sci. 2015;8(4):191–6.
29. Kurutas EB. The importance of antioxidants which play the role in cellular response against oxidative/nitrosative stress: current state. Nutr J. 2016;15(1):71. https://doi.org/10.1186/s12937-016-0186-5.
30. Lefièvre L, Jha KN, de Lamirande E, Visconti PE, Gagnon C. Activation of protein kinase A during human sperm capacitation and acrosome reaction. J Androl. 2002;23(5):709–16.
31. Lehti MS, Sironen A. Formation and function of sperm tail structures in association with sperm motility defects. Biol Reprod. 2017;97(4):522–36. https://doi.org/10.1093/biolre/iox096.
32. Leslie SW, Siref LE, Soon-Sutton TL, Khan MA. Male infertility. StatPearls [Internet]. StatPearls; 2021 Jan-. Available from: https://www.ncbi.nlm.nih.gov/books/NBK562258/
33. Lishko P, Kirichok Y. Signaling the differences between cilia. eLife. 2015;4:e12760. https://doi.org/10.7554/eLife.12760.
34. Loges NT, Antony D, Maver A, Deardorff MA, Güleç EY, Gezdirici A, Nöthe-Menchen T, Höben IM, Jelten L, Frank D, Werner C, Tebbe J, Wu K, Goldmuntz E, Čuturilo G, Krock B, Ritter A, Hjeij R, Bakey Z, Pennekamp P, et al. Recessive DNAH9 loss-of-function mutations cause laterality defects and subtle respiratory ciliary-beating defects. Am J Hum Genet. 2018;103(6):995–1008. https://doi.org/10.1016/j.ajhg.2018.10.020.
35. Lushchak VI. Free radicals, reactive oxygen species, oxidative stress and its classification. Chem Biol Interact. 2014;224:164–75. https://doi.org/10.1016/j.cbi.2014.10.016.
36. Manfrevola F, Chioccarelli T, Cobellis G, Fasano S, Ferraro B, Sellitto C, Marella G, Pierantoni R, Chianese R. CircRNA role and circRNA-dependent network (ceR-NET) in Asthenozoospermia. Front Endocrinol. 2020;11:395. https://doi.org/10.3389/fendo.2020.00395.
37. Miller MR, Kenny SJ, Mannowetz N, Mansell SA, Wojcik M, Mendoza S, Zucker RS, Xu K, Lishko PV. Asymmetrically positioned flagellar control units regulate human sperm rotation. Cell Rep. 2018;24(10):2606–13. https://doi.org/10.1016/j.celrep.2018.08.016.
38. Mirzaei Khorramabadi K, Reza Talebi A, Abbasi Sarcheshmeh A, Mirjalili A. Protective effect of vitamin E on oxidative stress and sperm apoptosis in diabetic Mice. Int J Reprod Biomed. 2019;17(2):127–34. https://doi.org/10.18502/ijrm.v17i2.3990.
39. Musset B, Clark RA, DeCoursey TE, Petheo GL, Geiszt M, Chen Y, Cornell JE, Eddy CA, Brzyski RG, El Jamali A. NOX5 in human spermatozoa: expression, function, and regulation. J Biol Chem. 2012;287(12):9376–88. https://doi.org/10.1074/jbc.M111.314955.
40. Naz RK, Rajesh PB. Role of tyrosine phosphorylation in sperm capacitation/acrosome reaction. Reprod Biol Endocrinol. 2004;2:75. https://doi.org/10.1186/1477-7827-2-75.

41. Nesci S, Spinaci M, Galeati G, Nerozzi C, Pagliarani A, Algieri C, Tamanini C, Bucci D. Sperm function and mitochondrial activity: an insight on boar sperm metabolism. Theriogenology. 2020;144 https://doi.org/10.1016/j.theriogenology.2020.01.004.
42. Nowicka-Bauer K, Nixon B. Molecular changes induced by oxidative stress that impair human sperm motility. Antioxidants (Basel). 2020;9(2):134. https://doi.org/10.3390/antiox9020134.
43. Nsota Mbango JF, Coutton C, Arnoult C, Ray PF, Touré A. Genetic causes of male infertility: snapshot on morphological abnormalities of the sperm flagellum. Basic Clin Androl. 2019;29(2) https://doi.org/10.1186/s12610-019-0083-9.
44. Omran H, Kobayashi D, Olbrich H, Tsukahara T, Loges NT, Hagiwara H, Zhang Q, Leblond G, O'Toole E, Hara C, Mizuno H, Kawano H, Fliegauf M, Yagi T, Koshida S, Miyawaki A, Zentgraf H, Seithe H, Reinhardt R, Watanabe Y, et al. Ktu/PF13 is required for cytoplasmic pre-assembly of axonemal dyneins. Nature. 2008;456(7222):611–6. https://doi.org/10.1038/nature07471.
45. Orta G, de la Vega-Beltran JL, Martín-Hidalgo D, Santi CM, Visconti PE, Darszon A. CatSper channels are regulated by protein kinase A. J Biol Chem. 2018;293(43):16830–41. https://doi.org/10.1074/jbc.RA117.001566.
46. Pereira R, Sá R, Barros A, Sousa M. Major regulatory mechanisms involved in sperm motility. Asian J Androl. 2017;19(1):5–14. https://doi.org/10.4103/1008-682X.167716.
47. Phaniendra A, Jestadi DB, Periyasamy L. Free radicals: properties, sources, targets, and their implication in various diseases. Indian J Clin Biochem. 2015;30(1):11–26. https://doi.org/10.1007/s12291-014-0446-0.
48. Pozdniakova S, Ladilov Y. Functional significance of the Adcy10-dependent intracellular cAMP compartments. J Cardiovasc Dev Dis. 2018;5(2):29. https://doi.org/10.3390/jcdd5020029.
49. Rahman N, Buck J, Levin LR. pH sensing via bicarbonate-regulated "soluble" adenylyl cyclase (sAC). Front Physiol. 2013;4:343. https://doi.org/10.3389/fphys.2013.00343.
50. Rauscher FM, Sanders RA, Watkins JB 3rd. Effects of coenzyme Q10 treatment on antioxidant pathways in normal and streptozotocin-induced diabetic rats. J Biochem Mol Toxicol. 2001;15(1):41–6. https://doi.org/10.1002/1099-0461(2001)15:1<41::aid-jbt5>3.0.co;2-z.
51. Roser M. Fertility rate. Published online at OurWorldInData.org. 2014.
52. Sabeti P, Pourmasumi S, Rahiminia T, Akyash F, Talebi AR. Etiologies of sperm oxidative stress. Int J Reprod Biomed. 2016;14(4):231–40.
53. Sapiro R, Kostetskii I, Olds-Clarke P, Gerton GL, Radice GL, Strauss JF III. Male infertility, impaired sperm motility, and hydrocephalus in mice deficient in sperm-associated antigen 6. Mol Cell Biol. 2002;22(17):6298–305. https://doi.org/10.1128/mcb.22.17.6298-6305.2002.
54. Schou KB, Morthorst SK, Christensen ST, Pedersen LB. Identification of conserved, centrosome-targeting ASH domains in TRAPPII complex subunits and TRAPPC8. Cilia. 2014;3:6. https://doi.org/10.1186/2046-2530-3-6.
55. Shamsi MB, Kumar R, Bhatt A, Bamezai RN, Kumar R, Gupta NP, Das TK, Dada R. Mitochondrial DNA mutations in etiopathogenesis of male infertility. Indian J Urol. 2008;24(2):150–4. https://doi.org/10.4103/0970-1591.40606.
56. Sironen A, Shoemark A, Patel M, Loebinger MR, Mitchison HM. Sperm defects in primary ciliary dyskinesia and related causes of male infertility. Cell Mole Life Sci. 2020;77:2029–48. https://doi.org/10.1007/s00018-019-03389-7.
57. Skoracka K, Eder P, Łykowska-Szuber L, Dobrowolska A, Krela-Kaźmierczak I. Diet and nutritional factors in male (in)fertility-underestimated factors. J Clin Med. 2020;9(5):1400. https://doi.org/10.3390/jcm9051400.
58. Tourmente M, Villar-Moya P, Rial E, Roldan ER. Differences in ATP generation via glycolysis and oxidative phosphorylation and relationships with sperm motility in mouse species. J Biol Chem. 2015;290(33):20613–26. https://doi.org/10.1074/jbc.M115.664813.
59. Wang C, Swerdloff RS. Limitations of semen analysis as a test of male fertility and anticipated needs from newer tests. Fertil Steril. 2014;102(6):1502–7. https://doi.org/10.1016/j.fertnstert.2014.10.021.

60. Wu H, Wang J, Cheng H, et al. Patients with severe asthenoteratospermia carrying SPAG6 or RSPH3 mutations have a positive pregnancy outcome following intracytoplasmic sperm injection. J Assist Reprod Genet. 2020;37(4):829–40. https://doi.org/10.1007/s10815-020-01721-w.
61. Xu K, Yang L, Zhao D, Wu Y, Qi H. AKAP3 synthesis is mediated by RNA binding proteins and PKA signaling during mouse spermiogenesis. Biol Reprod. 2014;90(6):119. https://doi.org/10.1095/biolreprod.113.116111.
62. Yang Y, Lu X. Drosophila sperm motility in the reproductive tract. Biol Reprod. 2011;84(5):1005–15. https://doi.org/10.1095/biolreprod.110.088773.
63. Zhou Q, Wu X, Liu Y, Wang X, Ling X, Ge H, Zhang J. Curcumin improves asthenozoospermia by inhibiting reactive oxygen species reproduction through nuclear factor erythroid 2-related factor 2 activation. Andrologia. 2020;52(2):e13491. https://doi.org/10.1111/and.13491.
64. Zuccarello D, Ferlin A, Cazzadore C, Pepe A, Garolla A, Moretti A, Cordeschi G, Francavilla S, Foresta C. Mutations in dynein genes in patients affected by isolated non-syndromic asthenozoospermia. Hum Reprod. 2008;23(8):1957–62. https://doi.org/10.1093/humrep/den193.

Chapter 13
Tale of Viruses in Male Infertility

Shreya Das, Arunima Mondal, Jayeeta Samanta, Santanu Chakraborty, and Arunima Sengupta

Abstract Male infertility is a condition where the males either become sterile or critically infertile. The World Health Organisation assessed that approximately 9% of the couple have fertility issues where the contribution of the male partner was estimated to be 50%. There are several factors that can amalgamate to give rise to male infertility. Among them are lifestyle factors, genetic factors and as well as several environmental factors. The causes of male infertility may be acquired, congenital or sometimes idiopathic. All these factors adversely affect the spermatogenesis process as well as they impart serious threats to male genital organs thus resulting in infertility. Viruses are submicroscopic pathogenic agents that rely on host for their replication and survival. They enter the host cell, hijack the host cell machinery to aid their own replication and exit the cell for a new round of infection. With the growing abundance of different types of viruses and the havoc they have stirred in the form of pandemics, it is very essential to decipher their route of entry inside the human body and understand their diverse functional roles in order to combat them. In this chapter, we will review how viruses invade the male genital system thus in turn leading to detrimental consequence on male fertility. We will discuss the tropism of various viruses in the male genital organs and explore their sexual transmissibility. This chapter will summarise the functional and mechanistic approaches employed by the viruses in inducing oxidative stress inside spermatozoa thus leading to male infertility. Moreover, we will also highlight the various antiviral therapies that have been studied so far in order to ameliorate viral infection in order to combat the harmful consequences leading to male infertility.

Keywords Virus · Infertility · Oxidative stress · Sperm · Antiviral

---

S. Das · A. Mondal · J. Samanta · A. Sengupta (✉)
Department of Life science and Biotechnology, Jadavpur University, Kolkata, West Bengal, India

S. Chakraborty
Department of Life sciences, Presidency University, Kolkata, India

13.1 Introduction

Since the beginning of life, living things have always been in a constant struggle for survival over the resources on earth for their nutrition, growth and reproduction. While some organisms are equipped with the ability to produce their own food known as autotrophs, but a vast majority of other organisms rely on some other organisms for their nutrition and propagation often being detrimental to their host known as the pathogens. The most notable mention of these pathogens that has left the world populace in a dramatic struggle over the decade is the emergence of various deadly viruses affecting millions of people not only to mortality but also various forms of deformities or disabilities.

Viruses are a group of microorganisms structurally consisting of a genetic core of nucleic acid, i.e., DNA or RNA and an outer protein coat called capsid. In some viruses, there is a presence of an outer envelope that often has some spike-like structure that helps in injecting the viral genetic material within their host. Because of the absence of any other cellular structures, viruses are generally hard to target and kill without posing potential risk to the host. The most effective solution to viral diseases till date is the vaccines and the body's own innate immune response. Viral diseases apart from the direct pathogenesis often leave several damages to the body of host organisms even after sometimes the disease is cured, like deafness, infertility, paralysis of limbs and many others.

13.1.1 Origin of Virus

Viruses are considered as the earliest form of living entity and there are several theories regarding how they came into existence. Till date, there are three major theories hypothesising the origin of viruses, namely:

Primordial Virus World/Virus Early Theory

This theory attempts to explain the origin of virus as directly emanating from the first pre-cellular genetic materials that eventually developed complex structures around the genetic core to form the virus in the course of evolution. This theory also hypothesises virus to give rise to cellular life forms although this has been of debate as all viruses require a host cell for survival and replication.

Regressive Virus Origin/Regression Theory

The regressive theory, based on the findings of morphological and genetic similarity between many parasitic bacteria and some recently discovered giant viruses, states that virus may have originated from the degeneration of cellular materials that have lost their autonomic life functions and become parasitic on host cells for propagation.

Escaped Genes Theory

The escaped theory says that virus may have originated from cellular genome from the 'escape' of selfish genes that have evolved to have their semi-self-replicative property within the host cells. The escaped theory is the most accepted theory of viral origin by far owing to the ability of modern-day virus to integrate cell's genes into their genetic materials [1, 2].

Viruses are obligate intracellular parasites as they lack the systems required for the essential life processes such as replication, translation and other independent metabolic activity. Hence, viruses use the host's metabolic machineries in an efficient way to infect and propagate within the host. It is important to know the interactions between the virus and its host's cell environments to understand the mode of infection a virus follows and ways to combat the viral infection.

13.1.2 Virus Life Cycle

The course of viral life cycle has five general stages: cell entry, translation of viral proteins, replication of the genome, assembly of the viral particle and egress from the cell (Fig. 13.1).

Cell Entry

After recognising a target host cell, virus generally attaches to the host cell's membrane by the interaction of viral factors with the specific receptors present on the cell surface, which then enters the cell cytoplasm through internalisation of the viral particle by endocytosis or by direct fusion with the host membrane in case of the viruses that lack capsid. Some enveloped virus also forms a pore-like structure within the host membrane to pass the viral genome into the cytoplasm [3, 4]. The viral receptors present on the cell determine which virus attacks which type of cells, for example, the neuraminidase and haemagglutinin protein present on the influenza virus's cell surface binds to the sialic acid-containing receptors on cell that are mostly found on the cells of the respiratory tract, the gp120 viral proteins on human immunodeficiency virus (HIV) binds to the CD4 receptors of T lymphocyte cells thus making the T lymphocyte cells of human immune systems susceptible to HIV infections [4, 5].

Following the entry of the virus, the capsid is uncoated and the viral genetic material is released into the host cell's cytoplasm for subsequent translation and replication.

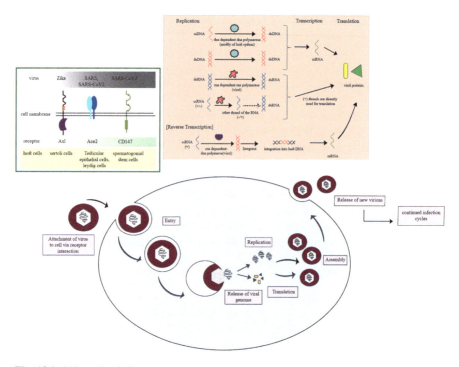

Fig. 13.1 Life cycle of virus inside spermatozoa. The life cycle of viruses can be characterised into six steps: attachment of virus to host cells, entry into host cell, replication of the viral genome and translation of the viral protein, assembly of the virus particles, release of the new formed virions. The attachment of the virus to the host is enhanced by various receptors like Axl, Ace2 and Cd147

Genome Replication, Transcription and Translation of Viral Proteins

Within the host cell viral genome follows various strategies to translate the viral proteins and replicate the genome based on their genetic contents. While DNA viruses enter the nucleus of host cell and most often use the host replication system to copy their genome, all RNA viruses replicate their genome within the cytoplasm using the RNA-dependent-RNA polymerase enzyme carried by the virus itself. Here it is important to note that viruses, prior to these genomic processes, often make various modulations to their surrounding environment in their favour for successful use of host systems and to escape from the host recognition and immune response. For example, hepatitis C virus (HCV) has been shown to increase the intracellular lipid and cholesterol level prior to replication, increased level of certain cell signalling mediating components like phosphatidylinositides (PIs) and so on; most of which is done by exploiting the host's metabolic system. Some virus also happens to downregulate the host's apoptotic pathway, antiviral immune response processes or expresses anti-apoptotic genes to escape the immune response-mediated apoptotic death of infected cells, e.g. HCV virus decreases the host's immunoproteasome activity responsible for degrading viral and displaying viral antigen forimmune cell recognition, herpes virus EBV uses the virally encoded deubiquitinating enzyme (DUB) functions, and human cytomegalovirus (HCMV) encodes its own DUBs to inhibit the activation of the antiviral innate immune response [3].

Cell Exit

The final stage of viral life cycle is to package the newly synthesised genomes and viral proteins into complete virions and release them into extracellular space for further infection and propagation. The newly synthesised virus particles attach to host cell membrane to eject from cell mainly by exocytosis [4]. Virus makes use of certain host systems or processes to execute their release from the cell-like herpes virus use the host's Rab-GTPase enzyme system to traffic between cell organelles and egress from the cell. Some viruses use the mitochondrial F1FO-ATPase system for budding off as new virions [3].

Apart from these, viruses use a multitude of tactics to evade from viral recognition and host's adaptive immune response, e.g. HIV infects the $CD4^+$ T_H cells, resulting in drastic immune suppression due to cell lysis; measles virus infects B cells, $CD4^+$, and $CD8^+$ memory T cells and monocytes, resulting in immune suppression; Epstein-Barr virus infects the B cells, affecting antigen recognition and antibody release and many others [6].

These viruses exert detrimental effects on male fertility by inducing oxidative stress which in turn results in various abnormalities in sperm morphology as well as sperm characteristics. Infection with these viruses often leads to decreased sperm count, sperm motility, sperm viability, increased leucocyte infiltration and development of orchitis [7–10]. Virus infection also leads to imbalance of the male hormone testosterone [10–13]. All these factors in concert result in the development of male infertility.

This chapter will elucidate the various detrimental effects of viral infection on male fertility. It will also highlight the development of antiviral therapy against these viruses in order to eliminate them which will indirectly assist in combating male infertility due to virus infection.

13.2 Classification of Viruses

Viruses, often microscopic in nature but vast in numbers, require a systematic grouping and characterisation for their easy scientific studies and controlling measures. In practice there are two ways of classifying viruses, one being direct morphological characteristics of the virus and the other focusing on the genetic makeup of the virus.

13.2.1 Morphological Characteristics

In the first way, viruses are known to have two major types based on the structure of their capsid along with other complex structural group of viruses. They are classified accordingly as:

Helical Virus

This group of viruses consists of nucleic acid coiled in the shape of a helix and the capsid proteins wound around the nucleic acid, forming a long tube or rod-like structure surrounded by a hollow cylindrical capsid. Most plant viruses have helical morphology and the common animal viruses having helical structure are influenza virus, measles virus, mumps virus, rabies virus and Ebola virus [14, 15].

Icosahedral

The most common capsid structure found within the animal viruses is icosahedral type. The genomes of icosahedral viruses are packaged within an icosahedral-shaped capsid that acts as a protein shell. An icosahedron is a geometric shape with 20 sides (or faces), each composed of an equilateral triangle. Several viral proteins form each face (small triangle) of the icosahedral capsid subunits. These subunits together form the structural unit that repeats to form the capsid of the virion. The common viruses having icosahedral structure are rotavirus, human papillomavirus (HPV), hepatitis B virus (HBV) and herpes viruses [16].

Complex Virus Structures

Apart from the helical and icosahedral structured capsid, present in most viruses, there are some viruses that have other complex structures like a modified icosahedral structure, presence of extra protein tails, or complex outer wall. Bacteriophages and poxvirus are some common examples of virus of this category [17].

13.2.2 Genetic Characteristics

Apart from the morphological categorisation, viruses can be essentially classified based on their genome characteristics and mode of transcription. Developed by Nobel Laureate David Baltimore in 1970, this system groups virus under seven categories:

DNA Virus

The DNA virus is classified as A) class I: dsDNA viruses: viral genome is double-stranded DNA molecule, mRNA is transcribed using RNA polymerase from host or sometimes their own and the negative DNA strand as the template for mRNA. Some of the major family of viruses in this group are adenovirus, herpes virus, papillomavirus, poxvirus and others [18]. B) class II: ssDNA viruses: The virus of this group contains single-stranded DNA as their genome. As the DNA is single-stranded the virus after entering the host cell first synthesises a double-stranded DNA from the existing DNA using DNA polymerase enzyme and then the mRNA is transcribed from the double-stranded DNA like dsDNA viruses. Parvovirus and anelloviridae are some of the important viral families of this group [18].

RNA Virus

The RNA virus is classified as A) class III: dsRNA viruses: The genetic content of this group of virus is a double-stranded RNA molecule. The virus upon entering the cell makes mRNA from the negative strand of the RNA using a RNA-dependent RNA polymerase. This mRNA is further used for other processes like translation and replication to generate the dsRNA genome of the virus. Rotavirus is one such common example of class III type virus [18]. B) class IV: positive-sense ssRNA viruses: viral genome is a single-stranded RNA molecule that has the positive or 'sense' directionality of the RNA sequence, hence, this can readily be translated into viral proteins once inside the cell. Possibly the group with the most number of infectious viruses to human is the common members of this group are hepatitis C, West Nile virus, dengue virus, Zika virus, rhinovirus, coronavirus and the Middle East respiratory syndrome (MERS), severe acute respiratory syndrome (SARS) and severe acute respiratory syndrome coronavirus 2 (SARS-CoV-2) [18]. C) class V: negative-sense ssRNA viruses: As this group of viruses contains the anti-sense strand RNA only as their genome so they require to synthesise the sense mRNA strand from the negative strand first using the RNA-dependent RNA polymerase which can then be used for other processes like translation. Some major pathogenic viruses of this type are Ebola virus, Hantavirus, influenza virus, the Lassa fever virus, and the rabies virus [18].

Reverse Transcribing Virus

The reverse transcribing viruses are classified as A) class VI: RNA viruses that reverse transcribe: These groups of viruses also known as retroviruses have the ability to make an intermediate DNA or RNA from viral RNA or DNA respectively using the viral reverse transcriptase enzyme system. The reverse transcriptase enzyme functions as both RNA-dependent RNA polymerase and then DNA-dependent DNA polymerase that makes a double-stranded DNA molecule from the single-stranded RNA molecule. This type of genome has the advantage of integrating the newly synthesised double-stranded DNA into the host cells' genomic DNA. During the host cells' transcription and replication the viral DNA is subsequently either transcribed into mRNA for further pathogenicity or can be used to replicate and produce multiple copies of the viral genome within the host cells for indefinite times as long as the host survives. The most notable member of this group is the HIV virus and the hepatitis B virus [18]. B) class VII: DNA viruses that reverse transcribe: DNA-RT viruses are a group of viruses that replicate through an RNA intermediate inside the host cell but while still inside the viral capsid. Hence, they require a viral reverse transcriptase enzyme just like RNA-RT viruses. The replicated genome is then transcribed using the host cells' transcription machinery once in the host cytoplasm. A common example of DNA-RT virus is hepadnavirus [18].

Often virus of various categories apart from their direct infectivity and disease induction can have severe consequences from a chronic infection, inflammation or oxidative stress caused in other organs. In recent years an increasing number of cases of occurrence of male infertility in patients suffering from major viral infection are being reported. The major viruses currently known to have a role in affecting male fertility are the classical human immunodeficiency virus, hepatitis C virus, and the more recently Zika virus and the SARS-CoV-2, among others.

13.3 Oxidative Stress and Male Infertility

Oxidative stress is a condition caused due to the increased accumulation of reactive oxygen species (ROS). During oxidative stress, the antioxidant system of the cell is compromised due to the overproduction of ROS. Oxidative stress has often been linked to male infertility. Oxidative stress-induced male infertility is caused by a plethora of factors including invasion of foreign particles like viruses (hepatitis B virus, human immunodeficiency virus, Severe Acute Respiratory Syndrome–Coronavirus-2, etc.), lifestyle-related factors, varicocele, obesity, metabolic disorders, etc. A prime source of ROS generation in sperm includes activated leucocyte in semen and mitochondria in the spermatozoa. These factors lead to the development of free radicals like superoxide anions, hydrogen peroxide and hydroxyl radicals. The free radicals in turn result in the fragmentation of sperm DNA thereby causing apoptosis and in turn male infertility (Fig. 13.2). Sperm capacitation is a process that requires low level of ROS generation in order to execute redox-sensitive processes. However, an augmented level of ROS often disrupts sperm membrane fluidity and permeability [19]. ROS generation was shown to cause axonemal damage. Presence of a single detection and repair mechanism of DNA damage makes spermatozoa more vulnerable to oxidative stress. Oxidative stress often leads to damage of sperm DNA, RNA and telomeres thus resulting in increased infertility [20]. This chapter stresses on the role of viruses in inducing male infertility with a special focus on oxidative stress. Many virus-like hepatitis B virus was shown to cause sperm damage by inducing oxidative stress. Malondialdehyde, a marker for oxidative stress, was found to be augmented in sperm during hepatitis infection [21]. Leucocyte infiltration which causes high oxidative stress was shown to be increased during SARS-Cov2 and Zika infection [22, 23]. HCV was also shown to induce ROS generation with the help of its proteins E1 and NS3 [24]. ROS generation often leads to DNA fragmentation which in turn causes loss in sperm integrity. Spermatozoa with damaged DNA leads to fall in the fertilisation capacity of the sperm and abnormalities in other sperm parameters thus resulting in male infertility. Orchitis is a phenomenon characterised by swollen testis and inflammation of the scrotum. Viral infection often leads to the development of orchitis which in turn leads to the development of oxidative stress due to heightened levels of pro-inflammatory cytokines. Mumps virus infection often leads to the development of orchitis. Mumps virus infection was shown to trigger Toll-like receptor 2 (TLR2)-mediated upregulation of pro-inflammatory cytokines TNF-α, IL-6 and chemokines CXCL10, monocyte chemoattractant protein-1 (MCP-1) [25]. These molecules in turn trigger the generation of oxidative stress leading to increased sperm cell apoptosis. Severe oxidative stress in turn leads to telomere attrition. The build-up of reactive oxygen species due to oxidative stress leads to the shortening of sperm telomere length [26]. HSV infection was shown to cause telomere attrition due to increased oxidative stress. HSV-1 was reported to cause the dissociation of shelterin from sperm telomere DNA. Shelterin is a protein that binds with telomere ends and protects it from DNA repair mechanism. HSV-1-encoded ICP8 protein was shown to cause the generation of single-stranded telomeric DNA which in turn causes loss of sperm telomere repeat DNA and subsequently leads to male infertility [27]. Viral infection also results in impairment of the antioxidant machinery due to excessive production of reactive oxygen species. Infection with SARS-CoV and influenza virus was shown to generate oxidised phospholipids (OxPLs). These viruses also led to the increase in the level of lipid peroxidation breakdown product malondialdehyde which is an indicative of oxidative stress [28]. Thus orchitis and surge in the inflammatory response results in the generation of oxidative stress. Oxidative stress in turn causes a debilitating effect on sperm by increasing sperm DNA fragmentation, mitochondrial DNA damage, telomere attrition and reduced antioxidant machinery thereby resulting in male infertility (Fig. 13.2).

13.4 Viruses and Male Infertility

Many viruses were shown to exert their detrimental effects on male fertility by inducing the generation of oxidative stress (Table 13.1). DNA viruses like human papillomavirus and herpes simplex virus, RNA viruses like SARS-CoV-2, Zika virus and hepatitis C and reverse transcribing viruses like hepatitis B virus and HIV were shown to cause male infertility (Fig. 13.3).

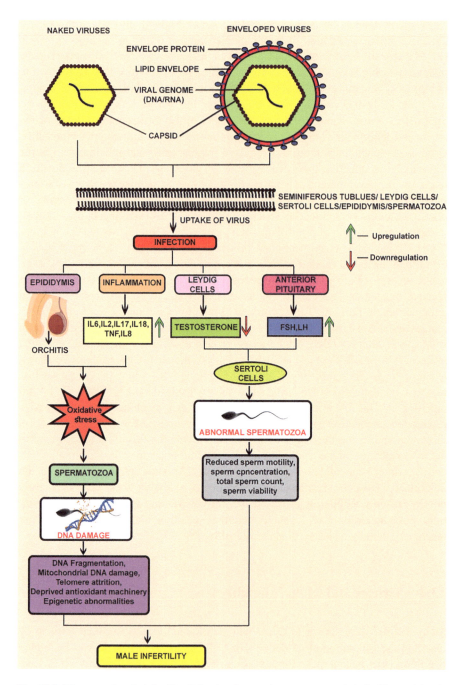

Fig. 13.2 Viruses and male infertility. Infection due to viruses causes male infertility mainly via two routes. The first route is through the development of oxidative stress due to orchitis and surge in the level of inflammatory molecules. Development of oxidative stress causes sperm DNA damage due to DNA fragmentation, mitochondrial DNA damage, telomere attrition and impaired antioxidant machinery. The second route of virus-dependent male infertility is via abnormal hormonal regulation. Decreased testosterone level and increased FSH and LH levels cause change in sperm characteristics leading to male infertility. Sperm motility, viability, total sperm count, and penetration capacity of sperms are decreased due to abnormal hormonal balance

Table 13.1 Adverse effect of viruses in inducing male infertility

Name of virus	Cells affected of male genital tract	Effects on male fertility	References
Human papillomavirus	Spermatozoa, epididymis, testis, vas deference	Asthenozoospermia Generates anti-sperm antibodies that perturb sperm motility, sperm oocyte interaction, disrupts sperm integrity Associates with mitochondrial membrane proteins, complex III, IV and V and increases the generation of ROS. E2 protein modifies the mitochondrial cristae morphology thereby increasing mitochondrial ROS generation. L1, capsid protein of the HPV interacts with AQP8 thus resulting in the inhibition of AQP8-mediated efflux of H_2O_2 from the sperms. It causes build-up of ROS inside the sperms leading to sperm dysfunction	[35] [36] [38]
Herpes simplex virus	Spermatozoa, epididymis, prostrate	Excessive build-up of oxidative stress due to HSV infection causes telomere attrition, a phenomenon that causes shortening of the telomere length Genital herpes, ameliorates sperm motility, motile sperm concentration, total motile sperm count, genital ulcer infection of epididymis and prostate gland	[26] [40] [42] [43] [44]
Hepatitis C virus	Spermatozoa, spermatid	Reduces sperm count, sperm volume, sperm motility and sperm viability, reduces fertility index Induces double-stranded DNA breaks by increasing nitric oxide production, HCV proteins E1 and NS3 increases ROS generation and ablates mitochondrial membrane potential Induces sperm diploidy and random aneuploidy	[13] [24] [55] [57]
West Nile virus	Spermatozoa, lymphocytes, Sertoli cells, interstitial cell, seminiferous tubules	Orchitis, thickens tubular basement of testis, reduces spermatogenesis Chronic inflammatory infiltration of lymphocytes, syncytial Sertoli cells, and multinucleated interstitial cell Causes tubular necrosis of seminiferous tubules	[7] [61]

(continued)

Table 13.1 (continued)

Name of virus	Cells affected of male genital tract	Effects on male fertility	References
Zika virus	Spermatozoa, Sertoli cells, epididymis	Haematospermia, oligospermia, painful ejaculation and penile discharge Leucocytes infiltration of the semen, low sperm concentration and motility, increases expression of FSH and inhibin B in the semen Orchitis and epididymitis	[71] [72] [73]
Severe acute respiratory syndrome virus	Spermatozoa, Sertoli cells, epididymis, Leydig cells, Germ cells, interstitial cells	Germ cell destruction Thickening of the basement membrane and peritubular fibrosis Leucocyte infiltration and vascular congestion of interstitial cells, increase apoptotic spermatogenetic cells Increases expression of $CD3^+$ T lymphocytes and $CD68^+$ macrophages, increases IgG immunoreaction deposits in seminiferous epithelium, interstitium, some degenerated germ cells and Sertoli cells	[23]
Severe acute respiratory syndrome–Coronavirus-2	Spermatozoa	Orchitis Leucocyte infiltration of the testis and atrophy of seminiferous tubules Increases the production of pro-inflammatory cytokines, oxidative stress, decreases sperm motility and spermatogenesis Causes testicular damage due to increased levels of FSH and LH and decreased levels of testosterone SARS-CoV-2 spike protein S binds with the ACE 2 protein This prevents the conversion of angiotensin II thus resulting in increased ROS and NADPH oxidase generation	[12] [23] [87] [89] [90] [94]
Ebola virus	Spermatozoa, epididymis, seminal vesicle, prostate	Erectile dysfunction, decreased libido Thrombocytopenia Intesticular haemorrhage of non-human primates	[105] [106] [107] [108]

(continued)

Table 13.1 (continued)

Name of virus	Cells affected of male genital tract	Effects on male fertility	References
Influenza virus	Spermatozoa	Decreases total sperm count, sperm motility Decreases sperm penetration capacity Develops pulverised chromosomes and displays chromosomal translocation Chromosomes display anomalous metaphases Generates ROS production which leads to the formation of DNA strand breaks thus ultimately leading to apoptosis. Excessive production of ROS due to influenza infection may also lead to the development of sperm DNA breakage thus in turn causing male infertility	[113] [114] [115] [116] [117]
Mumps virus	Spermatozoa, Leydig cells, seminiferous tubules	Gynecomastia Bilateral orchitis, testicular atrophy, oligospermia, hypofertility Mumps orchitis in turn results in the generation of oxidative stress due to accumulation of excessive ROS Decreases testosterone level and increases LH and FSH levels leading to decreased libido and impotence Leucocyte infiltration Germ cell degeneration, seminiferous tubule degeneration	[11] [121] [122] [123] [125]
Hepatitis B virus	Spermatozoa	Decreases sperm viability, alters sperm morphology, lowers fertility index Generates oxidative stress as indicated by increased malondialdehyde levels, ameliorates total antioxidant capacity, decreases matrix metalloprotease level leading to loss of mitochondrial integrity and increased sperm apoptosis Increases level of IL-17 and IL-18	[131] [132] [133]
Human immunodeficiency virus	Spermatozoa, seminal mononuclear cell	Hypogonadotropic hypogonadism Orchitis Azoospermia, hyalinisation of the boundary wall of seminiferous tubules, lymphocytic infiltration of the interstitium Perivasculitis, increases tubular wall thickness of the testes Decreases testosterone levels, increases the level of inflammatory cytokine thus increasing ROS generation	[147] [148] [149]

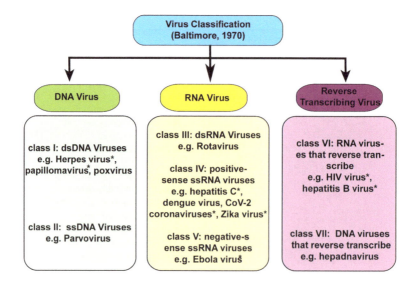

*marked are the viruses involved in male infertility induction in human

Fig. 13.3 Baltimore classification of viruses: Baltimore in 1970 classified viruses into three main groups according to their genetic material which was further subdivided into sub-groups. The first group is DNA virus which is further subdivided into double-stranded DNA (ds DNA) viruses (class I) and single-stranded DNA (ss DNA) viruses (class II). The second group is the RNA virus group which is subdivided into double-stranded RNA (ds RNA) viruses (class III), positive sense single-stranded RNA viruses (class IV) and negative sense single-stranded RNA viruses (class V). The third group is the reverse transcribing virus group. It is further subdivided into RNA viruses that reverse transcribe (class VI) and DNA viruses that reverse transcribe (class VII). *marked viruses like herpes virus, papillomavirus, hepatitis C virus, CoV-2 coronavirus, Zika virus, HIV, and hepatitis C virus was shown to impart debilitating effect on fertility of men giving rise to male infertility

13.4.1 Human Papillomavirus (HPV)

Human papillomavirus is a small, non-enveloped virus having a diameter of 52-55nm. Its genome contains a single double-stranded DNA molecule of about 800 bp in length that is bound to cellular histones. The genome is contained in a capsid composed of 72 capsomer proteins. The capsid consists of two structural proteins namely LI and L2 which are encoded by the virus itself. The genome consists of eight ORFs transcribed from a single strand of DNA. It is functionally divided into three types namely early (E) region encoding protein E1-E7 required for viral replication, late (L) region encoding structural proteins L1-L2 required for assembly of the virus and long control region (LCR) which is the non-coding region [29].

Human Papillomavirus (HPV) and Male Infertility

Human papillomavirus has been extensively tested in the prostate gland of males and is thought to be one of the main reasons for prostate cancer. In one study, Ostrow et al showed the presence of papillomavirus in the semen of three patients. Two of them had severe chronic wart disease. This study reported the notion that semen can act as an agent for sexual transmission of HPV [30]. Several studies have highlighted the presence of HPV in various male genital parts like epididymis, semen, testis and vas deference [31–34]. In patients with human papillomavirus-infected semen, the incidence of asthenozoospermia is considerably higher [35]. HPV was shown to induce infertility in men by generating anti-sperm antibodies that perturb sperm motility, sperm oocyte interaction and other sperm parameters [35]. It was also reported that HPV disrupts sperm integrity. HPV E2 protein was shown to induce the generation of ROS. It was shown to associate with mitochondrial membrane proteins, complex III, IV and V and augment the generation of ROS. E2 protein was also shown to modify the mitochondrial cristae morphology thereby increasing mitochondrial ROS generation [36]. HPV E6 protein was shown to decrease the levels of antioxidant proteins SOD 2 and GPx ½, thus increasing the generation of ROS [37]. Thus HPV infection disrupts the cellular redox balance which is sensed by redox sensor apoptosis signal-regulating kinase 1 (Ask1) thus in turn increasing apoptosis by upregulating the c-Jun N-terminal kinase (JNK) pathway. Aquaporins (AQP) are water channels that are present extensively in the sperms. They were shown to eliminate ROS by transporting hydrogen peroxide (H_2O_2) out of the spermatocyte. AQP3 is localised in sperm tails, head and midpiece whereas AQP8 is localised in the midpiece. L1, capsid protein of the HPV, was shown to interact with AQP8 thus resulting in the inhibition of AQP8 mediated efflux of H_2O_2 from the sperms [38]. This results in the build-up of ROS inside the sperms leading to sperm dysfunction.

13.4.2 Herpes Simplex Virus (HSV)

Herpesviruses have a unique four-layered structure: a core containing the large, double-stranded DNA genome is enclosed by an icosapentahedral capsid which is composed of capsomers. The capsid is surrounded by an amorphous protein coat called the tegument. It is encased in a glycoprotein-bearing lipid bilayer envelope. The size of the genome is approximately 240kbp. The nucleocapsid is embedded in a proteinaceous layer termed tegument. The viral envelope is embedded with viral glycoproteins B, glycoprotein C and glycoprotein D. They aid in viral attachment and entry into the host system [39].

Herpes Simplex Virus (HSV) and Male Infertility

HSV was shown to infect human spermatozoa. Herpes simplex virus was shown to transmit due to sexual intercourse [40]. Kotronias and co-workers demonstrated the connection of HSV and male infertility by in situ hybridisation technique [41]. They collected 60 sperm samples from men attending a clinic due to fertility issues. They detected HSV DNA in 37 of the 60 semen samples. HSV-1 was predominant in 21 cases and HSV-2 in 16 cases. The presence of HSV was accompanied by lower sperm count and sperm motility. HSV-2 was shown to be present in the genital tract and is one of the causative agents of genital herpes [40]. Csata et al. detected HSV in about 40% of patients suffering from infertility and the virus was mainly present in latent form in about 60% of sperms [42]. HSV infection was reported to have debilitating effects on male accessory sex organs and sperm parameters [43]. HSV imparts its harmful effects on fertility parameters of men by inducing oxidative stress. Excessive build-up of oxidative stress causes telomere attrition, a phenomenon that causes shortening of the telomere length [26]. Infection of HSV-1 was shown to cause sperm telomere attrition by dissociating shelterin protein from telomere DNA thus resulting in its gradual telomere degradation by DNA repair machinery and in turn increasing male infertility.

13.4.3 Hepatitis C Virus (HCV)

HCV belongs to the family Flaviviridae. It is a small enveloped RNA virus. The genome of HCV consists of single-stranded RNA with positive polarity. The genomic RNA is packed by the core protein and the complex is enveloped by a lipid bilayer. The lipid bilayer contains two viral glycoproteins E1 and E2. The HCV genome consists of three distinct regions namely a 5' untranslated region, a long ORF consisting of more than 9000 nucleotides and a 3' untranslated region. The ORF encodes a polypeptide that is arranged in the order of NH_2-C-E1-E2-p7-NS2-NS3-NS4A-NS4B-NS5A-NS5B-COOH. It is further processed co-translationally and posttranslationally by viral and host proteases [44].

Hepatitis C Virus (HCV) and Male Infertility

Many studies done so far have indicated the presence of HCV in the semen of infected males. In one study, Kotwal and Baroudy showed the presence of HCV antigen in the semen of infected patients [45]. Liou et al in another study showed the presence of HCV RNA in the semen of infected males by nested polymerase chain reaction. In 34 patients positive for HCV, 24% (4 of 17) showed the presence of HCV RNA in the semen. The patients do not have any history of parenteral exposure thus highlighting the possibility of sexual transmission of HCV virus [46]. Another study reported the presence of HCV RNA in supernatants of sperm and spermatid [47]. Although many studies highlighted the presence of HCV virus in the semen of infected patients, few other studies contradicted this notion [48–50]. Another study by Fiore et al showed the presence of HCV RNA in semen, however its prevalence was found to be low. One out of 11 patients was found to be positive for HCV RNA in the semen [51].

The sexual transmission of HCV virus was further corroborated by Zhao et al where they showed that heterosexual individuals with multiple partners have a higher prevalence of HCV infection as compared to healthy women with stable partners [52]. Another study confirmed the notion of sexual transmission of HCV. Nucleotide sequencing of HCV genome from 35 spouse pairs revealed that 33 (94%) of them have the same HCV genotype. Thirty couples (90%) showed 100% sequence homology and two had a high homology (>97%). Among the non-spouse pair studied, 27% showed same genotype and about less than 88% patients displayed low nucleotide homology [53, 130].

A study conducted by Hofny et al. showed 57 males with chronic HCV infection displayed decreased sperm count, sperm volume and sperm motility as compared to healthy controls. These males also displayed abnormal sperm morphology. The total testosterone level was also found to be lower as compared to control [13]. Lorusso et al. showed ameliorated sperm viability, sperm motility and increased deformation of sperms in patients with HCV infection [54]. Fluorescence in situ hybridisation of HCV-infected sperm nuclei revealed diploidy. The fertility index was reduced in HCV-infected males. Their sperms also displayed increased apoptosis and necrosis [55]. Machida et al. reported that HCV infection induces double-stranded DNA breaks by inducing nitric oxide (NO) production. They also reported increased ROS generation and ablated mitochondrial membrane potential due to HCV infection. The HCV proteins E1 and NS3 were found to be the main inducers of ROS generation [24]. HCV infection often leads to alteration of redox state of the cells due to the elevated levels of ROS. It was reported that infection with HCV resulted in increased NADPH-oxidase and reduced glutathione (GSH) levels which are indicative of oxidative stress [56]. HCV infection was also to increase random aneuploidy in infected patients [57].

13.4.4 West Nile Virus

The West Nile virus belongs to the family Flaviviridae. It is a positive-stranded RNA virus. Its genome is approximately 11kb in length and consists of a single ORF flanked by 5' and 3' untranslated regions. The virus consists of an envelope surrounding an icosahedral nucleocapsid of size 50nm. The 3000 amino acid polyprotein is cleaved into ten proteins by proteases. Three proteins constitute the structural components required for virus formation namely capsid protein (C), premembrane protein (prM) and envelope proteins (E). The seven other proteins are non-structural proteins (NS) required for replication of the viral genome [58].

West Nile Virus and Male Infertility

Till date very few studies have been done highlighting the presence of West Nile virus in the semen. Siemann et al have reported that primary human Sertoli cells (HSeCs) infected with West Nile virus have virus titres equivalent to that of Zika virus, which is known to have high infectivity in the semen [59]. Gorchakov and co-workers examined four semen samples for the presence of West Nile virus. One among the four semen samples was found to be positive with West Nile virus [60]. Another case study highlighted the association of West Nile virus outside the central nervous system. Immunohistochemistry of six post-mortem tissues of patients deceased due to West Nile virus infection revealed the presence of viral antigen in the testis and prostate gland of one out of six people. The infection in the testis was associated with inflammation.

Till date only one study has highlighted sexual transmission of West Nile virus. Kelley et al have shown a healthy middle-aged woman developed meningoencephalitis due to West Nile virus, two weeks after having unprotected sexual intercourse with her husband [61]. A study conducted by Smith et al reported orchitis in a post-mortem sample of a 43-year-old man with renal allograft who died of West Nile virus [7]. The testis sections were marked with thickened tubular basement. They also observed reduced spermatogenesis. They also observed chronic inflammatory infiltration of lymphocytes, syncytial Sertoli cells, and multinucleated interstitial cells. Chronic inflammatory infiltration leads to the development of severe oxidative stress. Tubular necrosis with karyorrhectic material in the seminiferous tubules was reported. West Nile virus infection was shown to induce the expression of heme oxygenase-1 (HO-1) and inducible NO synthase (iNOS), which are indicative of increased oxidative stress.

13.4.5 Zika Virus

Zika virus is a flavivirus first identified in febrile rhesus macaque monkeys in the Zika Forest of Uganda in the year 1947. Later it was also isolated from *Aedes africanus* mosquitoes from the same forest. Zika virus belongs to the family *Flaviviridae*. It is a positive-sense single-stranded RNA virus. Zika virus genome contains 10794 nucleotides and its open reading frame (ORF) encodes 3419 amino acids. The genome length of Zika virus is approximately 11kb. The genome encodes for a single polypeptide that is composed of structural proteins capsid (C), pre-membrane (prM) and envelope (E) and seven non-structural proteins namely NS1, NS2a, NS2b, NS3, NS4a, NS4b and NS5. The structural proteins aid in viral particle formation and the non-structural proteins aid in viral replication [62].

Zika Virus and Male Infertility

In a study carried out by Foy et al, they have reported that Zika virus present in the semen can be transmitted sexually [63]. In their study, two American scientists contacted the virus while they were staying at Senegal in 2008. The wife of one of the scientists also contacted the virus after his return to America. The most probable route for the transmission was thought to be through sexual route. In a study carried out by Mead et al, they have shown that Zika virus is often detected in the semen sample of symptomatic men [64]. They screened a total of 1327 semen samples from 184 men infected with Zika virus. They found 33% of the semen sample to be positive for Zika RNA, including 61% of those men who tested within one month of the disease onset. Infectious Zika virus was detected in 15% of Zika RNA positive semen samples, all obtained within one month of disease onset. A study conducted by Mansuy and colleagues reported Zika virus RNA to persist longer in the semen than in blood and urine in most of the studied cases [65]. A longitudinal follow up of a 32-year-old man with Zika RNA positive semen after his return from French Guyana revealed that the viral RNA persisted in the semen for 141 days after onset of symptoms. A study on five other symptomatic men showed Zika RNA to be present in the semen sample of two out of six men till days 69 and 115 after the onset of symptoms. Confocal and stimulated emission depletion microscopy revealed the presence of Zika virus antigen in the sperm of infected men. Since Sertoli cells extensively express Axl receptor which facilitates Zika virus entry, they hypothesised that Zika virus is transmitted to the spermatozoa through Sertoli cells as the spermatozoa lack receptor Axl. Zika virus RNA was shown to be present in the semen up to 188 days after onset of disease symptoms [66].

The presence of Zika virus RNA in the semen for a relatively longer period of time as compared to the other body fluids suggest that the male genital tract may serve as a potential reservoir for the virus. Zika virus was shown to be sexually transmitted from male to female, male to male and female to male both from semen as well as saliva. Hence Zika virus was shown to be transmitted during both anal as well as vaginal intercourse [67]. Counotte et al reported that Zika virus is more plausibly transmitted from men to women rather than women to men. A living systematic review conducted by them revealed that the median duration of the presence of Zika virus RNA in the semen is 40 days with a maximum of 370 days [68]. They reported that several factors are responsible for the persistence of Zika virus in the semen. Men who ejaculate more than 4 times a week was able to clear the viral RNA in more than 21 days earlier. Other factors that play role in determining the persistence of Zika in the semen are increased age, absence of joint pain and conjunctivitis. Zika virus was shown to replicate in other parts of male genital tract as vasectomised men displayed Zika virus RNA in bulbourethral glands, prostate and seminal vesicles [69].

A prospective cohort study was conducted by Huits et al in Belgian travellers with confirmed Zika virus infection who returned to America during the 2016 pandemic [70]. Ten of the 15 participants semen displayed haematospermia and 6 of them showed oligospermia. 11 participants showed the presence of leucocytes in the semen sample. Paz-Bailey et al observed painful ejaculation and penile discharge in Zika virus-positive men [71]. In an observational study by Joguet et al, semen samples were collected from acute Zika-infected patients at Pointe-à-Pitre University Hospital in Guadeloupe, French Caribbean [72]. Semen was collected at days 7, 11, 20, 30, 60, 90 and 120 post onset of symptoms. The study showed increased expression of follicle-stimulating hormone and inhibin B in the semen samples. The motile sperm count was found to be 50% lower on day 60 as compared to day 7. It reverted to normal by day 120. In a case study conducted by Avelino-Silva et al, burning sensation while urinating was reported in two of the total men [73]. One of them showed haematospermia. Among them cases of six participants were carried on forward. One out of six patients had a lower sperm concentration whereas 2 other men had concentration near the lower reference limit. Abnormal sperm motility was seen in all of the three participants.

Orchitis and epididymitis were shown to hamper testicular and epididymal function in a murine model of Zika virus infection. A study carried out by Governo et al showed abrogated spermatozoa number and sperm motility in Zika-infected male mice [8]. Zika virus infection was shown to infect Sertoli cells in both humans and mice [74, 75]. Orchitis and epididymitis in turn lead to the development of oxidative stress which may lead to male infertility. Zika infection was shown to augment the generation of ROS, malondialdehyde all of which are indicative of increased oxidative stress.

13.4.6 Severe Acute Respiratory Syndrome (SARS) Virus

The SARS genome consists of positive-sense single-stranded RNA. The genome is about 30 kb in length. The size of the virus ranges from 80 to 90nm. The viral envelope consists of lipid bilayer in which the spike (S), envelope (E) and membrane (M) proteins are embedded. Inside the envelope, the viral nucleocapsid is present. The virus consists of 14 ORF's which can overlap sometimes. The genome contains 5' methylated cap and 3' polyadenylated tail. ORFs 1a and 1b encode for replicase/transcriptase. ORF2, ORF4, ORF5 and ORF9a encode for proteins spike, envelope, membrane and nucleocapsid respectively. ORFs 3a to 9b encode for accessory proteins [76].

Severe Acute Respiratory Syndrome (SARS) and Male Infertility

The studies related to the presence of SARS virus in the semen was initiated by scientists after the discovery that angiotensin-converting enzyme 2 (ACE-2), a functional receptor of SARS virus, was expressed in the testicular tissues [77]. Zhao et al, while studying the autopsy sample by in-situ hybridisation and electron microscopy of 6 peoples who died of SARS virus, found the presence of SARS virus in testicular epithelial cells and Leydig's cells [78]. However, no further studies were able to reflect similar results. Ding and co-workers studied four autopsy samples of patients who died of SARS virus. In spite of the expression of high levels of ACE-2 in the testis of the samples, none of them was found to be positive for SARS RNA as well as N protein [79]. Another group of scientists examined eight autopsy samples of people with SARS infection. They reported testicular atrophy in the samples, however the parenchymal tissues were found to be SARS negative [80].

Being a respiratory virus, SARS is mainly transmitted through aerosols and droplets. No sexual transmission of SARS virus has been reported so far.

Xu et al reported the involvement of SARS virus with male infertility [23]. They examined autopsy samples of testis from Ditan Hospital, Beijing, China. All the samples were collected from people who were deceased due to SARS virus infection. All the autopsy samples displayed germ cell destruction. The tissue samples displayed thickening of the basement membrane. They also observed peritubular fibrosis. Leucocyte infiltration and vascular congestion which are characteristics of inflammatory response was observed in interstitial cells. Increased inflammatory response might result in increased oxidative stress. They also reported accentuated levels of apoptotic spermatogenetic cells in SARS-infected samples as compared to control samples which might be due to increased ROS generation leading to sperm DNA damage. The SARS-positive samples displayed increased expression of $CD3^+$ T lymphocytes and $CD68^+$ macrophages compared to control. In the SARS testis, extensive IgG immunoreaction deposits were observed in the seminiferous epithelium, interstitium, some degenerated germ cells and Sertoli cells.

13.4.7 Severe Acute Respiratory Syndrome–Coronavirus-2 (SARS-CoV-2)

The World Health Organisation in December 2019 declared about the outbreak of SARS-CoV-2 in Wuhan, Hubei Province, China. SARS-CoV-2 is a novel coronavirus causing the disease named coronavirus disease-19 (COVID-19) [81]. Later in March 2020, the World Health Organisation informed COVID-19 to be a global pandemic [82]. The SARS-CoV-2 belongs to the coronavirus family having a beta-coronavirus 2B lineage [83]. SARS-CoV-2 is closely related to virus strain BatCov RaTG13, with a sequence similarity of 96%. The genome size of coronavirus is about 27-32 kb. Its genome consist of single-stranded positive-sense RNA. The SARS-CoV-2 consists of four structural proteins namely spike protein (S), nucleocapsid protein (N), membrane protein (M), envelope protein (E) and 16 non-structural proteins (nsp1-nsp16). The two open reading frames (ORF) that distinguish SARS-CoV-2 from other viruses are ORF8 and ORF10. Of the total 103 SARS-CoV-2 genomes studied by reasearchers, they found 149 mutations. Among the 103 strains identified, 101 displays notable linkage between two single nucleotide polypeptides (SNPs). Majority of the SARS-CoV-2 types are the L-type and S-type. They are distinguished by the presence of two SNPs at sites 8,782 and

28,144. Among the 103 strains identified, the L type accounts for 70% and the S type for 30% [84]. The COVID-19 pandemic has stirred havoc in the current scenario. Although it infects people of all age groups, older people and people with co-morbidities are at a higher risk. Significant number of the deaths have occurred in individuals with age 65 and above.

Severe Acute Respiratory Syndrome–Coronavirus-2 (SARS-CoV-2) and Male Infertility

Few studies have claimed about the presence of SARS-CoV-2 in the semen. One such observational study carried out by Li et al with males having a history of COVID-19 and aged 15 years and older in Shangqiu Municipal Hospital. Initially, they identified 50 patients, however 12 of them were unable to provide semen samples due to erectile dysfunctions. Among the remaining 38 patients whose semen were collected for further testing, 23 people (60.5%) had already recovered and 15 people (39.5%) were in the stage of acute infection due to the virus. Semen from 6 patients (15.8%) were found to be positive for SARS-CoV-2 which included 4 of the 15 patients (26.7%) [85]. Massarotti et al reported the presence of SARS-CoV-2 in the semen of six male patients including two participants still recovering from the infection [86]. These reports were however contradicted by reports from various other studies indicating the absence of SARS-CoV-2 in the semen of infected males. Pan et al in their observational study with 34 male participants diagnosed with COVID-19 reported the absence of SARS-CoV-2 in their semen after a median of 31 days follow-up from diagnosis of the disease [87]. However, 6 patients (19%) reported scrotal discomfort due to orchitis. Another study by Song et al also corroborated the absence of SARS-CoV-2 in the semen of patients with COVID-19 disease. Paoli and co-workers also showed a similar result. They tested semen and urine samples of a 31-year-old man diagnosed with COVID-19 for the presence of viral E and S genes. Both the semen and urine sample was found to be negative for SARS-CoV-2 [88]. Due to discrepancy in the results obtained by different groups as well as a small sample size used in most of the studies, further research will be essential before reaching any conclusions.

Transmission of the SARS-CoV-2 through sexual route has not been studied extensively till date. Several discrepancies regarding the presence of SARS-CoV-2 in the semen sample have made it more elusive regarding the transmission of the viral particles through sexual route. Moreover, due to the high rate of infectivity of the SARS-CoV-2, it is always advisable to maintain caution during any intimate contact in order to prevent virus spread. SARS-CoV-2 is mainly known to be transmitted through respiratory droplets, mucus and faecal matters.

In a study by Pan et al with 34 male participants having SARS-CoV-2 infection, 6 of them (19%) reported scrotal discomfort around the time of the viral infection. The researchers concluded that the discomfort may be due to the occurrence of viral orchitis due to the infection [87]. Chen et al performed a structural analysis of receptor-binding domain (RBD) of the spike proteins present in SARS-CoV-2.

Molecular modelling studies revealed that RBD has strong interaction with angiotensin-converting enzyme 2 (ACE2). The RBD of SARS-CoV-2 has a definite loop with flexible glycyl residues. The presence of a unique phenylalanine F486 in the distinct loop permits it to penetrate deep into the hydrophobic pocket of ACE2. The presence of ACE2 in the testis, Leydig cells, renal tubular cells and intestines makes them a possible route for the transmission of the virus. The presence of ACE2 in the male genital tract also raises possibilities that the virus may have a debilitating effect on the male genital organs thus leading to infertility via the increased oxidative stress route. SARS-CoV-2 spike protein S was shown to bind with the ACE 2 protein and enter the sperm. This prevents the conversion of angiotensin II (Ang II) to angiotensin 1-7 due to the absence of free ACE2. Ang II in turn binds the angiotensin type 1 receptor (AT1R) and increase the production of ROS and NADPH oxidase [89]. Thus SARS-CoV-2 may cause sperm DNA damage and in turn result in infertility due to oxidative stress generation. The presence of CD147 receptors on the spermatogonial stem cells in the testis also raises a possibility of SARS-CoV-2 entry through it into the male genital tract [90]. Xu et al provided evidence of leucocyte infiltration in the testis as well as atrophy of seminiferous tubules in patients with SARS-CoV-2 infection [23]. A study by Shi et al reported a burst of pro-inflammatory cytokines in patients with SARS-CoV-2 infection [91]. The use of antiviral drug Ribavirin during SARS-CoV-2 infection was shown to augment oxidative stress, reduce the testosterone levels and impair spermatogenesis [12]. The increase in oxidative stress and pro-inflammatory cytokines was shown to have detrimental effects on sperm motility and spermatogenesis [92, 93]. Surge in inflammation level results in the production of ROS which in turn can lead to sperm DNA damage. Infection with SARS-CoV-2 was shown to produce oxidised phospholipids (OxPLs), which in turn leads to increased levels of lipid peroxidation breakdown product malondialdehyde which is an indicative of oxidative stress. The increased levels of FSH and LH which serve as markers of testicular damage during SARS-CoV-2 infection also raise the possibility that SARS-CoV-2 has a detrimental effect on the male genital system which could lead to infertility [94]. The low level of testosterone could also lead to the development of erectile dysfunction and infertility in males infected with SARS-CoV-2 [95]. Increasing evidence regarding the surge of cytokine productions as well as hormonal dysregulation during SARS-CoV-2 has made a possibility of a link between SARS-CoV-2 infection and male infertility. Hence, an extensive study in this respect is required in order to combat both the viral spread as well the detrimental effects it imparts on the male genital system.

13.4.8 *Ebola Virus*

Ebola virus belongs to the family Filoviridae. They are known to be highly pathogenic causing haemorrhagic fever [96]. It consists of a negative sense single-stranded RNA genome enclosed within a helical nucleocapsid. The RNA genome is non-segmented and 18-19 kb in size. The virus lacks 5' capping and 3' polyadenylation. The virus contains virally encoded glycoproteins that project as 7-10 nm long spikes from the lipid surface. They are filamentous viruses. The Ebola virus contains seven viral proteins namely NP (nucleoprotein), VP35, VP40, GP (glycoprotein), VP30, VP24 and an RNA-dependent RNA polymerase (L) [97].

Ebola Virus and Male Infertility

Numerous studies on the magnitude of viral ribonucleic acid (RNA) persistence in the body fluid after Ebola infection revealed the presence of viral RNA for a long time in males infected with the virus. Another study conducted by Liberia's Men's Health Screening Program (MHSP) revealed the presence of viral RNA in the serum of about 9% of Ebola survivors among the total number of participants studied. Among the total participants, 63% of males were found to have persistent Ebola virus RNA in their semen even after 12 months or more after disease recovery [98]. The longest period of time for the persistence of Ebola virus RNA in the semen of infected men after release from the Ebola treatment unit was shown to be 565 days [98]. It was also reported that men having age greater than 40 years are more likely to have a positive viral RNA in the semen as compared to younger men [98]. A 40-month follow-up of Ebola survivors in Guinea (PostEbogui) using the sensitive Ebola Xpert assay revealed the persistence of Ebola virus RNA in all the males studied for up to 3 months and 70% of the males had viral RNA in the semen for about 6 months [99]. Another report from Guinea revealed the presence of Ebola virus RNA in the semen of 1 survivor for up to 531 days [100]. However, Diallo et al showed that the persisting virus was different from the virus that affected the survivor initially. It was shown that the virus that persisted was transmitted sexually and it gave rise to a new strain of Ebola virus disease in Guinea and Liberia [100]. The persisting virus was transmitted about 470 days after the onset of symptoms. A longitudinal and modelling study by Sissoko et al revealed that 50% of the male survivors clear viral RNA from their semen at 115 days and 90% of men clear viral RNA at 294 days [101]. The longest period of time for the persistence of Ebola virus RNA in the semen was reported to be 965 days [102].

The presence of Ebola virus RNA was also detected in gonads of animals like macaque and rhesus monkey. A study conducted by Perry et al showed that four females and eight male macaques that perished of Ebola between 6 and 9 days after virus infection displayed Ebola infection in testis, seminal vesicle, epididymis, prostate gland, uterus and ovary [103]. The presence of Ebola virus in the reproductive tract led scientists to speculate the possibility of sexual transmission of Ebola virus. A study conducted by Zen et al reported Ebola virus replication harmonises with systematic inflammatory response in asymptomatic rhesus monkey that did not succumb to Ebola infection [104]. Macrophages were found to be the reservoirs of Ebola virus in the lumen of epididymis of these rhesus monkeys.

Although the mechanism of action of Ebola virus in connection to male infertility is not well comprehended, but few reports have linked Ebola virus disease to male infertility. A study conducted by Wadoum et al at Ebola survivor Mobile Health Clinic (MHC) in Sierra Leone between 7 February 2015 and 10 June 2016 reported 5.1% male survivors having erectile dysfunction and 10.2% male survivors having loss of sexual desire [105]. Another study in Liberia reported the virus to cause erectile dysfunction and decreased libido in male survivors [106]. In another study, immunohistochemical staining of Ebola virus antigen revealed the presence of fibrin clots in the blood vessel of testis of infected males. This led to the conclusion that thrombocytopenia might be a plausible outcome of platelet consumption due to fibrin clots [107]. Infection of non-human primates with the virus resulted in testicular haemorrhage [108]. Ebola virus infection was shown to cause increased levels of ROS, pro-inflammatory cytokines, chemokines which aid in the pathogenesis of Ebola [109]. Increased ROS generation in turn might lead to sperm DNA breakage and finally male infertility.

13.4.9 Influenza Virus

The influenza virus belongs to the family Orthomyxoviridae. The influenza virus is roughly a spherical virus of about 80-120 nm in diameter. It is an enveloped virus and its outer layer which is composed of lipid membrane is derived from the host. The influenza genome segmented and is organised into 8 pieces of single-stranded RNA (Influenza C has 7 pieces of RNA instead of 8). The RNA along with the nucleoprotein is packed in a helical ribonucleoprotein form. Glycolipids constitute the spikes which are embedded in the lipid membrane. The glycoproteins consist of proteins linked to sugars like HA (hemagglutinin) and NA (neuraminidase). About 80% of the spikes contains HA and the rest 20% contains NA.

Influenza Virus and Male Infertility

The presence of influenza virus in semen or their sexual transmissions has not been reported yet. Though no studies done so far have shown the presence of influenza receptor on human genital tract, a few studies have shed light on the presence of influenza receptors on the genital tract of turkeys [110, 111]. Sergerie et al carried out a study in which they reported parameters in semen sample and sperm deoxyribonucleic acid integrity following febrile influenza. By using sperm chromatin structure assay (SCSA) they reported decreased total sperm count at days 15, 37 and 58 after a 2-day fever of 39°C to 40°C. The count returned to normal after the fever subsided. Sperm DNA fragmentation measured by terminal deoxynucleotidyl transferase dUTP nick end labelling (TUNEL) assay increased from 17% before fever to 23% at day 15 which are indicative of increased ROS production. Sperm motility was remarkably abrogated at days 15 and 37 after the fever and bounced back to normal by day 58. High DNA stainability notably increased at day 37 after the fever [112]. Another study highlighted the variation in sperm penetration rate after febrile viral illness. By using sperm penetration assay, this study documented decreased sperm count and reduced sperm capacitation after febrile viral illness. The parameters like motility were reverted to normal at 4 weeks and sperm count was recovered at 8 weeks [113]. Another study by MacLEOD reported abrogated sperm motility and sperm count after febrile viral illness [114]. Kantorovich et al reported human diploid cell culture administered with influenza virus developed pulverised chromosomes as well as the displayed chromosomal translocation [115]. Influenza virus infection was shown to augment the generation of ROS which in turn induced RANTES production. The generation of ROS also leads to the formation of DNA strand breaks thus ultimately leading to apoptosis. Excessive production of ROS due to influenza infection may also lead to the development of sperm DNA breakage thus in turn causing male infertility.

In a murine model of influenza, Sharma and co-workers intraperitoneally injected mice with two doses of 1 and 2 hemagglutinating units (HA) of influenza A2 Hong Kong/68 virus [116]. The mice were shown to develop chromosomal aberrations. Mice administered with 1 and 2 HA U influenza virus displayed anomalous metaphases. Aneuploids percentile in mice inoculated with 1 and 2 HA virus were shown to plummet until 24 weeks post infection after an initial fluctuation between 1 and 4 weeks [116]. The euploidy percentage was also higher in virus-infected mice. X-Y dissociation in spermatocyte led to sterility in male mice. Inoculation of mice with influenza A2 Hong Kong/68 virus led to pulverisation [116]. Another study by Pathki and co-workers demonstrated that inoculation of male mice with 16HA of ultraviolet (UV) inactivated purified X-31 strain of influenza virus resulted in dominant lethal mutations in mice [117]. It was concluded that not only live virus but also inactivated virus gives rise to dominant lethality.

13.4.10 Mumps Virus

Mumps virus belongs to the Paramixoviridae family. It consists of a linear single-stranded RNA genome. The genome is surrounded by symmetric repeating protein units. The mumps virus is an irregular-shaped virus with a diameter ranging from 90 to 300nm. The virus contains an envelope that is three-layered and encloses nucleocapsid. The outer surface of the envelope consists of glycoproteins containing HA (hemagglutinin) and NA (neuraminidase). The middle layer consists of lipid bilayer and the innermost layer is a non-glycosylated membrane protein that forms the outer structure of the virus. The mumps genome codes for eight proteins namely haemagglutinin-neuraminidase protein (HN), fusion protein (F), nucleocapsid protein (NP), phosphoprotein (P), matrix protein (M), hydrophobic protein (SH), and L protein [118].

Mumps Virus and Male Infertility

Though no mechanistic evidence relating to mumps virus infection and male infertility have been deciphered so far, yet many reports have shown males having fertility issues after exposure to mumps virus infection. Aiman and co-workers first reported gynecomastia in males infected with mumps virus [119]. Mumps virus infection led to the development of bilateral orchitis which resulted in testicular atrophy. These men displayed decreased testosterone production and increased luteinising hormone (LH) and follicle-stimulating hormone (FSH) levels production thus highlighting the adverse effect of mumps virus on Leydig cell function. These men also displayed decreased libido and impotence. Another study showed mumps virus-induced bilateral orchitis that manifests in increased testicular pain and swelling thus leading to testicular atrophy and in turn infertility [120]. Bilateral orchitis was reported to induce oligospermia and testicular atrophy in 13% of the cases and it lead to the development of hypofertility [11]. Mumps associated with orchitis resulted in oedema of the interstitial tissues leading to congestion of blood vessel and accumulation of lymphocytes. As a consequence, the permeability of the blood vessels increases leading to interstitial haemorrhage [121, 122]. This also led to the extravasations of fibrin and leucocyte. The seminiferous tubules also degenerated. Repair of the scarring caused by the infection often leads to the deposition of collagen within the interstitial tissue resulting in fibrosis and atrophy. Mumps orchitis in turn results in the generation of oxidative stress due to the accumulation of excessive ROS. Generated ROS often results in sperm DNA fragmentation and apoptosis. Mumps virus infection was shown to trigger upregulation of pro-inflammatory cytokines TNF-α, IL-6 and chemokines CXCL10, (MCP-1) which in turn results in the generation of oxidative stress [26]. Mumps virus was also shown to have many indirect effects on germ cells. One of them is an increase in body temperature resulting from the viral infection thus leading to testicular dysfunction and in turn germ cell degeneration [123]. Another consequence is that of seminiferous tubule degeneration due to interstitial oedema thus leading to germ cell degeneration [124]. Adamopoulos and co-workers reported mumps virus to have an adverse effect on Leydig cells and testicular functions. They observed a dip in the testosterone level with a concomitant increase in luteinising hormone (LH) and follicle-stimulating hormone (FSH) levels in 27 patients infected with the virus [125].

13.4.11 Hepatitis B Virus (HBV)

HBV, a prototype of the Hepadnaviridae family, has been shown to infect approximately 2 billion people worldwide. More than 350 million people were shown to be chronic carriers of the virus [126]. According to the Global Burden of Disease study, the total mortality due to hepatitis B was estimated to be approximately 786000 in the year 2010 [127]. The virus can be transmitted through various sources including mother to child transmission during birth and delivery, blood or body fluids, sex with an infected partner, infected needles and syringes, etc. 257 billion people were estimated to be infected with HBV (defined as hepatitis B surface antigen-positive) in the year 2015 by World Health Organisation. In 2016 it was reported by World Health Organisation that of the more than 250 million people living with HBV, about 27 million people were aware of the infection and about 4.5 million people were on treatment.

HBV is classified into eight genotypes A to H each having a distinct geographical distribution. It consists of a circular, partially double-stranded DNA. Electron microscope shows three types of virus particles. Two of the virus particles consist of a small spherical structure with a diameter of 20nm and filament of width 22nm which are composed of hepatitis B surface antigen (HBsAg) and host-derived lipids. They are devoid of viral nucleic acid and are therefore non-infectious [128]. The third particle termed as the Dane particle is the infectious one having a spherical, double-shelled structure which is 42nm in diameter. The Dane particle consists of inner nucleocapsid composed of hepatitis B core antigen (HBcAg). The core is surrounded by lipid envelope containing HBsAg. The core is complexed with virally encoded polymerase and viral DNA genome. The genome of HBV is partially double-stranded and it consists of circular DNA of about 3.2 kilobase pairs. The HBV genome encodes four overlapping open reading frames (*S*, *C*, *P*, and *X)* [129]. The *S* ORF encodes for the viral surface envelope proteins HBsAg. The *C* ORF encodes either the viral nucleocapsid HBcAg or hepatitis B e antigen (HBeAg). The *P* ORF encodes for the polymerase protein. The polymerase protein is functionally divided into three domains namely the terminal protein domain, the reverse transcriptase domain and the ribonuclease domain. The *X* ORF encodes a 16.5-kd protein hepatitis b x antigen (HBxAg).

Hepatitis B Virus (HBV) and Male Infertility

HBV has been shown to be associated with male infertility. Recent studies have indicated that HBV infection results in increased male infertility. HBV was shown to pass the blood-testis barrier [130]. Males infected with HBV displayed lower fertility index [131]. In males infected with HBV, the sperm concentration was found to be significantly lower as compared to control [132]. Males infected with HBV have lower sperm viability as well as they exhibit morphological differences compared to the control [132]. A recent study has shown the involvement of hepatitis B virus S (HBs) protein in male infertility. Exposure of sperm cells with HBs resulted in the generation of oxidative stress. Malondialdehyde, an indicator of increased oxidative stress, was found to be high in spermatozoa treated with HBs [133]. These spermatozoa showed ameliorated total antioxidant capacity (TAC) [133]. HBs also resulted in heightened apoptosis of spermatozoa. Spermatozoa treated with HBs resulted in increased externalisation of phosphatidylserine (PS). They also showed an augmented expression level of apoptotic markers like caspase3, caspase8 and caspase 9[133]. Spermatozoa treated with HBs manifested decreased Matrix metalloprotease (MMP) leading to loss of mitochondrial integrity and increased apoptosis and reduced sperm motility [133]. In males with HBV infection, the level of IL-17 and IL-18 was found to be significantly higher. All these inflammatory molecules give rise to oxidative stress thus resulting in the generation of ROS which in turn leads to loss of sperm DNA integrity and in turn apoptosis.

13.4.12 Human Immunodeficiency Virus (HIV)

HIV belongs to the Retrovirus family in the genus of Lentivirus. HIV is approximately 100nm in diameter containing outer lipid envelope. Inside the lipid bilayer is embedded the transmembrane glycoprotein gp41. The surface glycoprotein gp120 is attached to the gp41. These two are responsible for the attachment of the virus to the host surface. The viral *gag* gene encodes for the matrix protein p17, core protein p6 and p24 and nucleocapsid protein p7. The viral core contains two identical single-stranded RNA molecule. The proviral DNA is formed by reverse transcription of the viral RNA genome into DNA followed by the subsequent degradation of the RNA and integration of the viral DNA in the human genome. The *pol* gene encodes for enzymes protease, reverse transcriptase and RNase H or reverse transcriptase plus RNase H (p66) and integrase. The virus also contains other genes that encode for various functional proteins like the *vif* gene encoding the viral infectivity protein, *vpr* gene encoding the viral protein R, *nef* gene encoding the negative regulatory factor, *vpu* gene encoding the virus protein unique, *tat* gene encoding the transactivator of transcription and *rev* gene encoding RNA splicing regulator. HIV-2 contains *vpx* gene encoding virus protein x instead of *vpu* [10, 134].

Human Immunodeficiency Virus (HIV) and Male Infertility

There are mainly two forms of HIV namely HIV-1 and HIV-2. Gupta et al reported the presence of HIV in the semen shortly after infection [135]. HIV virus can enter the semen via three routes namely as free virions, spermatozoa associated and virions compartmentalised inside the leucocytes [37]. Anderson et al have also reported enrichment of various cytokines and chemokines in the semen of patients infected with HIV [136]. This enrichment was thought to favour viral replication in the host. Asymptomatic carriers were shown to transmit the virus via their semen [137, 138]. Bagasra et al reported the presence of HIV-1 in the seminal cells of infected patients [139]. By using in situ polymerase chain reaction technique they observed the presence of HIV-1 in about 45% of the infected men at all stages of infection. Seminal mononuclear cell and sperm were shown to harbour HIV-1. A study conducted by Muciaccia et al highlighted the presence of HIV-1 DNA containing abnormal spermatozoa during ejaculation by infected males [140]. By using immunocytochemistry and in situ hybridisation techniques, Baccetti et al have shown that HIV-1 can bind and enter the spermatozoa. They detected viral particles, viral antigens and viral RNA in the sperm of infected patients [141].

The route of entry of the virus still remains unclear. The two co-receptors CXCR4 and CCR5 that facilitate the entry of HIV into the cells were not detected in the spermatozoa as was evidenced by flow cytometry data [142]. The CD4 receptor which is the main entry molecule for HIV was reported to be present in semen lymphocytes and monocytes but its presence in the spermatozoa was not confirmed in this study [143, 144]. Ashida and Scofield were the first to shed light on the interaction between the virus and the spermatozoa [145]. They highlighted the presence of a sperm ligand that was shown to react with the CD4 antibodies thus providing the first indirect clue for the presence of CD4 molecule in human sperms. Bergamo et al in their study highlighted the presence of CD4 molecule on the spermatozoa. They reported a 1kDa glycoprotein gp17 that was shown to interact with CD4+ T cells as well as to soluble CD4 recombinant protein. Western blotting of lysates from spermatozoa with monoclonal antibodies showed that proteins from the spermatozoa exhibited similar electrophoretic and antigenic properties as that of CD4 and gp17. Post acrosomal region of the spermatozoa also showed positive CD immunostaining in this study. The presence of CD4 molecules was also detected in the head region of sperm in mice [146]. Baccetti et al also detected a glycolipid on the surface of spermatozoon that was thought to bind HIV as well as aid its transmission [141].

HIV infection was shown to impart damage to the testes thus leading to infertility. HIV was reported to impart hypogonadotropic hypogonadism. Orchitis was shown to be a common phenomenon of HIV infection [147]. Orchitis in turn may lead to the generation of ROS which may cause loss of sperm DNA integrity. Pudney and Anderson carried out a study in which they tested autopsy samples from 43 male patients having acquired immune deficiency syndrome (AIDS) [148]. Azoospermia was observed in the samples. They also reported hyalinisation of the boundary wall of seminiferous tubules along with extensive infiltration of the interstitium with lymphocytes. They also reported the presence of CD4 positive white blood cells in the testis which is an indicative of the presence lymphocytes and macrophages for HIV. Paepe and Waxman carried out a study in which they examined autopsy samples of 57 patients with AIDS [9]. They observed reduced spermatogenesis in patients with AIDS. The autopsy samples of these patients also displayed fibrosis of the interstitium along with thickening of the basement membrane. Croxson et al reported that patients with AIDS have lower levels of testosterone and increased level of follicle-stimulating hormone (FSH) and luteinising hormone (LH) [149]. They concluded that a lower level of testosterone is due to impairment of the proper functioning of the testis due to HIV infection. Roders and Klatt carried out a retrospective case study in which they tested autopsy samples of testis from 100 AIDS patients. They observed increased perivasculitis, infiltration of lymphocytes, interstitial fibrosis and increased tubular wall thickness of the testes. This could lead to testicular atrophy thus resulting in male infertility. HIV infection was shown to increase the level of various inflammatory cytokines which in turn debilitate testosterone synthesis thus leading to male infertility.

13.5 Antiviral Therapies

Several antiviral therapies have been developed by different groups of scientists in order to combat the progression of viral infections (Table 13.2). Many chemically manufactured antiviral drugs have been developed so far in treating different kinds of viral infections. Most of the chemical drugs are nucleoside analogues. However, recently many natural compounds as well as interferon therapy have shown promising response to viral infections. They also incur less side effects as compared to the chemical drugs. Thus these modes of treatment can be effectively administered in order to ameliorate viral entry into the host cell. These antiviral agents also act to inhibit viral replication and the spread of the virus to other cells (Fig. 13.4). Thus antiviral therapies can be considered effective in preventing male infertility.

13.5.1 Natural Compounds

Flavonoids are a group of secondary metabolites derived from plants that were shown to have many antiviral properties. Flavonoids are further divided into various classes based on level of oxidation and substitution of the pyrene ring. Flavones are a class of flavonoids that consist of a 2-phenyl-1-benzopyran-4-one backbone. Mucsi et al showed a combination of flavones like apigenin, quercetin and quercitrin and acycloguanosine like acyclovir and Zovirax have antiviral action against herpes simplex virus types 1 and 2 in cell cultures [150]. Eliptica alba extracts and phytochemicals isolates containing wedelolactone, luteolin, and apigenin were shown to have antiviral properties. Eliptica alba and the phytochemicals abrogated HCV replication by inhibiting the RNA-dependent RNA polymerase function of HCV replicase in vitro [151]. Chiang et al in their comparative study have shown that purified components from *Ocimum basilicum* and purified components isolated from it like apigenin, linalool and ursolic acid have antiviral activities against DNA viruses like HSV, HBV and adenovirus as well as RNA viruses like coxsackievirus B1 and enterovirus 71 [152]. Apigenin was also shown to exert its antiviral properties against RNA viruses. Hakobyan et al in their study have shown that apigenin suppressed viral protein synthesis and viral factory formation of African swine fever virus. Apigenin was shown to be a very potent antiviral agent during the early stages of virus infection [153]. Apigenin downregulated the expression of mature microRNA122, which was shown to aid the replication of HCV [154]. Luteolin is another group of flavonoid that was shown to possess antiviral properties. Luteolin was shown to abrogate the Tat function of HIV thus inhibiting its reactivation during the latent phase. It inhibited clade B- and C-Tat-driven transactivation of LTR thus abrogating the reactivation of HIV [155]. Wu et al showed that luteolin has antiviral activity against EBV. It ablated the expression of proteins from EBV lytic genes. It also abrogated the EBV reactivating cell number by ameliorating the activities of the promoters of Zta (Zp) and Rta (Rp) genes by dysregulating the binding of

Table 13.2 Natural products and interferon therapy as antiviral agents

Name of the compound	Mode of action	References
Apigenin	Combination with Zovirax results in reduced multiplication of HSV-1 and HSV-2	[150]
	Inhibits RNA-dependent RNA polymerase function of HCV replicase, thus abrogating HCV replication	[151]
	Inhibits the secretion of HBsAg and HBeAg	[152]
	Suppress viral protein synthesis and viral factory formation of African swine fever virus	[153]
	Downregulates the expression of mature microRNA122 which aids in replication of HCV	[154]
Luteolin	Decreases clade B- and C-Tat-driven transactivation of LTR of HIV thus inhibiting the reactivation of the virus during the latent phase	[155]
	Inhibits the protein expression of the lytic genes of EBV	[156]
	Inhibits the binding of transcription factor Sp1 to promoters of Zta (Zp) and Rta (Rp) genes, thereby decreasing EBV reactivating cell number	[156]
	Inhibits the binding of SARS-CoV spike proteins with the host cell	[157]
Baicalein	Ablates the neuraminidase function of avian influenza H5N1 strain. Downregulates the influenza-induced cleavage of caspase-3 and nuclear export of the viral RNA particles	[158]
	Oral administration of baicalein to BALB/c mice with influenza infection increases mean time to death and decreases lung consolidation	[159]
	Disrupts the NS1-p85β binding influenza virus, increases interferon signalling, decreases (PI3K/Akt) activation thus preventing influenza infection	[160] [161]
	Inhibits the interaction between HIV-1 envelope proteins T cell tropic (X4), monocyte tropic (R5) and CD4/CXCR4 or CD4/CCR5, decreasing virus entry into the host. It blocks the replication of HIV during the early stages of viral infection	[163]
Quercitin	Inhibits the replication of influenza A virus subtypes (H1N1, H5N2, H7N3 and H9N2) by inducing the secretion of pro-inflammatory cytokines and type 1 interferon	[164]
	Inhibit binding, penetration and replication of HSV into the host cells	[165]
	Binds to the hemagglutinin subunit (HA2) of influenza A virus and inhibits its entry into the host, ablates the neuraminidase function of influenza A H1N1 and H7N9 viruses	[166] [167] [168]
	Binds with the nonstructural protein 3 (NS3) of HCV helicase and inhibit its replication, bind with HCV p7 protein and inhibit virus assembly	[169]

(continued)

Table 13.2 (continued)

Name of the compound	Mode of action	References
Kaempferol	Inhibits the neuraminidase activity of the influenza A virus	[172]
	Abrogates the reverse transcriptase activity of HIV	[173]
	Inhibits the function of NS3 thus inhibiting replication of HCV	[174]
	Inhibits SARS coronavirus infection by inhibiting the 3a ion channel of the virus thus limiting its release	[176]
Epigallocatechin gallate	Binds to haemagglutinin moieties of influenza virus and inhibiting its entry into host cells	[177]
	Prevents binding of CD4 with gp120 of HIV thereby inhibiting HIV infection, inhibits the reverse transcriptase activity of HIV	[179]
	Inhibits the lytic cycle of EBV by downregulating the expression of lytic cycle proteins Rta, Zta and EA-D, suppress EBV infection by inhibiting the activation of MEK/ERK1/2 and PI3-K/Akt pathways	[181]
	Ameliorates HSV infection by binding to the viral envelope glycoproteins gB and gD that are crucial for HSV infection	[183]
	Inhibits HBV replication by facilitating lysosomal acidification	[184]
	Reduces the formation of HBV covalently closed circular DNA, downregulates the synthesis of replicative intermediates of DNA thereby reducing HBV infection	[185]
	Inhibits HCV entry inside the host by preventing the attachment of the virus to the host cells, ameliorates the replication cycle of HCV	[187] [188]
	Reduces entry of ZIKA virus into host cells	[189]
Naringenin	Blocks the assembly of intracellular infectious HCV particles by activating peroxisome proliferator-activated receptor (PPAR)-alpha	[190]
Genistein	Inhibits HIV infection by dysregulating actin dynamics in resting CD4 T cells and macrophages	[192]
	Blocks HIV protein U thereby inhibiting the formation of cation permeable ion channel in infected cells	[193]
	Blocks the replication of HSV by inhibiting the transcription of viral particles	[194]
Curcumin	Downregulates the expression of PGC-1α, thereby abrogating HBV replication.	[195]
	Suppresses (SREBP-1)-Akt pathway, thereby inhibiting HCV replication	[196]
Chebulagic acid and punicalagin	Inactivates viral HCV particles and ablates HCV attachment and penetration into the host cell Competes with HSV-1 cell surface glycoprotein glycosaminoglycan (GAG) and inhibits viral entry and cell to cell spread	[197] [198]
Coumarin	Inhibits HIV replication by suppressing NF-κB activation	[199]

(continued)

Table 13.2 (continued)

Name of the compound	Mode of action	References
Type I interferon (IFN-α and IFN-β)	Releases myxovirus resistance (Mx) GTPases that interferes with viral nucleocapsid localisation Induces the production of 2'-5'-oligoadenylate synthetase (OAS), which activates nuclease RNaseL thereby degrading viral RNA Induces protein kinase RNA-activated (PKR), which disrupts recycling of guanidine diphosphate, thereby blocking translation of the viral RNA Induces the expression of gene 15 (ISG15) and tripartite motif (TRIM) that blocks the release of viral particles Increases the expression of APOBEC3, which in turn induces hypermutation of viral DNA Induces the expression of apoptotic genes including Fas ligand (FasL), PDL-1 and TRAIL PEG- IFN-α along with ribavirin is used for the treatment of HBV infection. It works by clearing serum hepatitis B e antigen (HBeAg) PEG- IFN-α is generally administered with nucleoside analogues lamivudine (LMV), adefovir (ADV), entecavir (ETV) and telbivudine (LdT) for better results	[200] [201] [203] [204] [205]
Type II interferon (IFN-γ)	Induces the expression of PKR, OAS and Mx GTPase Induces the expression of dsRNA-specific adenosine deaminase (ADAR), which exerts its effect by ablating viral replication Used in treatment against HIV-induced opportunistic infections with or without the combination of HAART	[206] [207] [208]
Type III interferon (IFN-λ)	Induces the expression of OAS and Mx protein Ablates HIV replication in macrophages and primary T cells by activating the JAK-STAT pathway-mediated innate immune response	[209]

transcription factor Sp1 [156]. Yi et al reported two small molecules, tetra-O-galloyl-beta-D-glucose (TGG) and luteolin to exert their antiviral properties by inhibiting the binding of SARS-CoV spike proteins with the host cell, thus preventing its entry [157]. Baicalein is another flavonoid which was shown to exert its antiviral activity against avian influenza H5N1 strain. Sithisarn et al showed that baicalein ablated the function of viral neuraminidase. It was also shown to downregulate the virus-induced cleavage of caspase-3 and nuclear export of the viral RNA particles. It also abrogated the expression of interleukin-6 and interleukin-8 [158]. Baicalein was also reported to harbour its antiviral activity during influenza infection in vivo. Oral administration of the compound to BALB/c mice with influenza infection resulted in increased mean time to death and decreased lung virus titre [159]. Lung consolidation was also inhibited. Baicalein was shown to exert its antiviral activity upon influenza infection by modulating the viral non-structural protein 1 (NS1) protein. Nayak et al showed that administration of baicalin during influenza infection disrupts the NS1- p85β binding, thus resulting in increased

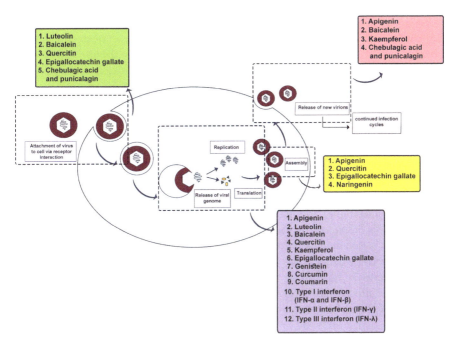

Fig. 13.4 Antiviral responses targeting different stages of viral infection of spermatozoa. Various natural antiviral agents and interferon signalling molecules were shown to target virus infection. Some of the viruses like luteolin, baicalein, quercitin, epigallocatechin gallate, chebulagic acid and punicalagin were shown to prevent viral attachment to host. Other molecules like apigenin, luteolin, baicalein, quercitin, kaempferol, genistein, curcumin, coumarin and Type I, Type II, Type III were shown to inhibit viral replication and translation of viral proteins. Apigenin, quercitin, epigallocatechin galate and naringenin were shown to inhibit viral assembly. Apigenin, baicalein, kaempferol, chebulagic acid and punicalagin were shown to inhibit the egress of newly formed virus thus inhibiting its spread

interferon (IFN) signalling and decreased phosphoinositide 3-kinase/Akt (PI3K/Akt) activation [160]. Baicalein was also shown to induce the production of IFN-γ in CD4+, CD8+ T-cells, natural killer (NK) cells of humans during influenza infection [161]. Synthetic analogues of baicalein with B ring-substituted bromine was found to be more effective than drugs like oseltamivir or ribavirin in treating Tamiflu-resistant H1N1 influenza virus [162]. Li et al reported that baicalein inhibited the interaction between HIV-1 Env proteins T cell tropic (X4), monocyte tropic (R5) and CD4/CXCR4 or CD4/CCR5 thus resulting in decreased virus entry into the host. It also blocks the replication of HIV during the early stages of viral infection [163]. Quercitin is another group of flavonoids having potent antiviral activity. Epimedium koreanum Nakai, a traditional medicinal plant, was shown to have antiviral activities against a range of viruses. Quercitin is a major component of the Epimedium koreanum plant. It was shown to ablate the replication of influenza A virus subtypes (H1N1, H5N2, H7N3 and H9N2) and HSV by inducing the secretion of pro-inflammatory cytokines and type 1 IFN [164]. Houttuynia cordata water extracts were shown to have antiviral properties against HSV-1, HSV-2 and acyclovir-resistant HSV-1 strain. Two compounds present in the extract quercetin and isoquercitrin were shown to inhibit HSV binding, penetration into the host cells as well as its replication [165]. They were also shown to inhibit the nuclear factor kappa B (NF-κB) pathway. Quercetin was shown to bind with the hemagglutinin subunit (HA2) of influenza A virus and inhibit its entry into the host [166]. Computational studies and molecular docking revealed that quercetin and chlorogenic acid ablates the neuraminidase function of influenza A H1N1 and H7N9 viruses [167, 168]. Fatima et al showed that quercetin, catechins, resveratrol and lutein bind with the nonstructural protein 3 (NS3) of HCV helicase and inhibit its replication [169]. Quercetin was also shown to bind with HCV p7 protein and inhibit virus assembly [170]. Kaempferol is another group of flavonoids with promising antiviral activity. It was shown to impart antiviral action against HSV-1 and HSV-2 [171]. Lee et al showed that a derivative of kaempferol, kaempferol 3-O-α-L-rhamnopyranoside, has antiviral action against influenza A virus. It was also shown to inhibit the neuraminidase activity of the virus at a much higher concentration [172]. Kaempferol and kaempferol-7-O-glucoside were found to have antiviral action during the early stage of HIV infection by abrogating its reverse transcriptase activity [173]. Another derivative of kaempferol, kaempferol-3,7-bisrhamnoside, was found to have antiviral activity against HCV by inhibiting NS3 serine protease of HCV [174]. Jeong et al showed that kaempferol exhibits its antiviral action against influenza strains H1N1 and H9N2 by inhibiting the neuraminidase activity of the virus [175]. Schwarz et al demonstrated that kaempferol inhibits SARS coronavirus infection by inhibiting the 3a ion channel of the virus thus limiting its release [176]. Catechin and its derivative epigallocatechin gallate (EGCG) were shown to have antiviral activity against influenza virus by binding to haemagglutinin of the virus and inhibiting its absorption [177]. EGCG was reported to compete with sialic acid residues for binding to influenza A virus thus inhibiting its attachment to the host [178]. Williamson et al showed that EGCG has a strong binding affinity for CD4 molecule. EGCG was reported to ameliorate HIV infection by binding itself with CD4, thus preventing binding of CD4 with gp120, an envelope protein of HIV-1 thus preventing HIV infection [179]. Epicatechin gallate and EGCG were shown to inhibit the reverse transcriptase activity of HIV [180]. EGCG was also found to be effective against EPV infection. EGCG was reported to inhibit the lytic cycle of EBV by ablating the expression of lytic proteins Rta, Zta and EA-D [181]. EGCG was shown to suppress EBV infection by inhibiting the activation of mitogen-activated protein kinase kinase (MEK)/extracellular signal-regulated kinase 1/2 (ERK1/2) and phosphatidylinositol 3-kinase (PI3-K)/Akt pathways [182]. EGCG was also found to be effective against HSV infection. EGCG was found to ameliorate HSV infection by binding to the viral envelope glycoproteins gB and gD that are crucial for HSV infection [183]. EGCG was shown to inhibit HBV replication by aiding in lysosomal acidification [184]. He et al showed EGCG to ameliorate HBV infection by reducing the formation of HBV covalently closed circular DNA as well as they downregulated the synthesis of replicative intermediates of DNA [185]. EGCG was found to inhibit the entry of HBV inside immortalised human primary hepatocytes with about 80% efficiency [186]. However, they did not observe any change in viral genome or virus secretion upon EGCG administration unlike the previous study. EGCG was also found to be an effective antiviral agent against HCV. It was shown to inhibit the virus entry inside the host by preventing the attachment of the virus to the host as well as ameliorated the replication cycle of HCV [187, 188]. Carneiro et al found EGCG to be effective against Zika virus infection. A study conducted by them during Zika virus outbreak in Brazil highlighted that administration of EGCG resulted in the entry of the virus into the host [189]. Naringenin is another flavonoid compound which was shown to be effective against HCV infection by blocking the assembly of intracellular infectious HCV particles by activating peroxisome proliferator-activated receptor (PPAR)-alpha [190]. Long-term administration of naringenin was shown to cause a rapid 1.4log reduction in the level of HCV [191]. Genistein is a group of isoflavons which was shown to inhibit HIV infection by interfering with the HIV-mediated actin dynamics in resting CD4 T cells and macrophages [192]. Genistein was reported to block the HIV protein U which is known to form cation permeable ion channel in infected cells [193]. Geninstein was reported to block HSV replication by inhibiting the transcription of viral particles [194]. Curcumin is another compound which was shown to have antiviral properties. It was shown to abrogate HBV replication by inhibiting the expression of peroxisome proliferator-activated receptor gamma coactivator 1-alpha (PGC-1α), which is a co-activator of transcription of HBV [195]. It was reported to suppress sterol regulatory element-binding protein-1 (SREBP-1)-Akt pathway, thereby inhibiting HCV replication [196]. Lin et al have reported two tannins, chebulagic acid and punicalagin, which inactivated viral HCV particles and inhibited HCV attachment and penetration into the host cell [197]. The two tannins were shown to compete with HSV-1 cell surface glycoprotein glycosaminoglycan (GAG) and inhibit viral entry and cell to cell spread [198]. Coumarins derived from *Calophyllum* species was shown to exhibit antiviral activities against HIV virus by suppressing nuclear factor-kappa B (NF-κB) activation and in turn inhibiting viral replication [199].

13.5.2 Interferon Therapy

Interferons (IFNs) are heterogenous group of signalling molecules that are effectively used by host cells in response to any viral threat. They are soluble glycoproteins. IFNs are classified into three types Type I, II and III based on the structure of the receptors. Type I interferons include IFN-α and IFN-β. They exert their antiviral activity by inducing a number of factors [200, 201]. Myxovirus resistance (Mx) GTPases released by interferons interferes with viral nucleocapsid localisation. Type I IFNs were shown to induce another protein 2'–5'-oligoadenylate synthetase (OAS) which in turn activates nuclease RNaseL that exerts its antiviral effect by degrading viral RNA. Protein kinase RNA-activated (PKR) is another enzyme induced by Type I IFN that disrupts recycling of guanidine diphosphate, thus blocking translation of the viral RNA. Type I interferons were also reported to induce two proteins, interferon-stimulated gene 15 (ISG15) and tripartite motif (TRIM) that blocks the release of viral particles. APOBEC3 is another protein induced by Type I IFNs that was shown to induce hypermutation of viral DNA. In response to viral infection, Type I IFN was shown to induce apoptotic genes including Fas ligand (FasL), PDL-1 and TRAIL. Type I IFN was shown to activate NK cells directly in response to viral infection. Type II interferon includes interferon-gamma (IFN-γ). During virus infection IFN-γ was shown to induce the expression of PKR, OAS and Mx GTPase. Along with these proteins IFN-γ also induces the expression of dsRNA-specific adenosine deaminase (ADAR), which exerts its effect by ablating viral replication. IFN-γ was also shown to directly convey the information of viral attack from innate to adaptive immune response. Type III interferon signalling includes interferon-lambda (IFN-λ). IFN-λ was shown to exert its antiviral response in target cells expressing IFN-λ receptors. It is structurally similar to interleukin-10. IFN-λ was shown to induce OAS and Mx protein [202].

Recombinant IFN-α has been used against HCV. However, due to their relatively shorter half-life and rapid absorption, IFN-α is generally attached with polyethylene glycol (PEG) moiety and administered to infected patients [203, 204]. Chronic HCV-infected patients are generally given on a weekly basis a subcutaneous injection of pegylated IFN-α (PEG- IFN-α) with a daily dose of oral ribavirin. Pegylated IFN-α2a (PEG- IFN-α2a) has been used for the treatment of HBV infection. PEG-IFN-α2a works against HBV by effective clearance of serum hepatitis B e antigen (HBeAg). Treatment with PEG-IFN-α2a was shown to have a long-lasting effect as compared to treatment with nucleoside analogues (NA), such as lamivudine (LMV), adefovir (ADV), entecavir (ETV) and telbivudine (LdT) [205]. PEG-IFN-α2a is generally administered along with NA for better results. IFN-β is used in patients with poor response to PEG-IFN-α/ribavirin treatment during HCV infection [206]. It was shown to work in patients with *IL-28* risk allele who were unable to respond to PEG- IFN-α/ribavirin treatment [207]. IFN-γ is used in treatment against HIV-induced opportunistic infections with or without the combination of highly active antiretroviral therapy (HAART) [208]. IFN-λ was shown to inhibit the replication of HBV and HCV viral genome [209]. IFN-λ was reported to ablate HIV replication in macrophages and primary T cells by activating the JAK-STAT pathway-mediated innate immune response [209]. Thus interferon therapy is a promising antiviral tool in ameliorating viral infection.

13.6 Conclusions

In this chapter, we have reviewed the impact of a plethora of viruses in inducing male infertility. A range of different viruses have been shown to exist in the semen of infected males and they were also shown to be transmitted sexually. These viruses were shown to induce male infertility by employing numerous patho-physiological mechanisms. The viruses were reported to impart detrimental consequences to sperm parameters leading to infertility. Sperm infections with viruses have been reported to ablate sperm motility, disrupt sperm morphology thus resulting in the development of infertility. Infection with viruses was also shown to induce aneuploidy and fragmentation of DNA. The mechanistic approach by which viruses induce male infertility is by inducing oxidative stress. ROS generated due to increased leucocyte infiltration and cytokine secretion as a result of viral infection leads to the development of oxidative stress inside spermatozoa leading to decreased fertility rate. Many chemical and natural products are emerging as promising therapeutic antiviral agents in ameliorating viral assembly and replication inside the host system. This review will shed light on the mechanism and extent of viral pathogenesis inside male genital tract, resulting in the development of infertility. It will summarise the available studies in light of antiviral therapies. Although many studies relating viruses and male infertility have been done so far yet a detailed mechanism relating to the development of viremia and male infertility have not been clearly elucidated. Hence, this review will shed light on the available research relating to the impact of viral infection on male fertility and the available antiviral therapies which could be modulated further in order to decrease male infertility.

References

1. Krupovic M, Dolja VV, Koonin EV. Origin of viruses: primordial replicators recruiting capsids from hosts. Nat Rev Microbiol. 2019;17(7):449–58.
2. Nasir A, Kim KM, Caetano-Anollés G. Viral evolution: primordial cellular origins and late adaptation to parasitism. Mob Genet Elements. 2012;2(5):247–52.
3. Desrochers GF, Pezacki JP. ABPP and Host-Virus Interactions. Curr Top Microbiol Immunol. 2019;420:131–54.
4. Ryu WS. Virus Life Cycle. Mol Virol Human Pathogenic Viruses. 2017:31–45.
5. Allen JD, Ross TM. H3N2 influenza viruses in humans: viral mechanisms, evolution, and evaluation. Hum Vaccin Immunother. 2018;14(8):1840–7.
6. Maarouf M, Rai KR, Goraya MU, Chen JL. Immune ecosystem of virus-infected host tissues. Int J Mol Sci. 2018;19(5):1379.
7. Smith RD, Konoplev S, DeCourten-Myers G, Brown T. West Nile virus encephalitis with myositis and orchitis. Hum Pathol. 2004;35(2):254–8.
8. Govero J, Esakky P, Scheaffer SM, Fernandez E, Drury A, Platt DJ, Gorman MJ, Richner JM, Caine EA, Salazar V, Moley KH, Diamond MS. Zika virus infection damages the testes in mice. Nature. 2016;540(7633):438–42.
9. De Paepe ME, Waxman M. Testicular atrophy in AIDS: a study of 57 autopsy cases. Hum Pathol. 1989;20(3):210–4.
10. Zumla A. Mandell, Douglas, and Bennett's principles and practice of infectious diseases. Lancet Infect Dis. 2010;10(5):303–4.
11. Casella R, Leibundgut B, Lehmann K, Gasser TC. Mumps orchitis: report of a mini-epidemic. J Urol. 1997;158(6):2158–61.
12. Narayana K, D'Souza UJ, Narayan P, Kumar G. The antiviral drug ribavirin reversibly affects the reproductive parameters in the male Wistar rat. Folia Morphol (Warsz). 2005;64(2):65–71.
13. La Vignera S, Condorelli RA, Vicari E, D'Agata R, Calogero AE. Sperm DNA damage in patients with chronic viral C hepatitis. Eur J Intern Med. 2012;23(1):e19–24.

14. Straus SK, Bo H. Filamentous bacteriophage proteins and assembly. Sub-cellular Biochem. 2018;88:261–79.
15. Louten J. Virus structure and classification. Essential Human Virol. 2016:19–29.
16. Wilson DP. Protruding features of viral capsids are clustered on icosahedral great circles. PLoS One. 2016;11(4):e0152319.
17. Long GW, Nobel J, Murphy FA, Herrmann KL, Lourie B. Experience with electron microscopy in the differential diagnosis of smallpox. Appl Microbiol. 1970;20(3):497–504.
18. Temin HM, Baltimore D. RNA-directed DNA synthesis and RNA tumor viruses. Adv Virus Res. 1972;17:129–86.
19. de Lamirande E, Gagnon C. Reactive oxygen species and human spermatozoa. I. Effects on the motility of intact spermatozoa and on sperm axonemes. J Androl. 1992;13(5):368–78.
20. Bisht S, Faiq M, Tolahunase M, Dada R. Oxidative stress and male infertility. Nat Rev Urol. 2017;14(8):470–85.
21. Kang X, Xie Q, Zhou X, Li F, Huang J, Liu D, Huang T. Effects of hepatitis B virus S protein exposure on sperm membrane integrity and functions. PLoS One. 2012;7(3):e33471.
22. Paz-Bailey G, Rosenberg ES, Doyle K, Munoz-Jordan J, Santiago GA, Klein L, Perez-Padilla J, Medina FA, Waterman SH, Gubern CG, Alvarado LI, Sharp TM. Persistence of Zika Virus in body fluids – final report. N Engl J Med. 2017;379(13):1234–43.
23. Xu J, Qi L, Chi X, Yang J, Wei X, Gong E, Peh S, Gu J. Orchitis: a complication of severe acute respiratory syndrome (SARS). Biol Reprod. 2006;74(2):410–6.
24. Machida K, Cheng KT, Lai CK, Jeng KS, Sung VM, Lai MM. Hepatitis C virus triggers mitochondrial permeability transition with production of reactive oxygen species, leading to DNA damage and STAT3 activation. J Virol. 2006;80(14):7199–207.
25. Wu H, Wang F, Tang D, Han D. Mumps Orchitis: clinical aspects and mechanisms. Front Immunol. 2021;12:582946.
26. Vasilopoulos E, Fragkiadaki P, Kalliora C, Fragou D, Docea AO, Vakonaki E, Tsoukalas D, Calina D, Buga AM, Georgiadis G, Mamoulakis C, Makrigiannakis A, Spandidos DA, Tsatsakis A. The association of female and male infertility with telomere length (Review). Int J Mol Med. 2019;44(2):375–89.
27. Deng Z, Kim ET, Vladimirova O, Dheekollu J, Wang Z, Newhart A, Liu D, Myers JL, Hensley SE, Moffat J, Janicki SM, Fraser NW, Knipe DM, Weitzman MD, Lieberman PM. HSV-1 remodels host telomeres to facilitate viral replication. Cell Rep. 2014;9(6):2263–78.
28. Imai Y, Kuba K, Neely GG, Yaghubian-Malhami R, Perkmann T, van Loo G, Ermolaeva M, Veldhuizen R, Leung YH, Wang H, Liu H, Sun Y, Pasparakis M, Kopf M, Mech C, Bavari S, Peiris JS, Slutsky AS, Akira S, Hultqvist M, Holmdahl R, Nicholls J, Jiang C, Binder CJ, Penninger JM. Identification of oxidative stress and Toll-like receptor 4 signaling as a key pathway of acute lung injury. Cell. 2008;133(2):235–49.
29. Zheng ZM, Baker CC. Papillomavirus genome structure, expression, and post-transcriptional regulation. Front Biosci. 2006;11:2286–302.
30. Ostrow RS, Zachow KR, Niimura M, Okagaki T, Muller S, Bender M, Faras AJ. Detection of papillomavirus DNA in human semen. Science. 1986;231(4739):731–3.
31. Foresta C, Patassini C, Bertoldo A, Menegazzo M, Francavilla F, Barzon L, Ferlin A. Mechanism of human papillomavirus binding to human spermatozoa and fertilizing ability of infected spermatozoa. PLoS One. 2011;6(3):e15036.
32. Gimenes F, Souza RP, Bento JC, Teixeira JJ, Maria-Engler SS, Bonini MG, Consolaro ME. Male infertility: a public health issue caused by sexually transmitted pathogens. Nat Rev Urol. 2014;11(12):672–87.
33. Lai YM, Yang FP, Pao CC. Human papillomavirus deoxyribonucleic acid and ribonucleic acid in seminal plasma and sperm cells. Fertil Steril. 1996;65(5):1026–30.
34. Rintala MA, Pollanen PP, Nikkanen VP, Grenman SE, Syrjanen SM. Human papillomavirus DNA is found in the vas deferens. J Infect Dis. 2002;185(11):1664–7.
35. Lai YM, Lee JF, Huang HY, Soong YK, Yang FP, Pao CC. The effect of human papillomavirus infection on sperm cell motility. Fertil Steril. 1997;67(6):1152–5.

36. Cruz-Gregorio A, Manzo-Merino J, Lizano M. Cellular redox, cancer and human papillomavirus. Virus Res. 2018;246:35–45.
37. Williams VM, Filippova M, Filippov V, Payne KJ, Duerksen-Hughes P. Human papillomavirus type 16 E6* induces oxidative stress and DNA damage. J Virol. 2014;88(12):6751–61.
38. Pellavio G, Todaro F, Alberizzi P, Scotti C, Gastaldi G, Lolicato M, Omes C, Caliogna L, Nappi RE, Laforenza U. HPV Infection Affects Human Sperm Functionality by Inhibition of Aquaporin-8. Cells. 2020;9(5):1241.
39. McElwee M, Vijayakrishnan S, Rixon F, Bhella D. Structure of the herpes simplex virus portal-vertex. PLoS Biol. 2018;16(6):e2006191.
40. Wald A, Matson P, Ryncarz A, Corey L. Detection of herpes simplex virus DNA in semen of men with genital HSV-2 infection. Sex Transm Dis. 1999;26(1):1–3.
41. Kotronias D, Kapranos N. Detection of herpes simplex virus DNA in human spermatozoa by in situ hybridization technique. Vivo. 1998;12(4):391–4.
42. Csata S, Kulcsár G. Virus-host studies in human seminal and mouse testicular cells. Acta Chir Hung. 1991;32(1):83–90.
43. Bezold G, Politch JA, Kiviat NB, Kuypers JM, Wolff H, Anderson DJ. Prevalence of sexually transmissible pathogens in semen from asymptomatic male infertility patients with and without leukocytospermia. Fertility Sterility. 2007;87(5):1087–97.
44. Li HC, Lo SY. Hepatitis C virus: virology, diagnosis and treatment. World J Hepatol. 2015;7(10):1377–89.
45. Kotwal GJ, Rustgi VK, Baroudy BM. Detection of hepatitis C virus-specific antigens in semen from non-A, non-B hepatitis patients. Dig Dis Sci. 1992;37(5):641–4.
46. Liou TC, Chang TT, Young KC, Lin XZ, Lin CY, Wu HL. Detection of HCV RNA in saliva, urine, seminal fluid, and ascites. J Med Virol. 1992;37(3):197–202.
47. Liu FH, Tian GS, Fu XX. Detection of plus and minus strand hepatitis C virus RNA in peripheral blood mononuclear cells and spermatid. Zhonghua Yi Xue Za Zhi. 1994;74(5):284–6.
48. Fried MW, Shindo M, Fong TL, Fox PC, Hoofnagle JH, Di Bisceglie AM. Absence of hepatitis C viral RNA from saliva and semen of patients with chronic hepatitis C. Gastroenterology. 1992;102:1306–8.
49. Hsu HH, Wright TL, Luba D, Martin M, Feinstone SM, Garcia G, Greenberg HB. Failure to detect hepatitis C virus genome in human secretions with the polymerase chain reaction. Hepatology. 1991;14:763–7.
50. Semprini AE, Persico T, Thiers V, Oneta M, Tuveri R, Serafini P, Boschini A, Giuntelli S, Pardi G, Brechot C. Absence of hepatitis C virus and detection of hepatitis G virus/GB virus C RNA sequences in the semen of infected men. J Infect Dis. 1998;177:848–54.
51. Fiore RJ, Potenza D, Monno L, Appice A, DiStefano M, Giannelli A, LaGrasta L, Romanelli C, DiBari C, Pastore G. Detection of HCV RNA in serum and seminal fluid from HIV-1 co-infected intravenous drug addicts. J Med Virol. 1995;46(4):364–7.
52. Zhao XP, Yang DL, Tang ZY, Huang YC, Hao LJ, Cheng L, Zhou W. Infectivity and risk factors of hepatitis C virus transmission through sexual contact. J Tongji Med Univ. 1995;15:147–50.
53. Kumar RM. Interspousal and intrafamilial transmission of hepatitis C virus: a myth or a concern? Obstet Gynecol. 1998;91(3):426–31.
54. Lorusso F, Palmisano M, Chironna M, Vacca M, Masciandaro P, Bassi E, Selvaggi Luigi L, Depalo R. Impact of chronic viral diseases on semen parameters. Andrologia. 2010;42(2):121–6.
55. Moretti E, Federico MG, Giannerini V, Collodel G. Sperm ultrastructure and meiotic segregation in a group of patients with chronic hepatitis B and C. Andrologia. 2008;40(3):173–8.
56. Anticoli S, Amatore D, Matarrese P, De Angelis M, Palamara AT, Nencioni L, Ruggieri A. Counteraction of HCV-induced oxidative stress concurs to establish chronic infection in liver cell cultures. Oxid Med Cell Longev. 2019;2019:6452390.
57. Goldberg-Bittman L, Kitay-Cohen Y, Hadari R, Yukla M, Fejgin MD, Amiel A. Random aneuploidy in chronic hepatitis C patients. Cancer Genet Cytogenet. 2008;180(1):20–3.
58. Rossi SL, Ross TM, Evans JD. West Nile virus. Clin Lab Med. 2010;30(1):47–65.

59. Siemann DN, Strange DP, Maharaj PN, Shi PY, Verma S. Zika Virus infects human sertoli cells and modulates the integrity of the *in vitro* blood-testis barrier model. J Virol. 2017;91(22):e00623–17.
60. Gorchakov R, Gulas-Wroblewski BE, Ronca SE, Ruff JC, Nolan MS, Berry R, Alvarado RE, Gunter SM, Murray KO. Optimizing PCR detection of West Nile Virus from body fluid specimens to delineate natural history in an infected human cohort. Int J Mol Sci. 2019;20(8):1934.
61. Kelley RE, Berger JR, Kelley BP. West Nile virus meningo-encephalitis: possible sexual transmission. J La State Med Soc. 2016;168(1):21–2.
62. Ayres CFJ, Guedes DRD, Paiva MHS, Morais-Sobral MC, Krokovsky L, Machado LC, Melo-Santos MAV, Crespo M, Oliveira CMF, Ribeiro RS, Cardoso OA, Menezes ALB, Laperrière-Jr RC, Luna CF, Oliveira ALS, Leal WS, Wallau GL. Zika virus detection, isolation and genome sequencing through Culicidae sampling during the epidemic in Vitória, Espírito Santo, Brazil. Parasit Vectors. 2019;12(1):220.
63. Foy BD, Kobylinski KC, Foy JLC. Probable non–vector-borne transmission of Zika virus, Colorado, USA. Emerging Infect Dis. 2011;17:880–2.
64. Mead PS, Duggal NK, Hook SA, Delorey M, Fischer M, Olzenak McGuire D, Becksted H, Max RJ, Anishchenko M, Schwartz AM, Tzeng WP, Nelson CA, McDonald EM, Brooks JT, Brault AC, Hinckley AF. Zika virus shedding in semen of symptomatic infected men. N Engl J Med. 2018;378(15):1377–85.
65. Mansuy JM, Suberbielle E, Chapuy-Regaud S, Mengelle C, Bujan L, Marchou B, Delobel P, Gonzalez-Dunia D, Malnou CE, Izopet J, Martin-Blondel G. Zika virus in semen and spermatozoa. Lancet Infect Dis. 2016;16(10):1106–7.
66. Mansuy JM, Dutertre M, Mengelle C, Fourcade C, Marchou B, Delobel P, Izopet J, Martin-Blondel G. Zika virus: high infectious viral load in semen, a new sexually transmitted pathogen? Lancet Infect Dis. 2016;16(4):405.
67. D'Ortenzio E, Matheron S, Yazdanpanah Y, de Lamballerie X, Hubert B, Piorkowski G, Maquart M, Descamps D, Damond F, Leparc-Goffart I. Evidence of sexual transmission of Zika virus. N Engl J Med. 2016;374(22):2195–8.
68. Counotte MJ, Kim CR, Wang J. Sexual transmission of Zika virus and other flaviviruses: a living systematic review. PLOS Medv. 2018;15:e1002611.
69. Mead PS, Duggal NK, Hook SA, Delorey M, Fischer M, Olzenak McGuire D, Becksted H, Max RJ, Anishchenko M, Schwartz AM, Tzeng WP, Nelson CA, McDonald EM, Brooks JT, Brault AC, Hinckley AF. Zika virus shedding in semen of symptomatic infected men. N Engl J Med. 2018;378(15):1377–85.
70. Huits R, De Smet B, Ariën KK, Van Esbroeck M, Bottieau E, Cnops L. Zika virus in semen: a prospective cohort study of symptomatic travellers returning to Belgium. Bull World Health Organ. 2017;95(12):802–9.
71. Paz-Bailey G, Rosenberg ES, Doyle K, Munoz-Jordan J, Santiago GA, Klein L, Perez-Padilla J, Medina FA, Waterman SH, Gubern CG, Alvarado LI, Sharp TM. Persistence of Zika virus in body fluids – final report. N Engl J Med. 2017;379(13):1234–43.
72. Oguet G, Mansuy JM, Matusali G, Hamdi S, Walschaerts M, Pavili L, Guyomard S, Prisant N, Lamarre P, Dejucq-Rainsford N, Pasquier C, Bujan L. Effect of acute Zika virus infection on sperm and virus clearance in body fluids: a prospective observational study. Lancet Infect Dis. 2017;17(11):1200–8.
73. Avelino-Silva VI, Alvarenga C, Abreu C, Tozetto-Mendoza TR, Canto CLMD, Manuli ER, Mendes-Correa MC, Sabino EC, Figueiredo WM, Segurado AC, Mayaud P. Potential effect of Zika virus infection on human male fertility? Rev Inst Med Trop Sao Paulo. 2018;60:e64.
74. Kumar A, Jovel J, Lopez-Orozco J, Limonta D, Airo AM, Hou S, Stryapunina I, Fibke C, Moore RB, Hobman TC. Human Sertoli cells support high levels of Zika virus replication and persistence. Sci Rep. 2018;8(1):5477.
75. Sheng ZY, Gao N, Wang ZY, Cui XY, Zhou DS, Fan DY, Chen H, Wang PG, An J. Sertoli cells are susceptible to ZIKV infection in mouse testis. Front Cell Infect Microbiol. 2017;7:272.

76. Snijder EJ, Bredenbeek PJ, Dobbe JC, Thiel V, Ziebuhr J, Poon LL, Guan Y, Rozanov M, Spaan WJ, Gorbalenya AE. Unique and conserved features of genome and proteome of SARS-coronavirus, an early split-off from the coronavirus group 2 lineage. J Mol Biol. 2003;331(5):991–1004.
77. Li W, Moore MJ, Vasilieva N, Sui J, Wong SK, Berne MA, Somasundaran M, Sullivan JL, Luzuriaga K, Greenough TC, Choe H, Farzan M. Angiotensin-converting enzyme 2 is a functional receptor for the SARS coronavirus. Nature. 2003;426(6965):450–4.
78. Zhao JM, Zhou GD, Sun YL, Wang SS, Yang JF, Meng EH, Pan D, Li WS, Zhou XS, Wang YD, Lu JY, Li N, Wang DW, Zhou BC, Zhang TH. Clinical pathology and pathogenesis of severe acute respiratory syndrome. Zhonghua Shi Yan He Lin Chuang Bing Du Xue Za Zhi. 2003;17(3):217–21.
79. Ding Y, He L, Zhang Q, Huang Z, Che X, Hou J, Wang H, Shen H, Qiu L, Li Z, Geng J, Cai J, Han H, Li X, Kang W, Weng D, Liang P, Jiang S. Organ distribution of severe acute respiratory syndrome (SARS) associated coronavirus (SARS-CoV) in SARS patients: implications for pathogenesis and virus transmission pathways. J Pathol. 2004;203(2):622–30.
80. Gu J, Gong E, Zhang B, Zheng J, Gao Z, Zhong Y, Zou W, Zhan J, Wang S, Xie Z, Zhuang H, Wu B, Zhong H, Shao H, Fang W, Gao D, Pei F, Li X, He Z, Xu D, Shi X, Anderson VM, Leong AS. Multiple organ infection and the pathogenesis of SARS. J Exp Med. 2005;202(3):415–24.
81. Sun P, Lu X, Xu C, Sun W, Pan B. Understanding of COVID-19 based on current evidence. J Med Virol. 2020;92(6):548–51.
82. Cucinotta D, Vanelli M. WHO Declares COVID-19 a Pandemic. Acta Biomed. 2020;91(1):157–60.
83. Lai CC, Shih TP, Ko WC, Tang HJ, Hsueh PR. Severe acute respiratory syndrome coronavirus 2 (SARS-CoV-2) and coronavirus disease-2019 (COVID-19): the epidemic and the challenges. Int J Antimicrob Agents. 2020;55(3):105924.
84. Tang X, Wu C, Li X, Song Y, Yao X, Wu X, Duan Y, Zhang H, Wang Y, Qian Z, Cui J, Lu J. On the origin and continuing evolution of SARS-CoV-2. Natl Sci Rev. 2020;036
85. Li D, Jin M, Bao P, Zhao W, Zhang S. Clinical characteristics and results of semen tests among men with coronavirus disease 2019. JAMA Netw Open. 2020;3(5):e208292.
86. Massarotti C, Garolla A, Maccarini E, Scaruffi P, Stigliani S, Anserini P, Foresta C. SARS-CoV-2 in the semen: where does it come from? Andrology. 2021;9(1):39–41.
87. Pan F, Xiao X, Guo J, Song Y, Li H, Patel DP, Spivak AM, Alukal JP, Zhang X, Xiong C, Li PS, Hotaling JM. No evidence of severe acute respiratory syndrome-coronavirus 2 in semen of males recovering from coronavirus disease 2019. Fertil Steril. 2020;113(6):1135–9.
88. Paoli D, Pallotti F, Colangelo S, Basilico F, Mazzuti L, Turriziani O, Antonelli G, Lenzi A, Lombardo F. Study of SARS-CoV-2 in semen and urine samples of a volunteer with positive naso-pharyngeal swab. J Endocrinol Invest. 2020;43(12):1819–22.
89. Suhail S, Zajac J, Fossum C, Lowater H, McCracken C, Severson N, Laatsch B, Narkiewicz-Jodko A, Johnson B, Liebau J, Bhattacharyya S, Hati S. Role of oxidative stress on SARS-CoV (SARS) and SARS-CoV-2 (COVID-19) infection: a review. Protein J. 2020;39(6):644–56.
90. Chen Z, Mi L, Xu J, Yu J, Wang X, Jiang J, Xing J, Shang P, Qian A, Li Y, Shaw PX, Wang J, Duan S, Ding J, Fan C, Zhang Y, Yang Y, Yu X, Feng Q, Li B, Yao X, Zhang Z, Li L, Xue X, Zhu P. Function of HAb18G/CD147 in invasion of host cells by severe acute respiratory syndrome coronavirus. J Infect Dis. 2005;191(5):755–60.
91. Shi Y, Wang Y, Shao C, Huang J, Gan J, Huang X, Bucci E, Piacentini M, Ippolito G, Melino G. COVID-19 infection: the perspectives on immune responses. Cell Death Differ. 2020;27:1–4.
92. Aitken RJ, Roman SD. Antioxidant systems and oxidative stress in the testes. Oxid Med Cell Longev. 2008;1(1):15–24.

93. Abd-Allah AR, Helal GK, Al-Yahya AA, Aleisa AM, Al-Rejaie SS, Al-Bakheet SA. Pro-inflammatory and oxidative stress pathways which compromise sperm motility and survival may be altered by L-carnitine. Oxid Med Cell Longev. 2009;2(2):73–81.
94. Ma L, Xie W, Li D, Shi L, Ye G, Mao Y, Xiong Y, Sun H, Zheng F, Chen Z, Qin J, Lyu J, Zhang Y, Zhang M. Evaluation of sex-related hormones and semen characteristics in reproductive-aged male COVID-19 patients. J Med Virol. 2021;93(1):456–62.
95. Mieusset R, Bujan L, Plantavid M, Grandjean H. Increased levels of serum follicle-stimulating hormone and luteinizing hormone associated with intrinsic testicular hyperthermia in oligospermic infertile men. J Clin Endocrinol Metab. 1989;68(2):419–25.
96. Geisbert TW: Marburg and Ebola hemorrhagic fevers (Filoviruses). Mandell, Douglas, and Bennett's Principles and Practice of Infectious Diseases 1995–1999.e1 (2015)
97. Bharat TA, Noda T, Riches JD, Kraehling V, Kolesnikova L, Becker S, Kawaoka Y, Briggs JA. Structural dissection of Ebola virus and its assembly determinants using cryo-electron tomography. Proc Natl Acad Sci U S A. 2012;109(11):4275–80.
98. Soka MJ, Choi MJ, Baller A, White S, Rogers E, Purpura LJ, Mahmoud N, Wasunna C, Massaquoi M, Abad N, Kollie J, Dweh S, Bemah PK, Christie A, Ladele V, Subah OC, Pillai S, Mugisha M, Kpaka J, Kowalewski S, German E, Stenger M, Nichol S, Ströher U, Vanderende KE, Zarecki SM, Green HH, Bailey JA, Rollin P, Marston B, Nyenswah TG, Gasasira A, Knust B, Williams D. Prevention of sexual transmission of Ebola in Liberia through a national semen testing and counselling programme for survivors: an analysis of Ebola virus RNA results and behavioural data. Lancet Glob Health. 2016;4(10):e736–43.
99. Keita AK, Vidal N, Toure A, Diallo MSK, Magassouba N, Baize S, Mateo M, Raoul H, Mely S, Subtil F, Kpamou C, Koivogui L, Traore F, Sow MS, Ayouba A, Etard JF, Delaporte E, Peeters M. PostEbogui study group. A 40-month follow-up of Ebola virus disease survivors in Guinea (PostEbogui) reveals long-term detection of Ebola viral ribonucleic acid in semen and breast milk. Open Forum Infect Dis. 2019;6(12):482.
100. Diallo B, Sissoko D, Loman NJ, Bah HA, Bah H, Worrell MC, Conde LS, Sacko R, Mesfin S, Loua A, Kalonda JK, Erondu NA, Dahl BA, Handrick S, Goodfellow I, Meredith LW, Cotten M, Jah U, Guetiya Wadoum RE, Rollin P, Magassouba N, Malvy D, Anglaret X, Carroll MW, Aylward RB, Djingarey MH, Diarra A, Formenty P, Keïta S, Günther S, Rambaut A, Duraffour S. Resurgence of Ebola virus disease in guinea linked to a survivor with virus persistence in seminal fluid for more than 500 days. Clin Infect Dis. 2016;63(10):1353–6.
101. Sissoko D, Duraffour S, Kerber R, Kolie JS, Beavogui AH, Camara AM, Colin G, Rieger T, Oestereich L, Pályi B, Wurr S, Guedj J, Nguyen THT, Eggo RM, Watson CH, Edmunds WJ, Bore JA, Koundouno FR, Cabeza-Cabrerizo M, Carter LL, Kafetzopoulou LE, Kuisma E, Michel J, Patrono LV, Rickett NY, Singethan K, Rudolf M, Lander A, Pallasch E, Bockholt S, Rodríguez E, Di Caro A, Wölfel R, Gabriel M, Gurry C, Formenty P, Keïta S, Malvy D, Carroll MW, Anglaret X, Günther S. Persistence and clearance of Ebola virus RNA from seminal fluid of Ebola virus disease survivors: a longitudinal analysis and modelling study. Lancet Glob Health. 2017;5(1):e80–8.
102. Fischer WA, Brown J, Wohl DA, Loftis AJ, Tozay S, Reeves E, Pewu K, Gorvego G, Quellie S, Cunningham CK, Merenbloom C, Napravnik S, Dube K, Adjasoo D, Jones E, Bonarwolo K, Hoover D. Ebola virus ribonucleic acid detection in semen more than two years after resolution of acute Ebola virus infection. Open Forum Infect Dis. 2017;4(3):155.
103. Perry DL, Huzella LM, Bernbaum JG, Holbrook MR, Jahrling PB, Hagen KR, Schnell MJ, Johnson RF. Ebola virus localization in the macaque reproductive tract during acute ebola virus disease. Am J Pathol. 2018;188(3):550–8.
104. Zeng X, Blancett CD, Koistinen KA, Schellhase CW, Bearss JJ, Radoshitzky SR, Honnold SP, Chance TB, Warren TK, Froude JW, Cashman KA, Dye JM, Bavari S, Palacios G, Kuhn JH, Sun MG. Identification and pathological characterization of persistent asymptomatic Ebola virus infection in rhesus monkeys. Nat Microbiol. 2017;2:17113.
105. Guetiya Wadoum RE, Samin A, Mafopa NG, Giovanetti M, Russo G, Turay P, Turay J, Kargbo M, Kanu MT, Kargbo B, Akpablie J, Cain CJ, Pasin P, Batwala V, Sobze MS, Potestà

M, Minutolo A, Colizzi V, Montesano C. Mobile health clinic for the medical management of clinical sequelae experienced by survivors of the 2013-2016 Ebola virus disease outbreak in Sierra Leone, West Africa. Eur J Clin Microbiol Infect Dis. 2017;36(11):2193–200.
106. de St Maurice A, Ervin E, Orone R, Choi M, Dokubo EK, Rollin PE, Nichol ST, Williams D, Brown J, Sacra R, Fankhauser J, Knust B. Care of ebola survivors and factors associated with clinical sequelae-monrovia, Liberia. Open Forum Infect Dis. 2018;5(10):239.
107. Martines RB, Ng DL, Greer PW, Rollin PE, Zaki SR. Tissue and cellular tropism, pathology and pathogenesis of Ebola and Marburg viruses. J Pathol. 2015;235(2):153–74.
108. Baskerville A, Fisher-Hoch SP, Neild GH, Dowsett AB. Ultrastructural pathology of experimental Ebola haemorrhagic fever virus infection. J Pathol. 1985;147(3):199–209.
109. El Sayed SM, Abdelrahman AA, Ozbak HA, Hemeg HA, Kheyami AM, Rezk N, El-Ghoul MB, Nabo MM, Fathy YM. Updates in diagnosis and management of Ebola hemorrhagic fever. J Res Med Sci. 2016;21:84.
110. Pantin-Jackwood M, Wasilenko JL, Spackman E, Suarez DL, Swayne DE. Susceptibility of turkeys to pandemic-H1N1 virus by reproductive tract insemination. Virol J. 2010;7:27.
111. Sid H, Hartmann S, Winter C, Rautenschlein S. Interaction of influenza A viruses with oviduct explants of different avian species. Front Microbiol. 2017;8:1338.
112. Sergerie M, Mieusset R, Croute F, Daudin M, Bujan L. High risk of temporary alteration of semen parameters after recent acute febrile illness. Fertil Steril. 2007;88(4):970.e1-7.
113. Buch JP, Havlovec SK. Variation in sperm penetration assay related to viral illness. Fertil Steril. 1991;55(4):844–6.
114. MacLEOD J. Effect of chickenpox and of pneumonia on semen quality. Fertil Steril. 1951;2(6):523–33.
115. Kantorovich RA, Sokolova NM, Gavrina EM, Solovieva AI, Shteiner AA. The action of influenza virus on human chromosomes. Byull Eksp Biol Med. 1974;77:99–102.
116. Sharma G, Polasa H. Cytogenetic effects of influenza virus infection on male germ cells of mice. Hum Genet. 1978;45(2):179–87.
117. Pathki VK, Polasa H. Dominant lethality in mice–a test for mutagenicity of influenza X-31 virus. Teratog Carcinog Mutagen. 1988;8(1):55–62.
118. Cox R, Green TJ, Purushotham S, Deivanayagam C, Bedwell GJ, Prevelige PE, Luo M. Structural and functional characterization of the mumps virus phosphoprotein. J Virol. 2013;87(13):7558–68.
119. Aiman J, Brenner PF, MacDonald PC. Androgen and estrogen production in elderly men with gynecomastia and testicular atrophy after mumps orchitis. J Clin Endocrinol Metab. 1980;50(2):380–6.
120. Masarani M, Wazait H, Dinneen M. Mumps orchitis. J R Soc Med. 2006;99(11):573–5.
121. Gall AE. The histopathology of acute mumps orchitis. Am J Pathol. 1947;23:637.
122. Charny CW, Meranze DR. Pathology of mumps orchitis. J Urol. 1948;60:140.
123. Dejucq N, Jegou B. Viruses in the mammalian male genital tract and their effects on the reproductive system. Microbiol Mol Biol Rev. 2001;65(2):208–31.
124. Manson AL. Mumps orchitis. Urology. 1990;36:355–8.
125. Adamopoulos DA, Lawrence DM, Vassilopoulos P, Contoyiannis PA, Swyer GI. Pituitary-testicular interrelationships in mumps orchitis and other viral infections. Br Med J. 1978;1:1177–80.
126. Liaw YF, Chu CM. Hepatitis B virus infection. Lancet. 2009;373:582–92.
127. Lozano R, Naghavi M, Foreman K, et al. Global and regional mortality from 235 causes of death for 20 age groups in 1990 and 2010: a systematic analysis for the Global Burden of Disease Study 2010. Lancet. 2012;380:2095–128.
128. Gavilanes F, Gonzalez-Ros JM, Peterson DL. Structure of hepatitis B surface antigen. Characterization of the lipid components and their association with the viral proteins. J Biol Chem. 1982;257(13):7770–7.
129. Liang TJ. Hepatitis B: the virus and disease. Hepatology. 2009;49(5 Suppl):S13–21.

130. Sun P, Zheng J, She G, Wei X, Zhang X, Shi H, Zhou X. Expression pattern of asialoglycoprotein receptor in human testis. Cell Tissue Res. 2013;352(3):761–8.
131. Moretti E, Federico MG, Giannerini V, Collodel G. Sperm ultrastructure and meiotic segregation in a group of patients with chronic hepatitis B and C. Andrologia. 2008;40(3):173–8.
132. Lorusso F, Palmisano M, Chironna M, Vacca M, Masciandaro P, Bassi E, Selvaggi Luigi L, Depalo R. Impact of chronic viral diseases on semen parameters. Andrologia. 2010;42(2):121–6.
133. Kang X, Xie Q, Zhou X, Li F, Huang J, Liu D, Huang T. Effects of hepatitis B virus S protein exposure on sperm membrane integrity and functions. PLoS One. 2012;7(3):e33471.
134. German Advisory Committee Blood (Arbeitskreis Blut), Subgroup 'Assessment of Pathogens Transmissible by Blood'. Human Immunodeficiency Virus (HIV). Transfus Med Hemother. 2016;43(3):203–22.
135. Gupta P, Leroux C, Patterson BK, Kingsley L, Rinaldo C, Ding M, Chen Y, Kulka K, Buchanan W, McKeon B, Montelaro R. Human immunodeficiency virus type 1 shedding pattern in semen correlates with the compartmentalization of viral Quasi species between blood and semen. J Infect Dis. 2000;182(1):79–87.
136. Anderson JA, Ping LH, Dibben O, Jabara CB, Arney L, Kincer L, Tang Y, Hobbs M, Hoffman I, Kazembe P, Jones CD, Borrow P, Fiscus S, Cohen MS, Swanstrom R. Center for HIV/AIDS Vaccine Immunology. HIV-1 Populations in Semen Arise through Multiple Mechanisms. PLoS Pathog. 2010;6(8):e1001053.
137. Anderson D. HIV can be transmitted from asymptomatic carriers. Am Fam Physician. 1991;44:936.
138. Araneta MR, Mascola L, Eller A, O'Neil L, Ginsberg MM, Bursaw M, Marik J, Friedman S, Sims CA, Rekart ML, et al. HIV transmission through donor artificial insemination. JAMA. 1995;273(11):854–8.
139. Bagasra O, Farzadegan H, Seshamma T, Oakes JW, Saah A, Pomerantz RJ. Detection of HIV-1 proviral DNA in sperm from HIV-1-infected men. AIDS. 1994;8(12):1669–74.
140. Muciaccia B, Corallini S, Vicini E, Padula F, Gandini L, Liuzzi G, Lenzi A, Stefanini M. HIV-1 viral DNA is present in ejaculated abnormal spermatozoa of seropositive subjects. Hum Reprod. 2007;22(11):2868–78.
141. Baccetti B, Benedetto A, Burrini AG, Collodel G, Ceccarini EC, Crisà N, Di Caro A, Estenoz M, Garbuglia AR, Massacesi A, et al. HIV-particles in spermatozoa of patients with AIDS and their transfer into the oocyte. J Cell Biol. 1994;127(4):903–14.
142. Kim LU, Johnson MR, Barton S, Nelson MR, Sontag G, Smith JR, Gotch FM, Gilmour JW. Evaluation of sperm washing as a potential method of reducing HIV transmission in HIV-discordant couples wishing to have children. AIDS. 1999;13(6):645–51.
143. Gil T, Castilla JA, Hortas ML, Molina J, Redondo M, Samaniego F, Garrido F, Vergara F, Herruzo A. CD4+ cells in human ejaculates. Hum Reprod. 1995;10(11):2923–7.
144. Gobert B, Amiel C, Tang JQ, Barbarino P, Béné MC, Faure G. CD4-like molecules in human sperm. FEBS Lett. 1990;261(2):339–42.
145. Ashida ER, Scofield VL. Lymphocyte major histocompatibility complex-encoded class II structures may act as sperm receptors. Proc Natl Acad Sci USA. 1987;84:3395–9.
146. Lavitrano M, Maione B, Forte E, Francolini M, Sperandio S, Testi R, Spadafora C. The interaction of sperm cells with exogenous DNA: a role of CD4 and major histocompatibility complex class II molecules. Exp Cell Res. 1997;233(1):56–62.
147. Poretsky L, Can S, Zumoff B. Testicular dysfunction in human immunodeficiency virus-infected men. Metabolism. 1995;44(7):946–53.
148. Pudney J, Anderson D. Orchitis and human immunodeficiency virus type 1 infected cells in reproductive tissues from men with the acquired immune deficiency syndrome. Am J Pathol. 1991;139(1):149–60.
149. Rogers C, Klatt EC. Pathology of the testis in acquired immunodeficiency syndrome. Histopathology. 1988;12(6):659–65.

150. Mucsi I, Gyulai Z, Béládi I. Combined effects of flavonoids and acyclovir against herpesviruses in cell cultures. Acta Microbiol Hung. 1992;39(2):137–47.
151. Manvar D, Mishra M, Kumar S, Pandey VN. Identification and evaluation of anti hepatitis C virus phytochemicals from Eclipta alba. J Ethnopharmacol. 2012;144(3):545–54.
152. Chiang LC, Ng LT, Cheng PW, Chiang W, Lin CC. Antiviral activities of extracts and selected pure constituents of Ocimum basilicum. Clin Exp Pharmacol Physiol. 2005;32(10):811–6.
153. Hakobyan A, Arabyan E, Avetisyan A, Abroyan L, Hakobyan L, Zakaryan H. Apigenin inhibits African swine fever virus infection in vitro. Arch Virol. 2016;161(12):3445–53.
154. Shibata C, Ohno M, Otsuka M, Kishikawa T, Goto K, Muroyama R, Kato N, Yoshikawa T, Takata A, Koike K. The flavonoid apigenin inhibits hepatitis C virus replication by decreasing mature microRNA122 levels. Virology. 2014;462-463:42–8.
155. Mehla R, Bivalkar-Mehla S, Chauhan A. A flavonoid, luteolin, cripples HIV-1 by abrogation of tat function. PLoS One. 2011;6(11):e27915.
156. Wu CC, Fang CY, Hsu HY, Chen YJ, Chou SP, Huang SY, Cheng YJ, Lin SF, Chang Y, Tsai CH, Chen JY. Luteolin inhibits Epstein-Barr virus lytic reactivation by repressing the promoter activities of immediate-early genes. Antiviral Res. 2016;132:99–110.
157. Yi L, Li Z, Yuan K, Qu X, Chen J, Wang G, Zhang H, Luo H, Zhu L, Jiang P, Chen L, Shen Y, Luo M, Zuo G, Hu J, Duan D, Nie Y, Shi X, Wang W, Han Y, Li T, Liu Y, Ding M, Deng H, Xu X. Small molecules blocking the entry of severe acute respiratory syndrome coronavirus into host cells. J Virol. 2004;78(20):11334–9.
158. Sithisarn P, Michaelis M, Schubert-Zsilavecz M, Cinatl J Jr. Differential antiviral and anti-inflammatory mechanisms of the flavonoids biochanin A and baicalein in H5N1 influenza A virus-infected cells. Antiviral Res. 2013;97(1):41–8.
159. Xu G, Dou J, Zhang L, Guo Q, Zhou C. Inhibitory effects of baicalein on the influenza virus in vivo is determined by baicalin in the serum. Biol Pharm Bull. 2010;33(2):238–43.
160. Nayak MK, Agrawal AS, Bose S, Naskar S, Bhowmick R, Chakrabarti S, Sarkar S, Chawla-Sarkar M. Antiviral activity of baicalin against influenza virus H1N1-pdm09 is due to modulation of NS1-mediated cellular innate immune responses. J Antimicrob Chemother. 2014;69(5):1298–310.
161. Chu M, Xu L, Zhang MB, Chu ZY, Wang YD. Role of Baicalin in anti-influenza virus A as a potent inducer of IFN-gamma. Biomed Res Int. 2015;2015:263630.
162. Chung ST, Chien PY, Huang WH, Yao CW, Lee AR. Synthesis and anti-influenza activities of novel baicalein analogs. Chem Pharm Bull (Tokyo). 2014;62(5):415–21.
163. Li BQ, Fu T, Dongyan Y, Mikovits JA, Ruscetti FW, Wang JM. Flavonoid baicalin inhibits HIV-1 infection at the level of viral entry. Biochem Biophys Res Commun. 2000;276(2):534–8.
164. Cho WK, Weeratunga P, Lee BH, Park JS, Kim CJ, Ma JY, Lee JS. Epimedium koreanum Nakai displays broad spectrum of antiviral activity in vitro and in vivo by inducing cellular antiviral state. Viruses. 2015;7(1):352–77.
165. Hung PY, Ho BC, Lee SY, Chang SY, Kao CL, Lee SS, Lee CN. Houttuynia cordata targets the beginning stage of herpes simplex virus infection. PLoS One. 2015;10(2):e0115475.
166. Wu W, Li R, Li X, He J, Jiang S, Liu S, Yang J. Quercetin as an antiviral agent inhibits influenza A virus (IAV) entry. Viruses. 2015;8(1):6.
167. Liu Z, Zhao J, Li W, Shen L, Huang S, Tang J, Duan J, Fang F, Huang Y, Chang H, Chen Z, Zhang R. Computational screen and experimental validation of anti-influenza effects of quercetin and chlorogenic acid from traditional Chinese medicine. Sci Rep. 2016;6:19095.
168. Liu Z, Zhao J, Li W, Wang X, Xu J, Xie J, Tao K, Shen L, Zhang R. Molecular docking of potential inhibitors for influenza H7N9. Comput Math Methods Med. 2015;2015:480764.
169. Fatima K, Mathew S, Suhail M, Ali A, Damanhouri G, Azhar E, Qadri I. Docking studies of Pakistani HCV NS3 helicase: a possible antiviral drug target. PLoS One. 2014;9(9):e106339.
170. Mathew S, Fatima K, Fatmi MQ, Archunan G, Ilyas M, Begum N, Azhar E, Damanhouri G, Qadri I. Computational docking study of p7 Ion channel from HCV Genotype 3 and Genotype 4 and its interaction with natural compounds. PLoS One. 2015;10(6):e0126510.

171. Yarmolinsky L, Huleihel M, Zaccai M, Ben-Shabat S. Potent antiviral flavone glycosides from Ficus benjamina leaves. Fitoterapia. 2012;83(2):362–7.
172. Ha SY, Youn H, Song CS, Kang SC, Bae JJ, Kim HT, Lee KM, Eom TH, Kim IS, Kwak JH. Antiviral effect of flavonol glycosides isolated from the leaf of Zanthoxylum piperitum on influenza virus. J Microbiol. 2014;52(4):340–4.
173. Behbahani M, Sayedipour S, Pourazar A, Shanehsazzadeh M. In vitro anti-HIV-1 activities of kaempferol and kaempferol-7-O-glucoside isolated from Securigera securidaca. Res Pharm Sci. 2014;9(6):463–9.
174. Yang L, Lin J, Zhou B, Liu Y, Zhu B. Activity of compounds from Taxillus sutchuenensis as inhibitors of HCV NS3 serine protease. Nat Prod Res. 2017;31(4):487–91.
175. Jeong HJ, Ryu YB, Park SJ, Kim JH, Kwon HJ, Kim JH, Park KH, Rho MC, Lee WS. Neuraminidase inhibitory activities of flavonols isolated from Rhodiola rosea roots and their in vitro anti-influenza viral activities. Bioorg Med Chem. 2009;17(19):6816–23.
176. Schwarz S, Sauter D, Wang K, Zhang R, Sun B, Karioti A, Bilia AR, Efferth T, Schwarz W. Kaempferol derivatives as antiviral drugs against the 3a channel protein of coronavirus. Planta Med. 2014;80(2-3):177–82.
177. Nakayama M, Suzuki K, Toda M, Okubo S, Hara Y, Shimamura T. Inhibition of the infectivity of influenza virus by tea polyphenols. Antiviral Res. 1993;21(4):289–99.
178. Colpitts CC, Schang LM. A small molecule inhibits virion attachment to heparan sulfate- or sialic acid-containing glycans. J Virol. 2014;88(14):7806–17.
179. Williamson MP, McCormick TG, Nance CL, Shearer WT. Epigallocatechin gallate, the main polyphenol in green tea, binds to the T-cell receptor, CD4: potential for HIV-1 therapy. J Allergy Clin Immunol. 2006;118(6):1369–74.
180. Kawai K, Tsuno NH, Kitayama J, Okaji Y, Yazawa K, Asakage M, Hori N, Watanabe T, Takahashi K, Nagawa H. Epigallocatechin gallate, the main component of tea polyphenol, binds to CD4 and interferes with gp120 binding. J Allergy Clin Immunol. 2003;112(5):951–7.
181. Nakane H, Ono K. Differential inhibitory effects of some catechin derivatives on the activities of human immunodeficiency virus reverse transcriptase and cellular deoxyribonucleic and ribonucleic acid polymerases. Biochemistry. 1990;29(11):2841–5.
182. Chang LK, Wei TT, Chiu YF, Tung CP, Chuang JY, Hung SK, Li C, Liu ST. Inhibition of Epstein-Barr virus lytic cycle by (-)-epigallocatechin gallate. Biochem Biophys Res Commun. 2003;301(4):1062–8.
183. Liu S, Li H, Chen L, Yang L, Li L, Tao Y, Li W, Li Z, Liu H, Tang M, Bode AM, Dong Z, Cao Y. (-)-Epigallocatechin-3-gallate inhibition of Epstein-Barr virus spontaneous lytic infection involves ERK1/2 and PI3-K/Akt signaling in EBV-positive cells. Carcinogenesis. 2013;34(3):627–37.
184. Isaacs CE, Wen GY, Xu W, Jia JH, Rohan L, Corbo C, Di Maggio V, Jenkins EC Jr, Hillier S. Epigallocatechin gallate inactivates clinical isolates of herpes simplex virus. Antimicrob Agents Chemother. 2008;52(3):962–70.
185. Zhong L, Hu J, Shu W, Gao B, Xiong S. Epigallocatechin-3-gallate opposes HBV-induced incomplete autophagy by enhancing lysosomal acidification, which is unfavorable for HBV replication. Cell Death Dis. 2015;6:e1770.
186. He W, Li LX, Liao QJ, Liu CL, Chen XL. Epigallocatechin gallate inhibits HBV DNA synthesis in a viral replication - inducible cell line. World J Gastroenterol. 2011;17(11):1507–14.
187. Huang HC, Tao MH, Hung TM, Chen JC, Lin ZJ, Huang C. (-)-Epigallocatechin-3-gallate inhibits entry of hepatitis B virus into hepatocytes. Antiviral Res. 2014;111:100–11.
188. Calland N, Albecka A, Belouzard S, Wychowski C, Duverlie G, Descamps V, Hober D, Dubuisson J, Rouillé Y, Séron K. (−)-Epigallocatechin-3-gallate is a new inhibitor of hepatitis C virus entry. Hepatology. 2012;55(3):720–9.
189. Chen C, Qiu H, Gong J, Liu Q, Xiao H, Chen XW, Sun BL, Yang RG. (−) Epigallocatechin-3-gallate inhibits the replication cycle of hepatitis C virus. Arch Virol. 2012;157(7):1301–12.
190. Carneiro BM, Batista MN, Braga ACS, Nogueira ML, Rahal P. The green tea molecule EGCG inhibits Zika virus entry. Virology. 2016;496:215–8.

191. Goldwasser J, Cohen PY, Lin W, Kitsberg D, Balaguer P, Polyak SJ, Chung RT, Yarmush ML, Nahmias Y. Naringenin inhibits the assembly and long-term production of infectious hepatitis C virus particles through a PPAR-mediated mechanism. J Hepatol. 2011;55(5):963–71.
192. Guo J, Xu X, Rasheed TK, Yoder A, Yu D, Liang H, Yi F, Hawley T, Jin T, Ling B, Wu Y. Genistein interferes with SDF-1- and HIV-mediated actin dynamics and inhibits HIV infection of resting CD4 T cells. Retrovirology. 2013;10:62.
193. Sauter D, Schwarz S, Wang K, Zhang R, Sun B, Schwarz W. Genistein as antiviral drug against HIV ion channel. Planta Med. 2014;80(8–9):682–7.
194. Argenta DF, Silva IT, Bassani VL, Koester LS, Teixeira HF, Simões CM. Antiherpes evaluation of soybean isoflavonoids. Arch Virol. 2015;160(9):2335–42.
195. Rechtman MM, Har-Noy O, Bar-Yishay I, Fishman S, Adamovich Y, Shaul Y, Halpern Z, Shlomai A. Curcumin inhibits hepatitis B virus via down-regulation of the metabolic coactivator PGC-1alpha. FEBS Lett. 2010;584(11):2485–90.
196. Kim K, Kim KH, Kim HY, Cho HK, Sakamoto N, Cheong J. Curcumin inhibits hepatitis C virus replication via suppressing the Akt-SREBP-1 pathway. FEBS Lett. 2010;584(4):707–12.
197. Lin LT, Chen TY, Lin SC, Chung CY, Lin TC, Wang GH, Anderson R, Lin CC, Richardson CD. Broad-spectrum antiviral activity of chebulagic acid and punicalagin against viruses that use glycosaminoglycans for entry. BMC Microbiol. 2013;13:187.
198. Lin LT, Chen TY, Chung CY, Noyce RS, Grindley TB, McCormick C, Lin TC, Wang GH, Lin CC, Richardson CD. Hydrolyzable tannins (chebulagic acid and punicalagin) target viral glycoprotein-glycosaminoglycan interactions to inhibit herpes simplex virus 1 entry and cell-to-cell spread. J Virol. 2011;85(9):4386–98.
199. Kudo E, Taura M, Matsuda K, Shimamoto M, Kariya R, Goto H, Hattori S, Kimura S, Okada S. Inhibition of HIV-1 replication by a tricyclic coumarin GUT-70 in acutely and chronically infected cells. Bioorg Med Chem Lett. 2013;23(3):606–9.
200. Boasso A. Type I interferon at the interface of antiviral immunity and immune regulation: the curious case of HIV-1. Scientifica (Cairo). 2013;2013:580968.
201. Stetson DB, Medzhitov R. Type I interferons in host defense. Immunity. 2006;25(3):373–81.
202. Brand S, Zitzmann K, Dambacher J, Beigel F, Olszak T, Vlotides G, Eichhorst ST, Göke B, Diepolder H, Auernhammer CJ. SOCS-1 inhibits expression of the antiviral proteins 2',5'-OAS and MxA induced by the novel interferon-lambdas IL-28A and IL-29. Biochem Biophys Res Commun. 2005;331(2):543–8.
203. Zeuzem S, Welsch C, Herrmann E. Pharmacokinetics of peginterferons. Semin Liver Dis. 2003;23(Suppl 1):23–8.
204. Aghemo A, Rumi MG, Colombo M. Pegylated interferons alpha2a and alpha2b in the treatment of chronic hepatitis C. Nat Rev Gastroenterol Hepatol. 2010;7(9):485–94.
205. Kwon H, Lok AS. Hepatitis B therapy. Nat Rev Gastroenterol Hepatol. 2011;8(5):275–84.
206. Ishikawa T, Kubota T, Abe H, Nagashima A, Hirose K, Togashi T, Seki K, Honma T, Yoshida T, Kamimura T. Efficacy of the regimen using twice-daily β-interferon followed by the standard of care for chronic hepatitis C genotype 1b with high viral load. Hepatol Res. 2012;42(9):864–9.
207. Itokawa N, Atsukawa M, Tsubota A, Kondo C, Hashimoto S, Fukuda T, Matsushita Y, Kidokoro H, Kobayashi T, Narahara Y, Nakatsuka K, Kanazawa H, Iwakiri K, Sakamoto C. Lead-in treatment with interferon-β/ribavirin may modify the early hepatitis C virus dynamics in pegylated interferon alpha-2b/ribavirin combination for chronic hepatitis C patients with the IL28B minor genotype. J Gastroenterol Hepatol. 2013;28(3):443–9.
208. Jarvis JN, Meintjes G, Rebe K, Williams GN, Bicanic T, Williams A, Schutz C, Bekker LG, Wood R, Harrison TS. Adjunctive interferon-γ immunotherapy for the treatment of HIV-associated cryptococcal meningitis: a randomized controlled trial. AIDS. 2012;26(9):1105–13.
209. Robek MD, Boyd BS, Chisari FV. Lambda interferon inhibits hepatitis B and C virus replication. J Virol. 2005;79(6):3851–4.

Chapter 14
Pathogenesis of Viral Infections and Male Reproductive Health: An Evidence-Based Study

Diptendu Sarkar, Shubham Dutta, Shubhadeep Roychoudhury, Preethi Poduval, Niraj Kumar Jha, Paltu Kumar Dhal, Shatabhisha Roychoudhury, and Kavindra Kumar Kesari

Abstract Viruses, being intracellular obligate parasites, can cause several congenital and sexually transmitted diseases. Depending on the site of infection, viruses can adopt various pathogenic mechanisms for their survival and to escape the host immune response. The male reproductive system is one of the attainable targets of many viruses including immunodeficiency virus (HIV), Zika virus (ZIKV), adenovirus, cytomegalovirus (CMV), and severe acute respiratory syndrome coronavirus

---

D. Sarkar
Department of Microbiology, Ramakrishna Mission Vidyamandira, Belur Math, West Bengal, India

S. Dutta
Department of Veterinary Microbiology, University of Saskatchewan, Saskatoon, Canada

S. Roychoudhury (✉)
Department of Life Science and Bioinformatics, Assam University, Silchar, India

P. Poduval
Department of Biotechnology, Dhempe College of Arts and Science, Miramar, Goa, India

N. K. Jha
Department of Biotechnology, School of Engineering and Technology (SET), Sharda University,
Greater Noida, India

P. K. Dhal
Department of Life Science and Biotechnology, Jadavpur University, Kolkata, India

S. Roychoudhury
Department of Microbiology, R. G. Kar Medical College and Hospital, Kolkata, India

Health Centre, Assam University, Silchar, India

K. K. Kesari
Department of Bioproducts and Biosystems, School of Chemical Engineering,
Aalto University, Espoo, Finland

© The Author(s), under exclusive license to Springer Nature Switzerland AG 2022
K. K. Kesari, S. Roychoudhury (eds.), *Oxidative Stress and Toxicity in Reproductive Biology and Medicine*, Advances in Experimental Medicine and Biology 1358, https://doi.org/10.1007/978-3-030-89340-8_14

2 (SARS-CoV-2), and infection with such viruses may cause serious health issues. Leydig cells and seminiferous tubules are the prime sites of mammalian testis for viral infection. The azoospermic condition is a common symptom of viral infection, wherein the hypothalamic-pituitary-testicular (HPT) axis can be disrupted, leading to decreased levels of luteinizing hormone (LH). Furthermore, oxidative stress (OS) is a major contributing factor to viral infection-associated male infertility. The likelihood of direct and indirect infection, as well as sex-based variability in the vulnerability pattern to viral infections, has been observed. However, there appears to be a long-term impact of viral infection on male reproductive performance due to testicular tissue pathogenicity – a process that requires thorough investigation. The present study aimed to explore how the viruses affect the male reproductive system, including their distribution in tissues and body fluids, possible targets as well as the effects on the endocrine system. We used the major electronic databases such as MEDLINE and SCOPUS. Google Scholar was also consulted for additional literature search related to the topic. Obtained literatures were sorted based on the content. The articles that reported the pathogenesis of viruses on male reproductive health and were published in the English language were included in the present study.

Keywords Viral infection · Epidemic · Sexual transmission · Male reproductive system · Reproductive endocrinology · Oxidative stress

14.1 Introduction

The discovery of the human immunodeficiency virus (HIV) in the 1980s sparked a growing interest in male reproductive research [1]. A number of viruses (e.g., adenovirus, Polyomavirus hominis 1 – BK virus, cytomegalovirus – CMV, parvovirus B-19, Zika virus – ZIKV) have been reported to cause viremia in human semen [2, 3]. The symptoms and outcomes of the virally infected pathogenesis can differ as the viruses belong to different families [3]. An infected partner may transfer the viruses into the semen through sexual interaction or by physical contact with the unprotected skin. Viruses may potentially be transmitted from the penile epithelia during the collection of samples from the infected patient, too [4, 5]. Sexual transmission may be one of the main reasons behind few epidemics, as there is a possibility that the viruses may be circulating across the world through the gametes and embryos of infected humans and/or animals [3].

Different pathogenic processes have been described depending on the site of infection on the human body – (a) a systemic, acute, or chronic infection can result in irreversible infertility, hormone imbalances, testicular dysfunction, and spermatogenesis disruption [4]; (b) testicular orchitis inhibits the sperm production [5]; and (c) male accessory gland (epididymis, prostate, and seminal vesicles) and urethral infections may impose serious complications on the male reproductive features and fertility. Such cases of virus-induced semen infection are believed to be increasing, especially in asymptomatic males associated with poor sperm quality [6]. Furthermore, another astonishing observation includes that with the increase in viral load, there is an absence of seminal leukocytes, which are the indicators of male genital tract infections [4, 7, 8]. Chronic sperm infections, and subsequent decline in fertility may particularly be brought about by the HIV, hepatitis B virus (HBV), and hepatitis C virus (HCV). More specifically, HIV contamination has been verified in causing persistent low genitourinary tract inflammation, sperm contamination, and decreased fertility [9]. It has been established that the HBV or HCV contamination in semen reduces sperm head motility [10] and increases the frequency of sperm aneuploidy and DNA fragmentation [11]. Human papillomavirus (HPV) has been demonstrated to exert a negative influence on sperm parameters [12–15]. Other viruses that may cause sexually transmitted diseases (STDs) are also capable of colonizing in the semen. This is a principal problem that scientists encompass especially with respect to the herpesvirus (HSV), human cytomegalovirus (HCMV), and adeno-related virus (AAV). Both the viral DNA and RNA at the sperm level can induce testicular damage and may exert negative impact on sperm parameters [16, 17]. In fact, little is known about the viral infectivity of the reproductive system at clinical as well as molecular levels. Data on the contamination capability of coronaviruses inside the human reproductive tract date back to the SARS-CoV epidemic of 2002. SARS-CoV2 has been purported to act in a really comparable way while it influences generative features. The male reproductive tissues express better degrees of the ACE2 receptor in comparison to that of females that can justify the higher vulnerability of male reproductive tract to the SARS-CoV and/or SARS-CoV2 infection relative to that of the female [4,5,8,]. SARS-CoV and SARS-CoV2 are usually not detected in the humoral samples of patients, despite the fact that ACE2 receptors are highly expressed in gametes, Leydig cells, and the Sertoli cells [4, 9, 18]. More importantly, the current scenario of COVID-19 pandemic raises questions about the pathogenesis of SARS-CoV2 particularly in relation to its potential impact on human reproduction and fertility. This evidence-based study presents the conceivable effect of viral infections on the regenerative capacity of the male. Alongside, the study also features the hypotheses that need in-depth research for the specific fundamental instruments as to how such infections including that of the coronaviruses are related to men's health and wellbeing.

14.2 Methodology

We primarily used major electronic databases such as MEDLINE and SCOPUS. The web search engine Google Scholar was also consulted for additional literature search related to the topic, using the keyword strings such as "testis and virus", "male reproductive organ and virus", "virus, semen and prostate", "testicular antiviral defense", etc. Obtained literature was sorted based on the content and relevance. Articles that did not fit to the purview of the theme according to our search key of the current topic were excluded. Even many papers were not considered as they were not in the English language. Every article that specifically highlighted the behavioral, histological, or any specific biochemical activity in mice or human due to various viral infections was kept back. Articles allied to the above issues on organisms besides human beings were excluded. Those articles reporting only pathogenesis of virus on male reproductive health published prior to the year 2010 have also been excluded.

14.3 Viruses and the Male Reproductive System

Different microorganisms, which include bacteria, viruses, and protozoa, can infect the male reproductive system and may affect fertility [19–23]. Among these, viruses play an important part in causing several congenital diseases (including mumps, measles, etc.) and severe STDs (such as HIV, HSV, HPV, etc.). Their infectivity could range from moderate to broad in the context of tropism (Table 14.1). Bacterial infections thrive in the urogenital tract and basically have an effect on the accessory glands and epididymis, which is comparable to the viral infections in the bloodstream and predominantly have an effect on the testes [13, 14].

Table 14.1 Viral infections of the male reproductive tract and their consequences

Virus	Site of infection	Disease/Effect
Mumps virus	Leydig cells (4), germ cells (14, 27, 36)	Orchitis (29, 34, 36), testicular atrophy (29, 34), sterility (3, 34, 58), decrease in androgen secretion (4), testicular cancer (34, 24)
HIV	Lymphocytes and macrophages (25, 37), germ cells (22, 24, 51)	Orchitis, interstitial fibrosis, lymphocyte infiltration, change in Leydig cell number, decrease in germ cell number (63, 22, 23, 64, 48), change in spermatogenesis (33, 42, 43, 52, 57, 54)
EBV	Semen (37, 79)	Orchitis (68, 73), testicular cancer (10, 26, 30)
Parvovirus B19	Testes (5, 27, 33, 56)	Testicular cancer (12, 54, 60)
HSV-2	Testes (22, 30,52)	Viral reservoir (5, 12, 24, 33–37)
HSV-1	Testes (35, 52)	Infertility (15, 38, 60)
Adenovirus	Semen (14, 21)	Infertility (45, 60, 93)
Coxsackie virus	Testes (15, 33, 36, 79)	Orchitis (21, 33, 36,56,79)
Influenza virus	Not available	Orchitis (56, 68,73)
Endogenous retroviruses	Germ cells (7, 25, 33, 59)	Testicular cancer (32, 40, 65)

14.3.1 Viruses in Semen

Infections can be explicitly communicated for up to 41 days [24]. A meta-analysis has determined the existence of 27 viruses in the human semen that infects the bloodstream [25]. Those encompass the viruses that cause Ebola virus disease, AIDS, HBV, and HSV [26]. Careful review of the literature revealed that at least 11 viruses can survive within the testes, especially in men infected with influenza, dengue, and SARS viruses [27]. HIV is the most dominant pathogen that is sexually transmitted. Its presence within the semen causes virus-induced cellular damage; however, the nature of the inflamed cells remains unclear [28]. The virus was discovered in the lymphoid fraction of 2 men's sperm samples in the initial period of the disease, which was principally spread by a HIV-1-positive individual [5, 15, 17, 29]. The virus is known to be transmitted through the semen of asymptomatic carriers [30]. HIV-1 has been found in both symptomatic and asymptomatic AIDS patients' cell-free seminal fluid [31]. This finding implies that HIV may be released into the epididymal fluid when epididymal epithelial cells become infected. In fact, the prostate and seminal vesicles are the viral reservoirs that release HIV into the sperm cells [32]. As a result, the virus is exposed to macrophages and lymphocytes that are not an unusual place within the sperm and are genuine targets for HIV [33]. Free virions, virions linked with sperm, and virions that have infiltrated leukocytes are the three principal sources of HIV-1 in the semen [14, 34]. The main vectors of HIV-1 in the semen have been identified as leukocytes, lymphocytes, monocytes, and macrophages from the precedent sources [35]. There are controversies regarding the internalization of HIV-1 by sperm that may be considered as passive carriers of HIV-1 facilitating attachment to the cell surface [36]. During sexual intercourse, individuals having HIV-induced inflammation can transmit the virus through their semen, which incorporates free-floating virus in addition to the inflamed leukocytes [19, 22]. Sperm cell expresses molecules that interact with HIV envelope proteins (e.g., heparan sulfate and mannose receptors) [37]. After the attachment, the viruses get transmitted from sperm to dendritic cells (DCs) as shown in culture. The CD4 and dendritic cell-specific intercellular adhesion molecule-3-grabbing non-integrin (DC-SIGN) are required for such a transmission [38]. This indicates that DCs become aware of the presence of the viruses by means of binding to sperm in place of consuming them. Da Silva et al. (2016) first confirmed the response of sperm ligand with CD4 antibodies and studied the interactions among sperm and HLA-DR high-quality cells which offer sturdy backing to the proof of expression of CD4 on sperm [34]. Immunofluorescence and Western blotting helped to detect the CD4 molecules on mouse sperm heads [21].

Transmission of ZIKV through sexual contact is one of the most pressing public health concerns. Infectious virus particles can survive in the sperm for up to 6 months after the infection and can be passed from male to female sexual partners during unprotected sexual contact. Reports of ZIKV RNA confirmed that ZIKV was able to infect the MRT, although the mechanism of invasion, replication, and persistence is unknown [22, 35]. Further research on how ZIKV enters the MRT and interacts with different cell types, which could aid in the development of vaccines and antiviral medicines, is needed. Furthermore, as ZIKV can be excreted repeatedly over several months, it poses a risk to the semen bank and human fertility efforts. Semen alterations have been reported, including a decreased sperm count with increasing sperm anomalies, specifically in patients with ZIKV RNA-positive seminal specimens [23]. However, the details of ZIKV infection in the genital tract are still not clear. Immunohistochemistry revealed the presence of ZIKV antigen in the heads of mature sperm in a chronic ZIKV positive patient within the sperm [34]. Post-infection studies of ZIKV RNA, by Atkinson et al., point toward the knowledge gap about the persistence of viruses in genital fluids during the replication of ZIKV2 [2, 5]. Despite having a pandemic potential, there is no strong evidence of semen infection by influenza virus; however, influenza virus infection was found to reduce sperm count in boars [20, 24].

On the other hand, human T-lymphotropic virus type I (HTLV-1) is sexually transmitted through the sperm, most likely by infected sperm lymphocytes [25]. Human herpesvirus 8 (HHV-8) has been considered as the principal pathogen behind Kaposis sarcoma that is integrally associated with HIV infection. Epidemiological studies revealed the transmission of the virus is mostly by sexual intercourse. A multicenter research published in 1999 concluded that HHV-8 DNA was found in the semen at a concentration that was insufficient for consistent detection [26]. Hence, the seminal viral load should be quantified to detect the transmission probability within the population. Another *Herpesviridae* family member HCMV has also been detected in infected body fluids including semen [27]. HCMV-infected sperm is thought to increase the risk of AIDS by activating CD4+ cells, making it one of the most common opportunistic infections in AIDS patients [28].

Recently, different cohort studies have assessed the risk of seminal infection by SARS-CoV2. It has been observed that mild infection did not affect testis or epididymis functionality but semen parameters were impaired after a moderate infection. Furthermore, viral RNA remained undetected by RT-PCR in the semen of acute COVID-19 patients suggesting the absence of sexual transmission of virus [29]. However, another group has detected the presence of SARS-CoV2 in semen, and even speculated that MRT may be seeded by SARS-CoV2 due to the lacunae of blood-testes/vas deferens/epididymis barriers in the presence of systemic local inflammation [30]. Testes are an organ that has the ACE2 receptor, according to research conducted using existing proteomic databases and an examination of sperm surface using monoclonal antibodies. In addition to ACE2, another protein called TMPRSS2 aids in the fusing of SARS-CoV2 with human sperm. This protease has been found in prostasomes, which are released into the seminal fluid by the prostate gland after ejaculation. A cautious scrutiny of the human sperm proteomic database additionally explained the associated proteases TMPRSS11B and TMPRSS12 in addition to FURIN in sperm cells – all of which province like activating proteases for viral contamination consisting of coronaviruses. The companionship of those activating proteases and ACE2 within the sperm plasma membrane increased the chances of sexual transmission of the virus. More studies would possibly help in discovering the sexual transmission of SARS in the future [7, 31, 32].

14.3.2 Viral Infection in the Testis

There are two main functional fundamental compartments within the mammalian testis [30]. The Leydig cells (liable for testosterone release) alongside macrophages, fibroblasts, blood, and lymphatic vessels are taken into consideration as interstitium (primary compartment). The second compartment consists of the seminiferous tubules that are surrounded by a multifaceted stratified epithelium that contains two distinct populations of cells: spermatogenic cells that develop into spermatozoa and Sertoli cells that serve a supportive and nutritive function. Infection is replicated in both these compartments in human males and other well-evolved life forms [5, 7, 32]. Figure 14.1 represents a mechanism of the viruses interaction and steps toward entering into the cells and causing seminal infection.

Most prevalent viruses that cause testicular infections include ZIKV, HIV, and mumps virus [30, 31]. Few investigations have confirmed the role of oncogenic viruses like herpes or papillomavirus in testicular cancer etiology. Different arboviruses such as ZIKV and bluetongue viruses have been documented to infect the testis, causing testicular atrophy [31]. Bluetongue virus can replicate within the endothelial cells of the peritubular area that surrounds the seminiferous tubules [6, 32]. Testicular degeneration results in azoospermia that ultimately lead to male infertility [33]. The symptoms associated with mumps are limited to infectious parotitis in adolescent

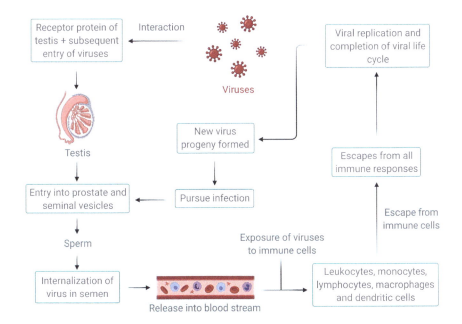

Fig. 14.1 Virus survival mechanism and virus-induced seminal infection. It represents the pathways of virus entry into cells, beginning with the bloodstream and progressing to leukocytes, lymphocytes, monocytes, and macrophages. This may allow the virus to accomplish its life cycle while evading the immune responses

males, while orchitis is the most common illness in men aged 15 to 30 years [3, 21]. Initially after infection, the virus immediately attacks the testis followed by destruction of testicular parenchyma and diminishing androgen production [34]. A link between HPV infection and testicular cancer has been confirmed in a recent meta-analysis. HPV DNA changes its nature inside the cell from episomal to host-genome integrated, influencing cellular genome transcription [32, 35].

In other situations, such as COVID-19, the similarity of coronaviruses indicates toward their sexual orientation, with males being more vulnerable than females. The ACE2 receptor is important in COVID-19 pathogenesis because it aids direct host cell damage. SARS-CoV2 can enter cells and replicate through binding to the ACE2 receptor [3]. Consequently, cells with higher ACE2 expression appear to be more vulnerable to SAR-CoV2 infection [3]. According to a recent study, the testes have the greatest levels of ACE2 mRNA and macromolecule expression as compared to other bodily tissues [3]. Spermatogonia, Leydig cells, and Sertoli cells are among the male reproductive gland cells that express ACE2 RNA [3–5]. Furthermore, ACE2 protein levels are significantly higher in male reproductive gland cells than in male sex glands [6, 7], supporting the idea that male ductless gland functions are more sensitive to SARS-CoV-2-mediated failure. The communication of ACE2 has also been linked to aging [5]. ACE2 receptor expression levels are the highest in patients between ages 15 and 40, and the lowest in those above 60 years of age [36]. COVID-19 has been demonstrated to render young males more susceptible to testicular injury than older patients. This also suggests that different secretory environments have a significant pathophysiological role in SARS-CoV2 infection in the male. The primary step required for SARS-CoV2 infectivity, as previously documented, is the proteases-mediated preparation of the viral spike proteins with the cell membrane interface, mostly through the transmembrane proteolytic enzyme TMPRSS2 [11]. TMPRSS2 is thought to cleave the ACE2 protein, allowing the virus to enter the host cell more easily [12]. According to an investigation, the sex hormone receptor activation is required to start the transcription of the TMPRSS2 gene. Augmented sex chromosome inheritance of genetic polymorphisms and, as a result, endogenous sex hormone activities could be a potential mechanism to explain male vulnerability to COVID-19. This is due to the fact that the gene loci for sex hormone receptor and ACE2 are situated on X chromosome [13].

14.3.3 Viral Infection in the Prostate

The predominant male steroid hormone is testosterone [3]. Androgenic hormone is important for the development of male reproductive tissues such as the testes and prostate, as well as secondary sexual traits such as increased muscle and bone mass [22–27]. Androgenic hormone is a steroid that belongs to the androstane type and has a keto and hydroxyl group at positions three and seventeen, respectively. It operates by attaching to the sex hormone receptor and causing it to be activated. Traditional sperm cell growth requires androgenic hormone [6–9, 11]. It activates genes in Sertoli cells that enhance spermatogonia demarcation. Under dominance demand, it synchronizes acute hypothalamic–pituitary–adrenal (HPA) axis paying back. Physiological stress in males has been linked to lower sperm cell quality and higher sexual dysfunction [38–42]. This could be due to repressive effects on the HPG axis, which affects the testosterone levels, causing Sertoli cell modifications and as a result, changing the blood–testis barrier, resulting in gametogenesis arrest. Proper discharge balance is a crucial necessity for male fertility. Endocrine dysfunction could impair the reproductive health. SARS-CoV2 is expected to cause inflammatory reactions that impair the hypothalamic–pituitary–testicular (HPT) axis operations, resulting in lower levels of luteinizing hormone (LH), cyst stimulating secretion (FSH), and androgenic hormones [42–48]. COVID-19 patients have been observed to contain lower blood serum androgenic hormone levels, higher gonadotropin levels, and a lower androgenic hormone to gonadotropin magnitude ratio than healthy men indicating that the present theory is flawed [45]. This reveals the missing links in the relationship between SARS-CoV2 infection and androgen regulation, which requires immediate attention in order to clarify the hypothesis regarding SARS-CoV2 infection in relation to male fertility. The male androgenic hormone secretion begins to drop around the age of 30, with signs and symptoms of androgenic hormone deficiency often appearing around the age of 65 [4]. Attenuated sexual desire, impotency, weariness, anxiety, diminished strength, bone loss, and increased belly fat are all hallmarks of age-related deficits in androgenic hormone production in the primary sector [49, 50].

Mumps virus causes a dramatic alteration in Leydig cell activity in the testis during the acute phase of the disease. Testicular failure is defined by a decrease in testosterone levels combined with an increase in LH and FSH levels [35]. Testicular shrinkage, libido loss, impotency, and gynecomastia have all been linked to a decrease in testosterone and a rise in LH and FSH levels in the male [2, 9–13]. As a result, it is possible that the mumps infection impairs Leydig cell function, leading to more serious consequences like testicular cancer, although it remains unclear whether this is due to a direct or indirect effect of the virus on the cells. Male testosterone levels have been observed to rise in the early stages of HIV infection, whereas testosterone levels have been found to fall in patients with AIDS. The number of Leydig cells decreases as a result of testosteronemia, which is a result of lymphocyte infiltration and interstitial tissue fibrosis [36].

Prostate cancer is one of the major leading causes of death in males throughout the world. HPV belongs to the class of oncogenic viruses that can cause prostate cancer [36]. Although viral DNA may stay inside the cells in episomal and integrated forms and cause a variety of prostate illnesses, it is still uncertain if HPV is genuinely involved in the etiology of a small percentage of prostate malignancies [38, 51]. It is controversial whether HSV in prostate has a notable relationship with prostate carcinoma or not [39]. HSV antigens are recognized in prostate cancer cells, as well as HSV antibodies in the serum [40–42]. However, a larger epidemiological investigation found no significant difference in HSV-2 antibody levels between individuals with prostate cancer and those with benign prostatic hyperplasia [43–47]. As a result, it is assumed that HSV-2 infection of the prostate is widespread in males, but there is no causative association between HSV-2 infection and prostate cancer, and more data in this area is needed [48]. When the tumor-associated

HSV-2 transformed hamster embryo cells, it resulted in the formation of human prostate cancer [49]. Teixeira et al. revealed that a human prostate cancer cell line can express HCMV-specific membrane antigen throughout their prostate cell culture study, and they presumed that HCMV was linked to prostate cancer in the light of their findings [50]. They also stated that males with prostate cancer have greater titers of anti-HCMV antibodies [51] than the controls with benign prostatic hyperplasia [52]. Patients who died of AIDS had HIV1-related proteins in several adjacent glandular epithelial cells in prostate sections, too [53]. However, morphological investigation revealed no significant abnormalities in the prostate [54]. Two investigations by the same group of researchers found HHV-8 in normal, hyperplastic, and neoplastic prostate glands from HIV-negative individuals [55, 56], while other reports contradicted these findings and found no link between HIV-negative men and HHV-8 in normal, hyperplastic, and neoplastic prostate glands [19, 21, 57–61]. This showed that the prostate gland would be a favored site for viral incubation and endurance. However, HHV-8 did not establish latent infection in prostate cancer cell lines, implying that it is unlikely to be implicated in prostate pathogenesis [62].

14.4 Gender-Based Susceptibility

A few investigations outlined the physiological contrasts among males and females leading to different responses to viral infection. Females showed reduced susceptibility to viral infections. Due to greater immunity against contaminations, with respect to viral weight inside an individual and transcendence number of polluted individuals inside a general population, viral illnesses are regularly lower for females than males [11, 46]. Nonetheless, there is a growing awareness that a significant proportion of the disease attributed to a viral infection is caused by an unusual host fuel [63–69]. As a result, greater antiviral, combustible, and cell-safe responses are generally observed in females. Regardless of the way the gender differences have been particularly recorded for viral defilements, there has been limited focus on differentiations among males and females in terms of prophylaxis and helpful medications against viral diseases [70]. Antibodies are the guideline precaution treatment for viral infections and have effectively diminished numerous sicknesses both in the male and the female. Viability of immunization depends on their capacity to initiate defensive resistance in either at a present moment (e.g., flu) or in long haul (e.g., measles) [70, 71].

There is developing proof that defensive resistant reactions and unfriendly responses to viral antibodies are higher in females than in males [7, 70]. Majority of antiviral medications are used to treat specific viral disorders, with these drugs typically preventing viral infections from replicating in host cells. Of late, it has been reported that the pharmacokinetics (i.e., ingestion, dispersion, digestion, and discharge of medication) and pharmacodynamics (i.e., the impact of the medication on physiological and biochemical cycles, both restorative and unfavorable) vary between the genders. SARS-CoV2 is a profoundly pathogenic infection that causes an intense respiratory trouble disorder which incites significant issues for worldwide wellbeing [2–7, 56, 72]. Studies have identified individual variances in the safe response to SARS-CoV2 tainting and triggering illnesses. Due to another natural barrier, steroid synthetics, and components connected to sex chromosomes, females are less vulnerable to viral contamination than males [5, 18]. Females with two X chromosomes, whether one is torpid or not, put their immune systems under strain. Female sexual direction has lower viral weight levels and less anxiety due to immune regulatory qualities encoded by the X chromosome, although CD4+ white platelets are higher and have a better safe reactivity than male sexual route [9, 32]. Females, on the other hand, tend to develop higher levels of antibodies that last longer. In comparison to the trigger of TLR7 and the manufacturing of IFN, the levels of genesis of immune cells are higher in women than in men. Level of TLR7 is higher in females than in males and its biallelic enunciation prompts higher resistant responses and assembles the resistance from viral infections [13, 73]. TLR7 defends against single-stranded RNA contamination by boosting the development of disease-specific immunoglobulin and lengthening the period of action for counter cytokines such as the IL-6 and IL-1 classes. Furthermore, after a viral infection, females produce less IL-6 burning than males, which is regularly linked to a shorter life span. There are also sites on the X chromosome that translate for features like FOXP3 that regulate safe cells, as well as a recall factor for infection and pathogenesis [6, 74]. The X chromosome affects the effective quality management system by circling back to proteins like TLR8, CD40L, and CXCR3, which can be overexpressed in females and affect the response to viral infection and vaccines. Regardless, the biallelic expression of X-associated qualities can lead to a terrifying insusceptible framework and flammable responses. Males are more likely to develop cardiovascular disorders, and those without cardiovascular problems who have been infected with SARS-CoV2 have a high prevalence of varying diseases, but these implications are still being investigated [66, 72, 75].

14.5 Virus-Induced Oxidative Damage and Male Infertility

Superoxide ($O\bullet^-_2$), hydrogen peroxide (H_2O_2), peroxyl radical ($ROO\bullet$), or hydroxyl ($\bullet OH$) are among the particles generated with a half-life of nanoseconds. The reactive nitrogen species (RNS), which include nitric oxide ($\bullet NO$), dinitrogen trioxide (N_2O_3), and peroxinitrite ($ONOO^-$), are less prevalent but still present in sperm cells [1–9, 56, 76]. When the balance between oxidants and reductants is disrupted due to excessive generation of reactive oxygen species (ROS), a working framework is triggered. Increased levels of oxygen radicals or ROS have been shown to influence sperm motility and architecture [61]. High ROS levels have been linked to DNA breakage and chromatin squeezing, which could endanger the sperm DNA integrity. Also, during extreme viremia, the ability to repair DNA is harmed, due to the modification of the nucleoprotein-intervened structure that the sperm possess initially [62, 77]. As a result, implantation rates could reduce further leading to an imbalanced pregnancy [63, 64]. SARS-CoV2 can initiate oxidant-sensitive pathways using provocative reactions, that could affect the human reproductive system in a manner discussed previously [42]. In addition to the sperm cells at various phases of development, human discharge comprises leukocytes, epithelial cells, and the monocytes from various phases of spermatogenesis. Two of these cells that produce a lot of free radicals are peroxidase-positive leukocytes and immature sperm [59]. Sperm are especially powerless to oxidative mischief on account of the presence of plentiful polyunsaturated fatty acids. These unsaturated fats are large biomolecules and provide a primary issue for multiple layers put together to facilitate the activities like acrosome response and sperm-egg interactions [60]. DNA bases are moreover arranged in order to withstand base alterations, strand-breaks, and chromatin cross-interfacing. Effect of superoxides and apoptosis are unquestionably important actions in generating DNA damage in the germline [64]. Oxidative stress (OS) is well accepted as a contributing factor in male infertility [57–60]. When the balance between oxidants and reductants (cancer-prevention agents) is upset due to an increase or decline in ROS levels (or even an unusual elevation or decrease in antioxidants), OS is triggered. Increased ROS can affect sperm auxiliary and usable integrity, such as motility, structure, count, and acceptability [61]. High levels of ROS have been linked to DNA fragmentation and chromatin squeezing, both of which have been linked to sperm DNA insights. Furthermore, the ability to repair sperm DNA damage is severely harmed by high viremia, which is due to the disruption of the original nucleoprotein-mediated defense framework of sperm [62, 69]. This, in turn, may reduce fertilization rates, implantation rates, embryogenesis, and may also enhance miscarriage [71, 72].

14.6 Testicular Antiviral Defense System

Infection of the testis with a virus has a negative impact on spermatogenesis and can develop into testicular cancer [6, 9, 11]. Hence, testicular immune system is crucial to the protection of male reproductive health. Interferon (IFN) and other proteins play a major role in the innate immune system immediately after any viral infection [63]. Leydig cells and Sertoli cells have the most antiviral capability of all the testicular cells [64, 65]. The IFNs bind to auto-expressed or other cell's expressed receptors followed by the induction of innate immune signaling cascade that involves toll-like receptors (TLRs), RIG-I-like receptors (RLRs), and several proteins (such as MAVS, IL, etc.) [66]. Rat seminiferous tubules were used to study the testicular defense system [3]. As infection of the germ cells in the seminiferous tubules was found to damage the germline or spread the virus or viral DNA to all germ cells, including the sperm, their protection became extremely important [67–70]. However, male germ cells have an autophagic mechanism to clear the virus. Moreover, male germ cells and macrophages were found to express beclin-1 and

LC3 which gives the impression of acquiring an autophagy machinery [36, 67, 71]. This autophagy plays an important role in recycling the intracellular components and removes dysfunctional cells to adapt to the stress so that cellular homeostasis is maintained [37, 72]. Autophagosomes may directly engulf the invading viral particle intracellularly [69, 73].

Regulation of the adaptive immune system and more specifically, the cell-mediated immunity has not been explored in detail within the testicular microenvironment in the context of viral infection [74]. Viruses are one of the most common causes of male infertility and infertile patients with a history of inflammation have a significantly greater antisperm antibody in their blood and sperm, implying that inflammatory response has a significant impact on male sexual health [40, 75]. Treg (along with Fox P3 molecules), γδ T cells, and helper T cells are involved in autoimmune orchitis, though their activity has not been investigated against viral infection. A considerable reduction in Leydig cells in the testes, as well as T lymphocyte invasion, was found in COVID-19-infected patients. Even more astonishingly, ACE2 receptor was found to be expressed at a higher degree in the testes, as stated earlier [41, 75, 76]. Nora et al. [8] reported similar T lymphocyte invasion, including CD3+ T cells and CD20+ B lymphocytes, and the analysis via functional genomics revealed an increased inflammatory response-related network in a separate study conducted on COVID-19 [18].

14.7 Conclusions

A variety of viruses, especially those infecting the human semen, can affect the male reproductive systems, as previously summarized in Table 14.1. The possible effects of viral infections on male reproduction are concluded in Fig. 14.2. Repercussions of such viral infections may be detrimental to varying degrees including organ integrity, and refashioning of the reproductive and endocrine systems. Viruses after penetrating the reproductive tissues cause an infection. To fight our immune system, viruses produce many inflammatory compounds like interleukin and tumor necrosis factors. Infection can also give rise to physiological stress. Ultimately, all these elements are responsible for producing ROS that could lead to OS in the process of reproduction. On the other hand, testosterone levels drop because of virus-induced damage in the Leydig cells that eventually pass on to the Sertoli cells. The disorder in Sertoli cells leads to the manufacturing of extraordinary sperm-mediated infertility in affected men. More research is needed to determine whether viruses attack and replicate in the adjacent glands and the urethral epithelium. Progress in these areas is necessary for the development of new therapeutic strategies to eradicate viruses and cure endocrine dysfunction caused by viruses.

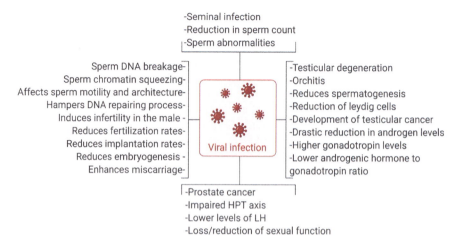

Fig. 14.2 Effects of viral infection on male reproduction. Role of viral infection on several reproductive organs, i.e., testis, semen, and prostate where virus induces oxidative stress and produces ROS. This may affect several sperm-related parameters and lead to cause infertility(?)

14.8 Future Perspectives

An increasing number of families of viruses have recently been discovered to colonize/infect the male reproductive system, and could potentially be life-threatening to the infected individual in particular and to the public health in general [2, 6, 23, 71]. An insight into the pathophysiology of these viruses in the reproductive system is important to help in the prevention of sexual transmission and associated reproductive problems in males [1, 17, 23, 78]. Preventive measures must be described with respect to the previous, current, and upcoming viral outbreaks, although many issues have remained unaddressed. Mechanisms underlying the dangerous effects of chronic viral infections on semen quality and male fertility remain widely unknown, discriminating the systemic from a more direct viral impact on the reproductive system [11, 25]. Greater understanding of the infection pathways and target cells of the male reproductive tract is essential to develop adequate treatment strategies [13, 79]. Semen as a vector of sexual and vertical viral transmission has to be examined and as a matter of fact, the simple presence of the viral nucleic acid levels in the semen may not be considered indicative of infection [27, 33, 79]. Finally, it is essential to draw attention to the complex role of certain viruses in the etiology and evolution of the reproductive cancers such as penile, prostatic, and testicular cancers. The presence of a human virome in the male genital system can lead to the discovery of viruses or combinations of viruses in previously undiagnosed asymptomatic males who may be withholding harmful conditions [33, 56, 59]. More studies on viruses can provide useful information for better understanding of the pathogenesis, infection pathways, and the target tissues as well. In addition, meaningful research can make it possible to elucidate the mechanisms of viral persistence in semen, sexual confirmation, and mechanisms for vertical transmission. Accordingly, appropriate and specific therapies can be developed to treat or prevent the potential implications of the male reproductive system for future generations.

References

1. Roychoudhury S, Das A, Jha NK, Kesari KK, Roychoudhury S, Jha SK, Kosgi R, Choudhury AP, Lukac N, Madhu NR, Kumar D, Slama P. Viral pathogenesis of SARS-CoV-2 infection and male reproductive health. Open Biol. 2021;11(1):200347. https://doi.org/10.1098/rsob.200347.
2. Roychoudhury S, Das A, Sengupta P, Dutta S, Roychoudhury S, Choudhury AP, Ahmed A, Bhattacharjee S, Slama P. Viral pandemics of the last four decades: pathophysiology, health impacts and perspectives. Int J Environ Res Public Health. 2020;17(24):9411. https://doi.org/10.3390/ijerph17249411.
3. Dutta S, Sengupta P. SARS-CoV-2 and male infertility: possible multifaceted pathology. Reprod Sci (Thousand Oaks, Calif.). 2021;28(1):23–6. https://doi.org/10.1007/s43032-020-00261-z.
4. Stanley KE, Thomas E, Leaver M, Wells D. Coronavirus disease (COVID-19) and fertility: viral ost entry protein expression in male and female reproductive tissues. Fertil Steril. 2020;114:33–43. https://doi.org/10.1016/j.fertnstert.2020.05.001.
5. Xu J, Qi L, Chi X, Yang J, Wei X, Gong E, et al. Orchitis: a complication of severe acute respiratory syndrome (SARS). Biol Reprod. 2006;74(2):410–6.
6. Pan F, Xiao X, Guo J, Song Y, Li H, Patel DP, et al. No evidence of SARS-CoV-2 in semen of males recovering from COVID-19. Fertil Steril. 2020;113:1135–9. https://doi.org/10.1016/j.fertnstert.
7. Nora H, Philippos E, Marcel A, Cornelius D, Dunja B-B, Ortwin A, et al. Assessment of SARS-CoV-2 in human semen-a cohort study. Fertil Steril. 2020; https://doi.org/10.1016/j.fertnstert.2020.05.028.
8. Hoffmann M, Kleine-Weber H, Schroeder S, Krüger N, Herrler T, Erichsen S, et al. SARS-CoV-2 cell entry depends on ACE2 and TMPRSS2 and is blocked by a clinically proven protease inhibitor. Cell. 2020;181:271–80. https://doi.org/10.1016/j.cell.2020.02.052.
9. Cascella M, Rajnik M, Cuomo A, Dulebohn SC, Napoli RD. Features, evaluation and treatment coronavirus (COVID-19). In: Statpearls [Internet]. Treasure Island: StatPearls Publishing; 2020. p. 1–16.
10. Roychoudhury S, Das A, Sengupta P, Dutta S, Roychoudhury S, Kolesarova A, Hleba L, Massanyi P, Salma P. Viral pandemics of twenty-first century. J Microbiol Biotechnol Food Sci. 2021;11(1):711–6. https://doi.org/10.015414/jmbfs.2021.10.4
11. Bloom DE, Cadarette D. Infectious disease threats in the twenty-first century: strengthening the global response. Front Immunol. 2019;10:549.
12. Alex P, Salam PWH. The breadth of viruses in human semen. J Med Microbiol. 2000;49(10):937–40. https://doi.org/10.1099/0022-1317-49-10-937.
13. Corona G, et al. SARS-CoV-2 infection, male fertility and sperm cryopreservation: a position statement of the Italian Society of Andrology and Sexual Medicine (SIAMS). J Endocrinol Investig. 2020;43:1153–7. https://doi.org/10.1007/s40618-020-01290-w.
14. Dejucq N, Jégou B. Viruses in the mammalian male genital tract and their effects on the reproductive system. Microbiol Mol Biol Rev. 2001;65(2):208–31. https://doi.org/10.1128/mmbr.65.2.208-231.2001.
15. La Vignera S, Vicari E, Condorelli RA, D'Agata R, Calogero AE. Male accessory gland infection and sperm parameters (review). Int J Androl. 2011;34(5PART 2) https://doi.org/10.1111/j.1365-2605.2011.01200.x.

16. Weidner W, Krause W, Ludwig M. Relevance of male accessory gland infection for subsequent fertility with special focus on prostatitis. Hum Reprod Update. 1999;5(5):421–32. https://doi.org/10.1093/humupd/5.5.421.
17. Bhushan S, Schuppe HC, Fijak M, Meinhardt A. Testicular infection: microorganisms, clinical implications and host-pathogen interaction. J Reprod Immunol. 2009;83(1–2):164–7. https://doi.org/10.1016/j.jri.2009.07.007.
18. Sarkar D. A Review on the Emerging Epidemic of Novel Coronavirus (nCoV- SARS CoV-2): Present Combating Strategies. Int J Pharm Sci. 2021;12(2):33–41. https://doi.org/10.22376/ijpbs.2021.12.2.b33-41.
19. Cardona-Maya W, Velilla PA, Montoya CJ, Cadavid Á, Rugeles MT. In vitro human immunodeficiency virus and sperm cell interaction mediated by the mannose receptor. J Reprod Immunol. 2011;92(1–2):1–7. https://doi.org/10.1016/j.jri.2011.09.002.
20. Dutta S, Thakare YR, Kshirsagar A, Sarkar D. A review on host genetic susceptibility to SARS CoV-2 related pneumonia. Int J Pharma Sci. 2021;12(2):b42–9. https://doi.org/10.22376/ijpbs.2021.12.2b42-49.
21. Nickel JC. Prostatitis and related conditions, orchitis, and epididymitis. In: Wein AJ, editor. Campbell–Walsh urology. 10th ed. Philadelphia: Elsevier Saunders; 2011. [chapter 11] [Google Scholar].
22. Wald A, Matson P, Ryncarz A, Corey L. Detection of herpes simplex virus DNA in semen of men with genital HSV-2 infection. Sex Transm Dis. 1999;26:1–3. [PubMed] [Google Scholar]
23. Salam AP, Horby PW. The breadth of viruses in human semen. Emerg Infect Dis. 2017;23:1922–4.
24. Dejucq-Rainsford N, Jegou B. Viruses in semen and male genital tissues—consequences for the reproductive system and therapeutic perspectives. Curr Pharm Des. 2004;10:557–75.
25. Murray K, Walker C, Herrington E, Lewis JA, McCormick J, Beasley DWC, Tesh RB, Fisher-Hoch S. Persistent infection with West Nile virus years after initial infection. J Infect Dis. 2010;201:2–4.
26. Bandeira AC, Campos GS, Rocha VF, Souza BS, Soares MB, Oliveira AA, Abreu YC, Menezes GS, Sardi S. Prolonged shedding of chikungunya virus in semen and urine: a new perspective for diagnosis and implications for transmission. IDCases. 2016;6:100–3.
27. Bornstein SR, Rutkowski H, Vrezas I. Cytokines and steroidogenesis. Mol Cell Endocrinol. 2004;215(1–2):135–41. https://doi.org/10.1016/j.mce.2003.11.022.
28. Wald A, Matson P, Ryncarz A, Corey L. Detection of herpes simplex virus DNA in semen of men with genital HSV-2 infection. Sex Transm Dis. 1999;26(1):1–3. https://doi.org/10.1097/00007435-199901000-00001.
29. Dejucq N, Jégou B. Viruses in the mammalian male genital tract and their effects on the reproductive system. Microbiol Mol Biol Rev. 2001;65:208–31.
30. Klimova RR, Chichev EV, Naumenko VA, et al. Herpes simplex virus and cytomegalovirus in male ejaculate: herpes simplex virus is more frequently encountered in idiopathic infertility and correlates with the reduction in sperm parameters. Vopr Virusol. 2010;55:27–31. [PubMed] [Google Scholar]
31. Mate SE, Kugelman JR, Nyenswah TG, Ladner JT, Wiley MR, Cordier-Lassalle T, Christie A, Schroth GP, Gross SM, Davies-Wayne GJ, et al. Molecular evidence of sexual transmission of Ebola virus. N Engl J Med. 2015;373:2448–54.
32. Deen GF, Broutet N, Xu W, Knust B, Sesay FR, McDonald SLR, Ervin E, Marrinan JE, Gaillard P, Habib N, et al. Ebola RNA persistence in semen of Ebola virus disease survivors—preliminary report. N Engl J Med. 2017;377:1428–37.
33. Dejucq N, Jégou B. Viruses in the mammalian male genital tract and their effects on the reproductive system. Microbiol Mol Biol Rev. 2001;65(2):208–31. https://doi.org/10.1128/MMBR.65.2.208-231.2001.
34. Casella R, Leibundgut B, Lehmann K, Gasser TC. Mumps orchitis: report of a mini-epidemic. J Urol. 1997;158(6):2158–61. https://doi.org/10.1016/S0022-5347(01)68186-2.

35. Lafferty WE, Downey L, Celum C, Wald A. Herpes simplex virus type 1 as a cause of genital herpes: impact on surveillance and prevention. J Infect Dis. 2000;181:1454–7. [PubMed] [Google Scholar]
36. Mikuz G, Damjanov I. Inflammation of the testis, epididymis, peritesticular membranes, and scrotum. Pathol Annu. 1982;17:101–28. [PubMed] [Google Scholar]
37. Kapranos N, Petrakou E, Anastasiadou C, Kotronias D. Detection of herpes simplex virus, cytomegalovirus, and Epstein-Barr virus in the semen of men attending an infertility clinic. Fertil Steril. 2003;79(Suppl 3):1566–70. https://doi.org/10.1016/s0015-0282(03)00370-4.
38. Musso D, Roche C, Robin E, Nhan T, Teissier A, Cao-Lormeau VM. Potential sexual transmission of Zika virus. Emerg Infect Dis. 2015;21(2):359–61. https://doi.org/10.3201/eid2102.141363.
39. Schuppe HC, Pilatz A, Hossain H, Diemer T, Wagenlehner F, Weidner W. Urogenital infection as a risk factor for male infertility. Dtsch Arztebl Int. 2017;114(19):339–46. https://doi.org/10.3238/arztebl.2017.0339.
40. Apari P, de Sousa JD, Muller V. Why sexually transmitted infections tend to cause infertility: an evolutionary hypothesis. PLoS Path. 2014;10(8):e1004111. https://doi.org/10.1371/journal.ppat.1004111Atkinson.
41. Nguyen O, Sheppeard V, Douglas MW, Tu E, Rawlinson W. Acute hepatitis C infection with evidence of heterosexual transmission. J Clin Virol. 2010;49(1):65–8. https://doi.org/10.1016/j.jcv.2010.06.008.
42. Sánchez-Montalvá A, Pou D, Sulleiro E, Salvador F, Bocanegra C, Treviño B, Rando A, Serre N, Pumarola T, Almirante B, et al. Zika virus dynamics in body fluids and risk of sexual transmission in a non-endemic area. Tropical Med Int Health. 2018;23:92–100.
43. Dejucq N, Jegou B. Viruses in the mammalian male genital tract and their effects on the reproductive system. Microbiol Mol Biol Rev. 2001;65(2):208–31. first and second pages, table of contents. https://doi.org/10.1128/MMBR.65.2.208-231.2001.
44. Harrower J, Kiedrzynski T, Baker S, Upton A, Rahnama F, Sherwood J, Huang QS, Todd A, Pulford D. Sexual transmission of Zika virus and persistence in semen, New Zealand, 2016. Emerg Infect Dis. 2016;22:1855–7.
45. Turmel JM, Abgueguen P, Hubert B, Vandamme YM, Maquart M, Le Guillou-Guillemette H, Leparc-Goffart I. Late sexual transmission of Zika virus related to persistence in the semen. Lancet. 2016;387:2501.
46. Couto-Lima D, Madec Y, Bersot MI, Campos SS, Motta MA, Santos FBD, Vazeille M, Vasconcelos PFDC, Lourenço-de-Oliveira R, Failloux AB. Potential risk of re-emergence of urban transmission of yellow fever virus in Brazil facilitated by competent Aedes populations. Sci Rep. 2017;7:4848.
47. Barbosa CM, Di Paola N, Cunha MP, Rodrigues-Jesus MJ, Araujo DB, Silveira VB, Leal FB, Mesquita FS, Botosso VF, Zanotto PMA, et al. Yellow fever virus RNA in urine and semen of convalescent patient. Brazil Emerg Infect Dis. 2018;24:176–8.
48. Venturi G, Zammarchi L, Fortuna C, Remoli ME, Benedetti E, Fiorentini C, Trotta M, Rizzo C, Mantella A, Rezza G, et al. An autochthonous case of Zika due to possible sexual transmission, Florence, Italy, 2014. Euro Surveill. 2016;21:30148.
49. Musso D, Roche C, Robin E, Nhan T, Teissier A, Cao-Lormeau VM. Potential sexual transmission of Zika virus. Emerg Infect Dis. 2015;21:2013–5.
50. Mansuy JM, Dutertre M, Mengelle C, Fourcade C, Marchou B, Delobel P, Izopet J, Martin-Blondel G. Zika virus: high infectious viral load in semen, a new sexually transmitted pathogen? Lancet Infect Dis. 2016;16:405.
51. Nguyen PV, Kafka JK, Ferreira VH, Roth K, Kaushic C. Innate and adaptive immune responses in male and female reproductive tracts in homeostasis and following HIV infection. Cell Mol Immunol. 2014;11:410–27.
52. Malolina EA, Kulibin AY, Naumenko VA, Gushchina EA, Zavalishina LE, Kushch AA. Herpes simplex virus inoculation in murine rete testis results in irreversible testicular damage. Int

J Exp Pathol. 2014;95(2):120–30. Published online 2014 Feb 23. https://doi.org/10.1111/iep.12071.
53. Musso D, Richard V, Teissier A, Stone M, Lanteri MC, Latoni G, Alsina J, Reik R, Busch MP. Detection of Zika virus RNA in semen of asymptomatic blood donors. Clin Microbio Infect. 2017;23:1001.e1–3.
54. Foresta C, Pizzol D, Moretti A, Barzon L, Palu G, Garolla A. Clinical and prognostic significance of human papillomavirus DNA in the sperm or exfoliated cells of infertile patients and subjects with risk factors. Fertil Steril. 2010;94(5):1723–7.
55. Prisant N, Bujan L, Benichou H, Hayot P-H, Pavili L, Lurel S, Herrmann C, Janky E, Joguet G. Zika virus in the female genital tract. Lancet Infect Dis. 2016;16:1000–1.
56. Davidson A, Slavinski S, Komoto K, Rakeman J, Weiss D. Suspected female-to-male sexual transmission of Zika virus—new York City, 2016. MMWR Morb Mortal Wkly Rep. 2016;65:716–7.
57. Foresta C, Patassini C, Bertoldo A, Menegazzo M, Francavilla F, Barzon L, Ferlin A. Mechanism of human papillomavirus binding to human spermatozoa and fertilizing ability of infected spermatozoa. PLoS One. 2011;6(3):e15036. https://doi.org/10.1371/journal.pone.0015036.
58. Gimenes F, Souza RP, Bento JC, Teixeira JJ, Maria-Engler SS, Bonini MG, Consolaro ME. Male infertility: a public health issue caused by sexually transmitted pathogens. Nat Rev Urol. 2014;11(12):672–87. https://doi.org/10.1038/nrurol.2014.285.
59. Laprise C, Trottier H, Monnier P, Coutlee F, Mayrand MH. Prevalence of human papillomaviruses in semen: a systematic review and meta-analysis. Hum Reprod. 2014;29(4):640–51. https://doi.org/10.1093/humrep/det453.
60. Croxson TS, Chapman WE, Miller LK, Levit CD, Senie R, Zumoff B. Changes in the hypothalamic-pituitary-gonadal axis in human immunodeficiency virus-infected homosexual men. J Clin Endocrinol Metab. 1989;68(2):317–21. https://doi.org/10.1210/jcem-68-2-317.
61. Poretsky L, Can S, Zumoff B. Testicular dysfunction in human immunodeficiency virus-infected men. Metabolism. 1995;44(7):946–53. https://doi.org/10.1016/0026-0495(95)90250-3.
62. Matheron S, D'Ortenzio E, Leparc-Goffart I, Hubert B, de Lamballerie X, Yazdanpanah Y. Long-lasting persistence of Zika virus in semen. Clin Infect Dis. 2016;63:ciw509.
63. Lai YM, Lee JF, Huang HY, Soong YK, Yang FP, Pao CC. The effect of human papillomavirus infection on sperm cell motility. Fertil Steril. 1997;67(6):1152–5. https://doi.org/10.1016/S0015-0282(97)81454-9.
64. Gimenes F, Souza RP, Bento JC, Teixeira JJ, Maria-Engler SS, Bonini MG, Consolaro ME. Male infertility: a public health issue caused by sexually transmitted pathogens. Nat Rev Urol. 2014;11(12):672–87. https://doi.org/10.1038/nrurol.2014.285.
65. Hills SL, Russell K, Hennessey M, Williams C, Oster AM, Fischer M, Mead P. Transmission of Zika virus through sexual contact with travelers to areas of ongoing transmission—continental United States, 2016. MMWR Morb Mortal Wkly Rep. 2016;65:215–6.
66. Gimenes F, Souza RP, Bento JC, Teixeira JJ, Maria-Engler SS, Bonini MG, Consolaro ME. Male infertility: a public health issue caused by sexually transmitted pathogens. Nat Rev Urol. 2014;11(12):672–87. https://doi.org/10.1038/nrurol.2014.285.
67. Huang JM, Huang TH, Qiu HY, Fang XW, Zhuang TG, Qiu JW. Studies on the integration of hepatitis B virus DNA sequence in human sperm chromosomes. Asian J Androl. 2002;4(3):209–12.
68. Jenabian MA, Costiniuk CT, Mehraj V, Ghazawi FM, Fromentin R, Brousseau J, Brassard P, Bélanger M, Ancuta P, Bendayan R, et al. Immune tolerance properties of the testicular tissue as a viral sanctuary site in ART-treated HIV-infected adults. AIDS. 2016;30:2777–86.
69. La Vignera S, Condorelli RA, Vicari E, D'Agata R, Calogero AE. Sperm DNA damage in patients with chronic viral C hepatitis. Eur J Intern Med. 2012;23(1):e19–24. https://doi.org/10.1016/j.ejim.2011.08.011.

70. Dutta S, Sengupta P. SARS-CoV-2 infection, oxidative stress and male reproductive hormones: can testicular-adrenal crosstalk be ruled-out? J Basic Clin Physiol. 2020;31:20200205. https://doi.org/10.1515/jbcpp-2020-0205.
71. Citil Dogan A, Wayne S, Bauer S, Ogunyemi D, Kulkharni SK, Maulik D, Carpenter CF, Bahado-Singh RO. The Zika virus and pregnancy: evidence, management, and prevention. J Matern Neonatal Med. 2017;30:386–96.
72. Yeniyol CO, Sorguc S, Minareci S, Ayder AR. Role of interferon-alpha-2B in prevention of testicular atrophy with unilateral mumps orchitis. Urology. 2000;55(6):931–3. https://doi.org/10.1016/S0090-4295(00)00491-X.
73. Klein SL, Dhakal S, Ursin RL, Deshpande S, Sandberg K, Mauvais-Jarvis F. Biological sex impacts COVID-19 outcomes. PLoS Pathog. 2020;16:e1008570.
74. Theas MS, Rival C, Jarazo-Dietrich S, Jacobo P, Guazzone VA, Lustig L. Tumour necrosis factor-alpha released by testicular macrophages induces apoptosis of germ cells in autoimmune orchitis. Hum Reprod. 2008;23(8):1865–72.
75. Alahmar AT. Role of oxidative stress in male infertility: an updated review. J Hum Reprod Sci. 2019;12:4–18. https://doi.org/10.4103/jhrs.JHRS_150_18.
76. Li N, Wang T, Han D. Structural, cellular and molecular aspects of immune privilege in the testis. Front Immunol. 2012;3:152.
77. Loveland KL, Klein B, Pueschl D, Indumathy S, Bergmann M, Loveland BE, Hedger MP, Schuppe H-C. Cytokines in male fertility and reproductive pathologies: immunoregulation and beyond. Front Endocrinol (Lausanne). 2017;8:307.
78. Mladinich MC, Schwedes J, Mackow ER. Zika virus persistently infects and is basolaterally released from primary human brain microvascular endothelial cells. mBio. 2017;8:e00952-17.
79. Wang Z, Xu X. scRNA-seq profiling of human testes reveals the presence of the ACE2 receptor, a target for SARS-CoV-2 infection in spermatogonia, Leydig and Sertoli cells. Cells. 2020;9:920.

Chapter 15
Sperm Redox System Equilibrium: Implications for Fertilization and Male Fertility

Lauren E. Hamilton, Richard Oko, Antonio Miranda-Vizuete, and Peter Sutovsky

Abstract Structural and regulatory requirements of mammalian spermatozoa in both development and function make them extremely unique cells. Looking at the complexity of spermatozoon structure and its requirements for both motility and quick breakdown within the post-fertilization environment, as well as its functional needs as an extremely streamlined cell with high energy requirements, demonstrate the high importance of oxidative-reductive processes. The oxidative state of the testis and epididymis during sperm development and maturation highly influences sperm structure, with a high dependence on disulfide bond formation, facilitated by thiol mediated processes. However, once functionally active, sperm transition to a new high-risk functional paradigm requiring low levels of reactive oxygen species (ROS) while also being highly susceptible to oxidative damage due to the high proportion of polyunsaturated fatty acids within the lipid bilayer of the plasmalemma and the lack of cytosolic antioxidant defenses. This chapter highlights how glutathione and thioredoxin systems mediate the oxidative environment of the male reproductive tract and facilitate the successful development, maturation and function of mammalian spermatozoa.

L. E. Hamilton · P. Sutovsky (✉)
Division of Animal Sciences, and Department of Obstetrics, Gynecology and Women's Health, University of Missouri, Columbia, MO, USA
e-mail: sutovskyp@missouri.edu

R. Oko
Department of Biomedical and Molecular Sciences, Queen's University, Kingston, ON, Canada

A. Miranda-Vizuete
Instituto de Biomedicina de Sevilla, Hospital Universitario Virgen del Rocío/CSIC/Universidad de Sevilla, Sevilla, Spain

© The Author(s), under exclusive license to Springer Nature Switzerland AG 2022
K. K. Kesari, S. Roychoudhury (eds.), *Oxidative Stress and Toxicity in Reproductive Biology and Medicine*, Advances in Experimental Medicine and Biology 1358, https://doi.org/10.1007/978-3-030-89340-8_15

Keywords Spermatozoa · ROS · Oxidative regulation · Glutathione transferase · Glutathione peroxidase · Testis thioredoxin system · Sperm maturation

15.1 Introduction

Mammalian spermatozoa function at the knife's edge of oxidative regulation, requiring low levels of reactive oxygen species (ROS) while also being highly susceptible to oxidative damage. The high proportion of polyunsaturated fatty acids within the lipid bilayer structure of the sperm plasmalemma and the lack of cytosolic antioxidant defenses within the streamlined cell structure result in a high-risk functional paradigm for mature spermatozoa. Protected by the antioxidant defenses of their surroundings in the testes and the epididymis during development and maturation, once functionally activated, spermatozoa have been, somewhat allegorically, described as following an apoptotic trajectory, with only a single cell successfully accomplishing its functional objectives, while the rest undergo cell death [1, 2]. Unlike somatic cells that need to continuously regenerate and maintain cellular integrity, sperm cells must only equip themselves with enough protection to survive the journey to the site of fertilization. This chapter will focus on how the glutathione and thioredoxin redox systems facilitate the protection, successful development, and maturation of mammalian spermatozoa.

15.2 The Glutathione Redox Cycle

Composed of glutamic acid, glycine, and cysteine amino acids, and harboring an active thiol group, glutathione (gamma-glutamyl-cysteinyl-glycine) has been shown to be the major nonprotein source of antioxidant defense within cells. Glutathione is synthesized in a series of six-enzyme catalyzed reactions facilitated by the actions of γ-glutamylcysteine synthetase (GCS) and glutathione synthetase (GS), as part of the γ-glutamyl cycle [3–6]. Present in both free and bound forms, the redox status of glutathione can be determined based on the ratio of reduced (GSH) and oxidized (GSSG) forms. Under physiologically normal cellular conditions, reduced glutathione (GSH) predominates where it can be oxidized to neutralize oxidative stress before being regenerated by glutathione reductases, a process facilitated by proton donations from NADPH and H$^+$ molecules [5]. The antioxidant capacity of GSH can be used directly but is also the driver of catalytic activity for many enzymes such as glutaredoxins, glutathione peroxidases, and glutathione transferases. Moreover, the functional implications of glutathione are not exclusive to oxidative cellular defense and have also been shown to be involved in amino acid transport mechanisms, nucleic acid and protein synthesis, and enzymatic modulation [4, 5].

In the male reproductive tract, GSH availability strongly influences both germ cell development and cell protection, acting through a wide range of enzymatic and direct interactions.

15.3 The Implications of Glutathione on Male Fertility from Testis to Embryo

Glutathione is the most abundant thiol in mammalian cells and acts both directly, in an antioxidant capacity, and indirectly, as a substrate in oxidative-reductive reactions [6–8]. Developmentally, reduced glutathione (GSH) concentrations have been shown to increase concomitantly with spermatogenetic cell development [9]. Beginning at birth and continuously rising until puberty, the presence of GSH within the testicular environment is proposed to mediate normal development and survival of male germ cells [9–12]. Throughout spermatogenesis, high levels of GSH have been recorded in early spermatogenic cells up to and including pachytene spermatocytes but dramatically decrease during the transition from spermatocyte to spermatid [9]. This suggests that as the elongating spermatids become more streamlined and compact, their dependence on extracellular antioxidant systems increases as their intracellular stores of GSH decrease. During the early stages of germ cell development, the synthesis and maintenance of reduced GSH is believed to be predominately facilitated by Sertoli cells, the somatic, germ cell nursing cells within the seminiferous epithelium [13–15]. Investigations into expression levels of glutathione reductases (GR), an essential component of the glutathione system, showed that Sertoli cells have approximately eightfold higher GR activity but half the concentration of GSH when compared to early, pre-meiotic spermatogenic cells [16]. These findings suggest that external GSH supplementation and the trafficking of GSH or its precursors to germ cells is required for germ cell sustenance, oxidative stress protection and proliferation [16]. The buffering of the oxidative environment within the basal compartment of the testicular epithelium by the surrounding somatic cells may also explain the increased resistance to oxidative injury shown by spermatogonia when compared to mature sperm cells [17–20].

As spermiogenesis commences, the accompanying rapid GSH depletion in spermatids is proposed to be due to their high production of hydrogen peroxide (H_2O_2), limited GSH regeneration, and the high enzymatic activity of glutathione peroxidases (GPX) such as GPX4, which acts in both oxidative defense and structural spermatid development [21]. The depletion of GSH within the maturing spermatids is speculated to be required for protein disulfide bond formation, in the structural formation of the mitochondrial capsule and the protamine compaction of the nucleus [21]. As sperm maturation continues within the epididymis, there is an increased dependence on extracellular reductive substrates, a trend that continues throughout the male reproductive tract, with high levels of antioxidant molecules being reported in epididymal secretions, seminal plasma, and prostate secretions [22, 23].

The concentration of GSH is also critical during fertilization events, specifically within early embryonic development [24–28]. The level of GSH within the oocyte has been suggested as an indicator of maturation, reaching its peak concentrations in the MII stage [24, 29–34]. Immediately after sperm-egg fusion, GSH is directly implicated in a variety of processes including the disassembly of sperm components, the decondensation of the sperm nucleus, the formation of the male pronucleus, the formation of the sperm aster, and the breakdown of the connecting piece [24, 25, 28, 35, 36]. Furthermore, the availability of GSH or its precursors has also been reported to enhance oocyte maturation and in vitro embryo development when supplemented into culture media [25, 27, 37, 38], demonstrating its continuous importance from testes to embryo and its overarching role in male reproductive success.

15.4 The Role of Glutathione Transferases in Sperm Development and Fertilization

Glutathione-transferases (GSTs) are multifunctional and ubiquitous proteins that function as detoxification enzymes through the conjugation of reduced glutathione (GSH) with the electrophilic centers of substrates [39]. Cytosolic GSTs all have the structurally conserved GST conical fold, but can be further categorized into different subclasses based on sequence and structural features. Thus far there are seven subclasses of soluble cytosolic GSTs in mammals: alpha, zeta, theta, mu, pi, sigma, and omega and one mitochondrial membrane-bound subtype, kappa. GSTs form both hetero- and homo-dimers within their subclasses and can facilitate different reactions through binding and/or catalytic activity [40]. GSTs are best-known for their detoxifying properties of both endogenous and exogenous toxic substrates mediated by the conjugation of glutathione (GSH), ligands, or other binding domains [41]. However, differences in their isoenzymatic architecture also facilitate different multifunctional capacities including glutathione peroxidase, dehydroascrobate reductase, and thiol transferase activity [39]. To date, cytosolic GSTs from the subclasses of alpha, mu, pi, theta, zeta, and omega have been identified and characterized throughout the male reproductive tract in detoxification processes as well as fertilization events [35, 42–50].

Transcripts of GSTs from all the soluble subclasses except pi have been reported to undergo alternate splicing events resulting in a highly heterogeneous and diverse group of enzymes. Furthermore, genetic variations within the GSTs of the male reproductive system have been thoroughly investigated in various ethnic populations, with relation to fertility status. Reportedly, GSTM1-null and GSTT1-null genotypes as well as a GST pi single nucleotide polymorphism (SNP), substituting Ile105Val, have been shown to be directly correlation with decreased sperm concentration, increased spermatozoon membrane damage, and an overall decrease in fertility [51–55]. Higher incidences of nonobstructive azoospermia, oglioasthenozoospermia, and asthenozoospermia have also been linked to these mutations [51–55]. Wu et al. (2013) have suggested that the null genotypes may remove binding sites for transcription factors [56], resulting in a change in gene expression related to sperm maturation and fertility; however, the exact mechanisms affected are not fully understood.

Mammalian testes have been reported to express the highest amount of total GSTs per milligram of cytosolic protein within the body [57], with the first documented GSTs of the male reproductive tract reported in 1978 [58]. Since then, isoenzymes of the GST Pi (GSTP), GST Mu (GSTM), GST Alpha (GSTA), GST Theta (GSTT), GST Zeta (GSTZ), and GST Omega (GSTO) have all been characterized to reside within the male reproductive system in various animal models. In the testis, immunostaining has identified GSTs as residents of Sertoli and Leydig cells, as well as germ cells. Investigations of specific GST expression have also shown that type A and type B spermatogonia strongly express GSTM2 specifically [59]. Furthermore, low expression of GSTA, GSTP, GSTT, and GSTO has also been reported in testicular cells [60]; however, GSTO expression in somatic cells appeared restricted to the Leydig cells [61]. High GST expression within the Leydig cells of the testicular interstitium is proposed to stem from the functional duality of certain GST subclasses, that harbor steroid isomerase activity in addition to electrophilic detoxification capabilities, implicating them in additional processes throughout the synthesis of testosterone and progesterone [62]. Furthermore, GSTZ1–1 immunoreactivity in the testis was shown to intensify concomitant with increased differentiation and maturation in both pre- and post-meiotic germ cells [63] and GSTO2 expression was shown to emerge post-meiotically, in the early stages of spermiogenesis before being shuttled to specific perinuclear theca compartments of the sperm head [48].

Within the epididymis, GSTs have regional specificity with high expression of GSTM2, GSTM5, and GSTO1 found in the caput epididymis, whereas GSTP1, GSTM3, and GSTO1 were identified in the corpus and caudal segments [64]. Light and electron microscopy investigations within the rat epididymis revealed cell-type specificity of GSTP enzymes in the basal cells of the corpus and proximal caudal regions, while GSTM reactivity was predominately found in the principal cells of the corpus and cauda epididymis [65]. Fluorescent immunohistochemical analysis in the mouse and boar epididymis showed evidence of GSTO2 within the luminal aspects of the epithelium, where it is proposed to be secreted into the luminal space via the apocrine secretory pathway [50]. Epididymal GSTs have been shown to regulate the oxidative-reductive environment of the epididymal epithelium itself, findings that supported not only by their presence but also by their differential regulation when exposed to xenobiotics [66, 67].

Biological activity of GSTs within the epididymis is further enhanced by their interactions with maturing spermatozoa as they transit through the luminal space. Initial investigations in goat were the first to show that GSTM and GSTP bind to the sperm surface using non-covalent interactions; however, further studies have shown their conserved surface localization in multiple mammalian species [43, 44, 66–68]. In ejaculated porcine spermatozoa, GSTP1 has been identified localizing to the midpiece, principal piece, and end piece of the tail as well as being present on the caudal aspect of the sperm head [47]. GSTM isoenzymes show a similar surface reactivity throughout the sperm tail but have also been shown to reside at the apical aspect of the sperm head in goats, co-localized with the acrosome [44]. Recent findings have similarly revealed that a subset of GSTO2 enzymes also associate with the sperm surface in both pigs and mice, and coat both the head and tail regions [50]. Functionally, these GST subclasses have been proposed to cooperate in facilitating detoxification processes to limiting lipid peroxidation on the sperm surface, which is of high importance for mammalian spermatozoa, as they have a high proportion of polyunsaturated fatty acids within the plasma membrane, making them more prone to oxidative damage [45–47, 50]. While GSTM and GSTP enzymes are proposed to utilize the traditional glutathione transferase activity to regulate oxidative stress, GSTO2, an enzyme shown to have high levels of dehydroascorbate reductase activity may exert its antioxidant effects through the replenishment of ascorbic acid within the surrounding environment as well as being able to neutralize oxidative stress via reduced glutathione conjugation [50]. Aside from detoxification roles facilitating plasma membrane oxidative homeostasis, both surface-bound GSTM and GSTP isoenzymes have also been shown to participate in fertilization events, irrespective of their catalytic functions, notably as sperm recognition factors in the oocyte zona pellucida binding at fertilization [43]. Furthermore, recent findings have also implicated GSTP1 as an inhibitor of the JNK kinases signal transduction pathway within porcine spermatozoa, a pathway associated with mitochondrial dysfunction and cell death in somatic cells [47].

While GST activity has been predominately characterized on the plasma membrane of spermatozoa, with isoforms of GSTM, GSTP, and GSTO subfamilies identified as plasma membrane residents, GSTO2 has also been localized within the perinuclear theca (PT) of the mammalian sperm head [48, 49]. Specifically, GSTO2 has been identified within the postacrosomal sheath region (PAS) of the PT in spatulate spermatozoa (e.g., bull, boar) and within both the PAS and perforatorial regions of the PT in falciform sperm (e.g., rat, mouse) [48, 49]. The PAS is a highly condensed cytosolic protein region that surrounds the caudal aspect of the sperm nucleus, and has previously been implicated as the reservoir of the sperm-borne oocyte activating factors (SOAF) for its quick solubility upon sperm-oocyte fusion [69, 70]. Through intracytoplasmic sperm injection (ICSI) studies in mice, Hamilton et al. (2019) demonstrated that sperm-borne GSTO2 inhibition causes a nuclear decondensation delay within developing embryos that is further exacerbated when GSTO enzymes within the oocyte are also inhibited [35]. Embryos that were conceived with spermatozoa that lacked GSTO2 activity showed a continuous and compounding delay that persisted throughout development and resulted in high rates of blastomere fragmentation and low blastocyst formation compared to vehicle controls [35]. Furthermore, dually inhibiting GSTO activity within both the female and male gametes resulted in a post-fertilization developmental arrest at the one-cell stage, suggesting a critical role for GSTOs in this early stage of zygotic development [35]. The multifaceted functions of GSTs support that GSTOs may be acting in a detoxifying manner during this highly active stage of embryonic development, while also facilitating the unpacking of the sperm head or nucleus at the onset of pronuclear development. Overall, the role of sperm-borne GSTO2 in post-fertilization events emphasizes the impact of sperm constituents on successful embryo development and highlights the importance of oxidative-reductive activity and sperm-borne contributions within early embryo development.

15.5 Glutathione Peroxidases of the Male Reproductive Tract

Glutathione peroxidases (GPX) are major scavengers of free radicals that facilitate the protection of cells against oxidative stress and lipid peroxidation. To date, the known GPXs can be divided into eight major subclasses (GPX1–8). In humans, the enzymatic activity of GPX1–4 and GPX6 is facilitated by the selenocysteine residue at their catalytic center; however, in rodents, the selenocysteine of GPX6 is replaced by solely cysteine in the catalytic center which may lead to differing enzymatic interactions and substrate specificity [71]. As a superfamily, their common functional denominator is the ability to catalyze the reduction of hydroperoxides using the reductive power of thiols. As their name suggests, GPX enzymes have a strong preference for GSH. GPX enzymes have also been described as a first line of cellular defense when cells respond to minor changes in H_2O_2 concentrations [72]. Furthermore, unlike other enzymes that facilitate cellular antioxidant functions, such as peroxiredoxins and catalase, GPXs are also able to metabolize specific organic molecules that have been affected by peroxidation due to hydroxyl radical production and facilitate the recycling of their components [73].

Within the male reproductive tract, ubiquitous expression of GPX1 can be found in both the testis and epididymis, whereas GPX3, GPX4, and GPX5 all show regional and sub-cellular specificities within the male reproductive tract [73]. GPX3 has been reported in the cytoplasmic compartment of the epithelium of the epididymis and vas deferens, and has been shown to be secreted in the lumen of the caput epididymis [74, 75]. Investigations into GPX3 expression have also found that while it is influenced by androgens within the corpus and caudal regions of the epididymis, the caput GPX3 expression is independent of androgenic control [74, 75]. GPX4 is more broadly expressed across the male reproductive tract, showing strong expression within the testicular environment, the germ cells, and the epididymis [73]. Conversely, GPX5 expression is exclusively reported in the principal cells of the epididymis where it has also been shown to be secreted into the lumen and loosely interact with spermatozoa [76].

Functionally, GPX1 has been characterized as an antioxidant, and facilitates the regulation of H_2O_2 within the epithelial environment of both the testis and the epididymis [73, 77]. GPX1 may also indirectly participate in the stress response through its regulation of hydrogen peroxide concentrations, as hydroperoxides have been shown to functionally mediate signal transduction pathways by activating the ubiquitination of signal-transducing proteins in the cytosol or by modulating thiol-containing transcription factors [78]. The functional implications of GPX1 expression may also exceed their roles within the reproductive epithelium. While no protein expression of GPX1 has been reported in germ cells, one study has suggested that GPX1 mRNA is shuttled into the newly formed zygote by the fertilizing spermatozoon and that its absence may have negative implications for the developing embryo [79].

GPX4 is a moonlighting enzyme that acts both in oxidative regulation and as a structural building block during sperm development. During spermatogenesis, GPX4 is heavily involved in the testicular defense against ferroptosis, a recently recognized type of non-apoptotic cell death triggered by a cell's inability to effectively protect itself against lipid peroxidation [80–82]. Ursini and Maiorino (2020) propose that traces of hydroperoxide derivatives of lipids are constantly being produced through aerobic metabolism, and consequently require the continuous reducing activity of GPX4 to avoid the accumulation of lipid peroxidation [80]. However, when GPX4 activity is inactivated or overwhelmed, and there is a build-up of reactive oxygen species derived from iron metabolism, ferroptosis can occur. To date, ferroptosis has been shown in the Sertoli cells of a mouse testicular ischemia-reperfusion model [81], and in the testes of mice treated with the chemotherapy/germ cell eradicant drug busulfan [82]. Furthermore, additional evidence demonstrated that ferroptosis induced testicular damage mimics phenotypes seen when GPX4 is inactivated or GSH is depleted, highlighting the vital role of GPX4 in the maintenance of the oxidative state of the membrane of testicular cells [82].

In germ cells, GPX4 is catalytically active from early spermatogenesis to late spermiogenesis, but does become inactivated in the mitochondrial sheath capsule of mature spermatozoa [21]. Three isoenzymes of GPX4 exist within mammalian spermatozoa due to alternative splicing, with two residing in the mitochondrial capsule of the sperm tail midpiece and accounting for approximately 50% of its protein content [83]. The third isoenzyme is located within the sperm nucleus where it retains its peroxidase activity and is believed to function during sperm maturation in the epididymis [21].

The functional contributions of each respective GPX4 isoenzyme have been investigated in regard to their implications on sperm development, structure, and fertility. Knockout models created in mice have targeted each of the differently expressed isoenzymes separately, to help deduce their specific functions. The expression of the mitochondrial GPX4 (mGPX4) begins during the pachytene spermatocyte stage of spermatogenesis and is continuous without interruption throughout sperm development [21]. At the onset of puberty, mGPX4 is suspected of maintaining the ROS concentrations surrounding the mitochondrial electron transport chain activity of developing spermatids, and specifically preventing ferroptosis, before being structurally incorporated into the mitochondrial capsule as a structural component [80, 84]. Conditional knockouts of the mGPX4 isoenzyme resulted in sperm-structural defects, with a high prevalence of hairpin folds (distal reflex) in the midpiece and a decrease in mitochondrial membrane potential, which ultimately led to decreased fertility due to altered motility [85]. The mitochondrial capsule is formed late in spermiogenesis when the level of GSH within maturing spermatids is low. Facilitated by the GSH depletion in elongating spermatids, mGPX4 is suggested to utilize surrounding protein thiols as alternate donors, with catalytic intermediates becoming cross-linked and inactivated, ultimately building the outer capsule of the mitochondrial sheath, which encircles the tail midpicce [21, 86, 87].

Investigations into sperm nuclear GPX4 (snGPX4) concluded that it was not essential to overall male fertility, but that knockout models did show defective DNA condensation during sperm maturation [88]. The lack of DNA compaction also resulted in higher rates of DNA oxidation, which was credited to a greater susceptibility to ROS-mediated damage during sperm transit [88]. These findings reinforce the previously proposed role for snGPX4 mediating the condensation of the sperm nucleus by acting as a protamine thiol peroxidase and crosslinking protamines through the formation of disulfide bridges [89]. Nuclear condensation not only conveys a level of protection to the paternal DNA in sperm cells as they travel toward the site of fertilization, but also acts as a structural modulator that creates a more streamlined cell with greater hydrodynamic properties to allow for improved performance.

The epididymis is the site of both conventional Se-dependent (GPX1, GPX3, and GPx4) and Se-independent (GPX5) GPX enzymes. The expression of GPX5 is conserved in various mammals and regulated by androgenic control in the caput epididymis region [90]. Functionally, GPX5 also differs from the other epididymal GPXs in that it is secreted into the luminal compartment where it has been documented to loosely interact with sperm cells [76, 91]. The mouse knockout model of GPX5 did not indicate any direct implications on the fertility status of young males; however, an increased incidence of late developmental defects was found in embryos conceived with spermatozoa from aged GPX5$^{-/-}$ mice [92]. Spermatozoa within the caput epididymis also presented with excessive chromatin compaction, which the authors suggested could be due to the increased availability of H_2O_2 and other ROS that could be utilized during the chromatin hypercondensation that occurs during sperm maturation [92]. Furthermore, regionally specific investigations of oxidative stress (OS) within the epididymis also revealed a heightened OS response in the cauda epididymis, which displayed an upregulation of the cytosolic GPX enzymes within the epithelium and an increase in the presence of catalase mRNA, both indicators of increased ROS [92]. Conversely, unlike caput spermatozoa, cauda epididymal spermatozoa of aged GPX5$^{-/-}$ knockout males showed decreased sperm chromatin compaction suggestive of decreased DNA integrity due to fragmentation and DNA breaks [92]. The increased expression of antioxidant enzymes within the epididymal epithelium and the increased DNA damage within luminal cauda epididymal spermatozoa were both suggestive of a heightened concentration of ROS within the caudal luminal environment. These findings were further supported through greater 8-OxodG reactivity in the nuclei of the cauda epididymal spermatozoa and increased lipid peroxidation of cell membranes [92]. Together these findings suggest that GPX5 is an important luminal ROS scavenger, specifically in the cauda epididymis, where the sperm reserve is stored until ejaculation, a role that is even more important in aged males that already have a decreased antioxidant capacity compared to younger counterparts.

15.6 The Influence of Selenium in the Male Reproductive Tract

Selenium (Se) has been well documented as a critical element for testicular function with comparatively high Se concentrations reported in the testes in relation to other organ systems [93, 94]. Past findings have shown that selenium is transported into the male reproductive system by selenoprotein P [95–97], which interacts with its receptor on Sertoli cells, facilitating the Se uptake required for spermatogenesis [96]. Radioactive Se tracking experiments suggest that up to 80% of testicular Se is incorporated into the midpiece of developing spermatids with the rest distributed throughout the epithelium of the reproductive tract [97]. Functionally, Se is incorporated into the active sites as a selenocysteine residue in most glutathione peroxidases, and thioredoxin reductases. Evolutionarily, the use of selenocysteine within the catalytic sites of proteins in the thioredoxin system is unique to vertebrates, with an exception being a cytosolic thioredoxin reductase in *C. elegans* [98]. Selenocysteine traditionally replaces a Cys residue in the active site, giving the proteins higher reactivity to a broader spectrum of substrates [99]. Se deficiency has been widely reported to produce testicular dysfunction in mammals, resulting in sperm morphological defects that mirror those of GPX4 knockouts, such as hairpin bends in the sperm midpiece, sperm head abnormalities, loss of motility, and disruptions to testosterone biosynthesis [93]. While GPX4 is shown to have very high affinity for Se and does take precedence over bioavailable Se when Se concentrations are limited, studies have suggested that it may not function in isolation. Instead, GPX4 cooperates in a disulfide bond formation and isomerization system with thioredoxin glutathione reductase (TGR), to promote the formation of structural components of the developing spermatid [100]. Besides working in close synergy with GPX4, TGR is also functionally driven by the selenoprotein at its catalytic center and is highly expressed in the testis [100]. Therefore, the dramatic consequences of Se deficiency that are reflective of GPX4 knockouts should not be looked at solely in isolation but also in the context of how the absence of GPX4 influences the functional capabilities of additional cooperative proteins that it interacts with during sperm development and maturation. Moreover, in fully mature sperm, specific selenoproteins, such as sperm mitochondrion-associated cysteine-rich protein (SMCP), which is highly expressed in the testis and localizes to the mitochondrial sheath of the sperm midpiece, have also been reported to have large implications on sperm fitness with $SMCP^{-/-}$ 129/Sv knockouts resulting in asthenozoospermia [101]. Therefore, selenium availability may also be a key regulator of fertilization success that could be investigated in individuals that demonstrate sperm motility disorders. These findings are also consistent with additional selenoprotein knockout models as well as clinical observations in Se deficient males, who show decreased sperm counts and higher incidence of oligoasthenzoospermia [93]. Excessive Se also contributes to negative fertility parameters, increased morphological defects, and decreased motility, demonstrating the need to moderate redox regulation in both developing spermatids and mature sperm [93]. Consequently, the upstream regulation of selenium and its mediating effects on selenoprotein activity have tremendous downstream implications on the male reproductive tract's oxidative homeostasis, the structural integrity of sperm cells, and overall fertility status.

15.7 The Mammalian Testis Thioredoxin System

Thioredoxins (TRX) are a class of small general protein disulfide reductases that are present in virtually all living organisms. Thioredoxins catalyze thiol-dependent thiol-disulfide exchange reactions in different types of proteins through the reversible oxidation of the cysteine residues of their conserve active site (WCGPC), impacting varied traits like transcription factor DNA binding activity, DNA synthesis, cell growth, antioxidant defense, immune response, or apoptosis [102]. Thioredoxins are maintained in their reduced active conformation by the flavoenzyme thioredoxin reductase (TRXR) that uses the reducing equivalents from NADPH, comprising the so-called thioredoxin system [99]. Initially identified in the early sixties as the electron donor required for ribonucleotide reductase activity in bacteria [103, 104], TRXR was later shown to be present in higher organisms when a human adult T-cell leukemia derived factor was found to be a mammalian thioredoxin [105]. While in bacteria there is only one thioredoxin system located in the cytoplasm, there are two main thioredoxin systems in eukaryotes, including TRX-1 and TRXR-1 in cytoplasm, and TRX-2 and TRXR-2 in the mitochondrial matrix [105–109]. Moreover, metazoans possess additional thioredoxin proteins and larger proteins containing one or several thioredoxin-like domains in different subcellular localizations (reviewed in [110]). Furthermore, some organisms contain a third thioredoxin reductase member, hybrid of a N-terminal glutaredoxin domain followed by thioredoxin reductase module (named TGR for thioredoxin glutathione reductase) that, in addition to its intrinsic thioredoxin reductase activity, can also act as a GSSG reductase thanks to the N-terminal glutaredoxin domain [100].

The first connection of the thioredoxin system with reproduction biology came through the identification of cytoplasmic TRX1 as a component of the early pregnancy factor, a pregnancy-dependent suppressor releasing hormone which is synthesized within the first hours post-fertilization to block the maternal immune response toward the fetal allograft [111]. Later on, two *Drosophila melanogaster* thioredoxins were found to be exclusively expressed in the germline. Firstly, the fly *deadhead* locus encodes an egg-specific thioredoxin required for meiotic oocyte progression [112] and for decondensing sperm chromatin at fertilization [113, 114]. Whereas, the fly *trxT* locus encodes a testis-specific thioredoxin of yet unknown function [115]. Curiously, these two *Drosophila* thioredoxins are arranged as a gene pair, transcribed in opposite directions and sharing a 471 bp regulatory region [115]. However, it was in mammals where the thioredoxin family flourished with the identification of several new members with exclusive or very high expression in testis and spermatozoa, which are discussed in more detail within the following sections.

The process of cellular differentiation that occurs in the male germ cell is a unique phenomenon not found in any other cell of the body. This highly coordinated and complex process transforms round spermatids into highly differentiated motile cells that must not only survive their transit to the site of fertilization but also be able to rapidly break apart in the oocyte cytoplasm for the paternal nucleus to successfully incorporate into the newly formed zygote. In mammalian spermatogenesis, this is achieved through extensive disulfide bond formation in both the nuclear and flagellar accessory structures, effectively creating strong, reversible stabilizing structures that can be rapidly degraded during fertilization. Together, the uniqueness of male germ cell differentiation and its high redox requirements due to the large amounts of disulfide bond formation help to explain the strong expression and presence of both traditional and testis-specific thioredoxin proteins in the male germline. Thioredoxins are critical facilitators that work cooperatively with other enzymes in disulfide bond formation and specifically participate in disulfide isomerization by reducing disulfide bonds to allow for their reoxidation by other oxidoreductases [100]. To date, TRX-1 and TRXR-1, part of the cytoplasmic thioredoxin system, TRX-2 a member of the mitochondrial thioredoxin system as well as TXL-2, TGR, and the spermatid-specific thioredoxins (SPTRX-1/TXNDC2, SPTRX-2/TXNDC3/NME8, and SPTRX-3/TXNDC8) have been reported to function in the testes [100, 116–121], also presented in Fig. 15.1.

Spermatid-specific thioredoxin 1 (SPTRX-1/TXNDC2) protein is first expressed in the post-meiotic round spermatids and peaks between stages 14–16 of the rat spermiogenesis cycle, commensurate with the onset and assembly of the sperm tail fibrous sheath (FS) [118, 122]. SPTRX-1 can act both as an oxidant or a reductant and has been co-localized associating with the longitudinal columns but not the transverse ribs of the FS during sperm tail formation; it shows diminished expression concomitant with the completion of FS assembly [118, 122]. Based on its temporal and spatial expression SPTRX-1 has been suggested as a possible component of the nucleation center of the FS longitudinal columns and a regulator of sperm tail morphogenesis [118, 122]. Furthermore, its redox duality also lends support to the possibility that it acts much like a protein disulfide isomerase (PDI), facilitating disulfide bond reshuffling to ensure the correct ultrastructural organization of the FS [118, 122]. Therefore, while it is not of structural importance in mature spermatozoa, it may be of significance when investigating the causes of structural tail malformations such as dysplasia of the fibrous sheath (DFS) for its role in regulating the development of sperm tail accessory structures.

Spermatid-specific thioredoxin 2 (SPTRX-2/TXNDC3/NME8) is the second of three thioredoxin proteins endemic to the testes. It harbors a thioredoxin catalytic domain in its N-terminus followed by three nucleotide diphosphate (NDP) kinase domains [119]. Due to sequence similarities and protein arrangement, SPTRX-2 has been suggested as the human homolog to the sea urchin protein IC1, a component of one of the three intermediate chains of the sperm axoneme outer dynein arms [123]. Its similarity to IC1, an axoneme-structural component, is further cemented by its exclusive expression within the sperm accessory tail structure of the FS [119]. Much like SPTRX-1, SPTRX-2 protein expression is restricted to spermiogenesis;

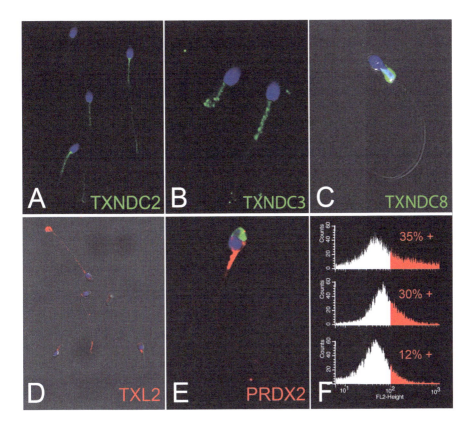

Fig. 15.1 Light microscopic (**a-e**) and flow cytometric detection of select redox proteins in human spermatozoa. Immunofluorescence detects thioredoxins TXNDC2/SPTRX1 (**a**) and TXNDC3/SPTRX2 (**b**) in the flagellar midpiece and principal piece of most morphologically normal and some aberrant spermatozoa, while TXNDC8/SPTRX3 accumulates in superfluous cytoplasm of defective spermatozoa (**c**). Thioredoxin TXL2 (**d**) is detected at varying intensity levels along the entire flagellum while peroxiredoxin PRDX2 (**e**) appears to be concentrated in the sperm tail midpiece/mitochondrial sheath, with lesser intensity in the sperm head postacrosomal sheath. (**f**) Differences in men's sperm content of TXNDC8, a candidate diagnostic marker of human male infertility and prognostic marker of ART outcomes are readily discernible in flow cytometric histograms of sperm samples labeled with fluorescent anti-TXNDC8 antibodies. Such individual differences are reflected by varying percentages of TXNDC8-positive spermatozoa in a teratospermic subject with extremely low sperm motility and concentration (top; two million sperm/ml, 1% motility), an oligo-asthenozoospermic subject (center) with low sperm motility (19%) and concentration (14 million/ml), and a normospermic donor (bottom; 84 million/ml, 85% motile)

however, its expression levels peak late in FS development, between steps 15–19 of rat spermiogenesis [119]. Moreover, unlike SPTRX-1, SPTRX-2 is not a transient component of flagellar development but rather has been shown to be an integral structural component of the FS that persists within the mature spermatozoa [119]. In in vitro studies using a physiological reducing system, *E. coil*-expressed human SPTRX-2 showed a lack of both thioredoxin and kinase activities, suggesting it may require posttranslational modifications or interactions with other co-factors or proteins to exert its functions [119]. Furthermore, SPTRX-2 does have several phosphorylation sites which could induce a required conformational change activating its catalytic functions [119]. These phosphorylation sites are of interest as phosphorylation events are integral to the sperm priming event of capacitation, which conveys sperm fertilizing ability within the female oviduct via changes in sperm motility patterns, and structural and membrane remodeling [119]. Lastly, immediately after fertilization occurs, the FS is one of the first sperm structures solubilized and recycled within the zygotic cytosol, superseding the breakdown of the paternal mitochondria and the outer dense fibers [124]. Therefore, the localization of SPTRX-2 as a structural component of the FS may be required to facilitating post-testicular functions such as the phosphorylation of the sperm tail substrates during capacitation and/or assisting in the rapid breakdown of the FS within the zygote after sperm-egg fusion [119]. Joint involvement of SPTRX1 and SPTRX2 in spermiogenesis and sperm anti-ROS defense can be inferred from a double knockout mouse model displaying the accelerated age-induced loss of sperm quality and motility [125].

Spermatid-specific thioredoxin 3 (SPTRX-3/TXNDC8) was the last of the trio of tissue-specific testicular thioredoxins discovered, and appears to have arisen from a gene duplication of the TRX-1 gene ancestor, with the only difference being an extra exon (exon V) present in the SPTRX-3 gene [121]. Clearly divergent from the two previously reported SPTRX proteins, SPTRX-3 does not contain any additional domains and is also uniquely localized in the Golgi apparatus of secondary spermatocytes and round spermatids rather than in the developing tail accessory structure of the elongating spermatids [121]. Surprisingly, SPTRX-3 lacks a clear signaling sequence to localize to the ER/Golgi system, and in fact has not been reported to associate with the ER in previous studies [116, 121]. However, SPTRX-3 does have additional cysteine residues within the exon V which suggest the possibility that Cys-acetylation serves as a signal for translocation to the Golgi apparatus, similar to the sorting mechanisms shown for GCP16 and PSKH1 [126, 127]. The expression of SPTRX-3 is highest in the Golgi apparatus of pachytene spermatocytes, before the commencement of proacrosomal granule formation and therefore is not suggested to have direct involvement in acrosome development [121]. Conversely, a functional role in regulating proteins through post-translational modifications has been suggested, as some proteins require phosphorylation during meiosis to be able to exert their functions, and previous studies have shown that TRX-1 does function in this capacity in other systems [128]. Additionally, SPTRX-3 may also be a candidate biomarker for male fertility, as SPTRX-3 overexpression and post-testicular retention via accumulation in sperm nuclear vacuoles and residual cytoplasm surrounding the flagellar midpiece have been reported in teratospermic patients [121] and have been associated with negative pregnancy outcomes in couples attending infertility clinics [129, 130].

Thioredoxin-like protein 2 (TXL-2) is a fusion protein that harbors a thioredoxin domain as well as one NDP domain, and has been found to share homology with SPTRX-2 [117]. TXL-2 also features an RCC1 signature that could suggest it interacts with Ran [117]. While not exclusive to testis, TXL-2 has a unique localization pattern that is tightly connected to microtubule-based structures, with its highest expression in the cilia of the lung airway epithelium and the manchette and axoneme of spermatids within the testicular seminiferous tubules [117]. While thioredoxins do not traditionally associate with microtubular structures in higher vertebrates, there are a few examples of such proteins in invertebrates such as LC14 and LC15 in *Chlamydomonas* flagellum and IC1 found in the dynein-containing intermediate filaments of sea urchin axonemal structures, previously discussed for its homology to SPTRX-2 [123, 131]. Furthermore, evidence suggests that there are members of the NDP kinase protein family that do interact with microtubules, giving some precedence for the unconventional localization patterns of TXL-2 [132–134]. The spermatid manchette is a hallmark structure of spermiogenesis that facilitates both nuclear shaping and the shuttling of proteins into the sperm head during spermatid elongation. Therefore, TXL-2 may act as a regulator of microtubule stability and maintenance within the manchette, using its functional disulfide reductase activity. Alternatively, TXL-2 could function to supply GTP for microtubule assembly or to facilitate the enzymatic activity of Ran through interactions with its previously stated RCC1 signature [117]. Clinically, TXL-2 may also be linked to aberrant sperm phenotypes as the gene maps to a region of the genome associated with primary ciliary dyskinesia (PCD), and one translocation in the gene has been associated with azoospermia and oligospermia [117]. Overall, the functional implications of TXL-2 activity could have large consequences on sperm morphology and energy regulation, specifically within the formation and function of the sperm axonemal structures.

As spermatogenesis progresses and GSH availability decreases, involvement of the glutathione system in the regulation of NADPH oxidation and antioxidant regulation most likely transitions to the peroxiredoxin-thioredoxin system. This is supported by the high number of sperm-specific enzymes endemic to the male germ line and the well-documented role of PRDXs as ROS regulators in spermatozoa [116, 118–121, 135]. However functionally, PRDXs require constant reduction to enact their role as H_2O_2 regulators and therefore must act synergistically with thioredoxins [136]. Once oxidized, PRDXs need to be reduced by a TRX which in turn is reduced by NADPH in the presence of a thioredoxin reductase [136]. Therefore, through the regulation of specific PRDX proteins, the TRX system may also be indirectly regulating the oxidative environment.

PRDX2 is the fastest regenerating redox protein and is highly efficient in neutralizing H_2O_2, providing better protection against H_2O_2-induced damage than catalase or glutathione peroxidases. It has recently been found to reside both in elongating spermatids and within seminal plasma [137, 138]. Put forth as a moonlighting enzyme, PRDX2 is suggested to act in a very similar capacity to and in cooperation with GPX4 [138]. PDRX2 is proposed to initially function in ROS regulation, via its peroxidase activity before being subsequently processed into an inactive form late in spermiogenesis, when it is structurally incorporated into the mitochondrial sheath and the highly compacted postacrosomal sheath region of the sperm head [138].

Regulation of the oxidative environment influences disulfide bond formation, through the availability of oxidative substrates, and in turn mediates the structural formation of key features during spermiogenesis. Therefore, hypercondensation of sperm chromatin and the formation of ultrastructural features of the sperm tail, two events that rely heavily on disulfide bond formation may also be highly regulated by a PRDX4, that functions as a newly proposed H_2O_2 sensor. While PRDX4 was originally documented in the testis/male germ line as a membrane-bound isoform localizing to the ER, a larger, alternatively transcribed cytosolic isoform (PRDX4-Long) has recently been shown to be restricted to spermatids of eutherian mammals [139]. Following a similar expression pattern of that of TGR, PRDX4-Long is proposed to act as an H_2O_2 sensor rather than in a peroxidative capacity, where it may influence TGR, GPX4, and PRDX2 [139]. This sensory function has been attributed to 2-Cys PRDXs in yeast and would functionally allow for disulfide bond formation without the production of damaging levels of H_2O_2 [140]. Therefore, PRDX4 may mediate the oxidative environment and act to signal for disulfide bond formation, ultimately conserving certain thiols under conditions that promote oxidation.

15.8 Conclusions and Perspectives

The glutathione and thioredoxin redox systems are well documented throughout the body but do show increased importance in the unique metamorphosis-like cellular differentiation and maturation processes of the male germ cells. While the glutathione-centered antioxidant system does have a heightened presence within the testicular environment and the early stages of spermatogenesis, antioxidant defense does appear to transition to a more peroxiredoxin-thioredoxin dominated system late in spermiogenesis and into sperm maturation, concomitant with the dramatic decrease in GSH availability. The shifting redox landscape within spermatogenesis may be explained by the unusual requirements of the male germ cell to be a structurally complex yet highly reducible cell. With such a high dependence on disulfide bond formation and reduction, and a limited cytosolic capacity for antioxidant enzymes, the sperm cell appears to have strategically incorporated inactive forms of antioxidants into its structural components, which could be re-activated once the sperm enters into the highly reductive environment of the zygotic cytosol. Furthermore, to ensure survival to the site of fertilization, a large amount of membrane-bound redox proteins have been identified coating the sperm surface which can interact with the high levels of antioxidant thiols in male reproductive system secretions, ultimately allowing the sperm cell to be streamlined yet protected. Therefore, sperm redox proteins such as glutathione transferases, glutathione peroxidases, thioredoxins, and peroxiredoxins should be regarded as critical for spermatozoon fitness and should be investigated as fertility markers for sperm function and structural integrity. Harnessing the knowledge of redox equilibrium in fertilization systems could produce large benefits in human-assisted reproductive therapies by optimizing oocyte and embryo culture media and stimulating various enzymatic activities in both male and female gametes. Maximizing the reducing power of human oocytes and selecting spermatozoa with high intrinsic reductase activities could promote pronuclear development via complete and even unpacking of the sperm nucleus, and timely release of oocyte activating sperm factors following intracytoplasmic sperm injection. Artificial insemination in agriculturally important livestock mammals may also benefit from the enrichment of semen diluents/extenders with antioxidants and reductase-stimulating ingredients to carefully balance the oxidative environment and preserve overall sperm function.

References

1. Aitken RJ, Baker MA, Nixon B. Are sperm capacitation and apoptosis the opposite ends of a continuum driven by oxidative stress? Asian J Androl. 2015;17(4):633.
2. Aitken RJ. The capacitation-apoptosis highway: oxysterols and mammalian sperm function. Biol Reprod. 2011;85(1):9–12.
3. Meister A, Larsson A. Glutathione synthetase deficiency and other disorders of the gamma-glutamyl cycle. Metab Mol Bases Inherit Dis. 1995;1:1461–95.
4. Meister A. Glutathione; metabolism and function via the γ-glutamyl cycle. Life Sci. 1974;15(2):177–90.
5. Pastore A, Federici G, Bertini E, Piemonte F. Analysis of glutathione: implication in redox and detoxification. Clin Chim Acta. 2003;333(1):19–39.
6. Meister A, Anderson ME. Glutathione. Annu Rev Biochem. 1983;52(1):711–60.
7. Sies H. Glutathione and its role in cellular functions. Free Radic Biol Med. 1999;27(9–10):916–21.
8. Dickinson DA, Forman HJ. Glutathione in defense and signaling: lessons from a small thiol. Ann N Y Acad Sci. 2002;973(1):488–504.
9. Grosshans K, Calvin HI. Estimation of glutathione in purified populations of mouse testis germ cells. Biol Reprod. 1985;33(5):1197–205.
10. Teaf CM, Harbison RD, Bishop JB. Germ-cell mutagenesis and GSH depression in reproductive tissue of the F-344 rat induced by ethyl methanesulfonate. Mutat Res Lett. 1985;144(2):93–8.
11. Mushahwar IK, Koeppe RE. Free amino acids of testes. Concentrations of free amino acids in the testes of several species and the precursors of glutamate and glutamine in rat testes in vivo. Biochem J. 1973;132(3):353–9.
12. Kochakian CD. Free amino acids of sex organs of the mouse: regulation by androgen. Am J Physiol Content. 1975;228(4):1231–5.
13. Gualtieri AF, Iwachow MA, Venara M, Rey RA, Schteingart HF. Bisphenol A effect on glutathione synthesis and recycling in testicular Sertoli cells. J Endocrinol Investig. 2011;34(5):e102–9.
14. Li L, Seddon AP, Meister A, Risley MS. Spermatogenic cell-somatic cell interactions are required for maintenance of spermatogenic cell glutathione. Biol Reprod. 1989;40(2):317–31.
15. Den Boer PJ, Mackenbach P, Grootegoed JA. Glutathione metabolism in cultured Sertoli cells and spermatogenic cells from hamsters. Reproduction. 1989;87(1):391–400.
16. Kaneko T, et al. The expression of glutathione reductase in the male reproductive system of rats supports the enzymatic basis of glutathione function in spermatogenesis. Eur J Biochem. 2002;269(5):1570–8.
17. Aitken RJ, Clarkson JS. Cellular basis of defective sperm function and its association with the genesis of reactive oxygen species by human spermatozoa. Reproduction. 1987;81(2):459–69.

18. Aruldhas MM, et al. Chronic chromium exposure-induced changes in testicular histoarchitecture are associated with oxidative stress: study in a non-human primate (Macaca radiata Geoffroy). Hum Reprod. 2005;20(10):2801–13.
19. Paul C, Teng S, Saunders PTK. A single, mild, transient scrotal heat stress causes hypoxia and oxidative stress in mouse testes, which induces germ cell death. Biol Reprod. 2009;80(5):913–9.
20. Celino FT, et al. Tolerance of spermatogonia to oxidative stress is due to high levels of Zn and Cu/Zn superoxide dismutase. PLoS One. 2011;6(2):e16938.
21. Roveri A, Ursini F, Flohé L, Maiorino M. PHGPx and spermatogenesis. Biofactors. 2001;14:213–22.
22. Potts RJ, Jefferies TM, Notarianni LJ. Antioxidant capacity of the epididymis. Hum Reprod. 1999;14(10):2513–6.
23. Colagar AH, Marzony ET. Ascorbic acid in human seminal plasma: determination and its relationship to sperm quality. J Clin Biochem Nutr. 2009;45(2):144–9.
24. Funahashi H, Cantley TC, Stumpf TT, Terlouw SL, Day BN. Use of low-salt culture medium for in vitro maturation of porcine oocytes is associated with elevated oocyte glutathione levels and enhanced male pronuclear formation after in vitro fertilization. Biol Reprod. 1994;51(March):633–9.
25. Sutovsky P, Schatten G. Depletion of glutathione during bovine oocyte maturation reversibly blocks the decondensation of the male pronucleus and pronuclear apposition during fertilization. Biol Reprod. 1997;56(6):1503–12.
26. Zuelke KA, Jeffay SC, Zucker RM, Perreault SD. Glutathione (GSH) concentrations vary with the cell cycle in maturing hamster oocytes, zygotes, and pre-implantation stage embryos. Mol Reprod Dev. 2003;64(1):106–12.
27. Kim IH, et al. Effect of exogenous glutathione on the in vitro fertilization of bovine oocytes. Theriogenology. 1999;52(3):537–47.
28. Yoshida M, Ishigaki K, Nagai T, Chikyu M, Pursel VG. Glutathione concentration during maturation and after fertilization in pig oocytes: relevance to the ability of oocytes to form male pronucleus. Biol Reprod. 1993;49(1):89–94.
29. Krisher RL, Bavister BD. Responses of oocytes and embryos to the culture environment. Theriogenology. 1998;49(1):103–14.
30. Eppig JJ. Coordination of nuclear and cytoplasmic oocyte maturation in eutherian mammals. Reprod Fertil Dev. 1996;8(4):485–9.
31. Furnus CC, De Matos DG, Moses DF. Cumulus expansion during in vitro maturation of bovine oocytes: relationship with intracellular glutathione level and its role on subsequent embryo development. Mol Reprod Dev. 1998;51(1):76–83.
32. de Matos DG, Furnus CC, Moses DF. Glutathione synthesis during in vitro maturation of bovine oocytes: role of cumulus cells. Biol Reprod. 1997;57(6):1420–5.
33. de Matos DG, Furnus CC. 2000-The importance of having high glutathione.pdf. Theriogenology. 2000;53(3):761–71.
34. Abeydeera LR, et al. Coculture with follicular shell pieces can enhance the developmental competence of pig oocytes after in vitro fertilization: relevance to intracellular glutathione. Theriogenology. 1998;58(9):1244–56.
35. Hamilton LE, et al. Sperm-borne glutathione-s-transferase omega 2 accelerates the nuclear decondensation of spermatozoa during fertilization in mice†. Biol Reprod. 2019;101(2):368–76.
36. Perreault SD, Wolff RA, Zirkin BR. The role of disulfide bond reduction during mammalian sperm nuclear decondensation in vivo. Dev Biol. 1984;101(1):160–7.
37. Furnus CC, et al. Metabolic requirements associated with GSH synthesis during in vitro maturation of cattle oocytes. Anim Reprod Sci. 2008;109(1):88–99.
38. Curnow EC, Ryan J, Saunders D, Hayes ES. Bovine in vitro oocyte maturation as a model for manipulation of the γ-glutamyl cycle and intraoocyte glutathione. Reprod Fertil Dev. 2008;20(5):579–88.

39. Hayes JD, Flanagan JU, Jowsey IR. Glutathione transferases. Annu Rev Pharmacol Toxicol. 2005;45:51–88.
40. Mannervik B, Helena Danielson U, Ketterer B. Glutathione transferases—structure and catalytic activit. Crit Rev Biochem. 1988;23(3):283–337.
41. Tew KD, Townsend DM. Glutathione-s-transferases as determinants of cell survival and death. Antioxid Redox Signal. 2012;17(12):1728–37.
42. Raijmakers MTM, et al. Glutathione and glutathione S-transferases A1-1 and P1-1 in seminal plasma may play a role in protecting against oxidative damage to spermatozoa. Fertil Steril. 2003;79(1):169–72.
43. Hemachand T, Gopalakrishnan B, Salunke DM, Totey SM, Shaha C. Sperm plasma-membrane-associated glutathione S-transferases as gamete recognition molecules. J Cell Sci. 2002;115(10):2053–65.
44. Petit FM, Serres C, Auer J. Moonlighting proteins in sperm–egg interactions. Biochem Soc Trans. 2014;42(6):1740–3.
45. Llavanera M, et al. Exploring seminal plasma GSTM3 as a quality and in vivo fertility biomarker in pigs—relationship with sperm morphology. Antioxidants. 2020;9(8) https://doi.org/10.3390/antiox9080741.
46. Gopalakrishnan B, et al. Studies on glutathione S-transferases important for sperm function: evidence of catalytic activity-independent functions. Biochem J. 1998;329(2):231–41.
47. Llavanera M, et al. Deactivation of the JNK pathway by GSTP1 is essential to maintain sperm functionality. Front Cell Dev Biol. 2021;9:627140.
48. Hamilton LE, Acteau G, Xu W, Sutovsky P, Oko R. The developmental origin and compartmentalization of glutathione-s-transferase omega 2 isoforms in the perinuclear theca of eutherian spermatozoa. Biol Reprod. 2017;97(4):612–21.
49. Protopapas N, et al. The perforatorium and postacrosomal sheath of rat spermatozoa share common developmental origins and protein constituents. Biol Reprod. 2019;100(6):1461–72.
50. Hamilton LE, et al. GSTO2 isoforms participate in the oxidative regulation of the plasmalemma in eutherian spermatozoa during capacitation. Antioxidants. 2019;8(12) https://doi.org/10.3390/antiox8120601.
51. Yu B, Huang Z. Variations in antioxidant genes and male infertility. Biomed Res Int. 2015:2015.
52. Tirumala Vani G, et al. Role of glutathione S-transferase Mu-1 (GSTM1) polymorphism in oligospermic infertile males. Andrologia. 2010;42(4):213–7.
53. Aydemir B, Onaran I, Kiziler AR, Alici B, Akyolcu MC. Increased oxidative damage of sperm and seminal plasma in men with idiopathic infertility is higher in patients with glutathione S-transferase Mu-1 null genotype. Asian J Androl. 2007;9(1):108–15.
54. Finotti AC, Costa E, Bordin BM, Silva CT, Moura KK. Glutathione S-transferase M1 and T1 polymorphism in men with idiopathic infertility. Genet Mol Res. 2009;8(3):1093–8.
55. Tang K, et al. Genetic polymorphisms of glutathione S-transferase M1, T1, and P1, and the assessment of oxidative damage in infertile men with varicoceles from northwestern China. J Androl. 2012;33(2):257–63.
56. Wu W, et al. GSTM1 and GSTT1 null polymorphisms and male infertility risk: an updated meta-analysis encompassing 6934 subjects. Sci Rep. 2013;3(1):1–11.
57. Listowsky I, et al. Human testicular glutathione S-transferases: insights into tissue-specific expression of the diverse subunit classes. Chem Biol Interact. 1998;111:103–12.
58. Mukhtar H, Lee IP, Bend JR. Glutathione S-transferase activities in rat and mouse sperm and human semen. Biochem Biophys Res Commun. 1978;83(3):1093–8.
59. Yu Z, et al. Gene expression profiles in different stages of mouse spermatogenic cells during spermatogenesis. Biol Reprod. 2003;69(1):37–47.
60. Klys HS, Whillis D, Howard G, Harrison DJ. Glutathione S-transferase expression in the human testis and testicular germ cell neoplasia. Br J Cancer. 1992;66(3):589–93.
61. Yin Z-L, Dahlstrom JE, Le Couteur DG, Board PG. Immunohistochemistry of omega class glutathione S-transferase in human tissues. J Histochem Cytochem. 2001;49(8):983–7.

62. Johansson A-S, Mannervik B. Human glutathione transferase A3-3, a highly efficient catalyst of double-bond isomerization in the biosynthetic pathway of steroid hormones. J Biol Chem. 2001;276(35):33061–5.
63. Lantum HBM, Baggs RB, Krenitsky DM, Board PG, Anders MW. Immunohistochemical localization and activity of glutathione transferase zeta (GSTZ1–1) in rat tissues. Drug Metab Dispos. 2002;30(6):616–25.
64. Li J, et al. Systematic mapping and functional analysis of a family of human Epididymal secretory sperm-located proteins*. Mol Cell Proteomics. 2010;9(11):2517–28.
65. Papp S, Robaire B, Hermo L. Immunocytochemical localization of the Ya, Yc, Yb1, and Yb2 subunits of glutathione S-transferases in the testis and epididymis of adult rats. Microsc Res Tech. 1995;30(1):1–23.
66. Sun Z, Wei R, Luo G, Niu R, Wang J. Proteomic identification of sperm from mice exposed to sodium fluoride. Chemosphere. 2018;207:676–81.
67. Sun Z, et al. Alterations in epididymal proteomics and antioxidant activity of mice exposed to fluoride. Arch Toxicol. 2018;92(1):169–80.
68. Fulcher KD, Welch JE, Klapper DG, O'Brien DA, Eddy EM. Identification of a unique μ-class glutathione S-transferase in mouse spermatogenic cells. Mol Reprod Dev. 1995;42(4):415–24.
69. Oko R, Sutovsky P. Biogenesis of sperm perinuclear theca and its role in sperm functional competence and fertilization. J Reprod Immunol. 2009;83(1–2):2–7.
70. Sutovsky P, Manandhar G, Wu A, Oko R. Interactions of sperm perinuclear theca with the oocyte: implications for oocyte activation, anti-polyspermy defense, and assisted reproduction. Microsc Res Tech. 2003;61(4):362–78.
71. Brigelius-Flohé R, Flohé L. Regulatory phenomena in the glutathione peroxidase superfamily. Antioxid Redox Signal. 2020;33(7):498–516.
72. Noblanc A, et al. Glutathione peroxidases at work on Epididymal spermatozoa: an example of the dual effect of reactive oxygen species on mammalian male fertilizing ability. J Androl. 2011;32(6):641–50.
73. Drevet JR. The antioxidant glutathione peroxidase family and spermatozoa: a complex story. Mol Cell Endocrinol. 2006;250(1–2):70–9.
74. Maser RL, Magenheimer BS, Calvet JP. Mouse plasma glutathione peroxidase. cDNA sequence analysis and renal proximal tubular expression and secretion. J Biol Chem. 1994;269(43):27066–73.
75. Schwaab V, Faure J, Dufaure J, Drevet JR. GPx3: the plasma-type glutathione peroxidase is expressed under androgenic control in the mouse epididymis and vas deferens. Mol Reprod Dev Inc Gamete Res. 1998;51(4):362–72.
76. Rejraji H, Vernet P, Drevet JR. GPX5 is present in the mouse caput and cauda epididymidis lumen at three different locations. Mol Reprod Dev Inc Gamete Res. 2002;63(1):96–103.
77. de Haan JB, et al. Mice with a homozygous null mutation for the most abundant glutathione peroxidase, Gpx1, show increased susceptibility to the oxidative stress-inducing agents paraquat and hydrogen peroxide. J Biol Chem. 1998;273(35):22528–36.
78. Morel Y, Barouki R. Repression of gene expression by oxidative stress. Biochem J. 1999;342(3):481–96.
79. Meseguer M, et al. Effect of sperm glutathione peroxidases 1 and 4 on embryo asymmetry and blastocyst quality in oocyte donation cycles. Fertil Steril. 2006;86(5):1376–85.
80. Ursini F, Maiorino M. Lipid peroxidation and ferroptosis: the role of GSH and GPx4. Free Radic Biol Med. 2020;152(March):175–85.
81. Li L, et al. Ferroptosis is associated with oxygen-glucose deprivation/reoxygenation-induced Sertoli cell death. Int J Mol Med. 2018;41(5):3051–62.
82. Zhao X, et al. Inhibition of ferroptosis attenuates busulfan-induced oligospermia in mice. Toxicology. 2020;440:152489.
83. Ursini F, et al. Dual function of the selenoprotein PHGPx during sperm maturation. Science (80-). 1999;285(5432):1393–6.
84. Flohe L (2007) Selenium in mammalian spermiogenesis.

85. Schneider M, et al. Mitochondrial glutathione peroxidase 4 disruption causes male infertility. FASEB J. 2009;23(9):3233–42.
86. Fisher HM, Aitken RJ. Comparative analysis of the ability of precursor germ cells and epididymal spermatozoa to generate reactive oxygen metabolites. J Exp Zool. 1997;277(5):390–400.
87. Shalgi R, Seligman J, Kosower NS. Dynamics of the thiol status of rat spermatozoa during maturation: analysis with the fluorescent labeling agent monobromobimane. Biol Reprod. 1989;40(5):1037–45.
88. Puglisi R, et al. The nuclear form of glutathione peroxidase 4 is associated with sperm nuclear matrix and is required for proper paternal chromatin decondensation at fertilization. J Cell Physiol. 2012;227(4):1420–7.
89. Pfeifer H, et al. Identification of a specific sperm nuclei selenoenzyme necessary for protamine thiol cross-linking during sperm maturation. FASEB J. 2001;15(7):1236–8.
90. Ghyselinck NB, et al. Structural organization and regulation of the gene for the androgen-dependent glutathione peroxidase-like protein specific to the mouse epididymis. Mol Endocrinol. 1993;7(2):258–72.
91. Vernet P, Faure J, Dufaure J, Drevet JR. Tissue and developmental distribution, dependence upon testicular factors and attachment to spermatozoa of GPX5, a murine epididymis-specific glutathione peroxidase. Mol Reprod Dev Inc Gamete Res. 1997;47(1):87–98.
92. Chabory E, et al. Epididymis seleno-independent glutathione peroxidase 5 maintains sperm DNA integrity in mice. J Clin Invest. 2009;119(7):2074–85.
93. Ahsan U, et al. Role of selenium in male reproduction—a review. Anim Reprod Sci. 2014;146(1–2):55–62.
94. Sprinker LH, Harr JR, Newberne PM, Whanger PD, Weswig PH. Selenium deficiency lesions in rats fed vitamin E supplemented rations. Nutr Rep Int. 1971;4(6):335–40.
95. Olson GE, Winfrey VP, NagDas SK, Hill KE, Burk RF. Selenoprotein P is required for mouse sperm development. Biol Reprod. 2005;73(1):201–11.
96. Olson GE, Winfrey VP, NagDas SK, Hill KE, Burk RF. Apolipoprotein E receptor-2 (ApoER2) mediates selenium uptake from selenoprotein P by the mouse testis. J Biol Chem. 2007;282(16):12290–7.
97. Kehr S, et al. X-ray fluorescence microscopy reveals the role of selenium in spermatogenesis. J Mol Biol. 2009;389(5):808–18.
98. Gladyshev VN, et al. Selenocysteine-containing Thioredoxin reductase in C. elegans. Biochem Biophys Res Commun. 1999;259(2):244–9.
99. Arnér ESJ, Holmgren A. Physiological functions of thioredoxin and thioredoxin reductase. Eur J Biochem. 2000;267(20):6102–9.
100. Su D, et al. Mammalian Selenoprotein Thioredoxin-glutathione reductase. J Biol Chem. 2005;280(28):26491–8.
101. Nayernia K, et al. Asthenozoospermia in mice with targeted deletion of the sperm mitochondrion-associated cysteine-rich protein (Smcp) gene. Mol Cell Biol. 2002;22(9):3046–52.
102. Lee S, Kim SM, Lee RT. Thioredoxin and thioredoxin target proteins: from molecular mechanisms to functional significance. Antioxidants Redox Signal. 2013;18(10):1165–207.
103. Moore EC, Reichard P, Thelander L. Synthesis of Deoxyribonucleotides. October. 1964;239(10):3436–44.
104. Laurent T, Moore EC, Reichard P. Synthesis of Deoxyribonucleotides. October. 1964;239(10):3436–44.
105. Yodoi J, Tagaya Y, Masutani H, Maeda Y, Kawabe T. IL-2 receptor and Fc epsilon R2 gene activation in lymphocyte transformation: possible roles of ATL-derived factor. Princess Takamatsu Symp. 1988;19:73–86.
106. Gasdaska PY, Gasdaska JR, Cochran S, Powis G. Cloning and sequencing of a human thioredoxin reductase. FEBS Lett. 1995;373(1):5–9.
107. Pedrajas JR, et al. Identification and functional characterization of a novel mitochondrial thioredoxin system in Saccharomyces cerevisiae. J Biol Chem. 1999;274(10):6366–73.

108. Miranda-Vizuete A, Damdimopoulos AE, Pedrajas JR, Gustafsson J-Å, Spyrou G. Human mitochondrial thioredoxin reductase. Eur J Biochem. 1999;261(2):405–12.
109. Spyrou G, Enmark E, Miranda-Vizuete A, Gustafsson JÅ. Cloning and expression of a novel mammalian thioredoxin. J Biol Chem. 1997;272(5):2936–41.
110. Hanschmann E-M, Godoy JR, Berndt C, Hudemann C, Lillig CH. Thioredoxins, glutaredoxins, and peroxiredoxins--molecular mechanisms and health significance: from cofactors to antioxidants to redox signaling. Antioxid Redox Signal. 2013;19(13):1539–605.
111. Clarke FM, et al. Identification of molecules involved in the 'early pregnancy factor' phenomenon. J Reprod Fertil. 1991;93(2):525–39.
112. Salz HK, et al. The Drosophila maternal effect locus deadhead encodes a thioredoxin homolog required for female meiosis and early embryonic development. Genetics. 1994;136(3):1075–86.
113. Tirmarche S, Kimura S, Dubruille R, Horard B, Loppin B. Unlocking sperm chromatin at fertilization requires a dedicated egg thioredoxin in Drosophila. Nat Commun. 2016;7:135–9.
114. Emelyanov AV, Fyodorov DV. Thioredoxin-dependent disulfide bond reduction is required for protamine eviction from sperm chromatin. Genes Dev. 2016;30(24):2651–6.
115. Svensson MJ, Chen JD, Pirrotta V, Larsson J. The ThioredoxinT and deadhead gene pair encode testis- and ovary-specific thioredoxins in Drosophila melanogaster. Chromosoma. 2003;112(3):133–43.
116. Jiménez A, et al. Absolute mRNA levels and transcriptional regulation of the mouse testis-specific thioredoxins. Biochem Biophys Res Commun. 2005;330(1):65–74.
117. Sadek CM, et al. Characterization of human thioredoxin-like 2: a novel microtubule-binding thioredoxin expressed predominantly in the cilia of lung airway epithelium and spermatid manchette and axoneme. J Biol Chem. 2003;278(15):13133–42.
118. Jiménez A, et al. Human spermatid-specific thioredoxin-1 (Sptrx-1) is a two-domain protein with oxidizing activity. FEBS Lett. 2002;530(1–3):79–84.
119. Sadek CM, et al. Sptrx-2, a fusion protein composed of one thioredoxin and three tandemly repeated NDP-kinase domains is expressed in human testis germ cells. Genes Cells. 2001;6(12):1077–90.
120. Miranda-Vizuete A, et al. The mammalian testis-specific Thioredoxin system. Antioxidants Redox Signal. 2004;6(1):25–40.
121. Jiménez A, et al. Spermatocyte/spermatid-specific thioredoxin-3, a novel golgi apparatus-associated thioredoxin, is a specific marker of aberrant spermatogenesis. J Biol Chem. 2004;279(33):34971–82.
122. Yu Y, Oko R, Miranda-Vizuete A. Developmental expression of spermatid-specific thioredoxin-1 protein: transient association to the longitudinal columns of the fibrous sheath during sperm tail formation. Biol Reprod. 2002;67(5):1546–54.
123. Ogawa K, et al. Is outer arm dynein intermediate chain 1 multifunctional? Mol Biol Cell. 1996;7(12):1895–907.
124. Sutovsky P, Navara CS, Schatten G. Fate of the sperm mitochondria, and the incorporation, conversion, and disassembly of the sperm tail structures during bovine Fertilization. Biol Reprod. 1996;55(6):1195–205.
125. Smith TB, Baker MA, Connaughton HS, Habenicht U, Aitken RJ. Functional deletion of Txndc2 and Txndc3 increases the susceptibility of spermatozoa to age-related oxidative stress. Free Radic Biol Med. 2013;65:872–81.
126. Ohta E, et al. Identification and characterization of GCP16, a novel Acylated Golgi protein that interacts with GCP170*. J Biol Chem. 2003;278(51):51957–67.
127. Brede G, Solheim J, Stang E, Prydz H. Mutants of the protein serine kinase PSKH1 disassemble the Golgi apparatus. Exp Cell Res. 2003;291(2):299–312.
128. Saitoh M, et al. Mammalian thioredoxin is a direct inhibitor of apoptosis signal-regulating kinase (ASK) 1. EMBO J. 1998;17(9):2596–606.

129. Ahlering P, et al. Sperm content of TXNDC8 reflects sperm chromatin structure, pregnancy establishment, and incidence of multiple births after ART. Syst Biol Reprod Med. 2020;66(5):311–21.
130. Buckman C, et al. Semen levels of spermatid-specific thioredoxin-3 correlate with pregnancy rates in ART couples. PLoS One. 2013;8(5):e61000.
131. Patel-King RS, Benashski SE, Harrison A, King SM. Two functional Thioredoxins containing redox-sensitive vicinal dithiols from the Chlamydomonas outer dynein arm (∗). J Biol Chem. 1996;271(11):6283–91.
132. Roymans D, et al. Identification of the tumor metastasis suppressor Nm23-H1/Nm23-R1 as a constituent of the centrosome. Exp Cell Res. 2001;262(2):145–53.
133. Pinon VP-B, et al. Cytoskeletal Association of the A and B Nucleoside Diphosphate Kinases of Interphasic but not mitotic human carcinoma cell lines: specific nuclear localization of the B Subunit. Exp Cell Res. 1999;246(2):355–67.
134. Nickerson JA, Wells WW. The microtubule-associated nucleoside diphosphate kinase. J Biol Chem. 1984;259(18):11297–304.
135. O'Flaherty C. Peroxiredoxins: hidden players in the antioxidant defence of human spermatozoa. Basic Clin Androl. 2014;24(1):1–10.
136. Rhee SG, Woo HA. Multiple functions of Peroxiredoxins: peroxidases, sensors and regulators of the intracellular messenger H2O2, and protein chaperones. Antioxid Redox Signal. 2010;15(3):781–94.
137. Chevallet M, et al. Regeneration of Peroxiredoxins during recovery after oxidative stress: only some overoxidized peroxiredoxins can be reduced during recovery after oxidative stress. J Biol Chem. 2003;278(39):37146–53.
138. Manandhar G, et al. Peroxiredoxin 2 and peroxidase enzymatic activity of mammalian spermatozoa. Biol Reprod. 2009;80(6):1168–77.
139. Yim SH, et al. Identification and characterization of alternatively transcribed form of peroxiredoxin IV gene that is specifically expressed in spermatids of postpubertal mouse testis. J Biol Chem. 2011;286(45):39002–12.
140. Veal EA, et al. A 2-Cys Peroxiredoxin regulates peroxide-induced oxidation and activation of a stress-activated MAP kinase. Mol Cell. 2004;15(1):129–39.

Index

A

Abasic sites, 79, 83
Acquired immune deficiency syndrome (AIDS), 304
Acrolein, 43
Adenosine triphosphate (ATP) synthesis, 46
Adipokine, 50
Alcohol consumption, 188, 189
Alkylating agents, 80, 82
Alkylating reaction, 82
Alkylation, 82
Alternative NHEJ pathway, 89
Angiotensin converting enzyme-2 (ACE2), 332
Antibiotics, 129
Antioxidants
    male infertility, 248
    MOSI, 193, 194
        cryodamage, 195
        DNA damage prevention, 195
        sperm motility, 195
Antiviral therapies
    male infertility, 305–309
        interferons, 311, 312
        natural compounds, 305, 309–311
Apigenin, 305, 306
Apoptosis, 167, 170, 172, 192, 266
    OS and male infertility, 170
Apoptotic markers, 84
Apurinic/apyrimidinic (AP) site, 87
Assisted reproductive technique (ART), 193
    clinical studies, 97, 98, 100–102
    DNA methylation profiling, 95
    embryo quality, 92–94
    fertilization rates, 92
    pregnancy, 92
Asthenozoospermia, 43, 258, 259
Athenozoospermia
    therapeutic approach for, 267, 268
Autophagy, 170
Azoospermia, 304
Azoospermia factor (AZF) locus, 85

B

Bacteria, 42, 116
Bacterial infection
    bacteriospermia, 119
    Chlamydia trachomatis, 120
    DNA fragmentation and fertilization rates, 119
    E. coli, 120
    gonorrhea, 121
    mycoplasma, 122
    Staphylococcus, 121
    treatment, 125
    U. urealyticum, 121
Bacteriospermia, 127, 142, 144, 155
    bacteria, 152
    E. coli, 149
    hydrophilic amino acids, 152
    infection and inflammatory responses, 145
    leukocytes, 146
    LPO, 153
    in male, 148
    male urogenital tract, 153

Bacteriospermia (*cont.*)
　management, 154
　prevalence, 145
　role, 146
　ROS, 152
Baicalein, 306, 308
Baltimore classification of viruses, 288
Base excision repair (BER), 87
Bioflavonoid, 227
Bisphenol A, 80
Blastulation, 169
Bluetongue virus, 331
Butylated hydroxyanisole (BHA), 23

C

Caloric restriction, 248
cAMP/PRKA pathway, 262
cAMP response element binding protein (CREB), 26
Carotenoids, 194
Catechin, 310
CCAAT/enhancer-binding protein beta (CEBPB), 171
C/EBP Homologous Protein (CHOP), 171
Chebulagic acid, 307
Chinese herbal medicine, 227, 228
Chlamydia trachomatis, 120, 150
Chlamydial infection, 150
Chromatin, 86
　in dynamic and plastic process, 86
　and histone proteins, 86
　packaging, 87
　remodeling, 86
　SCD, 92
　SCSA, 92
　stability, 89
Chronic inflammation, 117
Chronic OS, 172
Chronic sperm infections, 327
Cilia, 259
Cinnoxicam, 220
Clastogenic effect, 83
Coenzyme Q10 (CoQ10), 194, 268
Colchicine, 249
Comet assay, 93
Connective tissue cells, 173
Corpus luteum, 23, 24
Coumarin, 307
COVID-19, 332
Cranberry, 155
Creatine kinase (CK), 188
Cryopreserved sperm, 96
Curcumin, 267, 268, 307

Cyclic adenosine monophosphate (cAMP), 24
Cyclic guanosine monophosphate (cGMP), 47
Cyclobutane pyrimidine dimers, 80, 83
Cyclooxygenase activity, 22
Cystitis, 146
Cytokines, 42, 172
Cytosine, 81

D

Database for Annotation Visualization and Integrated Discovery (DAVID), 67
Deamination, 83
Difference gel electrophoresis (DIGE), 65
Differentially expressed proteins (DEPs), 67
DNA damage, *see* Sperm DNA damage
DNA damage response (DDR), 79
DNA double-strand break repair, 89, 90
DNA fragmentation, 83, 84
DNA fragmentation index (DFI), 96, 97, 99
DNA helix, 81, 82, 84, 88
DNA hydroxylation, 167
DNA methylation, 86, 87
DNA methylation profiling, 95
　NGS, 95
　pyrosequencing, 95
DNA polymerase, 79, 81
DNA repair, 79, 84
　molecular markers, 95, 96
DNA replication, 79
DNA sequencing, 95
DNA strand breaks, 167
DNA synthesis, 89
DNA virus, 281
Double-strand break (DSB) repair, 79, 80, 89, 90, 98
Double-stranded DNA, 89

E

Early paternal effect, 169
Ebola virus, 286, 297–299
Electron transfer chain complexes (ETCs), 261
Electron transport chain (ETC), 10
Electron transport system, 17
Eliptica alba, 305
Embryo implantation, 98–100
Endocrine disrupting chemicals (EDCs), 166
Endocrine-disrupting compounds (ECDs), 244
Endogenous DNA damage, 79, 80
Endometrial stromal cells, 17
Endometrium, 17, 18
Endonucleases, 84
Endoplasmic reticulum, 17

Epididymis, 90
Epididymitis, 118
Epigallocatechin gallate, 307
Epigenetic abnormalities, 86, 87
Epigenetic factors, 88
Epigenetics mechanisms, 86, 87
Epigenetics regulation, 98
Epithelial cells, 18
Epithelial tissue cells, 173
Erectile dysfunction (ED), 47–49
Escaped genes theory, 277
Eukaryotic cells, 88, 89
Excess residual cytoplasm (ERC), 19
Excessive/uncontrolled protracted inflammation, 171
Excision repair
    BER, 87
    NER, 87, 88
Exercise
    oxidative stress, male infertility, 247, 248
Exogenous antioxidants
    varicocele-associated male infertility, 220
        as adjunct therapy, 228–230
        sole therapy, 220, 222, 223, 225–227
Exogenous DNA damage, 80
EXOI, 88

F
Fallopian tube, 16
False discovery rates (FDRs), 67
Female reproductive system, 170
Fertility
    in men with varicocele, 212, 213, 215
    varicocele-associated OS and, 210, 213
Fertilization, 4, 5
    and embryo quality, 93
    in vitro, 91
    oocyte
        clinical studies in ART, 97, 98
        in vitro and animal studies, 96, 97, (*see also* Oocyte fertilization)
    and pregnancy (*see* Pregnancy)
    and sperm DNA damage (*see* Sperm DNA damage)
Flap endonuclease-1 (FEN-1), 87
Flavonoids, 305
Follicle stimulating hormone (FSH), 26
Follicular fluid, 15, 16
Folliculogenesis, 21
Fourier-transform ion cyclotron (FTIC), 67
Free radicals, 11, 143

G
Ganoderma tsugae, 54
GCHI (guanosine triphosphate cyclohydrolase I), 174
Gel electrophoresis, 65
Gender-based susceptibility, 334, 335
Gene Ontology (GO), 67
Genistein, 307, 311
Genotoxic agents, 80
Germ cell apoptosis, 168
Global genome NER (GG-NER), 87
Glucose-6-phosphate dehydrogenase enzyme (G6PD), 187
Glutathione, 347, 360
    availability, 348
    concentration, 347, 348
Glutathione peroxidases (GPX), 351
    enzymes, 351
    expression, 351, 352
    GPX4, 352
    isoenzymes, 352
    mouse knockout model, 353
    sperm nuclear, 353
Glutathione redox cycle
    glutamic acid, 346
Glutathione S-transferase omega-2 (GSTO2), 209
Glutathione-transferases (GSTs), 348
    activity, 350
    biological activity, 349
    cytosolic, 348
    enzymes, 350
    epididymal, 349
    epididymis, 349
    transcripts, 348
Glycolysis, 261
Gonorrhea, 121
Graafian follicles, 15
Guanine-cytosine, 81

H
Hayflick limit, 85
Heat Shock Protein Family D (HSPD), 171
Helical virus, 280
Hepatitis B virus (HBV), 287, 327
    and male infertility, 301–303
Hepatitis C virus (HCV), 285, 290, 327
    and male infertility, 290, 291
Herbal medicine, 155
    male infertility, 248, 249
Herpes simplex virus (HSV), 285, 289
    and male infertility, 290

Highly active antiretroviral therapy (HAART), 312
High-performance liquid chromatography (HPLC), 65–66
Homologous end-joining, 89
Homologous recombination (HR), 89
Human immunodeficiency virus (HIV), 287, 303, 326
    male infertility, 303, 304
Human papillomavirus (HPV), 123, 285, 288, 289
Human T-lymphotropic virus type I (HTLV-1), 330
Hydrogen peroxide (H_2O_2), 2, 13
Hydrolysis, 82
4-Hydroxynonenal (4-HNE), 43
8-Hydroxyguanosine (8-OHdG), 84
Hydroxyl radicals (OH•), 12
Hyperactivation, 143
Hyperglycaemia, 243

I

Icosahedral, 280
Idiopathic male infertility, 183, 184, 195
Idiopathic oligoasthenoteratozoospermia (IOAT), 183
Immotile cilia syndrome, 259
Immunofluorescence detects, 357
Inactive sperm DNA, 101
Infections, 117, 168, 172, 173
Infectious inflammatory diseases, 117
Infertile men
    with or without varicocele, 212, 214
Infertility, 78, 116, 145, 149, 258
Inflammation, 241
    inflammatory mediators, 173
    inflammatory transcription factors, 173
    leydig cells, 173, 174
    MAP kinases, 173
    mechanisms, 172
    in metabolic dysfunction, 244–246
    MyD88, 173
    and OS, 184, 185
    PAMPs, 173
    pathogen-specific molecules, 173
    Sertoli cell functions, 173
    TLRs, 173, 174
    vicious loop, 171, 172
Inflammatory cells, 171
Inflammatory mediators, 173
Inflammatory response, 145
Inflammatory transcription factors, 173
Influenza virus, 287, 299, 300

Ingenuity pathway analysis (IPA), 67
Intercellular adhesion molecule-1 (ICAM-1), 171
Interferon regulatory factor 3 (IRF3), 173
Interferons (IFNs), 311, 312, 336
    IFN-α, 308
    IFN-β, 308
    IFN-γ, 308
    IFN-λ, 308
Interferon therapy, 306–308
Interleukin (IL)-8, 19
Intermembrane space (IMS), 171
Internal spermatic vein (ISV), 45
Interstrand crosslinks, 83
Intracytoplasmic sperm injection (ICSI), 98–100
Intrauterine insemination (IUI), 100
In vitro and animal studies, 97
    oocyte fertilization, 96, 97
    pregnancy, 98–100
In vitro fertilization (IVF), 98, 100
In vitro model, 91
Ionizing radiation, 80

J

Jingling oral liquid, 228

K

Kaempferol, 307
Klebsiella pneumoniae, 152
Kyoto Encyclopedia of Genes and Genomes (KEGG), 67

L

Leucocytes, 171
Leukocytes, 18, 19, 126, 128, 168, 187, 188
Leukocytospermia, 126, 128–130
Levofloxacin, 129
Leydig cells, 173, 331, 336
Lipid peroxidation (LPO), 45, 167–169, 190
Lipid peroxyl radicals, 190
Liquid chromatography-tandem mass spectrometry (LC-MS/MS), 65
Long PCR, 93, 94
Long-patch BER, 87
Luteinizing hormone (LH), 22
Luteolin, 306

M

Macrophages, 19, 174
Male accessory gland infection (MAGI), 19

Male accessory tract infection (MAGI), 147, 148
Male factor infertility, 182
Male germ, 85
Male germ cells, 170, 171
Male infertility, 78, 79, 102, 142, 155, 182, 183, 276
   antioxidant treatment and effect, 72
   assessment of oxidative stress, 246
      herbal medicines, 248, 249
      management, 247
      nutrition and exercise, 247, 248
      nutritional supplements and antioxidants, 248
      pharmaceuticals, 249, 250
   biological functions, 64, 67
   causes and risk factors, 43
   cellular processes, 64
   clinical pregnancy, 64
   EAU, 116
   epididymitis, 118
   exogenous factors, 43
   gel-based methods, 65
   genital tract infections, 126
   genome, 64
   leukocytes, 64, 126
   liver cirrhosis, 43
   metabolic dysregulation and sperm motility
      asthenozoospermia, 258, 259
      sperm motility, cellular and molecular factors, 259–266
   OS
      aetiology, 168
      leukocytes, 168
      LPO, 168, 169
      MOSI, 168
      NADPH, 168
      SDF, 169
   and oxidative stress, 64, 282–284
   pathophysiology and oxidative stress
      alcohol consumption, 188, 189
      environmental sources, 189, 190
      immature sperm, 187
      inflammation and, 184, 185
      leukocyte, 187, 188
      obesity, 186
      radiation, 189
      ROS, sources of, 187
      smoking, 188
      varicocele and, 185, 186
   prostatitis, 118
   proteins, 64, 65
   proteomics, 64
   pyelonephritis, 117
   in semen, 116
   seminal plasma, 65, 69–71
   sexual dysfunctions, 43
   sexual intercourse, 116
   sperm, 68, 69
   sperm abnormalities, 69
   sperm DNA, 44
   sperm motility, 44
   sperm structural and functional integrity, 43
   spermatozoa, 64, 65
   testicular cancer, 71, 72
   types, 43
   urethritis, 118
   urinary and genital tract infections, 116
   urosepsis, 117
   UTIs, 117
   varicocele, 44–46, 71
   and viruses, 283, 284, 288
      antiviral therapies, 305–312
      classification of, 279–282
      ebola virus, 297–299
      HBV, 301–303
      hepatitis C virus, 290, 291
      herpes simplex virus, 289, 290
      HIV, 303, 304
      human papillomavirus, 288, 289
      influenza virus, 299, 300
      mumps virus, 300, 301
      origin of, 276, 277
      severe acute respiratory syndrome–coronavirus-2, 295–297
      severe acute respiratory syndrome virus, 294, 295
      viral life cycle, 277–279
      West Nile virus, 291, 292
      Zika virus, 292–294
   virus-induced oxidative damage and, 335, 336
   western blotting, 68
Male oxidative stress infertility (MOSI), 168, 192, 193
   antioxidants, 193, 194
      cryodamage, 195
      DNA damage prevention, 195
      sperm motility, 195
Male reproduction
   antioxidant therapy, 54
   ayurvedic herbs, 54
   erectile dysfunction (ED), 42, 47–49
   ethanol extract, 54
   free radicals, 41
   male infertility (*see* Male infertility)

Male reproduction (*cont.*)
   oxidative stress, 42
   oxygen metabolism, 42
   peroxiredoxins, 51
   pluripotent stem cells, 54
   polysaccharide, 54
   prostate cancer, 42
      androgen and androgenic hormones, 50
      carcinogens, 50
      chromosome 8q24 variants, 49
      chronic inflammation, 48
      development and growth, 50
      effects of ROS, 51–53
      fat-rich diet, 50
      growth factors, 48
      mechanisms, 51
      mineral supplements, 48
      mitochondria, 50
      myeloid-derived suppressor cells, 50
      nourishes and transports sperm, 48
      obesity, 50
      oxidative stress, 50
      post-translational DNA modification, 50
      prevalence, 49
      prostatitis, 48
      sex hormones, 50
      vasectomy, 50
      vitamin, 48
   sexual function, 54
   spermatozoa, 42
   varicocele, 54
   viral infection on, 338
Male reproductive health
   and male reproductive system in testis, 331, 332
   pathogenesis of
      gender-based susceptibility, 334, 335
      methodology, 327, 328
      testicular antiviral defense system, 336, 337
      viruses and male reproductive system, 328–334
      virus-induced oxidative damage and male infertility, 335, 336
   viruses and
      in semen, 329, 330
Male reproductive tissues, 174
Male urogenital infections, 130, 145
Malondialdehyde (MDA), 43, 102, 169
Mammalian spermatozoa function, 346
Mammalian testes, 349
Massachusetts Male Ageing Study (MMAS), 47

Matrix-assisted laser desorption/ionization time-of-flight (MALDI-TOF), 65
Metabolic dysfunction
   pathophysiology, of inflammation and oxidative stress in, 244–246
Metabolic dysregulation
   asthenozoospermia, 258, 259
Metabolic syndrome
   clinical assessment of, 239, 240
   impact of, 243, 244
   pathophysiology, inflammation and oxidative stress, 241–243
Metabolic syndrome (MetS), 238, 239
MetaCore™, 67
Metformin, 249
Methylcytosine, 81
Microbial infections, 127
Microdeletions, 85, 86
Microhomology-mediated end-joining pathway (MMEJ), 89, 90
MicroRNAs (miRNAs), 103
Microsatellite sequences, 81
MiOXSYS, 195
Mismatch repair, 88, 89
Mitochondrial DNA (mtDNA) copy number, 93, 94
Mitochondrial GPX4 (mGPX4), 352
Mitogen-activated protein (MAP), 173, 175
Mobile phone radiation, 189
Molecular markers
   γH2AX, 95, 96
   sperm DNA damage, 102–104
   XRCC1, 95, 96
Molecular mechanisms, sperm DNA repair
   DNA double-strand break repair/recombinational repair, 89, 90
   excision repair
      BER, 87
      NER, 87, 88
   mismatch repair, 88, 89
Molecular tests, DNA damage
   detection of DNA repair, 95, 96
   DNA methylation profiling, 95
   long PCR, 93, 94
   PCR, Y microdeletions, 94
   real-time PCR, 93, 94
   SDF testing, 92, 93
   telomere length measurement, 94
mtDNA integrity, 93, 94
Mumps virus, 122, 287, 300, 333
   and male infertility, 301
Murine in vitro system, 91
Murine model, 86
MutSα, 88

Mycoplasma, 122
Mycoplasma genitalium, 151
Myeloid differentiation primary response (MyD88), 173–175
Myo-inositol, 267

N
N-acetyl-L-cysteine (NAC), 23
Naringenin, 307, 311
Natural pregnancies
    clinical studies, 100–102
Neisseria gonorrhoeae, 150
Neutrophils, 19
Next generation sequencing (NGS), 95
Nicotinamide adenine dinucleotide phosphate (NADPH), 21, 168, 187
Nitric oxide (NO•), 12, 28, 174
Nitric oxide synthase (NOS), 12
Nitrotyrosine, 167
Non-adrenergic non-cholinergic (NANC), 47
Non-allelic homologous recombination (NAHR), 86
Nonhomologous end-joining (NHEJ), 89
Non-steroidal anti-inflammatory drugs, 249
Normozoospermia, 43
Nuclear chromatin decondensation test (NCD), 245
Nuclear Respiratory Factor 1 (NRF1), 171
Nucleotide excision repair (NER), 87, 88
Nutrition
    oxidative stress, male infertility, 247, 248
Nutritional supplements
    male infertility, 248

O
Obesity, 238, 239
    and OS, 186
    clinical assessment of, 239, 240
    impact of, 243, 244
    pathophysiology, inflammation and oxidative stress, 241–243
Ocimum basilicum, 305
Okazaki fragments, 84
Oligozoospermia, 43
Oocyte, 91
Oocyte fertilization
    clinical studies in ART, 97, 98
    in vitro and animal studies, 96, 97
Orchitis, 304
Ovulation, 22, 23
Oxidant-sensitive-inflammatory pathways
    OS (*see* Oxidative stress (OS))
    oxidative tissue injury, 167
    sperm mitochondrial response to stress, 171
    stress response pathways, 170, 171
Oxidation, 81, 82
Oxidation-reduction potential (ORP), 193
Oxidative environment, 360
Oxidative phosphorylation (OXPHOS), 261
Oxidative stress, 84, 85, 182
    antioxidants, 3, 4, 193, 194
        cryodamage, 195
        DNA damage prevention, 195
        sperm motility, 195
    cellular oxidative damage, 2
    environmental pollution, 3
    generation, 144
    herbal medicine, 3
    human infertility treatment, 5
    and inflammation (*see* Inflammation)
    levels, 144
    and male infertility, 143, 282–284
        aetiology, 168
        apoptosis, 170
        leukocytes, 168
        LPO, 168, 169
        MOSI, 168
        NADPH, 168
        SDF, 169
    MOSI and diagnosis, 192, 193
    lycopene, 3
    male infertility, 246
        herbal medicines, 248, 249
        management, 247
        nutritional supplements and antioxidants, 248
        nutrition and exercise, 247, 248
        pharmaceuticals, 249, 250
    in metabolic dysfunction, 244–246
    metabolism and motility, 2
    pathophysiology of
        alcohol consumption, 188, 189
        environmental sources, 189, 190
        immature sperm, 187
        inflammation and, 184, 185
        leukocyte, 187, 188
        obesity and, 186
        radiation, 189
        ROS, sources of, 187
        smoking, 188
        varicocele and, 185, 186
    physiological functions, 3, 4
    physiological roles, 3, 4
    reproductive system, 3
    and ROS, 166, 167

Oxidative stress (cont.)
  sex determination, 3
  source, 143
  on spermatozoa, 190, 191
    apoptosis, 192
    sperm DNA fragmentation, 191, 192
  sperm motility, 144
  traditional and herbal medicine, 3
  varicocele-associated male infertility, 206
    in adolescent, 215
    exogenous antioxidants, 220, 222–230
    and fertility, 210, 213
    fertility in men with, 212, 213, 215
    grade and, 210–212
    infertile men with or without
      varicocele, 212, 214
    pathophysiology of, 206–209
    and secondary male infertility, 216
    varicocelectomy, 216–220
  vitamins, 3
Oxidative stress (OS), 182, 353
Oxidative tissue injury, 167
Ozone (O_3), 14

P
Paramyxoviridae family, 122
Pathogen-associated molecular patterns
  (PAMPs), 173
Pathogen-specific molecules, 173
Pattern recognition receptors (PRRs), 173, 175
PCR-based assays, 130
Peroxidase activity, 22
Peroxynitrite, 49, 167
Pharmaceuticals
  male infertility, 249, 250
Phosphatidylinositides (PIs), 278
Phosphodiesterase 5 (PDE 5), 47
Phospholipase A2 (PLA2) activity, 29
Phthalates, 190
Physiological stress, 333
Poly (ADP-ribose) polymerase (PARP), 102
Polymerase chain reaction (PCR), 130
  Y microdeletions, 94
Polymorphonuclear leukocytes (PMNs), 19
Polyunsaturated fatty acids (PUFAs), 43, 168
Pregnancy
  clinical studies
    ART, 100–102
    in natural pregnancies, 100–102
    in vitro and animal studies, 98–100
Primary ciliary dyskinesia (PCD), 259
Pro-inflammatory mediators, 172
Proliferating cell nuclear antigen (PCNA), 87

Prostate cancer, 333
Prostatitis, 118
Protein kinase C (PKC), 29
Protein–protein interaction (PPI), 67
Proteomics, 64, 65, 67, 72
Pseudomonas aeruginosa, 149
Punicalagin, 307
Pyelonephritis, 117
Pyrimidine dimers, 83
Pyrimidines, 81
Pyrosequencing, 95

Q
Quantitative fluorescent in situ hybridization
  (Q-FISH), 94
Quercetin, 310
Quercitin, 306

R
Radiation, 189
Radiation-induced DNA damage, 99
Radical and non-radical ROS, 167
Ras-proximate-1 (RAP1), 84
Reactive chlorine species (RCS), 166
Reactive nitrogen species (RNS), 42, 166
Reactive oxygen species (ROS), 2, 80, 85,
    126, 187, 238, 346
  aerobic organisms, 11
  alcohol consumption, 188, 189
  antioxidant defence mechanism, 10
  cellular aerobic metabolism, 10
  chlorine-containing reactive species, 11
  endogenous sources
    immature sperm, 187
    leukocyte, 187
  endometrium, 17, 18
  endoplasmic reticulum, 10
  environmental sources, 189, 190
  enzymatic pathways, 10
  female reproductive system
    corpus luteum, 23, 24
    folliculogenesis, 21
    implantation, 24
    maintenance of pregnancy, 24, 25
    ovarian steroidogenesis, 22
    ovulation, 22, 23
    parturition, 25
  free radicals, 11
  fundamental chemical properties, 10
  hydroxyl radicals (OH•), 12
  immature sperm, 187
  leukocyte, 187, 188

male reproductive system
  acrosome reaction, 28
  capacitation, 27, 28
  human spermatozoa, 17
  immature spermatozoa, 19, 20
  leukocytes, 19
  mitochondrial oxidoreductase system, 18
  NADPH oxidases, 17
  NOX5, 18
  spermatogenesis, 26, 27
  sperm maturation, 26, 27
  sperm–oocyte fusion, 29
  varicocele, 20, 21
nitric oxide (NO•), 12
O_2-free radicals, 11
and OS, 166, 167
oxidative stress, 10
oxidative stress biomarkers, 30
peritoneal cavity, 16, 17
peroxisomes, 10
physiological activities, 10
radiation, 189
signalling molecules, 10
smoking, 188
spermatozoa, 29
sperm motility, 263, 264
superoxide Anion ($O_2^{•-}$), 11, 12
Reactive species, 166
Reactome, 67
Real-time PCR, 94
  mtDNA copy number, 93, 94
Receptor-tyrosine kinase (RTK) family, 26
Recombinant IFN-α, 312
Recombinational repair, 89, 90
Reductive stress, 2–4
Regression theory, 276–277
Reproduction
  DNA integrity, 79
Reproductive system, 145
Reverse transcribing viruses, 281, 282
RNA virus, 281

S
Search Tool for the Retrieval of Interacting Genes/Proteins (STRING), 67
Secondary male infertility
  varicocele and, 216
Selenium (Se), 354
  activity, 355
  GPX4, 354
  SMCP$^{-/-}$ 129/Sv knockouts, 354
Semen candidiasis, 125

Semen parameter, 206, 216
Seminal leukocytes, 129
Seminiferous epithelium, 173
Sequence-tagged site polymerase chain reaction (STS-PCR), 94
Sertoli cell cytoskeleton, 173
Sertoli cell functions, 173
Sertoli cells, 174, 175, 336
Severe acute respiratory syndrome (SARS) virus, 286, 294, 295
Severe acute respiratory syndrome coronavirus 2 (SARS-CoV-2), 124, 286, 295–297
Sexual transmission, 326, 330
Sexually transmitted diseases (STDs), 117, 327
Sexually transmitted infections (STIs), 148
Shelterin complex, 84
Short-patch BER, 87
Signaling pathway
  sperm motility, 262, 263
Single cell gel electrophoresis (Comet) assay, 92
Single copy nuclear gene, 93
Single-strand break (SSB), 79, 80, 98
Single-stranded DNA (ssDNA), 89
Singlet oxygen ($^1O_2$), 13
Smoking, 188, 266
SOD metalloenzymes, 194
Sodium dodecyl sulphate-polyacrylamide gel electrophoresis (SDS-PAGE), 65
Sole therapy, 220, 222, 223, 225–227
Somatic cells, 85
Sperm, 145, 279, 282, 283
Sperm chromatin dispersion (SCD) assay, 92
Sperm chromatin structure assay (SCSA), 92, 100
Sperm DNA damage
  abasic sites, 82
  alkylation, 82
  base modifications, 82
  causes, 79–81
  deamination, 83
  dimers of pyrimidines, 83
  DNA fragmentation, 83, 84
  endogenous DNA damage, 79, 80
  epigenetic abnormalities, 86, 87
  exogenous DNA damage, 80
  fertilization (see Fertilization)
  hydrolysis, 82
  interstrand crosslinks, 83
  and male infertility, 78, 79
  mismatched bases, 81, 82
  molecular markers, 102–104

Sperm DNA damage (*cont.*)
  molecular tests (*see* Molecular tests, DNA damage)
  oocyte, 91
  oxidation, 81, 82
  and pregnancy (*see* Pregnancy)
  telomeres, 84, 85
  Y chromosome microdeletions, 85, 86
Sperm DNA fragmentation (SDF), 92, 93, 98, 101, 168, 191, 192
  OS and male infertility, 169
Sperm DNA repair
  molecular mechanisms (*see* Molecular mechanisms, sperm DNA repair)
  spermatogenesis, 90
Sperm functions, 2
Sperm maturation, 347, 349
Sperm mitochondrial response to stress, 171
Sperm motility
  antioxidants, 195
  asthenozoospermia, 258, 259
  cellular and molecular factors for
    cilia role, 259
    energy driving, 261, 262
    metabolic pathways and gene, 265
    molecular pathways, and lifestyle factors, 266
    reactive oxygen species, 263, 264
    signaling pathway and genes, 262, 263
  cryodamage, 195
  DNA damage prevention, 195
  metabolic dysregulation and sperm motility in athenozoospermia, therapeutic approach for, 267, 268
Sperm parameters, 150
Spermatid-specific thioredoxin 1 (SPTRX-1/TXNDC2) protein, 356
Spermatid-specific thioredoxin 2 (SPTRX-2/TXNDC3/NME8), 356
Spermatid-specific thioredoxin 3 (SPTRX-3/TXNDC8), 358
Spermatogenesis, 79, 84–86, 90, 96, 98, 102, 104, 244
Spermatogonia, 85
Spermatozoa, 18, 44, 86, 90, 92, 96, 97, 99
  oxidative stress on, 190, 191
    apoptosis, 192
    sperm DNA fragmentation, 191, 192
Staphylococcus aureus, 151
Staphylococcus sp., 121
Stem cells, 85
Steroidogenic acute regulatory protein (StAR), 174

Stress response pathways in male germ cells, 170, 171
STS markers, 94
Superoxide Anion ($O_2^{\bullet-}$), 11, 12

T
Telomerase reverse transcriptase (TERT), 85
Telomerase RNA component (TERC), 85
Telomere length measurement, 94
Telomere shortening, 84, 85
Telomeres, 84, 85
Telomeric repeat-binding factor 1 (TRF1), 84
Telomeric repeat-binding factor 2 (TRF2), 84
Termination, 169
Testicular antiviral defense system, 336, 337
Testicular interstitial tissue, 174
Testis
  viral infection in, 331, 332
Testis thioredoxin system, 355
Testosterone, 249
Tetratozoospermia, 43
Thioredoxin-like protein 2 (TXL-2), 358
Thioredoxin system, 355
Thioredoxins (TRX), 355
Thymine-adenine, 81
Toll-like receptors (TLRs), 173, 174
Topoisomerase enzymes, 79
Topoisomerases, 79
Total antioxidant capacity (TAC), 209
Total fertility rate (TFR), 258
Traditional herbs, 156
Transcription-coupled NER (TC-NER), 87, 88
Transcription factor II H (TFIIH), 88
Trichomonas vaginalis, 124
Tripeptidyl peptidase 1 (TPP1), 84
TTAGGG repetitions, 84
Tumor necrosis factor-alpha (TNF-α), 50
TUNEL assay, 92
Two-dimensional gel electrophoresis (2D-GE), 65
Type 2 diabetes mellitus (T2DM), 238, 239
  clinical assessment of, 239, 240
  impact of, 243, 244
  pathophysiology, inflammation and oxidative stress, 241–243
Type I interferon, 308
Type II interferon, 308
Type III interferon, 308

U
Ubiquitin-proteasome system, 170
Unfolded protein response (UPR), 170

UniProt, 67
Ureaplasma urealyticum, 121
Ureplasma, 151
Urethritis, 118, 148
Urinary tract infections (UTIs), 116, 146
    bacterial species, 147
    cystitis, 147
Urogenital tract, 116
Uropathogens, 147
Urosepsis, 117

V
Varicocele, 20, 21
    and OS, 185, 186
Varicocele-associated male infertility, 206
    OS
        in adolescent, 215
        exogenous antioxidants, 220, 222–230
        and fertility, 210, 213
        fertility in men with, 212, 213, 215
        grade and, 210, 212
        infertile men with or without varicocele, 212, 214
        pathophysiology of, 206–209
        and secondary male infertility, 216
        varicocelectomy, 216–220
Varicocelectomy
    in adolescent, 217, 220
    in adult men, 216–219
Vascular endothelial growth factor (VEGF), 208
Vicious loop, 171, 172
Viral infections, 122
    fertility, 122
    HIV, 123
    HPV infections, 123
    infection, 123
    MuV infection, 122
    pathogenesis of
        gender-based susceptibility, 334, 335
        methodology, 327, 328
        testicular antiviral defense system, 336, 337
        viruses and male reproductive system, 328–334
        virus-induced oxidative damage and male infertility, 335, 336
    protozoal infections, 124
    SARS-CoV-2, 124
    semen candidiasis, 125
    Zika virus, 124
Virus early theory, 276
Viruses, 276

and male infertility, 283, 284, 288
    antiviral therapies, 305–312
    classification of, 279–282
    ebola virus, 297–299
    HBV, 301–303
    hepatitis C virus, 290, 291
    herpes simplex virus, 289, 290
    HIV, 303, 304
    human papillomavirus, 288, 289
    influenza virus, 299, 300
    mumps virus, 300, 301
    origin of, 276, 277
    severe acute respiratory syndrome–coronavirus-2, 295–297
    severe acute respiratory syndrome virus, 294, 295
    viral life cycle, 277–279
    West Nile virus, 291, 292
    Zika virus, 292–294
and male reproductive system, 328
    in semen, 329, 330
    in testis, 331, 332
Virus-induced oxidative damage
    and male infertility, 335, 336
Virus life cycle, 277, 278
    cell entry, 277
    cell exit, 279
    genome replication, transcription and translation of viral proteins, 278, 279
Vitamin C, 194, 228
Vitamin E, 194, 267

W
West Nile virus, 285, 291, 292
World Health Organization (WHO), 42, 145

X
Xenobiotics, 42
XRCC1, 95, 96

Y
Y chromosome microdeletions, 85, 86
Y microdeletions, 94, 98, 101

Z
Zika virus, 124, 286, 292–294
Zona pellucida glycoprotein 3 (ZP3), 28
Zygotic cytosol, 360

Printed by Printforce, the Netherlands